Pounder's
Marine Diesel Engines and
Gas Turbines

Eighth edition

Pounder's Marine Diesel Engines and Gas Turbines

Eighth edition

Edited by
Doug Woodyard

ELSEVIER
BUTTERWORTH
HEINEMANN

AMSTERDAM   BOSTON   HEIDELBERG   LONDON   NEW YORK   OXFORD
PARIS   SAN DIEGO   SAN FRANCISCO   SINGAPORE   SYDNEY   TOKYO

Butterworth-Heinemann is an imprint of Elsevier
Linacre House, Jordan Hill, Oxford OX2 8DP, UK
30 Corporate Drive, Suite 400, Burlington, MA 01803, USA

First edition 1984
Reprinted 1991, 1992
Seventh edition 1998
Reprinted 1999
Eighth edition 2004
Reprinted 2005, 2006, 2007

British Library Cataloguing in Publication Data
Pounders's marine diesel engines and gas turbines – 8th edn.
 1. Marine diesel motors. 2. Marine gas-turbines
 I. Woodyard, D. F. (Douglas F.) II. Marine diesel engines and
 gas turbines
 628.8′723′6

Library of Congress Cataloging-in-Publication Data
A catalog record for this book is available from the Library of Congress

ISBN–13: 978-0-7506-5846-1
ISBN–10: 0-7506-5846-0

For information on all Butterworth-Heinemann publications
visit our website at books.elsevier.com

Printed and bound in *Great Britain*

07 08 09 10 10 9 8 7 6 5

Contents

Preface

Developments in two-stroke and four-stroke designs for propulsion and auxiliary power drives in the five years since the publication of the seventh edition of *Pounder's Marine Diesel Engines* call for an update. Rationalization in the engine design/building industry has also been sustained, with the larger groups continuing to absorb (and in some cases phase out) long-established smaller marques.

This eighth edition reflects the generic and specific advances made by marine engine designers and specialists in support technologies—notably turbocharging, fuel treatment, emissions reduction and automation systems—which are driven by: ship designer demands for more compactness and lower weight; shipowner demands for higher reliability, serviceability and overall operational economy; and shipbuilder demands for lower costs and easier installation procedures.

A revised historical perspective logs the nautical milestones over the first century of marine diesel technology, which closed with the emergence of electronically-controlled low speed designs paving the path for future so-called Intelligent Engines. Development progress with these designs and operating experience with the first to enter commercial service are reported in this new edition.

Increasing interest in dual-fuel and gas-diesel engines for marine and offshore applications, since the last edition, is reflected in an expanded chapter. The specification of dual-fuel medium speed machinery for LNG carriers in 2002 marked the fall of the final bastion of steam turbine propulsion to the diesel engine.

Controls on exhaust gas emissions—particularly nitrogen oxides, sulphur oxides and smoke—have tightened regionally and internationally, dictating responses from engine designers exploiting common rail fuel systems, emulsified fuel, direct water injection and charge air humidification. These and other solutions, including selective catalytic reduction systems, are detailed in an extended chapter.

Also extended is the chapter on fuels and lube oils, and the problems of contamination, which includes information on low sulphur fuels, new cylinder and system lubricants, and cylinder oil feed system developments.

A new chapter provides an introduction to marine gas turbines, now competing more strongly with diesel engines in some key commercial propulsion sectors, notably cruise ships and fast ferries.

The traditional core of this book—reviews of the current programmes of the leading low, medium and high speed engine designers—has been thoroughly updated. Details of all new designs and major refinements to established models introduced since the last edition are provided. Technically important engines no longer in production but still encountered at sea justify their continued coverage.

In preparing the new edition the author expresses again his gratitude for the groundwork laid by the late C.C. Pounder and to the editors of the sixth edition, his late friend and colleague Chris Wilbur and Don Wight (whose contributions are respectively acknowledged at the end of sections or chapters by C.T.W. and D.A.W.).

In an industry generous for imparting information on new developments and facilitating visits, special thanks are again due to MAN B&W Diesel, Wärtsilä Corporation, Caterpillar Motoren, ABB Turbo Systems, the major classification societies, and the leading marine lube oil groups. Thanks also to my wife Shelley Woodyard for support and assistance in the project.

Finally, the author hopes that this edition, like its predecessors, will continue to provide a useful reference for marine engineers ashore and at sea, enginebuilders and ship operators.

Doug Woodyard

Introduction: a century of diesel progress

Ninety years after the entry into service of *Selandia*, generally regarded as the world's first oceangoing motor vessel, the diesel engine enjoys almost total dominance in merchant ship propulsion markets. Mainstream sectors have long been surrendered by the steam turbine, ousted by low and medium speed engines from large containerships, bulk carriers, VLCCs and cruise liners. Even steam's last remaining bastion in the newbuilding lists—the LNG carrier—has now been breached by competitive new dual-fuel diesel engine designs arranged to burn the cargo boil-off gas.

The remorseless rise of the diesel engine at the expense of steam reciprocating and turbine installations was symbolized in 1987 by the steam-to-diesel conversion of Cunard's prestigious cruise liner *Queen Elizabeth 2*. Her turbine and boiler rooms were ignominiously gutted to allow the installation of a 95 600 kW diesel–electric plant.

The revitalized *QE2*'s propulsion plant was based on nine 9-cylinder L58/64 medium speed four-stroke engines from MAN B&W Diesel which provided a link with the pioneering *Selandia*: the 1912-built twin-screw 7400 dwt cargo/passenger ship was powered by two Burmeister & Wain eight-cylinder four-stroke engines (530 mm bore/730 mm stroke), each developing 920 kW at 140 rev/min. An important feature was the effective and reliable direct-reversing system.

Progress in raising specific output over the intervening 70 years was underlined by the 580 mm bore/640 mm stroke design specified for the *QE2* retrofit: each cylinder has a maximum continuous rating of 1213 kW.

Selandia was built by the Burmeister & Wain yard in Copenhagen for Denmark's East Asiatic Company and, after trials in February 1912, successfully completed a 20 000 mile round voyage between the Danish capital and the Far East. The significance of the propulsion plant was well appreciated at the time. On her first arrival in London the ship was inspected by Sir Winston Churchill, then First Lord of the Admiralty; and *Fiona*, a sistership delivered four months later by the same yard, so impressed the German Emperor that it was immediately arranged for the Hamburg Amerika Line to buy her.

INTRODUCTION

Figure I.1 One of two Burmeister & Wain DM8150X engines commissioned (1912) to power the first Selandia (MAN B&W Diesel)

A third vessel in the series, *Jutlandia*, was built by Barclay, Curle in Scotland and handed over to East Asiatic in May 1912. The Danish company's oceangoing motor ship fleet numbered 16 by 1920, the largest being the 13 275 dwt *Afrika* with twin six-cylinder B&W engines of 740 mm bore/1150 mm stroke developing a combined 3300 kW at 115 rev/min. Early steam-to-diesel conversions included three 4950 dwt vessels built in 1909 and repowered in 1914/15 by the B&W Oil Engine Co of Glasgow, each with a single six-cylinder 676 mm bore/1000 mm stroke engine developing 865 kW at 110 rev/min.

Selandia operated successfully for almost 30 years (latterly as *Norseman*) and maintained throughout a fully loaded service speed of 10.5 knots before being lost off Japan in 1942. The propulsion plant of the second *Selandia*, which entered service in 1938, demonstrated the advances made in diesel technology since the pioneering installation. The single, double-acting two-stroke five-cylinder engine of the 8300 dwt vessel delivered 5370 kW at 120 rev/min: three times the output of the twin-engined machinery powering the predecessor.

The performance of *Selandia* and other early motor ships stimulated East Asiatic to switch completely from steamers, an example followed by more and more owners. In 1914 there were fewer than 300 diesel-powered vessels in service with an aggregate tonnage of 235 000 grt;

Figure I.2 A 20 bhp engine built in 1898 by Burmeister & Wain to drawings supplied by Dr. Diesel, for experimental and demonstration purposes. MAN built the first diesel engine—a 250 mm bore/400 mm stroke design—in 1893

a decade later the fleet had grown to some 2000 ships of almost two million grt; and by 1940 the total tonnage had risen to 18 million grt embracing 8000 motor ships.

Between the two world wars the proportion of oil-engined tonnage in service thus expanded from 1.3 to 25 per cent of the overall oceangoing fleet. By 1939 an estimated 60 per cent of the total tonnage completed in world yards comprised motor ships, compared with only 4 per cent in 1920.

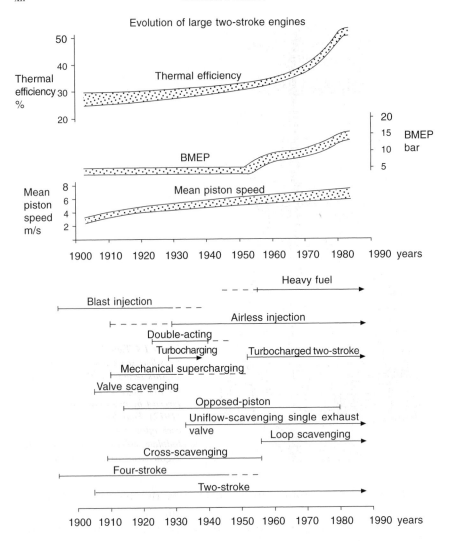

Figure I.3 Main lines of development for direct-drive low speed engines

In outlining the foundations of the diesel engine's present dominance in shipping other claimants to pioneering fame should be mentioned. In 1903 two diesel-powered vessels entered service in quick succession: the Russian naphtha carrier *Vandal*, which was deployed on the Volga, and the French canal boat *Petit Pierre*. By the end of 1910 there were 34 trading vessels over 30 m long worldwide with diesel propulsion, and an unknown number of naval vessels, especially submarines.

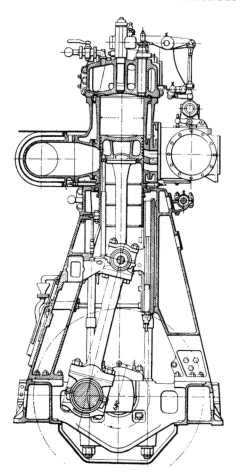

Figure I.4 Twin Sulzer 4S47 type cross-flow scavenged crosshead engines served the Monte Penedo, the first large oceangoing vessel powered by two-stroke engines (1912). Four long tie-rods secured each cylinder head directly to the bedplate, holding the whole cast iron engine structure in compression

The earliest seagoing motor vessel was the twin-screw 678 ton *Romagna*, built in 1910 by Cantieri Navali Riuniti with twin four-cylinder port-scavenged trunk piston engines supplied by Sulzer. Each 310 mm bore/460 mm stroke engine delivered 280 kW at 250 rev/min.

1910 also saw the single-screw 1179 dwt Anglo-Saxon tanker *Vulcanus* enter service powered by a 370 kW Werkspoor six-cylinder four-stroke crosshead engine with a 400 mm bore/600 mm stroke. The Dutch-built vessel was reportedly the first oceangoing motor ship to receive classification from Lloyd's Register.

In 1911 the Swan Hunter-built 2600 dwt Great Lakes vessel *Toiler* crossed the Atlantic with propulsion by two 132 kW Swedish Polar engines. Krupp's first marine diesel engines, six-cylinder 450 mm

Figure I.5 One of the two Sulzer 4S47 engines installed in the Monte Penedo (1912)

bore/800 mm stroke units developing 920 kW at 140 rev/min apiece, were installed the same year in the twin-screw 8000 dwt tankers *Hagen* and *Loki* built for the German subsidiary of the Standard Oil Co of New Jersey.

The following year, a few months after *Selandia*, Hamburg-South Amerika Line's 6500 dwt cargo/passenger ship *Monte Penedo* entered service as the first large oceangoing vessel powered by two-stroke diesel engines. Each of the twin four-cylinder Sulzer 4S47 crosshead units (470 mm bore/680 mm stroke) delivered 625 kW at 160 rev/min.

(The adoption of the two-stroke cycle by Sulzer in 1905 greatly increased power output and fostered a more simple engine. Port-

Figure I.6 The 6500 dwt cargo liner Monte Penedo (1912)

scavenging, introduced in 1910, eliminated the gas exchange valves in the cylinder cover to create a simple valveless concept that characterized the Sulzer two-stroke engine for 70 years: the change to uniflow scavenging only came with the RTA-series engines of 1982 because their very long stroke—required for the lower speeds dictated for high propeller efficiency—was unsuitable for valveless port scavenging.)

Another important delivery in 1912 was the 3150 dwt Furness Withy cargo ship *Eavestone*, powered by a single four-cylinder Carels two-stroke crosshead engine with a rating of 590 kW at 95 rev/min. The 508 mm bore/914 mm stroke design was built in England under licence by Richardsons Westgarth of Middlesbrough.

There were, inevitably, some failures among the pioneers. For example, a pair of Junkers opposed-piston two-stroke engines installed in a 6000 dwt Hamburg-Amerika Line cargo ship was replaced by triple-expansion steam engines even before the vessel was delivered. The Junkers engines were of an unusual vertical tandem design, effectively double-acting, with three pairs of cylinders of 400 mm bore and 800 mm combined stroke to yield 735 kW at 120 rev/min. More successful was Hapag's second motor ship, *Secundus*, delivered in 1914 with twin Blohm+Voss-MAN four-cylinder two-stroke single-acting engines, each developing 990 kW at 120 rev/min.

After the First World War diesel engines were specified for increasingly powerful cargo ship installations and a breakthrough made in large passenger vessels. The first geared motor ships appeared in 1921, and in the following year the Union Steamship Co of New Zealand ordered a 17 490 grt quadruple-screw liner from the UK's Fairfield yard. The four Sulzer six-cylinder ST70 two-stroke single-acting engines (700 mm bore/990 mm stroke) developed a total of 9560 kW at 127 rev/min—far higher than any contemporary motor ship—and gave *Aorangi* a speed of 18 knots when she entered service in December 1924.

Positive experience with these engines and those in other contemporary motor ships helped to dispel the remaining prejudices against using diesel propulsion in large vessels.

Swedish America Line's 18 134 grt *Gripsholm*—the first transatlantic diesel passenger liner—was delivered in 1925; an output of 9930 kW was yielded by a pair of B&W six-cylinder four-stroke double-acting 840 mm bore engines. Soon after, the Union Castle Line ordered the first of its large fleet of motor passengers liners, headed by the 20 000 grt *Caernarvon Castle* powered by 11 000 kW Harland & Wolff-B&W double-acting four-stroke machinery.

Figure I.7 Sulzer's 1S100 single-cylinder experimental two-stroke engine (1912) featured a 1000 mm bore

Another power milestone was logged in 1925 when the 30 000 grt liner *Augustus* was specified with a 20 600 kW propulsion plant based on four MAN six-cylinder double-acting two-stroke engines of 700 mm bore/1200 mm stroke.

It was now that the double-acting two-stroke engine began to make headway against the single-acting four-stroke design, which had enjoyed favour up to 1930. Two-stroke designs in single-and double-acting forms, more suitable for higher outputs, took a strong lead as ships became larger and faster. Bigger bore sizes and an increased number of cylinders were exploited. The 20 000 grt *Oranje*, built in 1939, remained the most powerful merchant motor ship for many years thanks to her three 12-cylinder Sulzer 760 mm bore SDT76 single-acting engines aggregating 27 600 kW.

The groundwork for large bore engines was laid early on. Sulzer, for example, in 1912 tested a single-cylinder experimental engine with a 1000 mm bore/1100 mm stroke. This two-stroke crosshead type 1S100 design developed up to 1470 kW at 150 rev/min and confirmed the effectiveness of Sulzer's valveless cross-scavenging system, paving the way for a range of engines with bores varying between 600 mm and 820 mm. (Its bore size was not exceeded by another Sulzer engine until 1968.)

At the end of the 1920s the largest engines were Sulzer five-cylinder 900 mm bore models (3420 kW at 80 rev/min) built under licence by John Brown in the UK. These S90 engines were specified for three twin-screw *Rangitiki*-class vessels of 1929.

GOODBYE TO BLAST INJECTION

It was towards the end of the 1920s that most designers concluded that the blast air–fuel injection diesel engine—with its need for large, often troublesome and energy-consuming high pressure compressors— should be displaced by the airless (or compressor-less) type.

Air-blast fuel injection called for compressed air from a pressure bottle to entrain the fuel and introduce it in a finely atomized state via a valve needle into the combustion chamber. The air-blast pressure, which was only just slightly above the ignition pressure in the cylinder, was produced by a water-cooled compressor driven off the engine connecting rod by means of a rocking lever.

Rudolf Diesel himself was never quite satisfied with this concept (which he called self-blast injection) since it was complicated and hence susceptible to failure—and also because the 'air pump' tapped as much as 15 per cent of the engine output.

Diesel had filed a patent as early as 1905 covering a concept for the solid injection of fuel, with a delivery pressure of several hundred atmospheres. A key feature was the conjoining of pump and nozzle and their shared accommodation in the cylinder head. One reason

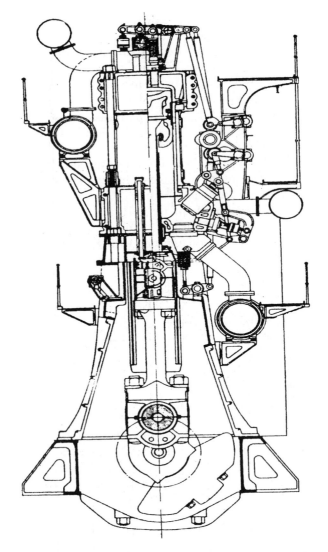

Figure I.8 A B&W 840-D four-stroke double-acting engine powered Swedish America Line's Gripsholm in 1925

advanced for the lack of follow-up was that few of the many engine licensees showed any interest.

A renewed thrust came in 1910 when Vickers' technical director McKechnie (independently of Diesel, and six months after a similar patent from Deutz) proposed in an English patent an 'accumulator system for airless direct fuel injection' at pressures between 140 bar

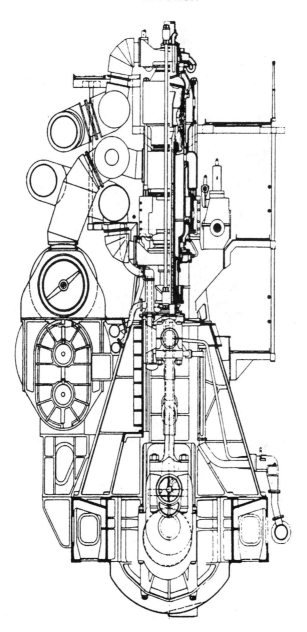

Figure I.9 A B&W 662-WF/40 two-stroke double-acting engine, first installed as a six-cylinder model in the Amerika (1929)

and 420 bar. By 1915 he had developed and tested an 'operational' diesel engine with direct injection, and is thus regarded as the main inventor of high intensity direct fuel injection. Eight years later it had become possible to manufacture reliable production injection pumps for high pressures, considerably expanding the range of applications.

The required replacement fuel injection technology thus had its roots in the pioneering days (a Doxford experimental engine was converted to airless fuel injection in 1911) but suitable materials and manufacturing techniques had to be evolved for the highly stressed camshaft drives and pump and injector components. The refinement of direct fuel injection systems was also significant for the development of smaller high speed diesel engines.

A BOOST FROM TURBOCHARGING

A major boost to engine output and reductions in size and weight resulted from the adoption of turbochargers. Pressure charging by various methods was applied by most enginebuilders in the 1920s and 1930s to ensure an adequate scavenge air supply: crankshaft-driven reciprocating air pumps, side-mounted pumps driven by levers off the crossheads, attached Roots-type blowers or independently driven pumps and blowers. The pumping effect from the piston underside was also used for pressure charging in some designs.

The Swiss engineer Alfred Büchi, considered the inventor of exhaust gas turbocharging, was granted a patent in 1905 and undertook his initial turbocharging experiments at Sulzer Brothers in 1911/15. It was almost 50 years after that first patent, however, before the principle could be applied satisfactorily to large marine two-stroke engines.

The first turbocharged marine engines were 10-cylinder Vulcan-MAN four-stroke single-acting models in the twin-screw *Preussen* and *Hansestadt Danzig*, commissioned in 1927. Turbocharging under a constant pressure system by Brown Boveri turboblowers increased the output of these 540 mm bore/600 mm stroke engines from 1250 kW at 240 rev/min to 1765 kW continuously at 275 rev/min, with a maximum of 2960 kW at 317 rev/min. Büchi turbocharging was keenly exploited by large four-stroke engine designers, and in 1929 some 79 engines totalling 162 000 kW were in service or contracted with the system.

In 1950/51 MAN was the forerunner in testing and introducing high pressure turbocharging for medium speed four-stroke engines for which boost pressures of 2.3 bar were demanded and attained.

Progressive advances in the efficiency of turbochargers and systems development made it possible by the mid-1950s for the major two-stroke enginebuilders to introduce turbocharged designs.

A more recent contribution of turbochargers, with overall efficiencies now topping 70 per cent, is to allow some exhaust gas to be diverted to a power recovery turbine and supplement the main engine effort or drive a generator. A range of modern power gas turbines is available to enhance the competitiveness of two-stroke and larger four-stroke engines, yielding reductions in fuel consumption or increased power.

HEAVY FUEL OILS

Another important step in strengthening the status of the diesel engine in marine propulsion was R&D enabling it to burn cheaper, heavier fuel oils. Progress was spurred in the mid-1950s by the availability of cylinder lubricants able to neutralize acid combustion products and hence reduce wear rates to levels experienced with diesel oil-burning. All low speed two-stroke and many medium speed four-stroke engines are now released for operation on low grade fuels of up to 700 cSt/50°C viscosity, and development work is extending the capability to higher speed designs.

Combating the deterioration in bunker quality is just one example of how diesel engine developers—in association with lube oil technologists and fuel treatment specialists—have managed successfully to adapt designs to contemporary market demands.

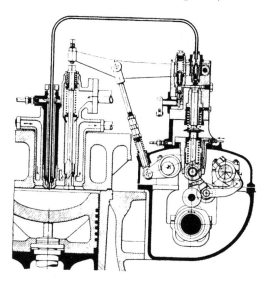

Figure I.10 Direct fuel injection system introduced by Sulzer in 1930, showing the reversing mechanism and cam-operated starting air valve. Airless fuel injection had been adopted by all manufacturers of large marine engines by the beginning of the 1930s: a major drawback of earlier engines was the blast injection system and its requirement for large, high pressure air compressors which dictated considerable maintenance and added to parasitic power losses

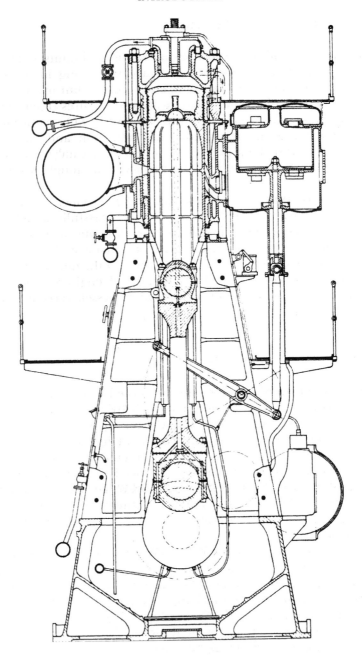

Figure I.11 Cross-section of Sulzer SD72 two-stroke engine (1943). Each cylinder had its own scavenge pump, lever driven off the crosshead. The pistons were oil cooled to avoid the earlier problem of water leaks into the crankcase

ENVIRONMENTAL PRESSURES

A continuing effort to reduce exhaust gas pollutants is another challenge for engine designers who face tightening international controls in the years ahead on nitrogen oxide, sulphur oxide, carbon dioxide and particulate emissions. In-engine measures (retarded fuel injection, for example) can cope with the IMO's NOx requirements while direct water injection, fuel emulsification and charge air humidification can effect greater curbs. Selective catalytic reduction (SCR) systems, however, are dictated to meet the toughest regional limits.

Demands for 'smokeless' engines, particularly from cruise operators in pollution-sensitive arenas, have been successfully addressed—common rail fuel systems playing a role—but the development of engines with lower airborne sound levels remains a challenge.

Environmental pressures are also stimulating the development and wider application of dual-fuel and gas-diesel engines, which have earned breakthroughs in offshore support vessel, ferry and LNG carrier propulsion.

LOWER SPEEDS, LARGER BORES

Specific output thresholds have been boosted to 6950 kW/cylinder by MAN B&W Diesel's 1080 mm bore MC/ME two-stroke designs. A single 14-cylinder model can thus deliver up to 97 300 kW for propelling projected 10 000 TEU-plus containerships with service speeds of 25 knots-plus. (The largest containerships of the 1970s typically required twin 12-cylinder low speed engines developing a combined 61 760 kW). Both MAN B&W Diesel and Wärtsilä Corporation (Sulzer) have extended their low speed engine programmes from the traditional 12-cylinder limit to embrace 14-cylinder models.

Power ratings in excess of 100 000 kW are mooted from extended in-line and V-cylinder versions of established low speed engine designs. V-configurations, although not yet in any official programme, promise valuable savings in weight and length over in-line cylinder models, allowing a higher number of cylinders (up to 18) to be accommodated within existing machinery room designs. Even larger bore sizes represent another route to higher powers per cylinder.

Engine development has also focused on greater fuel economy achieved by a combination of lower rotational speeds, higher maximum combustion pressures and more efficient turbochargers. Engine thermal efficiency has been raised to over 54 per cent and specific fuel consumptions can be as low as 155 g/kWh. At the same time, propeller efficiencies have been considerably improved due to minimum engine speeds reduced by more than 40 per cent to as low as 55 rev/min.

The pace and expense of development in the low speed engine sector have been such that only three designer/licensors remain active in the international market. The roll call of past contenders include names either long forgotten or living on in other fields: AEG-Hesselman, Deutsche Werft, Fullagar, Krupp, McIntosh and Seymour, Neptune, Nobel, North British, Polar, Richardsons Westgarth, Still, Tosi, Vickers, Werkspoor and Worthington. The most recent casualties were Doxford, Götaverken and Stork, some of whose products remain at sea in dwindling numbers.

These pioneering designers displayed individual flair within generic classifications which offered two-or four-stroke, single-or double-acting, and single-or opposed-piston arrangements. The Still concept even combined the Diesel principle with a steam engine: heat in the exhaust gases and cooling water was used to raise steam which was then supplied to the underside of the working piston.

Evolution has decreed that the surviving trio of low speed engine designers (MAN B&W, Mitsubishi and Sulzer) should all settle—for the present at least—on a common basic philosophy: uniflow-scavenged, single hydraulically-actuated exhaust valve in the head, constant pressure turbocharged, two-stroke crosshead engines exploiting increasingly high stroke/bore ratios (up to 4.2:1) and low operating speeds for direct coupling to the propeller. Bore sizes range from 260 mm to 1080 mm.

In contrast the high/medium speed engine market is served by numerous companies offering portfolios embracing four-stroke, trunk piston, uniflow- or loop-scavenged designs, and rotating piston types, with bore sizes up to 640 mm. Wärtsilä's 64 engine—the most powerful medium speed design—offers a rating of over 2000 kW/cylinder from the in-line models.

Recent years have seen the development of advanced medium and high speed designs with high power-to-weight ratios and compact configurations for fast commercial vessel propulsion, a promising market.

THE FUTURE

It is difficult to envisage the diesel engine being seriously troubled by alternative prime movers in the short-to-medium term but any

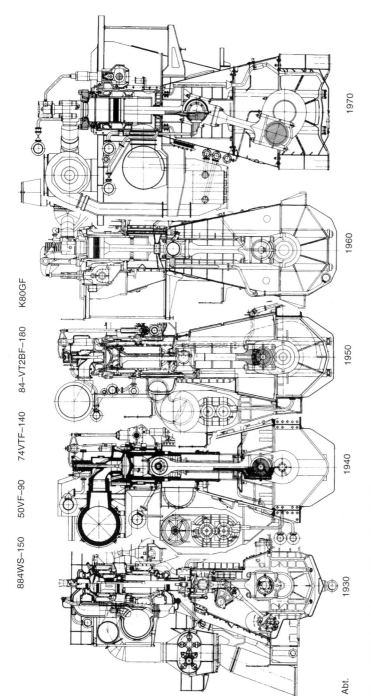

884WS–150 50VF–90 74VTF–140 84–VT2BF–180 K80GF

Abt.   1930   1940   1950   1960   1970

Figure I.12 Development of Burmeister & Wain uniflow-scavenged engine designs

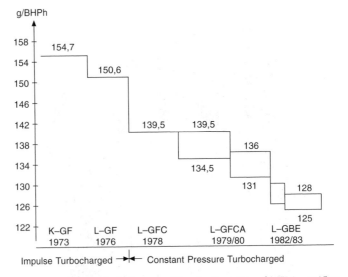

Figure I.13 Advances in specific fuel consumption by Burmeister & Wain uniflow-scavenged two-stroke engine designs (900 mm bore models)

regulation- or market-driven shift to much cleaner fuels (liquid or gas) could open the door to thwarted rivals.

As well as stifling coal- and oil-fired steam plant in its rise to dominance in commercial propulsion, the diesel engine shrugged off challenges from nuclear (steam) propulsion and gas turbines. Both modes found favour in warships, however, and aero-derived gas turbines have recently secured firm niches in fast ferry and cruise tonnage. A stronger challenge from gas turbines has to be faced (although combined diesel and gas turbine solutions are an option for high-powered installations) and diminishing fossil fuel availability may yet see nuclear propulsion revived in the longer term.

Diesel engine pioneers MAN B&W Diesel and Sulzer (the latter now part of the Wärtsilä Corporation) have both celebrated their centenaries since the last edition of Pounder's and are committed with other major designers to maintaining a competitive edge deep into this century. Valuable support will continue to flow from specialists in turbocharging, fuel treatment, lubrication, automation, materials, and computer-based diagnostic/monitoring systems and maintenance and spares management programs.

Key development targets aim to improve further the ability to burn low-grade bunkers (including perhaps coal-derived fuels and slurries) without compromising reliability; reduce noxious exhaust gas emissions; extend the durability of components and the periods

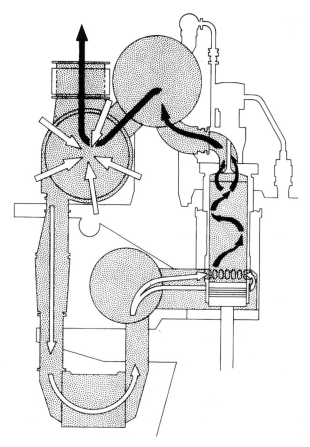

Figure I.14 Burmeister & Wain uniflow scavenging system

between major overhauls; lower production and installation costs; and simplify operation and maintenance routines.

Low speed engines with electronically-controlled fuel injection and exhaust valve actuating systems are entering service in increasing numbers, paving the way for future 'Intelligent Engines': those which can monitor their own condition and adjust key parameters for optimum performance in a selected running mode.

Potential remains for further developments in power and efficiency from diesel engines, with concepts such as steam injection and combined diesel and steam cycles projected to yield an overall plant efficiency of around 60 per cent. The Diesel Combined Cycle calls for a drastic change in the heat balance, which can be effected by

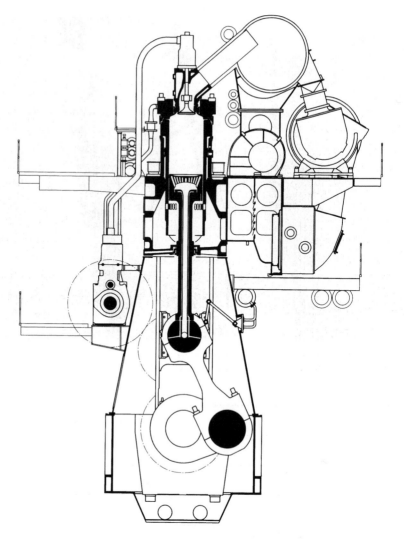

Figure I.15 Cross-section of a modern large bore two-stroke low speed engine, the Sulzer RTA96C

the Hot Combustion process. Piston top and cylinder head cooling is eliminated, cylinder liner cooling minimized, and the cooling losses concentrated in the exhaust gas and recovered in a boiler feeding high pressure steam to a turbine.

Acknowledgements to: ABB Turbo Systems, MAN B&W Diesel and Wärtsilä Corporation.

Figure I.16 The most powerful diesel engines in service are 12-cylinder MAN B&W K98MC low speed two-stroke models developing 68 640 kW

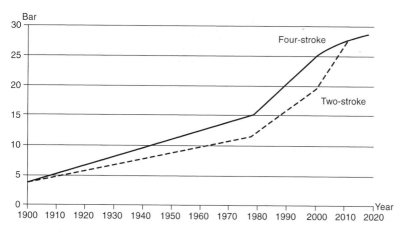

Figure I.17 Historical and estimated future development of mean effective pressure ratings for two-stroke and four-stroke diesel engines (Wärtsilä Corporation)

1 Theory and general principles

THEORETICAL HEAT CYCLE

In the original patent by Rudolf Diesel the diesel engine operated on the diesel cycle in which the heat was added at constant pressure. This was achieved by the blast injection principle. Today the term is universally used to describe any reciprocating engine in which the heat induced by compressing air in the cylinders ignites a finely atomized spray of fuel. This means that the theoretical cycle on which the modern diesel engine works is better represented by the dual or mixed cycle, diagrammatically illustrated in Figure 1.1. The area of the diagram, to a suitable scale, represents the work done on the piston during one cycle.

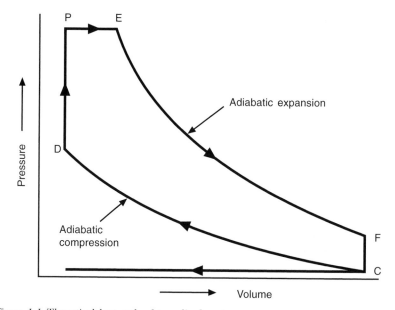

Figure 1.1 Theoretical heat cycle of true diesel engine

1

Starting from point C, the air is compressed adiabatically to a point D. Fuel injection begins at D, and heat is added to the cycle partly at constant volume as shown by vertical line DP, and partly at constant pressure, as shown by horizontal line PE. At the point E expansion begins. This proceeds adiabatically to point F when the heat is rejected to exhaust at constant volume as shown by vertical line FC.

The ideal efficiency of this cycle (i.e. of the hypothetical indicator diagram) is about 55–60 per cent: that is to say, about 40–45 per cent of the heat supplied is lost to the exhaust. Since the compression and expansion strokes are assumed to be adiabatic, and friction is disregarded, there is no loss to coolant or ambient. For a four-stroke engine the exhaust and suction strokes are shown by the horizontal line at C, and this has no effect on the cycle.

PRACTICAL CYCLES

While the theoretical cycle facilitates simple calculation, it does not exactly represent the true state of affairs. This is because:

1. The manner in which, and the rate at which, heat is added to the compressed air (the heat release rate) is a complex function of the hydraulics of the fuel injection equipment and the characteristic of its operating mechanism; of the way the spray is atomized and distributed in the combustion space; of the air movement at and after top dead centre (TDC); and to a degree also of the qualities of the fuel.

2. The compression and expansion strokes are not truly adiabatic. Heat is lost to the cylinder walls to an extent which is influenced by the coolant temperature and by the design of the heat paths to the coolant.

3. The exhaust and suction strokes on a four-stroke engine (and the appropriate phases of a two-stroke cycle) do create pressure differences which the crankshaft feels as 'pumping work'.

It is the designer's objective to minimize all these losses without prejudicing first cost or reliability, and also to minimise the cycle loss: that is, the heat rejected to exhaust. It is beyond the scope of this book to derive the formulae used in the theoretical cycle, and in practice designers have at their disposal sophisticated computer techniques which are capable of representing the actual events in the cylinder with a high degree of accuracy. But broadly speaking, the cycle efficiency is a function of the compression ratio (or more correctly the effective expansion ratio of the gas/air mixture after combustion).

The theoretical cycle (Figure 1.1) may be compared with a typical actual diesel indicator diagram such as that shown in Figure 1.2. Note that in higher speed engines combustion events are often represented on a crank angle, rather than a stroke basis, in order to achieve better accuracy in portraying events at the top dead centre (TDC), as in Figure 1.3. The actual indicator diagram is derived from it by transposition. This form of diagram is useful too when setting injection timing. If electronic indicators are used it is possible to choose either form of diagram.

An approximation to a crank angle based diagram can be made with mechanical indicators by disconnecting the phasing and taking a card quickly, pulling it by hand: this is termed a 'draw card'.

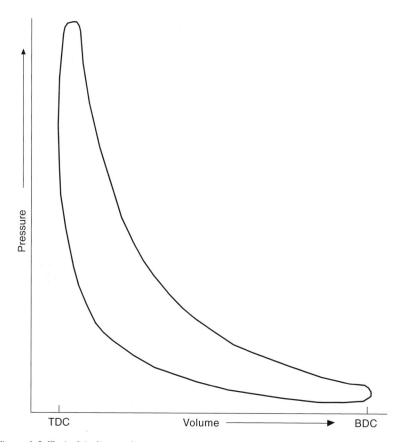

Figure 1.2 Typical indicator diagram (stroke based)

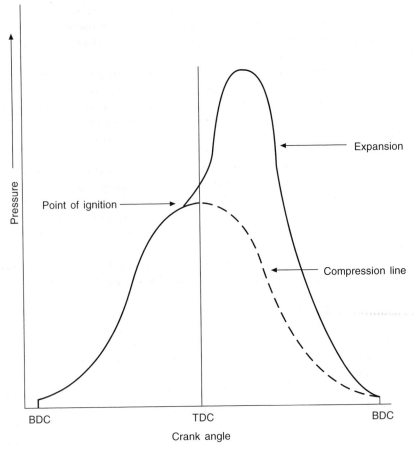

Figure 1.3 Typical indicator diagram (crank angle based)

EFFICIENCY

The only reason a practical engineer wants to run an engine at all is to achieve a desired output of useful work, which is, for our present purposes, to drive a ship at a prescribed speed, and/or to provide electricity at a prescribed kilowattage.

To determine this power he must, therefore, allow not only for the cycle losses mentioned above but for the friction losses in the cylinders, bearings and gearing (if any) together with the power consumed by engine-driven pumps and other auxiliary machines. He must also allow for such things as windage. The reckoning is further complicated by the fact that the heat rejected from the cylinder to exhaust is not necessarily totally lost, as practically all modern engines use up to 25 per cent of that heat to drive a turbocharger. Many use much of the remaining high temperature heat to raise steam, and low temperature heat for other purposes.

The detail is beyond the scope of this book but a typical diagram (usually known as a Sankey diagram), representing the various energy flows through a modern diesel engine, is reproduced in Figure 1.4. The right-hand side represents a turbocharged engine and an indication is given of the kind of interaction between the various heat paths as they leave the cylinders after combustion.

Note that the heat released from the fuel in the cylinder is augmented by the heat value of the work done by the turbocharger in compressing the intake air. This is apart from the turbocharger's function in introducing the extra air needed to burn an augmented quantity of fuel in a given cylinder, compared with what the naturally aspirated system could achieve, as in the left-hand side of the diagram.

It is the objective of the marine engineer to keep the injection settings, the air flow, coolant temperatures (not to mention the general mechanical condition) at those values which give the best fuel consumption for the power developed.

Note also that, whereas the fuel consumption is not difficult to measure in tonnes per day, kilograms per hour or in other units, there are many difficulties in measuring work done in propelling a ship. This is because the propeller efficiency is influenced by the entry conditions created by the shape of the afterbody of the hull, by cavitation, etc., and also critically influenced by the pitch setting of a controllable pitch propeller. The resulting speed of the ship is much dependent, of course, on hull cleanliness, wind and sea conditions, draught, and so on. Even when driving a generator it is necessary to allow for generator efficiency and instrument accuracy.

It is normal when talking of efficiency to base the work done on that transmitted to the driven machinery by the crankshaft. In a propulsion system this can be measured by a torsionmeter; in a generator it can be measured electrically. Allowing for measurement error, these can be compared with figures measured on a brake in the test shop.

THERMAL EFFICIENCY

Thermal efficiency (Thη) is the overall measure of performance. In absolute terms it is equal to:

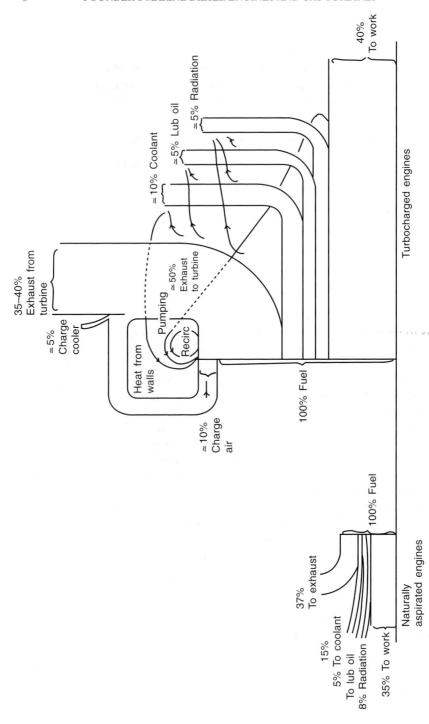

Figure 1.4 Typical Sankey diagrams

$$\frac{\text{heat converted into useful work}}{\text{total heat supplied}} \tag{1.1}$$

As long as the units used agree it does not matter whether the heat or work is expressed in pounds–feet, kilograms–metres, BTU, calories, kWh or joules. The recommended units to use are now those of the SI system.

$$\text{Heat converted into work per hour} = N \text{ kWh}$$

$$= 3600 \ N \text{ kJ}$$

where N = the power output in kW

$$\text{Heat supplied} = M \times K$$

where M = mass of fuel used per hour in kg
and K = calorific value of the fuel in kJ/kg

$$\text{therefore Th}\eta = \frac{3600 \ N}{M \times K} \tag{1.2}$$

It is now necessary to decide where the work is to be measured. If it is to be measured in the cylinders, as is usually done in slow-running machinery, by means of an indicator (though electronic techniques now make this possible directly and reliably even in high speed engines), the work measured (and hence power) is that indicated within the cylinder, and the calculation leads to the *indicated* thermal efficiency.

If the work is measured at the crankshaft output flange, it is net of friction, auxiliary drives, etc., and is what would be measured by a brake, whence the term *brake* thermal efficiency. [Manufacturers in some countries do include as output the power absorbed by essential auxiliary drives but some consider this to give a misleading impression of the power available.]

Additionally, the fuel is reckoned to have a higher (or gross) and a lower (or net) calorific value, according to whether one calculates the heat recoverable if the exhaust products are cooled back to standard atmospheric conditions, or assessed at the exhaust outlet. The essential difference is that in the latter case the water produced in combustion is released as steam and retains its latent heat of vaporization. This is the more representative case—and more desirable as water in the exhaust flow is likely to be corrosive. Today the net or lower calorific value (LCV) is more widely used.

Returning to our formula (Equation 1.2.), if we take the case of an engine producing a (brake) output of 10 000 kW for an hour using 2000 kg of fuel per hour having an LCV of 42 000 kJ/kg

$$\text{(Brake) Th}\eta = \frac{3600 \times 10\ 000}{2000 \times 42\ 000} \times 100\%$$

$$= 42.9\% \text{ (based on LCV)}$$

MECHANICAL EFFICIENCY

$$\text{Mechanical efficiency} = \frac{\text{output at crankshaft}}{\text{output at cylinders}} \qquad (1.3)$$

$$= \frac{\text{bhp}}{\text{ihp}} = \frac{\text{kW (brake)}}{\text{kW (indicated)}} \qquad (1.4)$$

The reasons for the difference are listed above. The brake power is normally measured with a high accuracy (98 per cent or so) by coupling the engine to a dynamometer at the builder's works. If it is measured in the ship by torsionmeter it is difficult to match this accuracy and, if the torsionmeter cannot be installed between the output flange and the thrust block or the gearbox input, additional losses have to be reckoned due to the friction entailed by these components.

The indicated power can only be measured from diagrams where these are feasible and they are also subject to significant measurement errors.

Fortunately for our attempts to reckon the mechanical efficiency, test bed experience shows that the 'friction' torque (that is, in fact, *all* the losses reckoned to influence the difference between indicated and brake torque) is not very greatly affected by the engine's torque output, nor by the speed. This means that the friction power loss is roughly proportional to speed, and fairly constant at fixed speed over the output range. Mechanical efficiency therefore falls more and more rapidly as brake output falls. It is one of the reasons why it is undesirable to let an engine run for prolonged periods at less than about 30 per cent torque.

WORKING CYCLES

A diesel engine may be designed to work on the two-stroke or on the four-stroke cycle: both of these are explained below. They should not be confused with the terms 'single-acting' or 'double-acting', which relate to whether the working fluid (the combustion gases) acts on one or both sides of the piston. (Note, incidentally, that the opposed piston two-stroke engine in service today is *single*-acting.)

The four-stroke cycle

Figure 1.5 shows diagrammatically the sequence of events throughout the typical four-stroke cycle of two revolutions. It is usual to draw such diagrams starting at TDC (firing), but the explanation will start at TDC (scavenge). Top dead centre is sometimes referred to as inner dead centre (IDC).

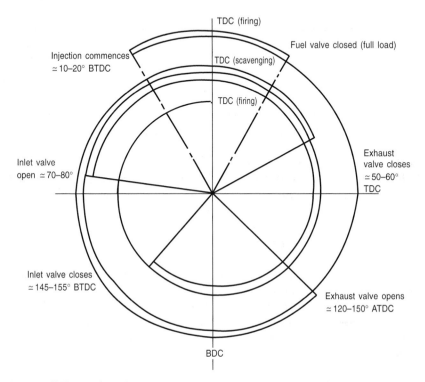

Figure 1.5 Four-stroke cycle

Proceeding clockwise round the diagram, both inlet (or suction) and exhaust valves are initially open. (All modern four-stroke engines have poppet valves.) If the engine is naturally aspirated, or is a small high speed type with a centripetal turbocharger, the period of valve overlap (i.e. when both valves are open) will be short, and the exhaust valve will close some 10° after top dead centre (ATDC).

Propulsion engines and the vast majority of auxiliary generator engines running at speeds below 1000 rev/min will almost certainly be turbocharged and will be designed to allow a generous throughflow of scavenge air at this point in order to control the turbine blade temperature. (See also Chapter 7.) In this case the exhaust valve will remain open until exhaust valve closure (EVC) at 50–60° ATDC. As the piston descends to outer or bottom dead centre (BDC) on the suction stroke, it will inhale a fresh charge of air. To maximize this, balancing the reduced opening as the valve seats against the slight ram or inertia effect of the incoming charge, the inlet (suction valve) will normally be held open until about 25–35° ABDC (145–155° BTDC). This event is called inlet valve closure (IVC).

The charge is then compressed by the rising piston until it has attained a temperature of some 550°C. At about 10–20° BTDC (firing), depending on the type and speed of the engine, the injector admits finely atomized fuel which ignites within 2–7° (depending on type again) and the fuel burns over a period of 30–50° while the piston begins to descend on the expansion stroke, the piston movement usually helping to induce air movement to assist combustion.

At about 120–150° ATDC the exhaust valve opens (EVO), the timing being chosen to promote a very rapid blow-down of the cylinder gases to exhaust. This is done: (a) to preserve as much energy as is practicable to drive the turbocharger, and (b) to reduce the cylinder pressure to a minimum by BDC to reduce pumping work on the 'exhaust' stroke. The rising piston expels the remaining exhaust gas and at about 70–80° BTDC the inlet valve opens (IVO) so that the inertia of the outflowing gas, plus the positive pressure difference, which usually exists across the cylinder by now, produces a through flow of air to the exhaust to 'scavenge' the cylinder.

If the engine is naturally aspirated the IVO is about 10° BTDC. The cycle now repeats.

The two-stroke cycle

Figure 1.6 shows the sequence of events in a typical two-stroke cycle, which, as the name implies, is accomplished in one complete revolution of the crank. Two-stroke engines invariably have ports to admit air when uncovered by the descending piston (or air piston where the engine has two pistons per cylinder). The exhaust may be via ports adjacent to the air ports and controlled by the same piston (loop scavenge) or via piston-controlled exhaust ports or poppet exhaust valves at the other end of the cylinder (uniflow scavenge). The principles of the cycle apply in all cases.

Starting at TDC, combustion is already under way and the exhaust opens (EO) at 110–120° ATDC to promote a rapid blow-down before the inlet opens (IO) about 20–30° later (130–150° ATDC). In this way the inertia of the exhaust gases—moving at about the speed of sound—

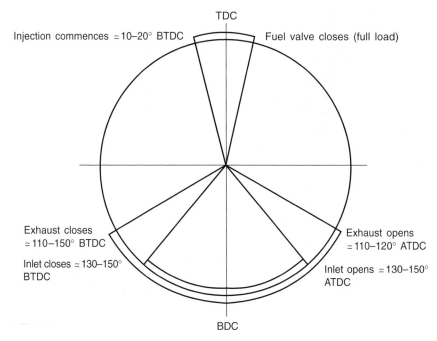

Figure 1.6 Two-stroke cycle

is contrived to encourage the incoming air to flow quickly through the cylinder with a minimum of mixing, because any unexpelled exhaust gas detracts from the weight of air entrained for the next stroke.

The exhaust should close before the inlet on the compression stroke to maximize the charge, but the geometry of the engine may prevent this if the two events are piston controlled. It can be done in an engine with exhaust valves, but otherwise the inlet and exhaust closure in a single-piston engine will mirror their opening. The inlet closure (IC) may be retarded relative to exhaust closure (EC) in an opposed-piston engine to a degree which depends on the ability of the designer and users to accept greater out-of-balance forces.

At all events the inlet ports will be closed as many degrees ABDC as they opened before it (i.e. again 130–150° BTDC) and the exhaust in the same region. Where there are two cranks and they are not in phase, the timing is usually related to that coupled to the piston controlling the air ports. The de-phasing is described as 'exhaust lead'.

Injection commences at about 10–20° BTDC depending on speed, and combustion occurs over 30–50°, as with the four-stroke engine.

HORSEPOWER

Despite the introduction of the SI system, in which power is measured in kilowatts (kW), horsepower cannot yet be discarded altogether. Power is the rate of doing work. In linear measure it is the mean force acting on a piston multiplied by the distance it moves in a given time. The force here is the mean pressure acting on the piston. This is obtained by averaging the difference in pressure in the cylinder between corresponding points during the compression and expansion strokes. It can be derived by measuring the area of an indicator diagram and dividing it by its length. This gives naturally the indicated mean effective pressure (imep), also known as mean indicated pressure (mip). Let this be denoted by 'p'.

Mean effective pressure is mainly useful as a design shorthand for the severity of the loading imposed on the working parts by combustion. In that context it is usually derived from the horsepower. If the latter is 'brake' horsepower (bhp), the mep derived is the brake mean effective pressure (bmep); but it should be remembered that it then has no direct physical significance of its own.

To obtain the total force the mep must be multiplied by the area on which it acts. This in turn comprises the area of one piston, $a = \pi d^2/4$, multiplied by the number of cylinders in the engine, denoted by N. The distance moved per cycle by the force is the working stroke (l), and for the chosen time unit, the total distance moved is the product of $l \times n$, where n is the number of *working* strokes in one cylinder in the specified time. Gathering all these factors gives the well-known 'plan' formula:

$$\text{power} = \frac{p \times l \times a \times n \times N}{k} \tag{1.5}$$

The value of the constant k depends on the units used. The units must be consistent as regards force, length and time.

If, for instance, SI units are used (newtons, metres and seconds) k will be 1000 and the power will be given in kW.

If imperial units are used (lb, feet and minutes) k will be 33 000 and the result is in imperial horsepower.

Onboard ship the marine engineer's interest in the above formula will usually be to relate mep and power for the engine with which he is directly concerned. In that case l, a, n become constants as well as k and

$$\text{power} = p \times C \times N \tag{1.6}$$

where $C = \dfrac{l \times a \times n}{k}$

Note that in an opposed-piston engine 'l' totals the sum of the strokes of the two pistons in each cylinder. To apply the formulae to double-acting engines is somewhat more complex since, for instance, allowance must be made for the piston rod diameter. Where double-acting engines are used it would be advisable to seek the builder's advice about the constants to be used.

TORQUE

The formula for power given in Equations (1.5) and (1.6) is based on the movement of the point of application of the force on the piston in a straight line. The inclusion of the engine speed in the formula arises in order to take into account the total distance moved by the force(s) along the cylinder(s): that is, the number of repetitions of the cycle in unit time.

Alternatively, power can be defined in terms of rotation.

If F = the effective resulting single force assumed to act tangentially at given radius from the axis of the shaft about which it acts

r = the nominated radius at which F is reckoned, and

n = revolutions per unit time of the shaft specified

then the circumferential distance moved by the tangential force in unit time is $2\pi\ rn$.

$$\text{Hence power} = \frac{F \times 2\,\pi rn}{K}$$

$$= \frac{Fr \times 2\,\pi n}{K} \tag{1.7}$$

The value of K depends on the system of units used, which, as before, must be consistent. In this expression $F \times r = T$, the torque acting on the shaft, and is measured (in SI units) in newton-metres.

Note that T is constant irrespective of the radius at which it acts. If n is in rev/min, and power is in kW, the constant $K = 1000$ so that

$$T = \frac{\text{power} \times 60 \times 1000}{2\pi n}$$

$$= \frac{30\,000 \times \text{power}}{\pi n} \text{ in newton-metres (Nm)} \tag{1.8}$$

If the drive to the propeller is taken through gearing, the torque acts at the pitch circle diameter of each of the meshing gears. If the pitch circle diameters of the input and output gears are respectively

d_1 and d_2 and the speeds of the two shafts are n_1 and n_2, the circumferential distance travelled by a tooth on either of these gears must be $\pi d_1 \times n_1$ and $\pi d_2 \times n_2$ respectively. But since they are meshed $\pi d_1 n_1 = \pi d_2 n_2$.

Therefore $\dfrac{n_1}{n_2} = \dfrac{d_2}{d_1}$

The tangential force F on two meshing teeth must also be equal. Therefore the torque on the input wheel

$$T_1 = \frac{Fd_1}{2}$$

and the torque on the output wheel

$$T_2 = \frac{Fd_2}{2}$$

or

$$\frac{T_1}{T_2} = \frac{d_1}{d_2}$$

If there is more than one gear train the same considerations apply at each. In practice there is a small loss of torque at each train due to friction, etc. This is usually of the order of 1.5–2 per cent at each train.

MEAN PISTON SPEED

This parameter is sometimes used as an indication of how highly rated an engine is. However, although in principle a higher piston speed can imply a greater degree of stress, as well as wear, etc., in modern practice the lubrication of piston rings and liner, as well as of other rubbing surfaces, has become much more scientific. It no longer follows that a 'high' piston speed is of itself more detrimental than a lower one in a well-designed engine.

Mean piston speed is simply $\dfrac{l \times n}{30}$ (1.9)

This is given in metres/sec if l = stroke in metres and n = revolutions per minute.

FUEL CONSUMPTION IN 24 HOURS

In SI units:

$$W = \frac{w \times \text{kW} \times 24}{1000}$$ (1.10)

$$w = \frac{1000\ W}{kW \times 24} \tag{1.11}$$

where w = fuel consumption rate, kg/kW h
        W = total fuel consumed per day in tonnes.
  (1 tonne = 1000 kg)

VIBRATION

Many problems have their roots in, or manifest themselves as, vibration. Vibration may be in any linear direction, and it may be rotational (torsional). Vibration may be resonant, at one of its natural frequencies, or forced. It may affect any group of components, or any one. It can occur at any frequency up to those which are more commonly called noise.

That vibration failures are less dramatic now than formerly is due to the advances in our understanding of it during the last 80 years. It can be controlled, once it is recognized, by design and by correct maintenance, by minimizing it at source, by damping, and by arranging to avoid exciting resonance. Vibration is a very complex subject and all that will be attempted here is a brief outline.

Any elastically coupled shaft or other system will have one or more natural frequencies which, if excited, can build up to an amplitude which is perfectly capable of breaking crankshafts. 'Elastic' in this sense means that a displacement or a twist from rest creates a force or torque tending to return the system to its position of rest, and which is proportional to the displacement. An elastic system, once set in motion in this way, will go on swinging, or vibrating, about its equilibrium position forever, in the theoretical absence of any damping influence. The resulting time/amplitude curve is exactly represented by a sine wave, i.e. it is sinusoidal.

In general, therefore, the frequency of torsional vibration of a single mass will be:

$$f = \frac{1}{2\pi} \sqrt{\frac{q}{I}} \ \text{cycles per second} \tag{1.12}$$

where q is the stiffness in newton metres per radian
and I is the moment of inertia of the attached mass in kg metres^2.

For a transverse or axial vibration

$$f = \frac{1}{2\pi} \sqrt{\frac{s}{m}} \ \text{cycles per second} \tag{1.13}$$

where s is the stiffness in newtons per metre of deflection and m is the mass attached in kg.

The essence of control is to adjust these two parameters, q and I (or s and m), to achieve a frequency which does not coincide with any of the forcing frequencies.

Potentially the most damaging form of vibration is the torsional mode, affecting the crankshaft and propeller shafting (or generator shafting). Consider a typical diesel propulsion system, say a six-cylinder two-stroke engine with a flywheel directly coupled to a fixed pitch propeller. There will be as many 'modes' in which the shaft can be induced to vibrate naturally as there are shaft elements: seven in this case. For the sake of simplicity, let us consider the two lowest: the one-node mode and the two-node mode (Figure 1.7 (a) and (b)).

In the one-node case, when the masses forward of the node swing clockwise, those aft of it swing anti-clockwise and vice versa. In the two-node case, when those masses forward of the first node swing clockwise, so do those aft of the second node, while those between the two nodes swing anti-clockwise, and vice versa.

The diagrams in Figure 1.7 show (exaggerated) at left: the angular displacements of the masses at maximum amplitude in one direction. At right: they plot the corresponding circumferential deflections from the mean or unstressed condition of the shaft when vibrating in that mode. The line in the right-hand diagrams connecting the maximum amplitudes reached simultaneously by each mass on the shaft system is called the 'elastic curve'.

A node is found where the deflection is zero and the amplitude changes sign. The more nodes that are present, the higher the corresponding natural frequency.

The problem arises when the forcing frequencies of the externally applied, or input, vibration coincide with, or approach closely, one of these natural frequencies. A lowish frequency risks exciting the one-node mode; a higher frequency will possibly excite the two-node mode, and so on. Unfortunately, the input frequencies or—to give them their correct name, the 'forcing frequencies'—are not simple.

As far as the crankshaft is concerned, the forcing frequencies are caused by the firing impulses in the cylinders. But the firing impulse put into the crankshaft at any loading by one cylinder firing is not a single sinusoidal frequency at one per cycle. It is a complex waveform which has to be represented for calculation purposes by a component at $1 \times$ cycle frequency; another, usually lower in amplitude, at $2 \times$ cycle frequency; another at $3 \times$ and so on up to at least 10 before the components become small enough to ignore. These components are

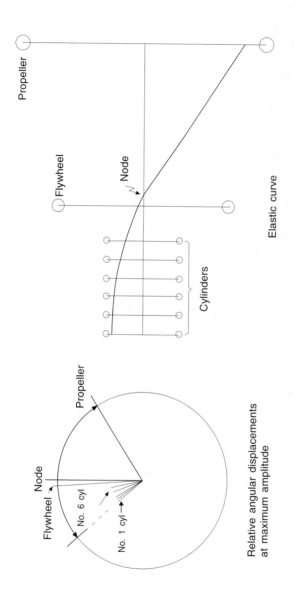

Figure 1.7(a) One-node mode

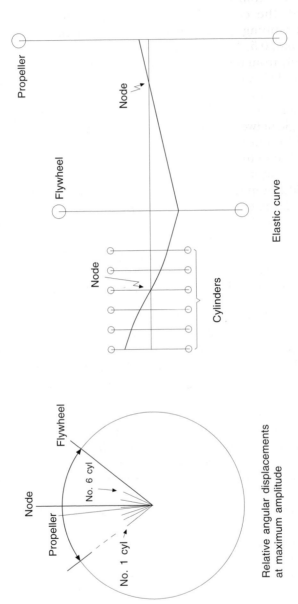

Figure 1.7(b) Two-node mode

called the 1st, 2nd, 3rd up to the 10th orders or harmonics of the firing impulse. For four-stroke engines, whose cycle speed is half the running speed, the convention has been adopted of basing the calculation on running speed. There will therefore be 'half' orders as well: for example, 0.5, 1.5, 2.5 and so on.

Unfortunately from the point of view of complexity, but fortunately from the point of view of control, these corresponding impulses have to be combined from all the cylinders according to the firing order. For the 1st order the interval between successive impulses is the same as the crank angle between successive firing impulses. For most engines, therefore, and for our six-cylinder engine in particular, the one-node 1st order would tend to cancel out, as shown in the vector summation in the centre of Figure 1.8. The length of each vector shown in the diagram is scaled from the corresponding deflection for that cylinder shown on the elastic curve, such as is in Figure 1.7.

On the other hand, in the case of our six-cylinder engine, for the 6th order, where the frequency is six times that of the 1st order (or fundamental order) to draw the vector diagram (right of Figure 1.8) all the 1st order phase angles have to be multiplied by 6. Therefore, all the cylinder vectors will combine linearly and become much more damaging.

If, say, the natural frequency in the one-node mode is 300 vibrations per minute (vpm) and our six-cylinder engine is run at 50 rev/min, the 6th harmonic ($6 \times 50 = 300$) would coincide with the one-node frequency and the engine would probably suffer major damage. 50 rev/min would be termed the '6th order critical speed' and the 6th order in this case is termed a 'major critical'.

Not only the engine could achieve this. The resistance felt by a propeller blade varies periodically with depth while it rotates in the water, and with the periodic passage of the blade tip past the stern post, or the point of closest proximity to the hull in the case of a multi-screw vessel. If a three-bladed propeller were used and its shaft run at 100 rev/min, a 3rd order of propeller excited vibration could also risk damage to the crankshaft (or whichever part of the shaft system was most vulnerable in the one-node mode).

The most significant masses in any mode of vibration are those with the greatest amplitude on the corresponding elastic curve. That is to say, changing them would have the greatest effect on frequency. The most vulnerable shaft sections are those whose combination of torque and diameter induce in them the greatest stress. The most significant shaft sections are those with the steepest change of amplitude on the elastic curve and therefore the highest torque. These are usually near the nodes but this depends on the relative shaft diameter. Changing

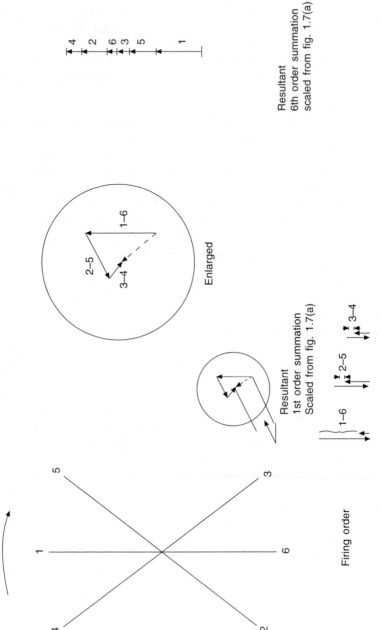

Figure 1.8 *Vector summations based on identical behaviour in all cylinders*

the diameter of such a section of shaft will also have a greater effect on the frequency.

The two-node mode is usually of a much higher frequency than the one-node mode in propulsion systems, and in fact usually only the first two or three modes are significant. That is to say that beyond the three-node mode the frequency components of the firing impulse that could resonate in the running speed range will be small enough to ignore.

The classification society chosen by the owners will invariably make its own assessment of the conditions presented by the vessel's machinery, and will judge by criteria based on experience.

Designers can nowadays adjust the frequency of resonance, the forcing impulses and the resultant stresses by adjusting shaft sizes, number of propeller blades, crankshaft balance weights and firing orders, as well as by using viscous or other dampers, detuning couplings and so on. Gearing, of course, creates further complications—and possibilities. Branched systems, involving twin input or multiple output gearboxes, introduce complications in solving them; but the principles remain the same.

The marine engineer needs to be aware, however, that designers tend to rely on reasonably correct balance among cylinders. It is important to realise that an engine with one cylinder cut out for any reason, or one with a serious imbalance between cylinder loads or timings, may inadvertently be aggravating a summation of vectors which the designer, expecting it to be small, had allowed to remain near the running speed range.

If an engine were run at or near a major critical speed it would sound rough because, at mid-stroke, the torsional oscillation of the cranks with the biggest amplitude would cause a longitudinal vibration of the connecting rod. This would set up in turn a lateral vibration of the piston and hence of the entablature. Gearing, if on a shaft section with a high amplitude, would also probably be distinctly noisy.

The remedy, if the engine appears to be running on a torsional critical speed, would be to run at a different and quieter speed while an investigation is made. Unfortunately, noise is not always distinct enough to be relied upon as a warning.

It is usually difficult, and sometimes impossible, to control all the possible criticals, so that in a variable speed propulsion engine it is sometimes necessary to 'bar' a range of speeds where vibration is considered too dangerous for continuous operation.

Torsional vibrations can sometimes affect camshafts also. Linear vibrations usually have simpler modes, except for those which are known as axial vibrations of the crankshaft. These arise because firing

causes the crankpin to deflect and this causes the crankwebs to bend. This in turn leads to the setting up of a complex pattern of axial vibration of the journals in the main bearings.

Vibration of smaller items, such as brackets holding components, or pipework, can often be controlled either by using a very soft mounting whose natural frequency is *below* that of the lowest exciting frequency, or by stiffening.

BALANCING

The reciprocating motion of the piston in an engine cylinder creates out-of-balance forces acting along the cylinder, while the centrifugal force associated with the crankpin rotating about the main bearing centres creates a rotating out-of-balance force. These forces, if not in themselves necessarily damaging, create objectionable vibration and noise in the engine foundations, and through them to the ship (or building) in which the engine is operating.

Balancing is a way of controlling vibrations by arranging that the overall summation of the out-of-balance forces and couples cancels out, or is reduced to a more acceptable amount.

The disturbing elements are in each case forces, each of which acts in its own plane, usually including the cylinder axis. The essence of balancing is that a force can be exactly replaced by a parallel force acting in a reference plane (chosen to suit the calculation) and a couple whose arm is the distance perpendicularly between the planes in which these two forces act (Figure 1.9). Inasmuch as balance is usually considered at and about convenient reference planes, all balancing involves a consideration of forces and of couples.

In multi-cylinder engines couples are present because the cylinder(s) coupled to each crankthrow act(s) in a different plane.

There are two groups of forces and couples. These relate to the revolving and to the reciprocating masses. (This section of the book is not concerned with balancing the power outputs of the cylinders. In fact, cylinder output balance, while it affects vibration levels, has no effect on the balance we are about to discuss.)

Revolving masses are concentrated at a radius from the crankshaft, usually at the crankpin, but are presumed to include a proportion of the connecting rod shank mass adjacent to the crankpin. This is done to simplify the calculations. Designers can usually obtain rotating balance quite easily by choice of crank sequence and balance weights.

Reciprocating masses are concentrated at the piston/crosshead, and similarly are usually assumed to include the rest of the connecting rod shank mass.

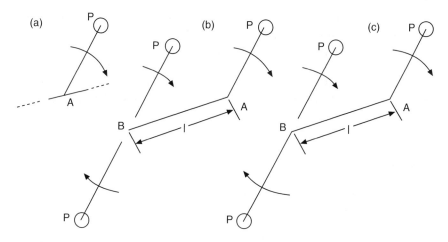

Figure 1.9 Principle of balancing
  (a) Force (P) to be balanced acting at A
  (b) Equal and opposite forces, of magnitude (P) assumed to act at B, a distance
      (I) arbitrarily chosen to suit calculation
  (c) System is equivalent to couple (P × I) plus force (P) now acting at B

Revolving masses turning at a uniform speed give rise to a single system of revolving forces and couples, each system comprising the related effects of each crank and of any balance weights in their correct phase relationship.

Reciprocating masses are more complicated, because of the effect of the obliquity of the connecting rod. This factor, particularly as the connecting rod length is reduced in relation to crank radius, causes the motion of the piston to depart from the simple sinusoidal time/ displacement pattern along the cylinder axis. It has instead to be represented by a primary (sinusoidal) component at a frequency corresponding to engine speed in rev/min, a secondary component at 2 × engine speed, a tertiary at 3 × engine speed, and so on. (It is seldom necessary to go beyond secondary.)

In practical terms, the reversal of direction involves greater acceleration at TDC than at BDC. This can be thought of as an average reversal force (the primary) at both dead centres, on which is superimposed an extra *inward* force at twice the frequency which adds to the primary at TDC and detracts from it at BDC. Reciprocating forces and couples always act in the plane of the cylinder. For V-cylinder engines they are combined vectorially. (The effect of side-by-side design for V-engine big ends is usually ignored.)

Most cylinder combinations of two- and of four-stroke engines lend themselves to complete balance of either primary forces, primary couples, secondary forces or secondary couples; some to any two, some to any three, and some to all of them.

Where one or more systems do not balance, the designer can contain the resulting vibration by low frequency anti-vibration mountings (as in the four-stroke four-cylinder automotive engine which normally has severe unbalanced secondary forces); or he can ameliorate it by adding extra rotational balance weights on a different shaft or shafts running at crankshaft speed (for primary unbalance) or at twice that speed (for secondary unbalance).

Note that a reciprocating unbalance force can be cancelled out if two shafts, each carrying a balance weight equal to half the offending unbalanced weight, and phased suitably, are mounted in the frame, running in opposite directions. The lateral effect of these two contra-rotating weights cancel out. This is the principle of Lanchester balancing used to cope with the heavy unbalanced secondary forces of four in-line and V8-cylinder four-stroke engines with uniform crank sequence.

In general it is nowadays dangerous to consider only the crankshaft system as a whole, because of the major effect that line-by-line balance has on oil film thickness in bearings. It is often necessary to balance internal couples much more closely, i.e. for each half of the engine or even line-by-line.

NOISE

Noise is unwanted sound. It is also a pervading nuisance and a hazard to hearing, if not to health itself. Noise is basically a form of vibration, and it is a complex subject; but a few brief notes will be given here.

To the scientist noise is reckoned according to its sound pressure level, which can be measured fairly easily on a microphone. Since the topic is essentially subjective, that is to say, it is of interest because of its effect on people via the human ear, the mere measurement of sound pressure level is of very limited use.

Noises are calibrated with reference to a sound pressure level of 0.0002 dynes per cm^2 for a pure 1000 Hz sine wave. A dyne is the force that gives 1 g an acceleration of 1 cm/s^2. It is 10^{-5} newtons. A linear scale gives misleading comparisons, so noise levels are measured on a logarithmic scale of bels. (1 bel means 10 times the reference level, 3 bel means 1000 times the reference level, i.e. 10^3.) For convenience the bel is divided into 10 parts, hence the decibel or db. This means that 80 db (8 bel) is 10^8 times the reference pressure level.

The human ear hears high and lower frequencies (over say 1 kHz or below 50 Hz) as louder than the middle frequencies, even when sound pressure levels are the same. For practical purposes, sounds are measured according to a frequency scale weighted to correspond to the response of the human ear. This is the 'A' scale and the readings are quoted in dBA.

The logarithmic-based scale is sometimes found confusing, so it is worth remembering that:

30 db corresponds to a gentle breeze in a meadow
70 db to an open office
100 db to a generator room.

Any prolonged exposure to levels of 85 db or above is likely to lead to hearing loss in the absence of ear protection; 140 db or above is likely to be physically painful.

The scope for reducing at source noise emitted by a diesel engine is limited, without fundamental and very expensive changes of design principle. However, the measures which lead to efficiency and economy—higher pressures, faster pressure rise and higher speeds—all, unfortunately, lead also to greater noise, and to noises emitted at higher frequencies and therefore more objectionably.

The noise from several point sources is also added in the logarithmic scale. Two point sources (e.g. cylinders) are twice as noisy as one. $\text{Log}_{10} 2 = 0.3010$, i.e. 0.301 bel or 3.01 db louder than one. Six cylinders, since $\log_{10} 6 = 0.788$, are 7.88 db louder than one.

If there is no echo, sound diminishes according to the square of the distance from the source (usually measured at a reference distance of 0.7 metres in the case of diesel engines). So moving ten times further away reduces the noise to 1/100th, or by 20 db.

Unfortunately a ship's engine room has plenty of reflecting surfaces so there is little benefit to be had from moving away. The whole space tends to fill with noise only a few db less than the source. Lining as much as possible with sound-absorbing materials does two things:

1. It reduces the echo, so that moving away from the engine gives a greater reduction in perceived noise.
2. It tends to reduce sympathetic (resonant) vibration of parts of the ship's structure, which, by drumming, add to the vibration and noise which is transmitted into the rest of the ship.

Anti-vibration mounts (see page 29) help in the latter case also but are not always practical. The only other measure which can successfully reduce noise is to put weight, particularly if a suitable cavity can be incorporated (or a vacuum), between the source and the observer.

A screen, of almost any material, weighing 5 kg/m^2 will effect a reduction of 10 db in perceived noise, and pro rata. Where weight is increased by increasing thickness it is more effective to do so in porous/flexible material than in rigid material. For the latter the db reduction in transmitted noise is proportional to the $\log_{10}$ of the weight, but for the former it is proportional to the thickness. A tenfold increase in the thickness of the rigid material of 5 kg/m^2 would double the attenuation; in porous material it would be tenfold.

The screen must be totally effective. A relatively small aperture will destroy much of the benefit. For example, most ships' enginerooms now incorporate a control room to provide some noise protection. The benefit will be appreciated every time the door is opened.

ACHIEVING QUIETER ENGINEROOMS

On the basis of engine noise measurements and frequency analyses, MAN B&W Diesel explains, it can be determined that noise emissions from two-stroke engines primarily originate from:

- The turbocharger, air and gas pulsations.
- Exhaust valves.
- Fuel oil injection systems.

The chain drive also contributes to a certain extent. The best way of reducing engine-related noise is, naturally, to reduce the vibrational energy at the source or, if this is neither feasible nor adequate, to attenuate the noise as close to its source as possible (Figure 1.10).

National and international standards for noise levels in ships have, in general, resulted in a considerable reduction of noise in newly-built tonnage, especially in accommodation spaces, but enginerooms remain a problem, suggests MAN B&W Diesel. Limits for maximum sound pressure levels in areas of a newbuilding are either defined specifically between shipowner/yard and enginebuilder or indirectly by referring to relevant national and international legislation.

Many owners refer to the German SBG (See-Berufsgenossenschaft) specifications or the IMO recommendations (see table below). On the bridge wing, where exhaust gas noise predominates, there are certain limitations as this area is regarded as a listening post. The requirement, depending on the noise standard to be met, is a maximum of 60–70 dB(A), which can always be satisfied by installing a suitable exhaust gas silencer. Ships built and registered in the Far East are often not equipped with exhaust gas silencers, while European-built

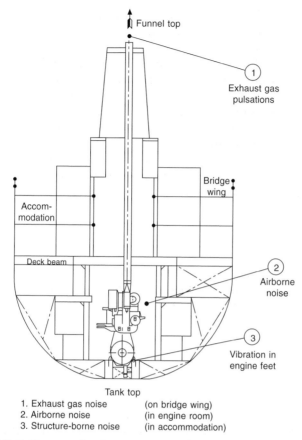

Figure 1.10 Typical sources of engine noise (MAN B&W Diesel)

tonnage is normally fitted with silencers in accordance with rules and regulations prevailing in Europe. In saving the cost of silencers, owners often do not specify compliance with European rules for ships built in the Far East; and even if such tonnage is later transferred to Europe compliance with the rules is not required.

Noise reductions have not been achieved in the engineroom, however, where airborne noise from the main engine dominates. The reason, MAN B&W Diesel explains, is that the acceptable noise limits for periodically-manned enginerooms have been set at around 110 dB(A) for many years; and the introduction of stricter requirements has not been realistic as noise emissions from engines have increased over the years with rising power ratings. The unchanged noise limit in itself seems to have constituted a serious limitation for enginebuilders. It is recognised, however, that a noise level of over 110 dB(A) can, in the long term, cause permanent damage to hearing, and any easing of this limit cannot be expected—rather to the contrary. Engine designers must therefore pay particular attention to reducing the airborne sound levels emitted from their future models.

Noise emitted by the main diesel engine's exhaust gas, and the structure-borne noise excited by the engine, are generally so low that keeping within the noise level requirements for the bridge wing and accommodation does not pose a problem. Airborne noise emitted from the engine, on the other hand, is so high that in some cases there is a risk that the noise limits for the engineroom cannot be met unless additional noise reduction measures are introduced. These measures may include:

- A Helmholtz resonator lining in the scavenge air pipe.
- External insulation of the scavenge air receiver.
- External insulation of the scavenge air cooler.
- Additional absorption material at the engine and/or at the engineroom walls (shipyard responsibility).
- Additional turbocharger intake silencer attenuation (turbocharger maker's responsibility).
- Additional attenuation material at the turbocharger inspection cover.
- Low noise diffusor for turbocharger compressor (if available).

Such measures may reduce the maximum noise levels by 3–5 dB(A) or more, depending on their extent, says MAN B&W Diesel. It would nevertheless be extremely difficult to meet stricter engineroom noise level requirements of, say, 105 dB(A) instead of 110 dB (A), especially in view of the influence of sound reverberations and the noise emitted by other machinery. The possibility of reducing the noise from an existing engine is greatly limited because the noise stems from many different sources, and because the noise transmission paths (through which vibrational energy is transferred from one area to another through the engine) are numerous. In principle, however, the transmission of airborne noise from the engineroom to other locations (accommodation quarters, for example) normally has no influence on the actual noise level in those locations.

IMO Noise Limits (Sound Pressure Level) in dB(A)

Workspaces

Machinery spaces (continuously unmanned)*	90
Machinery spaces (not continuously manned)*	110
Machinery control rooms	75
Workshops	85
Unspecified workspaces*	90

Navigation spaces

Navigating bridge and chartrooms	65
Listening posts (including bridge wings and windows)	70
Radio rooms (with radio equipment operating but not producing audio signals)	60
Radar rooms	65

Accommodation spaces

Cabins and hospital	60
Messrooms	65
Recreation rooms	65
Open recreation areas	75
Offices	65

*Ear protectors should be worn when the noise level is above 85 dB(A), and no individual's daily exposure duration should exceed four hours continuously or eight hours in total.

ANTI-VIBRATION MOUNTINGS

Over the past 25 years the use of elastomer-based mounting systems to suppress or attenuate noise and vibration in ships has advanced both in technical sophistication and growth of application. Rubber-to-metal bonded systems are the most commonly applied anti-vibration mountings, and also deliver the highest performance rating for a given size. Shipboard installations mainly benefit propulsion engines, gensets and diverse ancillary machinery, such as ventilation fans, compressors, pumps, sewage treatment units, refrigeration plant and instrument panels.

Any item of machinery with moving parts is subject to out-of-balance forces and couples created by the interaction of those parts. The magnitude of the forces and couples will vary considerably, depending on the configuration of the machine design. In theory, some machines will be totally balanced and there will be no external forces or couples (for example, an electric motor or a six-cylinder four-stroke engine). In practice, however, manufacturing tolerances may create out-of-balance excitations that can become significant in large elements of the machinery. The weights of pistons and connecting rods in large diesel engines may vary substantially, although to some extent this is allowed for by selective assembly and balancing. Uneven firing in large engines, due to variation in combustion between the cylinders, can also cause significant excitation.

The various excitations in a machinery element will either have discrete frequencies, based on a constant running speed, or there will be a frequency range to contend with (as in the case of a variable speed machine). In practical terms, it may not be possible or cost effective to eliminate vibration at source, and the role of anti-vibration mountings is therefore to reduce the level of transmitted vibration.

In an active isolation system the use of anti-vibration mountings beneath a machine will isolate the surrounding structure from the vibrating machine and equipment; in a passive system the mountings may be used to protect an item of sensitive equipment—such as an instrument panel—from external sources of vibration or shock. The amplitude of movement of the equipment installed on a flexible mounting system can be reduced by increasing the inertia of the vibrating machine, typically by adding mass in the form of an inertia block. Unfortunately, adding mass is not generally acceptable for marine and offshore applications because of the penalties of weight and cost.

A number of standard ranges of rubber/metal-bonded anti-vibration and suspension components are offered by specialist companies, but most installations differ and may call for custom-designed solutions. Computer analysis is exploited to study individual applications, predict the behaviour of various isolation systems and recommend an optimum solution. It is usually possible to design an anti-vibration system of at least 85 per cent efficiency against the worst possible condition, reports Sweden-based specialist Trelleborg.

LOW SPEED ENGINE VIBRATION: CHARACTERISTICS AND CURES

Ship machinery installations have two principal sources of excitation: the main engine(s) and the propeller(s). The two components are essentially linked by elastic shaft systems and may also embrace gearboxes and elastic couplings. The whole system is supported in flexible hull structures, and the forms of vibration possible are therefore diverse.

Problems in analysing ship machinery vibration have been compounded in recent years by the introduction of new fuel-efficient engine designs exploiting lower running speeds, longer stroke/bore ratios and higher combustion pressures. Another factor is the wider use of more complex machinery arrangements including, for example, power take-offs, shaft alternators, exhaust gas power turbines and multiple geared engines. The wider popularity of four- and five-cylinder low speed two-stroke engines for propulsion plant—reflecting attractive installation and operating costs—has also stimulated efforts by designers to counteract adverse vibration characteristics.

The greater complexity of vibration problems dictates a larger number of calculations to ensure satisfactory vibration levels from projected installations. For optimum results the vibration performance of the plant has to be investigated for all anticipated operational modes. In one case cited by Sulzer, in which two unequal-sized low speed engines were geared to a single controllable pitch propeller with a pair of PTO-driven generators, some eleven different operational configurations were possible.

Sophisticated tools such as computer software and measuring systems are now available to yield more detailed and accurate analysis of complex vibration problems. Even so, inaccuracies in determining shaft stiffness, damping effects, coupling performance and hull structure response make it advisable in certain borderline cases to allow, at the design stage, for suitable countermeasures to be applied after vibration measurements have been taken during sea trials.

Both MAN B&W Diesel and Wartsila (Sulzer) stress that proper consideration should be given to the vibration aspects of a projected installation at the earliest possible stage in the ship design process. The available countermeasures provide a good safety margin against potential vibration problems. Close collaboration is desirable between naval architects, machinery installation designers, enginebuilders and specialist component suppliers.

MAN B&W Diesel emphasises that the key factor is the interaction with the ship and not the mere magnitude of the excitation source. Excitations generated by the engine can be divided into two categories:

- *Primary excitations:* forces and moments originating from the combustion pressure and the inertia forces of the rotating and reciprocating masses. These are characteristics of the given engine which can be calculated in advance and stated as part of the engine specification with reference to a certain speed and power.
- *Secondary excitations:* stemming from a forced vibratory response

in a ship sub-structure. The vibration characteristics of sub-structures are almost independent of the remaining ship structure.

Examples of secondary excitation sources from sub-structures could be anything from transverse vibration of the engine structure to longitudinal vibration of a radar or light mast on top of the deckhouse. Such sub-structures of the complete ship might have resonance or be close to resonance conditions, resulting in considerable dynamically magnified reaction forces at their interface with the rest of the ship. Secondary excitation sources cannot be directly quantified for a certain engine type but must be calculated at the design stage of the specific propulsion plant.

The vibration characteristics of low speed two-stroke engines, for practical purposes, can be split into four categories that may influence the hull (Figures 1.11(a) and 1.11(b)):

- External unbalanced moments: these can be classified as unbalanced 1st and 2nd order external moments which need to be considered only for engines with certain cylinder numbers.

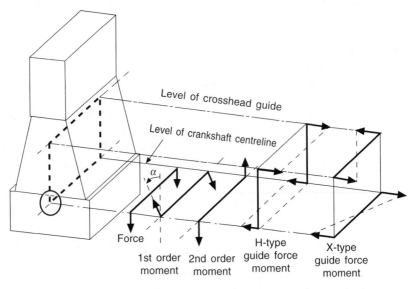

Figure 1.11(a) Forces and moments of a multi-cylinder low speed two-stroke engine. The firing order will determine the vectorial sum of the forces and moments from the individual cylinders. A distinction should be made between external forces and moments, and internal forces and moments. The external forces and moments will act as resultants on the engine and thereby also on the ship through the foundation and top bracing of the engine. The internal forces and moments will tend to deflect the engine as such (MAN B&W Diesel)

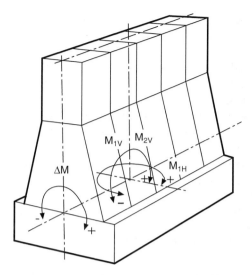

Figure 1.11(b) Free couples of mass forces and the torque variation about the centre lines of the engine and crankshaft (New Sulzer Diesel): M_{1V} is the 1st order couple having a vertical component. M_{1H} is the 1st order couple having a horizontal component. M_{2V} is the 2nd order couple having a vertical component. ΔM is the reaction to variations in the nominal torque. Reducing the 1st order couples is achieved by counterweights installed at both ends of the crankshaft

- Guide force moments.
- Axial vibrations in the shaft system.
- Torsional vibrations in the shaft system.

The influence of the excitation forces can be minimized or fully compensated if adequate countermeasures are considered from the early project stage. The firing angles can be customized to the specific project for nine-, ten-, eleven- and twelve-cylinder engines.

External unbalanced moments

The inertia forces originating from the unbalanced rotating and reciprocating masses of the engine create unbalanced external moments although the external forces are zero. Of these moments, only the 1st order (producing one cycle per revolution) and the 2nd order (two cycles per revolution) need to be considered, and then only for engines with a low number of cylinders. The inertia forces on engines with more than six cylinders tend, more or less, to neutralize themselves.

Countermeasures have to be taken if hull resonance occurs in the operating speed range, and if the vibration level leads to higher accelerations and/or velocities than the guidance values given by international standards or recommendations (for example, with reference to a special agreement between shipowner and yard). The natural frequency of the hull depends on its rigidity and distribution of masses, while the vibration level at resonance depends mainly on the magnitude of the external moment and the engine's position in relation to the vibration nodes of the ship.

1st order moments

These moments act in both vertical and horizontal directions, and are of the same magnitude for MAN B&W two-stroke engines with standard balancing. For engines with five cylinders or more, the 1st order moment is rarely of any significance to the ship but it can be of a disturbing magnitude in four-cylinder engines.

Resonance with a 1st order moment may occur for hull vibrations with two and/or three nodes. This resonance can be calculated with reasonable accuracy, and the calculation will show whether a compensator is necessary or not on four-cylinder engines. A resonance with the vertical moment for the two-node hull vibration can often be critical, whereas the resonance with the horizontal moment occurs at a higher speed than the nominal because of the higher natural frequency of the horizontal hull vibrations.

Four-cylinder MAN B&W MC two-stroke engines with bores from 500 mm to 980 mm are fitted as standard with adjustable counterweights (Figure 1.12). These can reduce the vertical moment to an insignificant

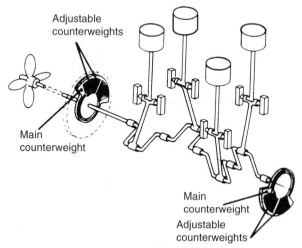

Figure 1.12 Four-cylinder engines of MAN B&W Diesel's 500 mm to 980 mm bore MC types are fitted with adjustable counterweights

value (although increasing, correspondingly, the horizontal moment) so this resonance is easily handled. A solution with zero horizontal moment is also available.

For smaller bore MAN B&W engines (S26MC, L35MC, S35MC, L42MC, S42MC and S46MC-C series) these adjustable counterweights can be ordered as an option.

In rare cases, where the 1st order moment will cause resonance with both the vertical and the horizontal hull vibration modes in the normal speed range of the engine, a 1st order compensator (Figure 1.13) can be introduced in the chain tightener wheel, reducing the 1st order moment to a harmless value. The compensator is an option and comprises two counter-rotating masses rotating at the same speed as the crankshaft.

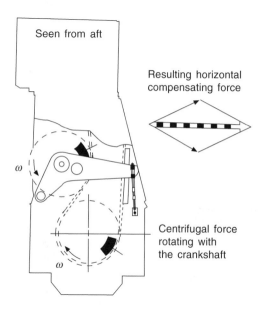

Figure 1.13 1st order moment compensator (MAN B&W Diesel)

With a 1st order moment compensator fitted aft, the horizontal moment will decrease to between 0 and 30 per cent of the value stated in MAN B&W Diesel tables, depending on the position of the node. The 1st order vertical moment will decrease to around 30 per cent of the value stated in the tables.

Since resonance with both the vertical and the horizontal hull vibration mode is rare the standard engine is not prepared for the fitting of such compensators.

2nd order moments

The 2nd order moment acts only in the vertical direction and precautions need only be considered for four-, five- and six-cylinder engines. Resonance with the 2nd order moment may occur at hull vibrations with more than three nodes.

A 2nd order moment compensator comprises two counter-rotating masses running at twice the engine speed. Such compensators are not included in the basic extent of MAN B&W Diesel engine delivery.

Several solutions are available to cope with the 2nd order moment (Figure 1.14), from which the most efficient can be selected for the individual case:

- No compensators, if considered unnecessary on the basis of natural frequency, nodal point and size of 2nd order moment.

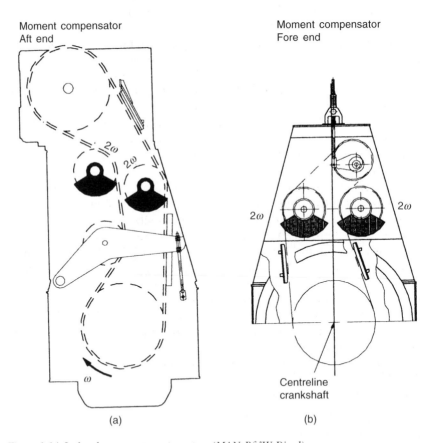

Figure 1.14 2nd order moment compensators (MAN B&W Diesel)

- A compensator mounted on the aft end of the engine, driven by the main chain drive.
- A compensator mounted on the fore end, driven from the crankshaft through a separate chain drive.
- Compensators on both aft and fore end, completely eliminating the external 2nd order moment.

Experience has shown, MAN B&W Diesel reports, that ships of a size propelled by the S26MC, L/S35MC and L/S42MC engines are less sensitive to hull vibrations. Engine-mounted 2nd order compensators are therefore not applied on these smaller models.

A decision regarding the vibrational aspects and the possible use of compensators must be taken at the contract stage. If no experience is available from sisterships (which would be the best basis for deciding whether compensators are necessary or not) it is advisable for calculations to be made to determine which of the recommended solutions should be applied.

If compensator(s) are omitted the engine can be delivered prepared for their fitting at a later date. The decision for such preparation must also be taken at the contract stage. Measurements taken during the sea trial, or later in service with a fully loaded ship, will show whether compensator(s) have to be fitted or not.

If no calculations are available at the contract stage, MAN B&W Diesel advises the supply of the engine with a 2nd order moment compensator on the aft end and provision for fitting a compensator on the fore end.

If a decision is made not to use compensators and, furthermore, not to prepare the engine for their later fitting, an electrically driven compensator can be specified if annoying vibrations occur. Such a compensator is synchronised to the correct phase relative to the external force or moment and can neutralize the excitation. The compensator requires an extra seating to be fitted—preferably in the steering gear compartment where deflections are largest and the effect of the compensator will therefore be greatest.

The electrically driven compensator will not give rise to distorting stresses in the hull but it is more expensive than the engine-mounted compensators mentioned above. Good results are reported from the numerous compensators of this type in service (Figure 1.15).

Power related unbalance

To evaluate if there is a risk that 1st and 2nd order external moments will excite disturbing hull vibrations the concept of 'Power Related Unbalance' (PRU) can be used as a guide, where:

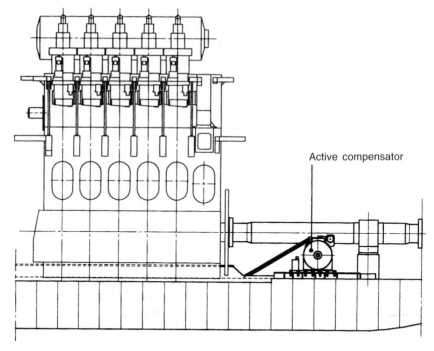

Figure 1.15 Active vibration compensator used as a thrust pulse compensator (MAN B&W Diesel)

$$PRU = \frac{\text{External moment}}{\text{Engine power}} \; Nm/kW$$

With the PRU value, stating the external moment relative to the engine power, it is possible to give an estimate of the risk of hull vibrations for a specific engine. Actual values for different MC models and cylinder numbers rated at layout point L1 are provided in MAN B&W Diesel charts.

Guide force moments

The so-called guide force moments are caused by the transverse reaction forces acting on the crossheads due to the connecting rod/crankshaft mechanism. These moments may excite engine vibrations, moving the engine top athwartships and causing a rocking (excited by H moment) or twisting (excited by X-moment) movement of the engine (Figure 1.16).

Guide force moments are harmless except when resonance vibrations occur in the engine/double bottom system. As this system is very

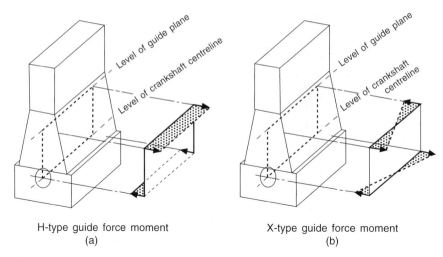

H-type guide force moment   X-type guide force moment
          (a)   (b)

Figure 1.16 H-type and X-type guide force moments (MAN B&W Diesel)

difficult to calculate with the necessary accuracy MAN B&W Diesel strongly recommends as standard that top bracing is installed between the engine's upper platform brackets and the casing side for all its two-stroke models except the S26MC and L35MC types.

The top bracing comprises stiff connections (links) either with friction plates which allow adjustment to the loading conditions of the ship or, alternatively, a hydraulic top bracing (Figure 1.17). With

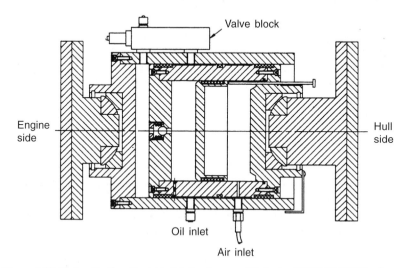

Figure 1.17 Hydraulically adjustable top bracing, modified type (MAN B&W Diesel)

both types of top bracing the above-mentioned natural frequency will increase to a level where resonance will occur above the normal engine speed.

AXIAL VIBRATIONS

The calculation of axial vibration characteristics is only necessary for low speed two-stroke engines. When the crank throw is loaded by the gas pressure through the connecting rod mechanism the arms of the crank throw deflect in the axial direction of the crankshaft, exciting axial vibrations. These vibrations may be transferred to the ship's hull through the thrust bearing.

Generally, MAN B&W Diesel explains, only zero-node axial vibrations are of interest. Thus the effect of the additional bending stresses in the crankshaft and possible vibrations of the ship's structure due to the reaction force in the thrust bearing are to be considered.

An appropriate axial damper is fitted to all MC engines to minimize the effects of the axial vibrations (Figure 1.18). Some engine types in

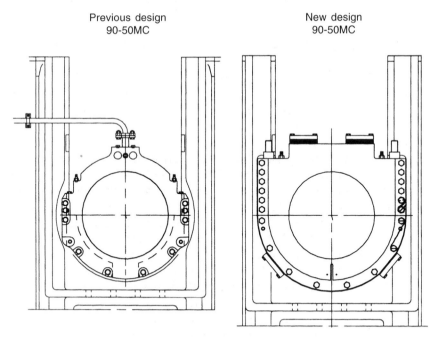

Figure 1.18 Mounting of previous and improved axial vibration damper for MAN B&W Diesel's 500 mm to 900 mm bore MC engines

five- and six-cylinder form with a power take-off mounted at the fore end require an axial vibration monitor for alarm and slow down. For the crankshaft itself, however, such a damper is only necessary for engines with large cylinder numbers.

The examination of axial vibration (sometimes termed longitudinal vibration) can be confined to the main shaftline embracing crankshaft, intermediate shaft and propeller shaft because the axial excitation is not transmitted through PTO gearboxes to branch shaftlines. The vibration system can be plotted in equivalent constant masses and stiffnesses, just as with torsional vibration.

The masses are defined by the rotating components as a whole and can be calculated very accurately with the aid of CAD programs. Axial stiffnesses of the crankshaft are derived, as with the torsional values, from an empirical formula and, if necessary, corrected by comparing measured and calculated axial natural frequencies. Alternatively, axial stiffnesses can be calculated directly by finite element analysis. Damping in the crankshaft is derived from measurements of axial vibration.

The dominating order of the axial vibration is equivalent to the number of cylinders for engines with less than seven cylinders. For engines with more than six cylinders the dominating order is equal to half the cylinder numbers. For engines with odd cylinder numbers the dominating orders are mostly the two orders closest to half the cylinder number.

When MAN B&W Diesel's MC series was introduced an axial vibration damper was only standard on engines with six or more cylinders. It was needed because resonance with the order corresponding to the cylinder number would otherwise have caused too high stresses in the crank throws.

An early case was experienced in which a five-cylinder L50MC engine installed in an LPG tanker recorded excessive axial vibration of the crankshaft during the trial trip. A closer analysis revealed that the crankshaft was not in resonance, and that the situation was caused by a coupled phenomenon. The crankshaft vibration was coupled to the engine frame and double bottom which, in turn, transferred vibration energy back to the crankshaft. As a result, both the whole engine and the superstructure suffered from heavy longitudinal vibration.

MAN B&W Diesel decided to tackle the problem from two sides. An axial vibration damper was retrofitted to the crankshaft while top bracing in the longitudinal direction was fitted on the aft end of the engine. Both countermeasures influenced the vibration behaviour of the crankshaft, the engine frame and the superstructure.

The axial vibration damper alone actually eliminated the problems; and the longitudinal top bracing alone reduced the vibration level in the deckhouse to below the ISO-recommended values. With both countermeasures in action, the longitudinal top bracing had only insignificant influence. The incident, together with experience from some other five-cylinder models, led MAN B&W Diesel to install axial vibration dampers on engines of all cylinder numbers, although those with fewer cylinders may not need the precaution.

TORSIONAL VIBRATIONS

Torsional vibration involves the whole shaft system of the propulsion plant, embracing engine crankshaft, intermediate shafts and propeller shaft, as well as engine running gear, flywheel, propeller and (where appropriate) reduction gearing, flexible couplings, clutches and power take-off drives.

The varying gas pressure in the cylinders during the working cycle and the crankshaft/connecting rod mechanism create a varying torque in the crankshaft. It is these variations that cause the excitation of torsional vibration of the shaft system. Torsional excitation also comes from the propeller through its interaction with the non-uniform wake field. Like other excitation sources the varying torque is cyclic in nature and thus subject to harmonic analysis. Such analysis makes it possible to represent the varying torque as a sum of torques acting with different frequencies which are multiples of the engine's rotational frequency.

Torsional vibration causes extra stresses, which may be detrimental to the shaft system. The stresses will show peak values at resonances: that is, where the number of revolutions multiplied by the order of excitation corresponds to the natural frequency.

Limiting torsional vibration is vitally important to avoid damage or even fracture of the crankshaft or other propulsion system elements. Classification societies therefore require torsional vibration characteristics of the engine/shafting system to be calculated, with verification by actual shipboard measurements. Two limits are laid down for the additional torsional stresses. The lower limit T1 determines a stress level which may only be exceeded for a short time; this dictates a 'barred' speed range of revolutions in which continuous operation is prohibited. The upper stress limit T2 must never be exceeded.

Taking a shaftline of a certain length, it is possible to modify its natural frequency of torsional vibration by adjusting the diameter: a small diameter results in a low natural frequency, a larger diameter in a high natural frequency. The introduction of a tuning wheel will also lower the natural frequency. Classification societies have also laid down rules determining the shaft diameter. An increase in the diameter is permissible but a reduction must be accompanied by the use of a material with a higher ultimate strength.

Based on its experience, MAN B&W Diesel offers the following guidelines for low speed engines:

Four-cylinder engines normally have the main critical resonance (4th order) positioned above but close to normal revolutions. Thus, in worst cases, these engines require an increased diameter of the shaftline relative to the diameters stipulated by classification society rules in order to increase the natural frequency, taking it 40–45 per cent above normal running range.

For five-cylinder engines the main critical (5th order) resonance is also positioned close to—but below—normal revolutions. If the diameter of the shafting is chosen according to class rules the resonance with the main critical will be positioned quite close to normal service speed, thus introducing a barred speed range. The usual and correct way to tackle this unacceptable position of a barred speed range is to mount a tuning wheel on the front end of the crankshaft; to design the intermediate shaft with a reduced diameter relative to the class diameter; and to use better material with a higher ultimate tensile strength. In some cases, however, an intermediate shaft of large diameter is installed in order to increase the resonance to above the maximum continuous rating speed level. This is termed 'overcritical running' (see section below).

Only torsional vibrations with one node generally need to be considered. The main critical order, causing the largest extra stresses in the shaftline, is normally the vibration with order equal to the number of cylinders: that is, five cycles per revolution on a five-cylinder engine. This resonance is positioned at the engine speed corresponding to the natural torsional frequency divided by the number of cylinders.

The torsional vibration conditions may, for certain installations, require a torsional vibration damper which can be fitted when necessary at additional cost. Based on MAN B&W Diesel's statistics, this need may arise for the following types of installation:

- Plants with a controllable pitch propeller.
- Plants with an unusual shafting layout and for special owner/ yard requirements.
- Plants with eight-, ten-, eleven- or twelve-cylinder engines.

Four-, five- and six-cylinder engines require special attention. To avoid the effect of the heavy excitation the natural frequency of the system with one-node vibration should be situated away from the normal operating speed range. This can be achieved by changing the masses and/or the stiffness of the system so as to give a much higher, or much lower, natural frequency (respectively termed 'undercritical' or 'overcritical' running).

Owing to the very large variety of possible shafting arrangements that may be used in combination with a specific engine only detailed torsional vibration calculations for the individual plant can determine whether or not a torsional vibration damper is necessary.

Undercritical running

The natural frequency of the one-node vibration is so adjusted that resonance with the main critical order occurs about 35–45 per cent above the engine speed at specified maximum continuous rating (mcr). The characteristics of an undercritical system are normally:

- Relatively short shafting system.
- Probably no tuning wheel.
- Turning wheel with relatively low inertia.
- Large diameter shafting, enabling the use of shafting material with a moderate ultimate tensile strength, but requiring careful shaft alignment (due to relatively high bending stiffness).
- Without barred speed range.

When running undercritical, significant varying torque at mcr conditions of about 100–150 per cent of the mean torque is to be expected. This torque (propeller torsional amplitude) induces a significant varying propeller thrust which, under adverse conditions, might excite annoying longitudinal vibrations on engine/double bottom and/or deckhouse. Shipyards should be aware of this and ensure that the complete aft body structure of the ship, including the double bottom in the engineroom, is designed to cope with the described phenomena.

Overcritical running

The natural frequency of the one-node vibration is so adjusted that resonance with the main critical order occurs about 30–70 per cent below the engine speed at specified mcr.

Such overcritical conditions can be realized by choosing an elastic shaft system, leading to a relatively low natural frequency. The characteristics of overcritical conditions are:

- Tuning wheel may be necessary on crankshaft fore end.
- Turning wheel with relatively high inertia.
- Shafts with relatively small diameters, requiring shafting material with a relatively high ultimate tensile strength.
- With barred speed range of about +/− 10 per cent with respect to the critical engine speed.

Torsional vibrations in overcritical conditions may, in special cases, have to be eliminated by the use of a torsional vibration damper which can be fitted when necessary at extra cost.

For six-cylinder engines the normal procedure is to adopt a shaftline with a diameter according to the class rules and, consequently, a barred speed range. Excitations associated with engines of seven or more cylinders are smaller, and a barred speed range is not normally necessary.

MAN B&W Diesel cites a series of tankers, each powered by a five-cylinder L80MC engine and featuring a shaft system of a larger diameter than that required by the classification society, and with no tuning wheel to avoid a barred speed range. The torsional vibration-induced propeller thrust was about 30 per cent of the mean thrust. During sea trials heavy longitudinal vibration of the engine frame as well as of the superstructure (excited by the varying thrust) was experienced.

A replacement of the entire shaft system was considered virtually impossible (expensive and time-consuming) so efforts to restrict the heavy longitudinal vibration were concentrated on longitudinal top bracing. After a few attempts it became evident that the steelwork of the deck in way of the fore end of the engine had to be strengthened to yield sufficient rigidity. Vibration levels became acceptable after this strengthening had been executed.

HULL VIBRATION

The natural frequencies of ship hulls are relatively difficult to predict accurately and are also influenced by the loading condition. MAN B&W Diesel and Sulzer two-stroke engines do not generate any free forces; and free moments only are generated for certain numbers of cylinders.

Hull vibration can only be excited by a Sulzer RTA low speed engine if it is located on or near a node of critical hull vibration, and if the frequency of this vibration mode coincides with the first or second harmonic of the engine excitation. Furthermore, the magnitude of the troublesome free moment has to exceed the stabilising influence of the natural damping in the hull structure.

If resonance is foreseen, Sulzer notes the following solutions (Figures 1.19 and 1.20):

- Lanchester balancers, either on the engine or electrically driven units usually located on the steering gear flat, compensate for ship vibration caused by the 2nd order vertical moment.
- Counterweights on the crankshaft often represent a simple and effective solution for primary unbalance, which is only relevant to four-cylinder engines.

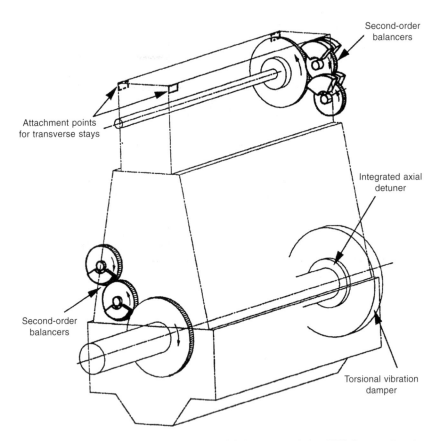

Figure 1.19 Provisions for vibration control and balancing on Sulzer RTA low speed engine

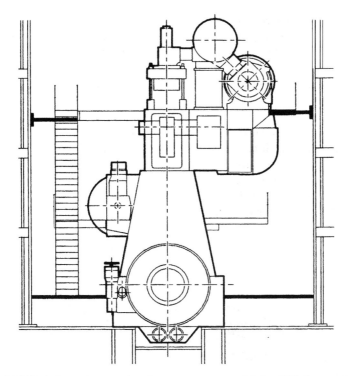

Figure 1.20 Typical attachment points for transverse stays of Sulzer RTA low speed engine

- Combined primary/secondary balancers are available for RTA engines to counteract entirely both primary and secondary unbalance.
- Side stays can be fitted between the engine top and hull structure to reduce vibration caused by lateral moments, a situation mainly arising for four-, eight- and twelve-cylinder engines. Hydraulic-type side stays are preferable.

Acknowledgements are made to MAN B&W Diesel and Wartsila Corporation (Sulzer) in the preparation of this chapter. Both designers have produced valuable papers detailing the theory of vibration in shipboard machinery installations, analysis techniques and practical countermeasures.

2 Gas-diesel and dual-fuel engines

Growing opportunities for dual-fuel and gas-diesel engines in land and marine power markets have stimulated designs from leading medium speed and low speed enginebuilders. Development is driven by the increasing availability of gaseous fuels, the much lower level of noxious exhaust emissions associated with such fuels, reduced maintenance and longer intervals between overhauls for power plant. A healthy market is targeted from floating oil production vessels and storage units, rigs, shuttle tankers, offshore support vessels and LNG carriers; LNG-fuelled RoPax ferries have also been proposed with dual-fuel diesel propulsion.

Valuable breakthroughs in mainstream markets have been made since 2000 with the specification of LNG-burning engines for propelling a small Norwegian double-ended ferry (Mitsubishi high speed engines), offshore supply vessels and a 75 000 cu.m. LNG carrier (Wärtsilä medium speed engines).

Natural gas is well established as a major contributor to the world's energy needs. It is derived from the raw gas from onshore and offshore fields as the dry, light fraction and mainly comprises methane and some ethane. It is available directly at the gas field itself, in pipeline systems, condensed into liquid as LNG or compressed as CNG.

Operation on natural gas results in very low emissions thanks to the clean-burning properties of the fuel and its low content of pollutants. Methane, the main constituent, is the most efficient hydrocarbon fuel in terms of energy content per amount of carbon; operation on natural gas accordingly reduces emissions of another key pollutant—carbon dioxide—by over 20 per cent compared with operation on diesel fuel. Natural gas has very good combustion characteristics in an engine and, because it is lighter than air and has a high ignition temperature, is also a very safe fuel. The thermal efficiencies of various prime movers as a function of load are illustrated in Figure 2.1.

Wärtsilä's dual-fuel (DF) four-stroke engines can be run in either gas mode or liquid-fuelled diesel mode. In gas mode the engines work according to the lean-burn Otto principle, with a lean pre-

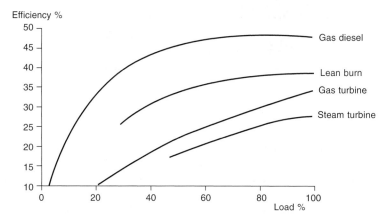

Figure 2.1 Thermal efficiency of diesel and gas-diesel engines as a function of load (Wärtsilä)

mixed air-gas mixture in the combustion chamber. (Lean burn means the mixture of air and gas in the cylinder has more air than is needed for complete combustion, reducing peak temperatures). Less NOx is produced and efficiency increases during leaner combustion because of the higher compression ratio and optimized injection timing. A lean mixture is also necessary to avoid knocking (self-ignition).

The gas is fed into the cylinder in the air inlet channel during the intake stroke (Figure 2.2). Instead of a spark plug for ignition—normally used in lean-burn gas engines—the lean air-gas mixture is ignited by a small amount of diesel fuel injected into the combustion

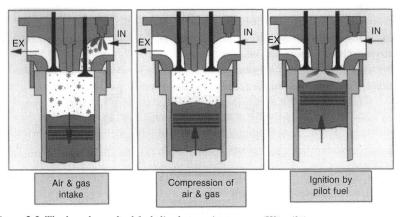

Figure 2.2 The lean burn dual-fuel diesel operating system (Wärtsilä)

chamber. This high energy source ensures reliable and powerful ignition of the mixture, which is needed when running with a high specific cylinder output and lean air-gas mixture. To secure low NOx emissions it is essential that the amount of injected diesel fuel is very small. The Wärtsilä DF engines therefore use a 'micro-pilot' injection, with less than 1 per cent diesel fuel injected at nominal load, to achieve NOx emissions approximately one-tenth those of a standard diesel engine.

When the DF engine is running in gas mode with a pre-mixed air-gas mixture the combustion must be closely controlled to prevent knocking and misfiring. The only reliable way to effect this, says Wärtsilä, is to use fully electronic control of both the pilot fuel injection and the gas admission on every cylinder head. The global air-fuel ratio is controlled by a wastegate valve, which allows some of the exhaust gases to bypass the turbine of the turbocharger. This ensures that the ratio is of the correct value independent of changing ambient conditions, such as the temperature (Figure 2.3).

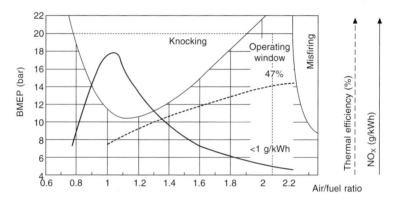

Figure 2.3 A special electronic system for the Wärtsilä DF engine controls combustion in each cylinder, and optimizes efficiency and emissions under all conditions by keeping each cylinder within the operating window

The quantity and timing of the injected pilot fuel are adjusted individually together with the cylinder-specific and global air-fuel ratio to keep every cylinder at the correct operating point and within the operating window between the knock and misfire limits. This is the key factor for securing reliable operation in gas mode, automatically tuning the engine according to varying conditions, Wärtsilä explains.

In diesel mode the engine works according to the normal diesel concept using a traditional jerk pump fuel injection system: diesel fuel is injected at high pressure into the combustion chamber just before the top dead centre. Gas admission is de-activated but the pilot fuel remains activated to ensure reliable pilot ignition when the engine is transferred to gas operation.

The gas pressure in the engine is less than 4 bar at full load, making a single-wall pipe design acceptable if the engineroom is arranged with proper ventilation and gas detectors. The gas valve on every cylinder head is a simple and robust, electronically-controlled solenoid valve, promising high reliability with long maintenance intervals.

The pilot fuel system is a common rail system with one engine-mounted high pressure pump supplying the pilot fuel to every injection valve at a constant pressure of 900 bar. Due to the high pressure a double-walled supply system is used with leak fuel detection and alarms. The injection valve is of twin-needle design, with the pilot fuel needle electronically controlled by the engine control system. It is important that the injection system can reliably handle the small amount of pilot fuel with minimum cycle-to-cycle variation. The main needle design is a conventional system in which mechanically-controlled pumps control the injection.

Other main components and systems are similar to designs well proven over a decade on Wärtsilä standard diesel engines, further underwriting high reliability of the dual-fuel engines.

DUAL-FUEL ENGINE OPERATION

Wärtsilä DF engines can be run on gas or on light fuel (marine diesel or gas oil) and heavy fuel oil. When running in gas mode the engine instantly switches over to diesel operation if the gas feed is interrupted or component failure occurs. The switch-over takes less than one second and has no effect on the engine speed and load during the process. Transferring from diesel to gas operation, however, is a gradual process: the diesel fuel supply is slowly reduced while the amount of gas admitted is increased. The effect on the engine speed and load fluctuations during transfer to gas is minimal, Wärtsilä reports.

Switching from gas to heavy fuel oil dictates some minor modifications on the engine, although the necessary changes can be executed in one day. Using HFO also demands another type of lubricating oil quality (in gas and light fuel oil operation the same lube oil quality can be used).

The DF engine is normally started only with the pilot fuel injection enabled to a speed of approximately 60 per cent of the nominal engine speed. When ignition is detected in all cylinders the gas admission is activated and the engine speed increased to the nominal speed; this safety measure ensures that no excess unburned fuel is fed directly into the exhaust gas system.

Wärtsilä has supplemented its portfolio with dual-fuel versions of the 320 mm bore and 500 mm bore medium speed designs, which have respectively earned contracts for offshore supply vessel and LNG carrier propulsion projects, both burning natural gas. These 32DF and 50DF engines—covering a power band from 2010 kW to 17 100 kW with six-cylinder to V18-cylinder models—extended a programme which also embraces gas-diesel (GD) variants of the W32 and W46 engines as well as smaller spark-ignited gas engines.

The Wärtsilä 32DF engine was introduced for marine applications in 2000 to meet the requirements of a new safety class for installations with a gas pressure of less than 10 bar in a single-pipe arrangement. It provided an alternative to the 32GD (gas-diesel) engine (see below) which had been successful in offshore markets. An early application called for shipsets of four 6L32DF LNG-burning genset engines to power diesel-electric offshore supply vessels; each vessel has capacity for 220 cu.m. of LNG stored at a pressure of 5.5 bar. The larger 50DF engine was officially launched in the following year, four 6-cylinder models (each developing 5700 kW) being specified to drive the main gensets of a diesel-electric LNG carrier.

Wärtsilä DF engine data	32DF	50DF
Bore	320 mm	500 mm
Stroke	350 mm	580 mm
Engine speed (rev/min)	720/750	500/514
Mean piston speed (m/s)	8.4/8.75	9.7/9.9
Mean effective pressure (bar)	19.9	20/19.5
Output/cylinder (kW)	335/350	950
Cylinder numbers	6,9L/12,18V	6,8,9L/12,16,18V
Power band	2010–6300 kW	5700–17 100 kW

WÄRTSILÄ 50DF ENGINE DETAILS

Based on the well proven Wärtsilä 46 diesel engine, but with special systems for dual-fuel operation, the 50DF is produced in six-cylinder in-line to V18-cylinder configurations yielding 950 kW per cylinder at a speed of 500 or 514 rev/min. A thermal efficiency of 47 per cent (higher than any other gas engine) is claimed, and the engine delivers the same output whether running on natural gas, light fuel oil or heavy fuel oil.

The air-fuel ratio is very high, typically 2.2:1. Since the same specific heat quantity released by combustion is used to heat up a larger mass of air, the maximum temperature and hence NOx formation are lower. The mixture is uniform throughout the cylinder since the fuel and air are pre-mixed before introduction into the cylinders, which helps to avoid local NOx formation points within the cylinder.

Engine starting is always in diesel mode, using both main diesel and pilot fuel. At 300 rev/min the main diesel injection is disabled and the engine transferred to gas mode. Gas admission is activated only when combustion is stable in all cylinders, ensuring safe and reliable starting. In gas running mode, the pilot fuel amounts to less than 1 per cent of the full load fuel consumption, the quantity of pilot fuel controlled by the Wärtsilä Engine Control System.

Natural gas is fed to the engine through a valve station, the gas first filtered to ensure a clean supply. The gas pressure—dependent on engine load—is controlled by a valve located near the engine driving end. The full load pressure is 3.6 bar (g) for a lower heating value of 36 MJ/m^3N; for a lower LHV the pressure has to be increased. The system incorporates the necessary shut-off and venting valves to secure a safe and trouble-free gas supply. Gas is supplied through large common rail pipes running along the engine, each cylinder then served by an individual feed pipe to the gas admission valve on the cylinder head.

The diesel fuel oil supply on the engine is divided into two systems: one for the pilot fuel and one for the back-up fuel (Figure 2.4). The pilot fuel is raised to the required pressure by a pump unit incorporating duplex filters, pressure regulator and engine-driven radial piston-type pump. The high pressure pilot fuel is then distributed through a common rail pipe to the injection valves at each cylinder; pilot fuel is injected at approximately 900 bar pressure and the timing and duration are electronically controlled.

Back-up fuel is fed to a normal camshaft-driven injection pump, by which it is pumped at high pressure to a spring-loaded injection valve of standard design for a diesel engine. The larger needle of the twin-needle injection valve is used in diesel engine mode, and the smaller needle for pilot fuel oil when the engine is running in gas mode. Pilot injection is electronically controlled; main diesel injection is hydro-mechanically controlled.

The individually-controlled solenoid valve allows optimum timing and duration of pilot fuel injection into every cylinder when the engine is running in gas mode. Since NOx formation depends greatly on the amount of pilot fuel, the system secures a very low NOx

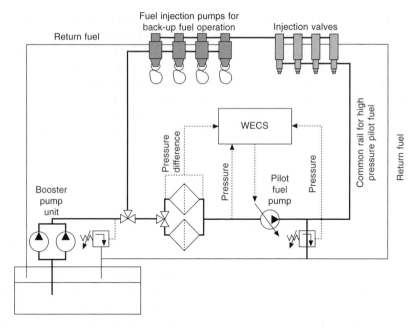

Figure 2.4 Diesel fuel oil supply system for the Wärtsilä DF engine (WECS = Wärtsilä Engine Control System)

creation while retaining a stable and reliable ignition source for the lean air-gas mixture in the combustion chamber.

Gas is admitted to the cylinders just before the air inlet valve. The gas admission valves are electronically actuated and controlled by the engine control system to deliver exactly the correct amount of gas to each cylinder. Combustion in each cylinder can thus be fully and individually controlled. Since the admission valve can be timed independently of the inlet valves, the cylinder can be scavenged without risk of gas being fed directly to the exhaust system. Independent gas admission ensures the correct air-fuel ratio and optimized operating point with respect to efficiency and emissions; it also fosters reliable performance without shutdowns, knocking and misfiring. The gas admission valves have a short stroke and embody special materials to underwrite low wear and long maintenance intervals.

A well proven Wärtsilä-developed monobloc fuel injection pump is specified to withstand the high pressures involved, with a constant pressure relief valve provided to avoid cavitation and a plunger featuring a wear-resistant coating. The fuel pump is ready for operation at all times.

The pilot fuel pump is a stand-alone unit comprising a radial piston pump and necessary filters, valves and control system. It receives the signal and required pressure level from the engine control unit and independently sets and maintains the pressure at that level, transmitting the prevailing fuel pressure to the engine control system. High pressure fuel is delivered to each injection valve via a common rail pipe which acts as a pressure accumulator and damper against pressure pulses in the system. The fuel system is of double-wall design with alarm for leakage.

The 50DF engine can be switched automatically from fuel oil to gas operation at loads below 80 per cent of the full load. The transfer takes place automatically after the operator's command, without load changes. During the switchover—which lasts for about one minute— the fuel oil is gradually substituted by gas. Trip from gas to fuel oil operation occurs instantaneously and automatically (for example, in the event of gas supply interruption). A correct air-fuel ratio under any operating conditions is essential for optimum performance and emissions, a status addressed by an exhaust gas wastegate valve. Part of the exhaust gases bypass the turbocharger through this valve, which adjusts the air-fuel ratio to the correct value independently of the varying conditions under high engine loads. The wastegate is actuated electro-pneumatically.

A flexible cooling system, optimized for different cooling applications, has two separate circuits: high temperature and low temperature. The HT circuit controls the cylinder liner and cylinder head temperatures, while the LT circuit serves the lubricating oil cooler. Both circuits are also connected to the respective parts of the two-stage charge air cooler; and both HT and LT water pumps are engine driven as standard.

All engine functions are controlled by the Wärtsilä Engine Control System (WECS), a microprocessor-based distributed control system mounted on the engine. The various WECS modules are dedicated and optimized for certain functions and communicate with each other via a databus. Diagnostics are built in for easy trouble shooting. The core of the WECS is the main control module, which is responsible for keeping the engine at optimum performance in all operating conditions, such as varying ambient temperature and methane number. The main module reads the information sent by all the other modules and applies it in adjusting the engine speed and load control by determining reference values for the main gas admission, air-fuel ratio and pilot fuel amount and timing. It also automatically controls the start and stop sequences of the engine and the safety system, and communicates with the plant control system.

Each cylinder control module is arranged to monitor and control three cylinders; it controls the cylinder-specific air-fuel ratio by adjusting the gas admission individually for each cylinder. The module measures the knock intensity (uncontrolled combustion in the cylinder), this information being applied in adjusting the cylinder-specific pilot fuel timing and gas admission. Light knocking leads to automatic adjustment of the pilot fuel timing and cylinder-specific air-fuel ratio; heavy knocking leads to load reduction or a gas trip.

GAS-DIESEL ENGINES

Wärtsilä gas-diesel (GD) technology differs from the dual-fuel diesel concept in using direct, high pressure injection of gas and 3 per cent pilot fuel for its ignition. GD medium speed engines can be operated on heavy fuel and diesel oils, crude oil directly from the well or natural gas; supply can be switched instantly and automatically from one fuel to another without shutdown. Conversions of existing engines from normal heavy fuel mode to natural gas/diesel oil operation can be executed with small modifications. Containerized engineroom packages have been delivered along with the associated gas compressors and other ancillaries.

The gas-diesel engine was developed in the late 1980s mainly as a prime mover for the offshore industry. High pressure direct injection for the gaseous fuel maintains the diesel process, making the concept insensitive to the methane content of the fuel. Such a feature makes the GD engine especially suitable for mobile oil production processing plants, where the gas composition may change with the location of the vessel and the stage in the production process.

Initially, the gas injection was controlled by a mechanical/hydraulic system using a cam-actuated jerk pump to generate the control pressure for the gas needle. Although reliable, Wärtsilä explains, this system did not offer the required flexibility and was later replaced by an electro-hydraulic system based on a constant rail pressure to control the gas injection via solenoid valves. Gas-specific monitoring and safety functions were all integrated, and the system allowed adjustment of individual cylinders. This represented the first step towards fully computerized systems for engine control, and valuable experience was gained in both software and hardware requirements.

Rolls-Royce is another specialist in the technology, its Norwegian subsidiary Bergen Diesel having started development of a lean-burn gas-fuelled version of the 250 mm bore K-series medium speed diesel engine in the mid-1980s. Numerous engines of this type have been sold for land power and co-generation installations, reportedly demonstrating high efficiency and very low emissions. The company has also been keen to demonstrate the potential in gas-fuelled ferries in an era of concern over atmospheric coastal pollution and the increasing availability of gas from domestic pipeline terminals.

Rolls-Royce Bergen spark-ignition lean burn gas engine technology has matured through four generations of engines and thousands of operating hours in shoreside power installations, running on various gas fuels including natural gas and biogas derived from landfill. The 250 mm bore KV-G4 engine features variable geometry turbocharging and individual electronic gas and air controls for each cylinder to yield a shaft efficiency of 44 per cent on a brake mean effective pressure of 16 to 18 bar. Very low NOx emissions (1.1 g/kWh) are reported from the engine, which is available with ratings up to 3600 kW. A gas engine version of the 320 mm bore Bergen B32:40 medium speed diesel, exploiting the proven features of the KV-G4 series, was planned to offer outputs up to around 6000 kW.

Medium and low speed gas-burning engine designs are offered by the MAN B&W Diesel group, citing successful experience with a Holeby 16V28/32-GI (gas injection) four-stroke model (Figure 2.5),

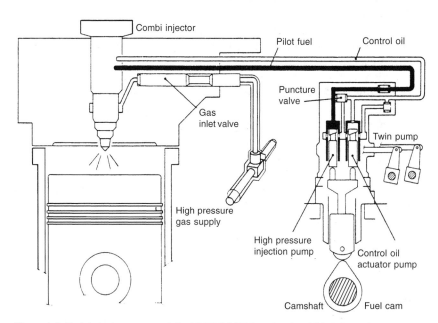

Figure 2.5 Fuel injection system of the MAN B&W Diesel 16 V 28/32-GI high pressure gas-injection engine

and a 6L35MC two-stroke model. A high pressure gas-injection system for MC low speed engines was jointly developed by the group and its key Japanese licensee Mitsui, which delivered a 41 000 kW 12-cylinder K80MC-GI engine for power generation in an environmentally sensitive region (Figures 2.6 and 2.7).

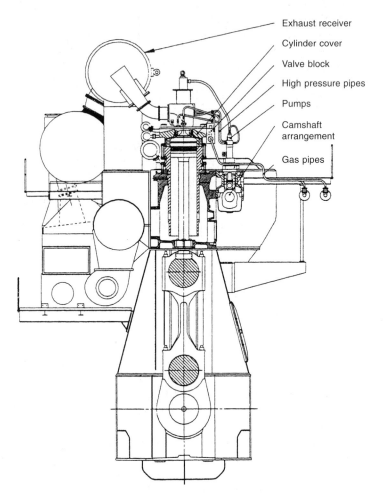

Figure 2.6 MAN B&W Diesel's MC-GI low speed dual fuel engine, indicating new or modified components compared with the standard diesel version

The prime marine target for large gas-fuelled engines is LNG carrier propulsion, which allows the cargo boil-off gas to be tapped, but a modified high pressure gas-injection version of the MC low

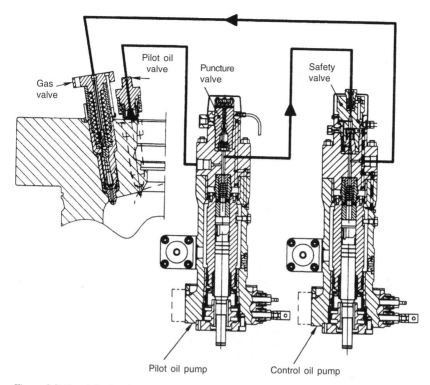

Figure 2.7 Gas injection is stopped immediately on the first failure to inject the pilot oil by the patented safety valve incorporated in the control oil pump of the MAN B&W MC-GI engine

speed engine is at the heart of MAN B&W Diesel's system for burning volatile organic compound (VOC) discharges from offshore shuttle tankers. Oil vapours or VOCs are light components of crude oil evaporated mainly during tanker loading and unloading but also during a voyage when cargo splashes around the tanks.

VOC discharge represents a significant loss of energy as well as an environmental problem (the non-methane part of the VOC released to the atmosphere reacts in sunlight with nitrogen oxide and may create a toxic ground-level ozone and smog layer). The VOC-burning system was first applied at sea to the twin 6L55GUCA main engines of the 125 000 dwt shuttle tanker *Navion Viking* in 2000, an onboard recovery system collecting the discharges and liquefying the product by compressing and cooling. The resulting condensate is stored and burned as fuel for propulsion, yielding substantial savings in heavy fuel oil consumption (Figure 2.8).

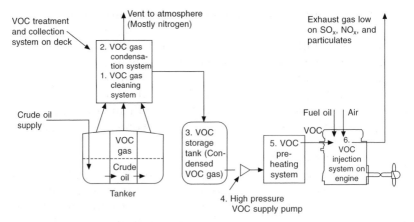

Figure 2.8 Schematic representation of system for burning volatile organic compound (VOC) discharges in the main engines of a shuttle tanker (MAN B&W Diesel)

Engines for such applications must be able to operate on ordinary heavy fuel oil if VOC fuel is unavailable, and they must also accept any possible composition of VOC condensate since it is not possible to apply any fuel specification: whatever is collected must be burned. Shoreside tests by MAN B&W Diesel confirmed that its dual-fuel MC engines could use any VOC/HFO ratio between 92/8 per cent and 0/100 per cent, as well as any relevant VOC quality.

REVIEWING THE DESIGN FEATURES OF GAS-BURNING ENGINES, MAN B&W DIESEL NOTES:

The fuel type will depend on the demands of the plant and on fuel availability. For large engines, the dual fuel system with a liquid pilot fuel is necessarily applied, providing the plant with redundancy to operate on liquid fuel alone in the event of shortage or failure of the gas supply. In smaller plants, the engine may exploit the same principles or be optimized for using a single gaseous fuel or even a number of different gases.

The combustion chamber design may be the single chamber configuration used on all large engines or it may be a divided chamber design featuring a pre-chamber. In the latter case the pre-chamber normally operates with an over-rich fuel/air mixture, provided either by a separate gas supply to the chamber or by pilot fuel injection. The aim in both cases is to secure stable ignition in the pre-chamber and to minimize nitrogen oxides (NOx) formation (due to the lack of oxygen). Combustion in the main chamber may then use a lean mixture effectively ignited by the hot combustion products emerging from the pre-chamber. The lean mixture burns at a lower temperature which, in turn, minimizes the formation of NOx in the main chamber.

Charge exchange may be with or without supercharging and the medium may be pure air or a gas/air mixture. Small, simple plants are normally naturally aspirated with a gas/air mixture while larger plants, for cost reasons, are turbocharged. Four-stroke engines may use air or a gas/air mixture for the charge exchange but two-stroke engines are restricted to air.

The air/fuel ratio may be selected according to engine type and application. A rich mixture can be exploited to secure a high output and stable running from small engines without emission restraints. Such applications will become scarce in the future: a stoichiometric air/fuel ratio controlled by a sensor and combined with a three-way catalyst to comply with emission regulations is considered a more suitable choice.

Except for such small engines, lean burn or stratified charge is the principle being applied in most new developments, and this may be used without a pre-chamber. The reasons are the favourable emission characteristics and high engine efficiency. Mixture formation, defined as the mixing process of air and gas, may take place in the cylinder by direct gas injection at low pressure (after valve closure but before compression start) or by high pressure gas injection near TDC. Two-stroke engines need to apply one of these alternatives to avoid excessive gas losses which would promote high HC-emissions and reduced efficiency.

With low pressure direct injection, the gas/air mixture is exposed to the compression process and hence increasing pressure and temperature. This may lead to self-ignition and severe knocking, excluding its use in new engine developments.

In the case of late-cycle high pressure gas injection the gas is only admitted to the combustion chamber at the desired time for combustion. This is initiated either by a pilot fuel injection or by directing the gas jets to a glow plug.

The high pressure gas injection system is relatively costly but it renders the engine practically insensitive to fuel gas self-ignition properties. For this reason, MAN B&W Diesel suggests, the system can be considered as underwriting true multi-fuel engines, provided that the auxiliary systems for fuel supply have been designed accordingly.

For most four-stroke gas-burning engines the mixture formation occurs outside the cylinder (an exception is the high pressure gas-injection engine). The gas may be mixed with air before the turbocharger, yielding a very uniform mixture but also a rather large volume of ignitable gas in the intake system of the engine. Alternatively, it may be added to the intake air in the cylinder head or the intake air manifold, close to the intake valve, by a combined intake valve/gas valve or by a separate gas valve.

Ignition sources take various forms but are all, basically, just high temperature sources. The highest ignition energy, MAN B&W Diesel explains, is provided by pilot injection which also makes it possible to locate multiple ignition points deep inside the gas/air mixture: beneficial for the fast and efficient combustion of lean mixtures. This solution is reliable and yields extended times-between-overhaul but it is more costly than a spark plug or glow plug arrangement. Pilot injection, by its nature, dictates the presence of a liquid fuel which, in some cases, is considered a disadvantage. In other cases, this is viewed positively as a means of fuel redundancy.

GAS ENGINE TYPES AND CHARACTERISTICS

Small engines are often derived from automotive gasoline and diesel engines, covering an output range from a few kW up to 500 kW. They are generally spark-ignited without a pre-chamber, operating with stoichiometric mixture and with mixture formation in the intake system. Naturally aspirated as well as turbocharged engines are found in this category, some equipped with a three-way catalyst and a sensor for emission control.

Spark-ignited lean burn engines may either be developed solely for co-generation purposes or derived from small bore marine diesel engines. Such engines cover an output range from 500–1000 kW up to around 7000 kW. They are often turbocharged lean burn engines with a pre-chamber and with external mixture formation. Engine efficiency can be high and NOx emissions so low that rather stringent regulations can be met without the use of a selective catalytic reduction (SCR) system. CO emissions, if regulated, may require the use of an oxidation catalyst.

Dual-fuel engines are normally derived from marine diesel engines; most are turbocharged, without pre-chamber and with external mixture formation for the gas/air mixture. The pilot oil amount is typically some 5 per cent if the engine must also be able to operate as a diesel engine. If this facility is not required a special small injection system allows for pilot oil injection amounts of 1–2 per cent, which has a beneficial effect on NOx emissions. These are generally lower than for the corresponding diesel engine but the engine output is often some 10–20 per cent lower, depending on the methane number of the actual gas.

High pressure gas-injection engines are derived from marine diesel engines and have the same performance and output. An output range of 2000 kW to 80 000 kW can be covered by these turbocharged engines. NOx emissions are somewhat lower than those from corresponding diesel engines (around 20 per cent); an SCR system is required if stringent limits are to be satisfied. The pilot fuel amount is 5–8 per cent and, if necessary, full output can be achieved in the oil-only mode, giving fuel redundancy.

Gas-injection engines are relatively complicated and need a high pressure gas compressor. In the lower output range, therefore, the costs are hardly justified except where special gases are to be used. But for large engines, and for special gases, the high pressure gas-injection engine is considered the only option. Long term service experience has confirmed that such engines can provide reliable and safe operation.

3 Exhaust emissions and control

Marine engine designers in recent years have had to address the challenge of tightening controls on noxious exhaust gas emissions imposed by regional, national and international authorities responding to concern over atmospheric pollution.

Exhaust gas emissions from marine diesel engines largely comprise nitrogen, oxygen, carbon dioxide and water vapour, with smaller quantities of carbon monoxide, oxides of sulphur and nitrogen, partially reacted and non-combusted hydrocarbons and particulate material (Figures 3.1 and 3.2).

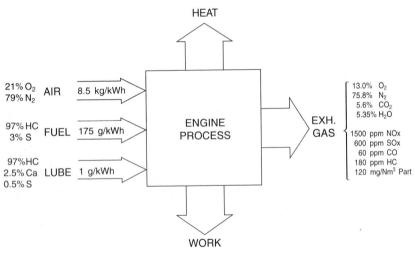

Figure 3.1 Typical exhaust emissions from a modern low speed diesel engine

Nitrogen oxides (NOx)—generated thermally from nitrogen and oxygen at high combustion temperatures in the cylinder—are of special concern since they are believed to be carcinogenic and contribute to photochemical smog formation over cities and acid rain (and hence excess acidification of the soil). Internal combustion engines primarily generate nitrogen oxide but less than 10 per cent

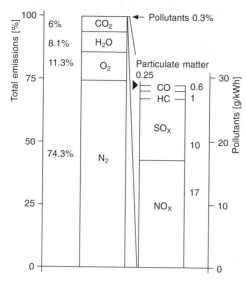

Figure 3.2 Typical composition of the exhaust gas products of a medium speed diesel engine burning fuel with an average 3 per cent sulphur content. Some 6 per cent of the total emission is carbon dioxide, with the 'real' pollutants representing only a 0.3 per cent share (MAN B&W Diesel)

of that oxidizes to nitrogen dioxide the moment it escapes as exhaust gas.

Sulphur oxides (SOx)—produced by oxidation of the sulphur in the fuel—have an unpleasant odour, irritate the mucus membrane and are a major source of acid rain (reacting with water to form sulphurous acid). Acid deposition is a trans-boundary pollution problem: once emitted, SOx can be carried over hundreds of miles in the atmosphere before being deposited in lakes and streams, reducing their alkalinity.

Sulphur deposition can also lead to increased sulphate levels in soils, fostering the formation of insoluble aluminium phosphates which can cause a phosphorous deficiency. Groundwater acidification has been observed in many areas of Europe; this can lead to corrosion of drinking water supply systems and health hazards due to dissolved metals in those systems. Forest soils can also become contaminated with higher than normal levels of toxic metals, and historic buildings and monuments damaged.

Hydrocarbons (HC)—created by the incomplete combustion of fuel and lube oil, and the evaporation of fuel—have an unpleasant odour, are partially carcinogenic, smog forming and irritate the mucus membrane (emissions, however, are typically low for modern diesel engines.)

Carbon monoxide (CO)—resulting from incomplete combustion due to a local shortage of air and the dissociation of carbon dioxide—is highly toxic but only in high concentrations.

Particulate matter (PM) is a complex mixture of inorganic and organic compounds resulting from incomplete combustion, partly unburned lube oil, thermal splitting of HC from the fuel and lube oil, ash in the fuel and lube oil, sulphates and water. More than half of the total particulate mass is soot (inorganic carbonaceous particles), whose visible evidence is smoke. Soot particles (unburned elemental carbon) are not themselves toxic but they can cause the build-up of aqueous hydrocarbons, and some of them are believed to be carcinogens. Particulates constitute no more than around 0.003 per cent of the engine exhaust gases.

Noxious emissions amount to 0.25-0.4 per cent by volume of the exhaust gas, depending on the amount of sulphur in the fuel and its lower heat value, and the engine type, speed and efficiency. Some idea of the actual pollutants generated is provided by MAN B&W Diesel, which cites an 18 V 48/60 medium speed engine in NOx-optimized form running at full load on a typical heavy fuel oil with 4 per cent sulphur content. A total of approximately 460 kg of harmful compounds are emitted per hour out of around 136 tonnes of exhaust gas mass per hour. Of the 0.35 per cent of the exhaust gas formed by pollutants, NOx contributes 0.17 per cent, sulphur dioxide 0.15 per cent, hydrocarbons 0.02 per cent, carbon monoxide 0.007 per cent and soot/ash 0.003 per cent.

Carbon dioxide: some 6 per cent of the exhaust gas emissions from this engine is carbon dioxide. Although not itself toxic, carbon dioxide contributes to the greenhouse effect (global warming) and hence to changes in the Earth's atmosphere. The gas is an inevitable product of combustion of all fossil fuels, but emissions from diesel engines—thanks to their thermal efficiency—are the lowest of all heat engines. A lower fuel consumption translates to reduced carbon dioxide emissions since the amount produced is directly proportional to the volume of fuel used, and therefore to the engine or plant efficiency. As a rough guide, burning one tonne of diesel fuel produces approximately three tonnes of carbon dioxide.

International concern over the atmospheric effect of carbon dioxide has stimulated measures and plans to curb the growth of such emissions, and the marine industry must be prepared for future legislation. (A switch from other transport modes—air, road and rail—to shipping would nevertheless yield a substantial overall reduction in emissions of the greenhouse gas because of the higher efficiency of diesel engines.)

The scope for improvement by raising the already high efficiency level of modern diesel engines is limited and other routes have to be pursued: operating the engines at a fuel-saving service point; using marine diesel oil or gas oil instead of low sulphur heavy fuel oil; adopting diesel-electric propulsion (the engines can be run continuously at the highest efficiency); or exploiting a diesel combined cycle incorporating a steam turbine. The steam-injected diesel engine is also promising.

Compared with land-based power installations, fuel burned by much of shipping has a very high sulphur content (up to 4.5 per cent and more) and contributes significantly to the overall amount of global sulphur oxide emissions at sea and in port areas. Studies on sulphur pollution showed that in 1990 SOx emissions from ships contributed around 4 per cent to the total in Europe. In 2001 such emissions represented around 12 per cent of the total and could rise to as high as 18 per cent by 2010.

SOx emissions in diesel engine exhaust gas—which mostly comprise sulphur dioxide with a small amount of sulphur trioxide—are a function of the amount of sulphur in the fuel and cannot be controlled by the combustion process. If the fuel contains 3 per cent sulphur, for example, the volume of SOx generated is around 64 kg per tonne of fuel burned; if fuel with a 1 per cent sulphur content is used SOx emissions amount to around 21 kg per tonne of fuel burned.

Chemical and washing/scrubbing desulphurization processes can remove SOx from the exhaust gases but are complex, bulky and expensive for shipboard applications, and increase overall maintenance costs. The most economical and simplest approach is thus to burn bunkers with a low sulphur content. (If a selective catalytic reduction system is installed to achieve the lowest NOx emission levels—see section below—then low sulphur fuels are dictated anyway to avoid premature fouling of the system's catalyst package.)

A global heavy fuel oil sulphur content cap of 4.5 per cent and a 1.5 per cent fuel sulphur limit in certain designated Sulphur Emissions Control Areas (SECAs)—such as the Baltic Sea, North Sea and English Channel—is sought by the International Maritime Organisation (IMO) to reduce SOx pollution at sea and in port. The European Union strategy for controlling air pollution calls for all ships in EU ports to burn fuel with a maximum sulphur content of 0.2 per cent, which would force uni-fuel ships to carry low sulphur fuel specifically for this purpose. (See also the Fuels and Lubes chapter.)

CONTROLLING NOx EMISSIONS

The global approach to NOx emission controls was taken by the IMO whose Annex VI (Regulations for the Prevention of Air Pollution from Ships) to Marpol 73/78 was adopted at a diplomatic conference in 1997. Ships burning marine diesel oil and heavy fuel oil at that time were reportedly responsible for around 7 per cent of global NOx emissions, around 4 per cent of global sulphur dioxide emissions and 2 per cent of global carbon dioxide emissions.

Annex VI will enter into force 12 months after the date on which not less than 15 states, together constituting not less than 50 per cent of the gross tonnage of the world's merchant fleet, have ratified it. Ships constructed after 1 January 2000 (date of keel laying) were nevertheless required to comply. The annex addresses engines with a power output of more than 130 kW installed in new ships constructed from that date and engines in existing ships that undergo a major modification.

Engines have to fulfil the NOx emission limits set by the IMO curve, which is related to engine speed. To show compliance, an engine has to be certified according to the NOx Technical Code and delivered with an EIAPP (Engine International Air Pollution Prevention) letter of compliance. The certification process includes NOx measurement for the engine type concerned, stamping of components that affect NOx formation and a Technical File that is delivered with the engine.

IMO's current maximum allowable NOx emission levels depend on the speed category of the engine and range from 17 g/kWh for engines of speed <130 rev/min to 9.84 g/kWh for engines of speed >2000 rev/min (Figure 3.3). Much tougher curbs on NOx and other emissions are set by regional authorities such as California's Air Resources Board; and Sweden has introduced a system of differentiated port and fairway dues making ships with higher NOx emissions pay higher fees than more environment-friendly tonnage of a similar size.

With stricter controls planned by the IMO, the reduction of NOx emissions remains a priority for engine designers whose concern is to secure environmental acceptability without compromising the impressive gains in engine fuel economy and reliability achieved in recent years. (Advances in thermal efficiency have, ironically, directly contributed to a rise in NOx emissions.) Considerable progress has been made and is projected (Figure 3.4).

Dominating influences in the formation of NOx are temperature and oxygen concentration: the higher the temperature and the higher

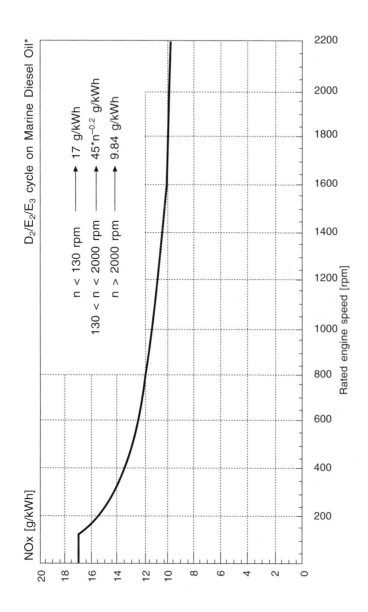

NOx [g/kWh]

$D_2/E_2/E_3$ cycle on Marine Diesel Oil*

n < 130 rpm   ⟶ 17 g/kWh
130 < n < 2000 rpm ⟶ $45*n^{-0.2}$ g/kWh
n > 2000 rpm   ⟶ 9.84 g/kWh

Rated engine speed [rpm]

* Testcycles in accordance with ISO 8178 Part 4

Figure 3.3 Maximum allowable NOx emissions for marine diesel engines (IMO)

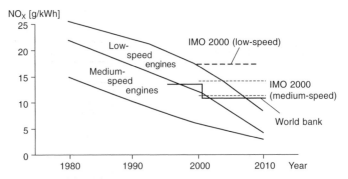

Figure 3.4 NOx emission trends for typical two-stroke and four-stroke diesel engines compared with IMO requirements (Wärtsilä)

the residence time at high temperature in the cylinder, the greater the amount of thermal NOx that will be created. A longer combustion timespan means that low speed two-stroke engines therefore generate higher NOx emissions than medium and high speed four-stroke engines of equivalent output. Apart from the use of alternative fuels, such as methanol, two main approaches can be pursued in reducing NOx emission levels:

Primary (in-engine) measures aimed at reducing the amount of NOx formed during combustion by optimizing engine parameters with respect to emissions (valve timing, fuel injection and turbocharging). Emission levels can be reduced by 30–60 per cent.

Secondary measures designed to remove NOx from the exhaust gas by downstream cleaning techniques. Emission reductions of over 95 per cent can be achieved.

De-NOx technology options are summarized in Figure 3.5.

The primary NOx reduction measures can be categorised as follows:

- Water addition: either by direct injection into the cylinder or by emulsified fuel.
- Altered fuel injection: retarded injection; rate-modulated injection; and a NOx-optimized fuel spray pattern.
- Combustion air treatment: Miller supercharging; turbocooling; intake air humidification; exhaust gas recirculation; and selective non-catalytic reduction. (Miller supercharging and turbocooling are covered in the Pressure Charging chapter.)
- Change of engine process: compression ratio; and boost pressure.

The basic aim of most of these measures is to lower the maximum temperature in the cylinder since this result is inherently combined with a lower NOx emission.

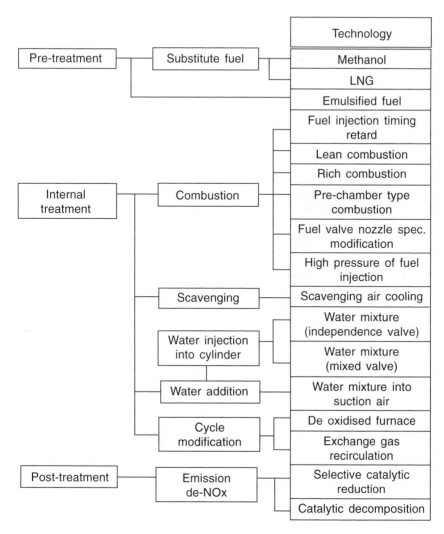

Figure 3.5 Methods of reducing NOx emissions from marine diesel engines (Mitsubishi Heavy Industries)

New generations of medium speed engines with longer strokes, higher compression ratios and increased firing pressures addressed the NOx emission challenge (Figure 3.6). The low NOx combustion system exploited by Wärtsilä throughout its medium speed engine programme, for example, is based on an optimized combination of compression ratio, injection timing and injection rate. The engine parameters affecting the combustion process are manipulated to secure

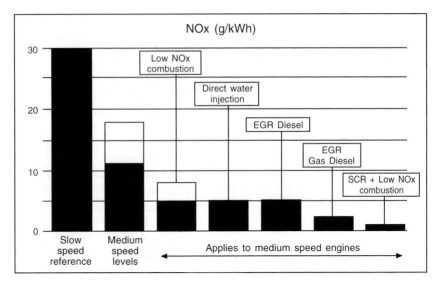

Figure 3.6 Emission control measures and their potentials in reducing NOx output. EGR = exhaust gas recirculation; SCR = selective catalytic reduction (Wärtsilä Diesel)

a higher cylinder pressure by increasing the compression ratio. The fuel injection equipment is optimized for late injection with a short and distinct injection period. NOx reductions of up to 50 per cent are reported without compromising thermal efficiency, achieving an NOx rate of 5–8 g/kWh compared with the 15 g/kWh of a typical conventional engine with virtually no effect on fuel consumption.

Low NOx combustion is based on:

- A higher combustion air temperature at the start of injection, which significantly reduces the ignition delay.
- A late start of injection and shorter injection duration to place combustion at the optimum point of the cycle with respect to efficiency.
- Improved fuel atomization and matching of combustion space with fuel sprays to facilitate air and fuel mixing.

Some Wärtsilä Vasa 32 engines were equipped with a 'California button' to meet the US regional authority's strict NOx control. A planetary gear device allows the necessary small injection timing retard to be effected temporarily while the engine is running when the ship enters the Californian waters.

WATER-BASED TECHNIQUES

Water–fuel emulsions injected via the fuel valve achieve a significant reduction in NOx production. The influence varies with the engine type but generally 1 per cent of water reduces NOx emissions by 1 per cent. Water can also be added to the combustion space through separate nozzles or by the stratified segregated injection of water and fuel from the same nozzle.

Direct Water Injection (DWI) is effective in reducing NOx by adding mass and stealing heat from the combustion process when the water is evaporated. Wärtsilä developed DWI for its medium speed engines after initially investigating the merits of injecting ammonia into the cylinder during the expansion stroke. It was found that injecting water into the combustion chamber during the compression stroke could achieve the same degree of NOx reduction as ammonia.

Wärtsilä medium speed engines with DWI systems feature a combined injection valve and nozzle for injecting water and fuel oil into the cylinder (Figure 3.7). The nozzle has two separate needles that are also controlled separately, such that neither mode (water on/water off) will affect operation of the engine. The engine can be transferred to non-water operational mode at any load, the transfer in alarm situations being automatic and instantaneous. Water injection takes place before fuel injection, resulting in a cooler combustion space and hence lower NOx generation; water injection stops before fuel oil is injected into the cylinder so that the ignition and combustion process is not disturbed.

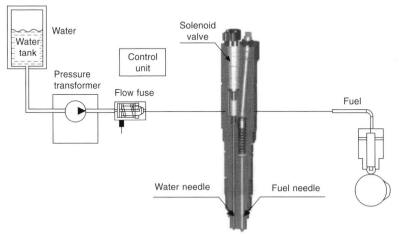

Figure 3.7 Wärtsilä's Direct Water Injection system features a combined injection valve and nozzle for water and fuel oil

Clean water is fed to the cylinder at a pressure of 210–400 bar (depending on the engine type) by a common rail system, the pressure generated in a high pressure pump module; a low pressure pump is also necessary to ensure a sufficiently stable water flow to the high pressure pump. The water is filtered before the low pressure pump to remove all solid particles; the pumps and filters are built into modules to ease installation.

A flow fuse installed on the cylinder head side acts as a safety device, shutting off the water flow into the cylinder if the water needle gets stuck; immediate water shut-off is initiated in the event of excessive water flow or leakage. Water injection timing and duration are electronically controlled by the control unit, which receives its input from the engine output; timing and duration can be optimized conveniently via a keyboard for different applications. Space requirements for the equipment are minimal, facilitating retrofit applications.

Water and fuel are injected in a typical water-to-fuel ratio of 0.4–0.7:1, reducing NOx emissions by 50–60 per cent without adversely affecting power output or engine components. NOx emissions are typically 4–6 g/kWh when the engine is running on marine diesel oil, and 5–7 g/kWh when heavy fuel oil is being burned. NOx reduction is most efficient from 40 per cent load and higher of nominal engine output.

Numerous applications of the above DWI system have benefited Wärtsilä 32 and 46 engines powering diverse ship types. Some W46 'EnviroEngines' (see section below) combine electronically-controlled common rail fuel injection with DWI, the water then injected into the combustion chamber separately from the side of the cylinder head through a dedicated valve.

Mitsubishi Heavy Industries developed a stratified fuel-water injection (SFWI) system using a common valve to inject 'slugs' of fuel/water/fuel sequentially into the combustion chamber. The system reportedly worked in a stable condition throughout extensive trials on a ship powered by a Mitsubishi UEC 52/105D low speed engine, NOx emission reduction being proportional to the amount of water injected.

Charge air humidification

Another way of introducing water into the combustion zone is by humidifying the scavenge air: warm water is injected and evaporated in the air intake, whose absolute humidity is thereby increased. An early drawback was that too much water in the air can be harmful to the cylinder condition but the introduction of the anti-polishing ring in the liner (see the chapter Medium Speed Engines—Introduction) has allowed a much higher humidity to be accepted.

In the Combustion Air Saturation System (CASS), which resulted from co-operation between Wärtsilä and the Finnish company Marioff Oy, special Hi-Fog nozzles introduce water directly into the charge air stream after the turbocharger in the form of very small droplets. Only a few microns in size, the droplets evaporate very quickly in an environment of more than 200°C and 75 m/s air velocity. Further heat for evaporation is provided by the air cooler (but acting now as a heater), resulting in combustion air with a humidity of around 60 g per kg of air. With such an amount of water, Wärtsilä reports, experiments demonstrated it was possible to secure NOx levels of less than 3 g/kWh (assuming the starting value is 10–15 g/kWh).

MAN B&W Diesel has also pursued NOx reduction by increasing the humidity of the charge air with water vapour. The process is based on the Humid Air Motor (HAM) system developed by the German company Munters Euroform (Figure 3.8). Compressed and heated air from the turbocharger is passed through a cell that humidifies and cools the air with evaporated water, the distillation process making it possible to use sea water rather than fresh water. The relative humidity of the inlet air can be kept constant at 99 per cent, yielding a reported maximum NOx reduction potential of around 70 per cent, exploiting all the waste heat of the engine.

Test results from an SEMT-Pielstick 3PC2.6 medium speed engine equipped with a prototype HAM system showed a NOx reduction from 13.5 g/kWh to 3.5 g/kWh. There was no significant influence

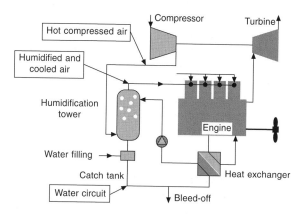

Figure 3.8 Schematic diagram of Humid Air Motor (HAM) system for NOx emission reduction

on specific fuel consumption, no significant increase in carbon monoxide and hydrocarbon emissions, and no smoke deterioration. These results—along with no trace of water in the lube oil, no corrosion and cleaner engine internals—were repeated in subsequent seagoing installations. In retrofit projects, the HAM system replaced the original intercoolers.

Exhaust gas recirculation

Exhaust gas recirculation (EGR) is a method of modifying the inlet air to reduce NOx emissions, an approach widely used in automotive applications. Some of the exhaust gas is cooled and cleaned before recirculation to the scavenge air side. Its effect on NOx formation is partly due to a reduction of the oxygen concentration in the combustion zone, and partly due to the content of water and carbon dioxide in the exhaust gas. The higher molar heat capacities of water and carbon dioxide lower the peak combustion temperature which, in turn, curbs the formation of NOx.

EGR is a very efficient method of reducing NOx emissions (by 50–60 per cent) without affecting the power output of the engine but is considered more practical for engines burning cleaner bunkers such as low sulphur and low ash fuels, alcohol and gas. Engines operating on high sulphur fuel might invite corrosion of turbochargers, intercoolers and scavenging pipes.

Wärtsilä acknowledges that EGR leads to reductions in NOx emissions but suggests its practical application is limited by the fact that it quite soon starts to adversely affect combustion, leading to increased fuel consumption, particulates, unburned hydrocarbons and carbon monoxide. A worse drawback is that, even with an insignificant sulphur content in the fuel, cooling down of the exhaust gas results in a sticky liquid of sulphur acid products, water and soot matter. This makes it difficult to maintain the functionality of the cooling device and, above all, results in a difficult residual product, says Wärtsilä, which abandoned EGR for diesel engine applications.

For low speed engines, especially those with electronically-controlled exhaust valve timing, Wärtsilä suggests that the Water Cooled Rest Gas (WaCoReG) technique offers an opportunity for NOx reduction. In this concept some of the exhaust gas is left in the cylinder, which in normal circumstances leads to increased thermal loading and inferior combustion. These drawbacks are largely avoided, however, if the rest gas is cooled by an internal spray of water. As the surrounding combustion space components are rather hot, there will be no condensation of acid products on the metal surfaces.

Fuel nozzles

Different fuel nozzle types and designs have a significant impact on NOx formation, and the intensity of the fuel injection also has an influence.

The increased mean effective pressure ratings of modern engines require increased flow areas throughout the fuel valve which, in turn, leads to increased sac volumes in the fuel nozzle itself and a higher risk of after-dripping. Consequently, more fuel from the sac volume may enter the combustion chamber and contribute to the emission of smoke and unburned hydrocarbons as well as to increased deposits in the combustion chamber. The relatively large sac volume in a standard design fuel nozzle thus has a negative influence on the formation of soot particles and hydrocarbons.

The so-called 'mini-sac' fuel valve introduced by MAN B&W Diesel incorporates a conventional conical spindle seat as well as a slide inside the fuel nozzle. The mini-sac leaves the flow conditions in the vicinity of the nozzle holes similar to the flow conditions in the conventional fuel nozzle. But its much reduced sac volume—only about 15 per cent that of the conventional fuel valve—has demonstrated a positive influence on the cleanliness of the combustion chamber and exhaust gas outlet ducts. Such valves also reduce the formation of NOx during combustion.

A new type of fuel valve—essentially eliminating the sac volume—was subsequently developed and introduced by MAN B&W Diesel as standard to its larger low speed engines (Figure 3.9). The main advantages of this slide-type fuel valve are reduced emissions of NOx, CO, smoke and unburned hydrocarbons as well as significantly fewer deposits inside the engine. A positive effect on the cylinder condition in general is reported.

Applying slide fuel valves to a 12K90MC containership engine yielded a 40 per cent reduction in smoke (BSN10) compared with the mini-sac valved engine, while hydrocarbons and CO were reduced by 33 per cent and 42 per cent, respectively, albeit from a low level. NOx was reduced by 14 per cent, while the fuel consumption remained virtually unchanged and with a slight reduction at part load.

SELECTIVE CATALYTIC REDUCTION

Primary methods are generally adequate for the IMO NOx emission limits but tougher regional controls may dictate the use of secondary methods—exhaust gas treatment techniques—either alone or in

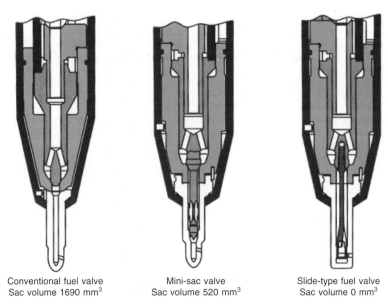

Conventional fuel valve   Mini-sac valve   Slide-type fuel valve
Sac volume 1690 mm^3 Sac volume 520 mm^3 Sac volume 0 mm^3

Figure 3.9 Evolution of fuel injection valve design for MAN B&W Diesel MC two-stroke engines; smoke and NOx emissions were lowered by reducing or eliminating the sac volume

combination with engine modifications. Here the focus is on selective catalytic reduction (SCR) systems, developed from land-based power station installations for shipboard applications, which can cut NOx reductions by well over 90 per cent.

In an SCR system the exhaust gas is mixed with ammonia (preferably in the form of a 40 per cent solution of urea in water) before passing through a layer of special catalyst at a temperature between 290°C and 450°C. The lower limit is mainly determined by the sulphur content of the fuel: at temperatures below 270°C ammonia and SOx will react and deposit as ammonium sulphate. At excessive temperatures, the catalyst will be degraded.

The NOx is reduced to the harmless gaseous waste products water and nitrogen; and part of the soot and hydrocarbons in the exhaust is also removed by oxidation in the SCR process reactor.

Among the desirable qualities of a catalyst are: a low inertia, which means that the ammonia slip (the quantity slipping past the catalyst) is extremely low, even during transient operations; a low pressure drop and a short heating-up time; and a low fouling tendency, ensuring minimal deterioration in performance over time.

The catalytic conversion rate of SCR systems is highly dependent on the amount of urea dosed: increased dosage yields increased conversion. However, excessive urea causes ammonia slip downstream of the reactor, which is detrimental both to the process and operational economy. Urea dosing must therefore be very accurate under different load conditions. The set point for dosing is derived primarily from the engine speed and load; in addition, control is adjusted by measuring the residual NOx level after the reactor. The NOx measurement data are used for tuning the stoichiometric ammonia/ NOx ratio and maintaining the ammonia slip at a constant level.

The temperature requirement of the process generally dictates that the SCR reactor serving a low speed two-stroke engine is installed before the turbocharger; the post-turbocharger exhaust gas temperature of a four-stroke engine, however, is sufficient for the catalytic process. If an exhaust gas boiler is specified it should be installed after the SCR unit. The temperature window has been the subject of catalyst development.

The impressive de-Noxing efficiency of SCR technology has been demonstrated in deepsea and coastal vessel installations since early 1990. NOx emission reductions of up to 95 per cent are yielded by pioneering marine SCR plant serving the MAN B&W 6S50MC low speed engines of bulk carriers on a dedicated trade between Korea and California where strict environmental regulations have to be satisfied. The amount of ammonia injected into the exhaust gas duct is controlled by a process computer which arranges dosing in proportion to the NOx produced by the engine as a function of engine load (Figure 3.10).

Typically, a ship on a regular trade to California will bypass the SCR reactor on the ocean leg of the voyage, complying with the IMO emission rules by using primary methods for controlling NOx. Approaching the Californian-regulated waters, the engine feed is switched from heavy fuel to fuel complying with California Air Resources Board rules (gas oil). The exhaust gas is then gradually passed through the SCR reactor and, when the temperature has been raised to the right level, ammonia dosing is started to effect near-complete NOx control.

An early SCR installation serving medium speed marine engines was commissioned in 1992 on a diesel–electric ferry plying short crossings between Denmark and Sweden. Its initial NOx reduction efficiency of 96 per cent was increased to 98 per cent after 12 000 running hours.

The number of SCR systems in service has proliferated as shipowners anticipate tougher regulations and seek NOx emission levels of 2 g/kWh or lower from the main and auxiliary engines of diverse tonnage, including fast car/passenger ferries. Applications in

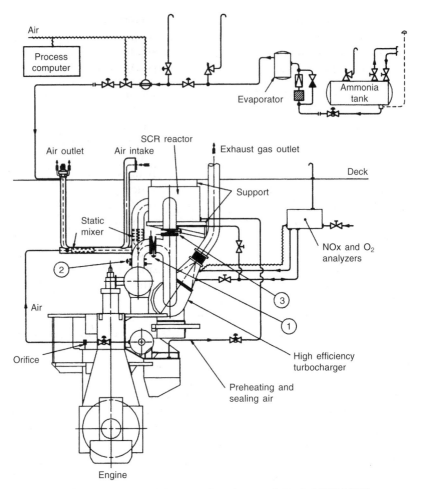

Figure 3.10 Schematic layout of SCR system for a low speed diesel (MAN B&W)

newbuilding and retrofit projects have been facilitated by the efforts of system designers to reduce reactor space requirements and offer compact solutions. The reactors can now be installed separately in the engineroom or integrated in the exhaust manifolds of both four-stroke and two-stroke engines, doubling as an efficient silencer.

Siemens' SINOx SCR system is designed to reduce the emissions of different exhaust gas pollutants simultaneously—notably NOx, hydrocarbons and soot—as well as to achieve a sound attenuation effect. The following performance can be expected, according to the German designer:

- NOx reduction.............................90–99 per cent at MCR
- Hydrocarbon/CO reduction............80–90 per cent at MCR
- Soot reduction............................30–40 per cent at MCR
- Noise reduction...........................30–35 dB(A).

A system investment cost of US$40–70 per kW and an operating cost of US$3–4 per MWh were quoted in 2002, with some 15–20 litres/ hour per MW of 40 per cent urea solution consumed.

Important factors in designing such a system are the exhaust gas temperature and the amount, composition and required purity level of the cleaned gas. The control system ensures optimum process supervision and correct dosing of the reducing agent; a special algorithm secures a fully automatic supply of the correct amount of aqueous urea solution in line with the NOx emission levels from the engine. Reducing agent consumption is thus minimized and emissions are reliably reduced below the specified limits. The exhaust gas emissions are measured during plant commissioning at different engine loads and the derived values programmed into the control system to ensure correct dosing at various loads.

The SINOx system is based on a fully ceramic, fine-celled honeycomb catalyst made of titanium dioxide and containing vanadium pentoxide as the active substance. The honeycomb elements are packed in prefabricated steel modules and can be replaced quickly and individually. The catalytic converter is designed chemically and physically to suit the particular conditions of a given installation, the catalyst lifespan varying from 10 000 to 40 000 hours depending on the application. The converters are characterized by high catalytic activity, high selectivity, high resistance to erosion and to chemicals such as sulphur, as well as insensitivity to particulate deposits and high mechanical and thermal loads.

R&D by SCR system designers targets increased compactness, low weight, no waste products, extended catalyst lifetimes, and lower operating and maintenance costs. Catalysts themselves remain the subject of continuing development to improve performance and longevity.

Results from a research programme on a Sulzer 6RTA38 low speed engine led Wärtsilä to the following conclusions on SCR systems:

- NOx reductions greater than 90 per cent can be achieved commercially.
- The catalyst housing (reactor) including insulation has a volume of 2–5 cu.m. per MW of engine power, depending on the make of catalyst; the size of the housing is more or less independent of the input NOx concentration.

- The exhaust gas back pressure imposed by an SCR plant is typically between 15 and 25 mbar.
- The reactor can be designed in such a way that it serves as a silencer with achievable noise reductions of more than 25 dBA.
- Some 30 litres of 40 per cent urea solution (corresponding to 15 kg of urea granulates) are needed per MWh.
- If the SCR plant is only for temporary use a burner is absolutely necessary to heat the catalyst before the engine is started; otherwise, the catalyst will inevitably become clogged by ammonium sulphates.

The schematic arrangement of an SCR system installed to serve the Sulzer 7RTA52U two-stroke engine of a RoRo paper carrier is shown in Figure 3.11; the SCR converter is installed before the turbine of the turbocharger.

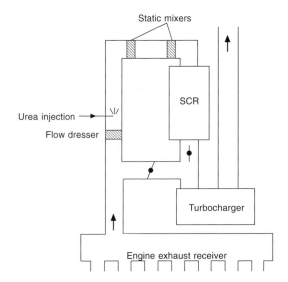

Figure 3.11 Schematic arrangement of SCR system installed to serve a Sulzer low speed engine

PARTICULATES, SOOT AND SMOKE

Particulate generation in diesel engines is a complex process depending on numerous factors (such as engine type, speed, engine setting, operating mode, load, fuel and even weather conditions), explains MAN B&W Diesel.

Particulates comprise all solid and liquid exhaust gas components which, after cooling by exhaust gas dilution with filtered particulate-free ambient air to a temperature below 51.7°C, are collected on specified filters (dilution tube sampling). The so-called PM fraction represents a broad mixture of: partly-burned or unburned hydrocarbons; sulphate bound water; sulphates; ash; and elemental carbon (soot).

Soot and ash are therefore only part of the total particulates. At high loads, ash and soot might contribute less than 20 per cent to the particulates, but at low loads and idling this percentage can be much higher.

The specific mass of all particulates from modern MAN B&W medium speed diesel engines averages around 0.6 g/kWh at maximum continuous rating, assuming heavy fuel oil with a 2 per cent sulphur content is burned. Lower particulate values can be achieved, however, the company citing a PM value of 0.2 g/kWh from the four 18V48/60 engines of a floating barge.

Much of the elemental carbon formed in a diesel engine is oxidized during combustion and only the remainder leaves the combustion space with the exhaust gas as soot, which becomes visible as a dark smoke plume emanating from the funnel. There is a clear correlation between the level of soot formation and the type of fuel used; heavy fuel oil combustion generates substantially greater volumes of particles (and soot) than the burning of cleaner fuels, such as marine diesel and marine gas oils. While soot particles themselves are not toxic, notes MAN B&W Diesel, the potential hazard posed by the build-up of liquid hydrocarbons onto them is viewed critically by many.

At very high engine loads, combustion in a state-of-the-art medium speed diesel engine can be modelled to give invisible smoke. At low service loads, however, and especially during rapid start-up manoeuvres and load changes, the turbochargers deliver less intake air than the engine needs for complete combustion and the engine 'smokes'.

Smoke as the visible manifestation of soot production is highly undesirable in all types of ships, but particularly in passenger ferries and cruise liners. In sensitive waters, such as the glacier regions and bays of Alaska or the Galapagos Islands, a vessel producing excessive amounts of visible soot (smoke) can even be banned from cruising there. Achieving zero emissions may dictate shutting down all engines and boilers during a port stay, and plugging into a land-based electrical supply system.

MAN B&W Diesel designed and successfully field-tested a package of measures to suppress the formation of soot in highly turbocharged medium speed engines, even during long periods of slow steaming.

The full package for these IS (invisible smoke) near-zero soot engines with low NOx emissions comprises:

- Turbocharger optimized for part load, with a waste gate.
- Charge air bypass below 65 per cent engine load.
- Charge air preheating (80°C) below 20 per cent load.
- Smaller injection bores.
- Nozzles with short sac holes.
- Auxiliary blower for operation below 20 per cent load*.
- Fuel-water emulsification (15–20 per cent water)**.
- Retarded fuel injection below 80 per cent engine load.

With smoke readings of 0.3 maximum on the Bosch scale, at very low engine (MAN B&W L/V 48/60IS) loads and even idling, soot production at steady operating conditions is drastically reduced and kept at an invisible level over the entire operating range. Smoke reduction was by a factor of seven at idling and by a factor of five at 10 per cent load, respectively. The mass of soot produced during diesel combustion is reduced by even higher factors than these due to the non-linear correlation between smoke emission and soot mass.

Elements of the IS package have become standard for all MAN B&W medium speed engines; the other measures are only specified when extremely strict requirements on soot and NOx emissions have to be met. The package is available for new engines and by retrofit to existing engines.

A different approach to smokeless medium speed engines was adopted by Wärtsilä, which explains that smoke in a large diesel engine can be formed in two different ways: if the fuel spray touches a metal surface and there is not enough remaining combustion time to burn away the soot formed (this is consequently a low load problem); and when more fuel is injected than can be burned in the air amount in the cylinder (this can easily happen at load variations with a conventional injection system).

A common rail fuel system can cure both these problems, Wärtsilä asserts, because it is possible to maintain high injection pressures independently of engine load and thus ensure good spray atomization even at very low loads. If the combustion space is also optimized there will be no risk of low load smoke. The risk of over-injection of fuel can also be avoided by a common rail system because its

* Can be dispensed with if load below 20 per cent is not required.
** For NOx cycle values between 6.7–8 g/kWh; can be dispensed with if NOx emission levels meeting the IMO limit curve are sufficient.

computer—supplied with air temperature, pressure and rotational speed data—can calculate the amount of air trapped in the cylinder. Based on empirical maps, the computer can decide how much fuel can be injected. A common rail fuel-injected engine can have a faster load response than a conventionally injected engine and without smoke. (The common rail system for Wärtsilä medium speed engines is detailed in the Fuel Injection chapter.)

Wärtsilä's EnviroEngine concept for its W46 and W32 medium speed designs exploits an electronically-controlled common rail fuel injection system to achieve 'smokeless' performance (Figure 3.12). The package embraces:

- Camshaft-driven high pressure pumps.
- Accumulators for eliminating pressure waves.
- Hot box enclosure of all elements for maximum safety.
- Engine-driven control oil pump for easy 'black' start.
- Suction control of fuel flow for high efficiency.
- Low cam load for high reliability.

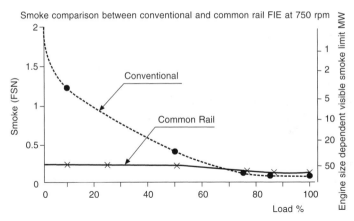

Figure 3.12 Smoke test results from a Wärtsilä 32 medium speed engine with and without common rail fuel injection

Such a system offers the freedom to choose the fuel injection pressure and timing completely independently of the engine load, while computerized control allows several key engine parameters to be considered and the injection and combustion optimized for each load condition. The ability to maintain fuel injection pressures sufficiently high at all engine loads and speeds (even at the lowest levels and during starting and transient load changes) contributes to clean combustion with no visible smoke emissions. Four 16V46 EnviroEngines were specified as the diesel element of the CODAG-electric propulsion plant of Cunard's *Queen Mary 2*.

There is limited scope to significantly reduce particulate emissions by improving combustion but methods of removing particulates from the exhaust gas are available, Wärtsilä notes. Compact ceramic filters can be used with some success on smaller engines but appear not to work when operating with heavy fuel-burning engines because they become clogged by the metal matter. Bag filters work with a certain efficiency, but for most applications are too bulky and hence impracticable. Electrostatic precipitators are efficient in removing particulates, but they are also extremely bulky and too expensive.

Conventional scrubbers have proved to be quite inefficient for the task because the surface contact between the exhaust gas and the water is limited. Promising results were reported by Wärtsilä, however, for a scrubber based on ultrafine water droplets created by Hi-Fog nozzles which yield a much greater surface area between gas and liquid.

SUMMARY OF OPTIONS

Summing up the various options for noxious exhaust emission reduction, Wärtsilä suggests:

- The first choice is engine tuning modifications that can achieve up to 39 per cent reductions in NOx emission levels compared with those of standard engines in 1990.
- For further NOx reductions, separate water injection is considered the most appropriate solution; the technique has proved its ability on the testbed to reduce emission levels by some 60 per cent compared with today's standard engines.
- Exhaust gas after-treatment by selective catalytic reduction (SCR) has proved an effective solution in reducing NOx by 90 per cent or more, despite the special difficulties imposed by using high sulphur fuel oils.
- Requirements for lower emissions of sulphur oxides (SOx) are most favourably met by using fuel oils with reduced sulphur contents. Although SOx reduction by exhaust gas after-treatment is technically feasible, the de-sulphurization process inevitably imposes a disposal problem which would not be acceptable for shipping.
- Carbon monoxide and hydrocarbons can, if necessary, be reduced by an oxidation catalyst housed within an SCR reactor.

- Particulates reduction, with the engine running on heavy fuel, poses a challenge. Technical solutions are available (electrostatic precipitators, for example) but involve either great space requirements or great expense. Particulate emissions are reduced by 50–90 per cent, however, through a switch to distillate fuel oils.

Acknowledgements to MAN B&W Diesel, Wärtsilä Corporation and Siemens.

4 Fuels and lubes: chemistry and treatment

Fuel remains one of the highest single cost factors in running a ship and also the source of the most potent operating problems. The reason for this is that new refining techniques, introduced as a result of political developments in the Middle East in 1973/74, have meant that fluid catalytic cracking and vis breaking have produced a more concentrated residual fuel of very poor quality. This residual fuel is the heavy fuel oil traditionally supplied to ships as bunkers and used in the majority of motor ships of a reasonable size for the main engine. Despite the high cost of these poor quality residual fuels owners generally have no alternative but to burn them, though some still prefer to use even more expensive intermediate grades produced as a result of mixing residual fuel oil with distillate.

The problems are reflected in the effects on the engine in terms of wear and tear and corrosion resulting from harmful components in the fuel. It is the duty of the ship's engineer to be aware of these harmful constituents, their effect on the operation of the engines and the solutions available to counter the harmful properties.

PROBLEMS WITH HEAVY FUELS

The problems of present and future heavy residual fuels can be categorized as:

1. Storage and handling.
2. Combustion quality and burnability.
3. Contaminants, resulting in corrosion and/or damage to engine components: for example, burnt out exhaust valves.

Storage problems

The problems of storage in tanks of bunker fuel result from build-up of sludge leading to difficulties in handling. The reason for the increase in sludge build-up is because heavy fuels are generally blended from

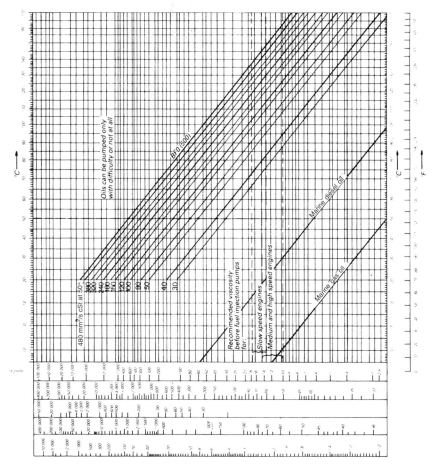

Figure 4.1 A typical temperature/viscosity chart for heavy fuels

a cracked heavy residual using a lighter cutter stock resulting in a problem of incompatibility. This occurs when the asphaltene or high molecular weight compound suspended in the fuel is precipitated by the addition of the cutter stock or other diluents. The sludge which settles in the bunker tanks or finds its way to the fuel lines tends to overload the fuel separators with a resultant loss of burnable fuel, and perhaps problems with fuel injectors and wear of the engine through abrasive particles.

To minimize the problems of sludging the ship operator has a number of options. He may ask the fuel supplier to perform stability checks on the fuel that he is providing. Bunkers of different origins should be kept segregated wherever possible and water contamination

Table 4.1 Effects of heavy fuels

Properties		Present H.O.	Future H.O.	Effect on engine
Viscosity (Red 1 at 37°C Heating temp.)	pumping centrifuging injection	3500 50 95 110–120	5200 65 98 115–130	Increased fuel heating required
Density at 15°C		0.98	0.99	Water elimination becomes more difficult
Pour point °C		30	30	
Noxious element Carbon residue %		6–12	15–22	Fouling risk of components Increased combustion delay
Asphaltenes %		4–8	10–13	Hard asphaltene producing hard particles Soft asphaltene giving sticky deposits at low output Increased combustion delay with defective combustion and pressure gradient increases
Cetane number		30–55	25–40	High pressure gradients and starting problems
Sulphur %		2–4	5	Wear of components due to corrosion below dew point of sulphuric acid (about 150°C)
Vanadium ppm		100–400	120–500	Burning of exhaust valves at about 500°C
Sodium ppm		18–25	35–80	Lower temp. in case of high Na content
Silicon and aluminium (CCF slurries)				Wear of liners, piston grooves, rings, fuel pump and injectors

kept to a minimum. Proper operation of the settling tanks and fuel treatment plant is essential to prevent sludge from entering the engine itself. A detergent-type chemical additive can be used to reduce the formation of sludge in the bunker tanks.

Water in the fuel

Water has always been a problem because it finds its way into the fuel during transport and storage on the ship. Free water can seriously damage fuel injection equipment, cause poor combustion and lead to excessive cylinder liner wear. If it happens to be seawater, it contains sodium which will contribute to corrosion when combined with vanadium and sulphur during combustion.

Water can normally be removed from the bunkers by proper operation of separators and properly designed settling and daily service tanks. However, where the specific gravity of the fuel is the same or greater than the water, removal of the water is difficult—or indeed not possible—and for this reason the maximum specific gravity of fuel supplied for ship's bunkers has generally been set at 0.99.

Burnability

The problems that are related to poor or incomplete combustion are many and complex and can vary with individual engines and even cylinders. The most significant problem, however, is the fouling of fuel injectors, exhaust ports and passages and the turbocharger gas side due to failure to burn the fuel completely. Fuels that are blends of cracked residual are much higher in aromatics and have a high carbon to hydrogen ratio, which means they do not burn as well. Other problems arising from these heavier fuels are engine knocking, after burning, uneven burning, variation in ignition delay, and a steeper ignition pressure gradient. These factors contribute to increased fatigue of engine components, excessive thermal loading, increased exhaust emissions, and critical piston ring and liner wear. The long-term effects on the engine are a significant increase in fuel consumption and component damage. The greatest fouling and deposit build-up will occur when the engine is operated at reduced or very low loads.

The fuel qualities used to indicate a fuel's burnability are: Conradson carbon residue; asphaltenes; cetane value; and carbon to hydrogen ratio.

To ensure proper fuel atomization, effective use of the centrifugal separators, settling tanks and filters is essential; and the correct fuel viscosity must be maintained by heating, adequate injection pressure and correct injection timing. The maintenance of correct running temperatures according to the engine manufacturer's recommendations is also important, particularly at low loads. Additives which employ a reactive combustion catalyst can also be used to reduce the products of incomplete combustion.

High-temperature corrosion

Vanadium is the major fuel constituent influencing high-temperature corrosion. It cannot be removed in the pre-treatment process and it combines with sodium and sulphur during the combustion process to form eutectic compounds with melting points as low as 530°C. Such molten compounds are very corrosive and attack the protective oxide layers on steel, exposing it to corrosion.

Exhaust valves and piston crowns are very susceptible to high temperature corrosion. One severe form is where mineral ash deposits form on valve seats, which, with constant pounding, cause dents leading to a small channel through which the hot gases can pass. The compounds become heated and then attack the metal of the valve seat.

As well as their capacity for corrosion, vanadium, sulphur and sodium deposit out during combustion to foul the engine components and, being abrasive, lead to increased liner and ring wear. The main defence against high temperature corrosion has been to reduce the running temperatures of engine components, particularly exhaust valves, to levels below that at which the vanadium compounds are melted. Intensively cooled cylinder covers, liners, and valves, as well as rotators fitted to valves, have considerably reduced these problems. Special corrosion resistant coatings such as Stellite and plasma coatings have been applied to valves.

Low-temperature corrosion

Sulphur is generally the cause of low-temperature corrosion. In the combustion process the sulphur in the fuel combines with oxygen to form sulphur dioxide (SO_2). Some of the sulphur dioxide further combines to form sulphur trioxide (SO_3). The sulphur trioxide formed during combustion reacts with moisture to form sulphuric acid vapours, and where the metal temperatures are below the acid dew point (160°C) the vapours condense as sulphuric acid, resulting in corrosion.

The obvious method of reducing this problem is to maintain temperatures in the engine above the acid dew point through good distribution and control of the cooling water. There is always the danger that an increase in temperatures to avoid low temperature corrosion may lead to increased high temperature corrosion. Attack on cylinder liners and piston rings as a result of high sulphur content fuels has been effectively reduced by controlled temperature of the cylinder liner walls and alkaline cylinder lubricating oils.

Abrasive impurities

The normal abrasive impurities in fuel are ash and sediment compounds. Solid metals such as sodium, nickel, vanadium, calcium and silica can result in significant wear to fuel injection equipment, cylinder liners, piston rings and ring grooves. However, a comparatively new contaminant is the metallic catalyst fines composed of very hard and abrasive alumina and silica particles which are a cause for much concern. These particles carry over in the catalytic cracking refinery process and remain suspended in the residual bottom fuel for extended periods. It has been known for brand new fuel pumps to be worn out in a matter of days, to the point where an engine fails to start through insufficient injection pressure, as a result of catalyst fines in the fuel.

The only effective method of combating abrasive particles is correct fuel pre-treatment. Separator manufacturers recommend that the separators should be operated in series (a purifier followed by a clarifier) at throughputs as low as 20 per cent of the rated value. See section on fuel oil treatment later in this chapter.

PROPERTIES OF FUEL OIL

The quality of a fuel oil is generally determined by a number of specific parameters or proportions of metals or impurities in a given sample of the particular fuel. Such parameters include: viscosity; specific gravity; flash point; Conradson carbon; asphaltenes content; sulphur content; water content; vanadium content; and sodium content. Two parameters of traditional importance have been the calorific value and viscosity. Viscosity, once the best pointer to a fuel's quality or degree of heaviness, is now considered as being only partially a major quality criterion because of the possible effects of constituents of a fuel.

Calorific value

The calorific value or heat of combustion of a fuel oil is a measure of the amount of heat released during complete combustion of a unit mass of the fuel, expressed kJ/kg. Calorific value is usually determined by a calorimeter but a theoretical value can be calculated from the following:

Calorific value in kcal/kg of fuel

$$= \frac{8100\ C + 34\ 000\ (H - \frac{O}{8})}{100}$$

where C, H and O are percentages of these elements in one kilogram of fuel. Carbon yields 8080 kcal/kg when completely burned, hydrogen 34 000; the oxygen is assumed to be already attached to its proportion of hydrogen—that is, an amount of hydrogen equal to one-eighth the weight of the oxygen is nullified. The sulphur compounds are assumed to have their combustion heat nullified by the oxy-nitrogen ones.

Another variant of the formula is:

Calorific value is kcal/kg of fuel

$$= 7500 \, C + 33 \, 800 \, (H - \tfrac{0}{8})$$

C, H, O being fractions of a kilogram per kilogram of fuel.

To convert the kcal/kg values to SI units as widely used:

$$1 \, \text{kcal/kg} = 4.187 \, \text{kJ/kg}$$

Example: A fuel of 10 500 kcal/kg in SI units

$$10 \, 500 \, \text{kcal/kg} = 10 \, 500 \times 4.187 \, \text{kJ/kg}$$

$$= 44 \, 000 \, \text{kJ/kg}$$

The calorific value as determined by a bomb calorimeter is the gross or 'higher' value which includes the latent heat of water vapour formed by the combustion of the hydrogen. The net or 'lower' calorific value is that obtained from subtracting this latent heat. The difference between the gross and net values is usually about 600–700 kcal/kg, depending upon the hydrogen percentage.

A formula that can be used to calculate the net value is:

$$CVn = CVg - 25 \, (f + w) \, \text{kJ/kg}$$

where: CVn = net calorific value in kJ/kg
$\quad\quad\;\; CVg$ = gross calorific value in kJ/kg
$\quad\quad\quad$ f = water content of the fuel in percentage by weight
$\quad\quad\quad$ w = water percentage by weight generated by combustion of the hydrogen in the fuel.

Typical gross calorific values for different fuels are:

diesel oil   10 750 kcal/kg
gas oil   10 900 kcal/kg
boiler fuel   10 300 kcal/kg

When test data is not available the specific gravity of a fuel will provide an approximate guide to its calorific value:

Specific gravity at 15°C	= 0.85	0.87	0.91	0.93
Gross CV kcal/kg	= 10 900	10 800	10 700	10 500

Viscosity

The viscosity of an oil is a measure of its resistance to flow which decreases rapidly with increase in temperature. Heating is necessary to thin the heavy fuels of high viscosity in current common use and ease their handling.

Over the years different units have been used for viscosity (Engler Degrees, Saybolt Universal Seconds, Saybolt Furol Seconds and Redwood No. 1 Seconds). The majority of marine fuels today are internationally traded on the basis of viscosity measured in centistokes ($1 \text{ cSt} = 1 \text{ mm}^2/\text{sec}$). When quoting a viscosity it must be accompanied by the temperature at which it is determined. Accepted temperatures for the viscosity determination of marine fuels are 40°C for distillate fuels and 100°C for residuals. If a fuel contains an appreciable amount of water, testing for viscosity determination at 100°C becomes impossible and therefore many laboratories may routinely test fuels at a lower temperature (80–90°C) and calculate the viscosity at 100°C.

Due to the variability of residual fuel composition, warns Det Norske Veritas Petroleum Services (DNVPS), calculations of viscosity measured at one temperature to that at another temperature may not be accurate. Calculated results should therefore be treated with caution.

A complication noted by DNVPS is that fuel is traded on the basis of viscosity at 50°C but the two main standards (ISO/CIMAC) give limits at 100°C. By reference to ISO 8217 a buyer could order a 180 cSt fuel at 50°C to comply with ISO RME 25. If the fuel was tested at 100°C and the viscosity result was 25 centistokes at 100°C this could be equivalent (by calculation) to 225 centistokes at 50°C. Clearly, DNVPS explains, the fuel in this case would meet the ISO requirements for viscosity but would be 45 centistokes above the 180 cSt ordered viscosity. This anomaly could be easily removed if trading of marine fuels moved to viscosity at 100°C.

The maximum admissible viscosity of the fuel that can be used in an installation depends on the onboard heating and fuel preparation outfit. As a guide, the necessary pre-heating temperature for a given nominal viscosity can be taken from the viscosity/temperature chart in the engine instruction manual. Wärtsilä's recommended viscosity range for its Sulzer low speed engines is: 13–17 cSt or 60–75 sec Redwood No. 1.

(Details of the fuel handling/treatment requirements of enginebuilders are provided in a later section of this chapter.)

Cetane number

The cetane number of a fuel is a measure of the ignition quality of the oil under the conditions in a diesel engine. The higher the cetane number, the shorter the time between fuel injection and rapid pressure rise. A more usable pointer of ignition quality is the diesel index, expressed as:

$$\text{diesel index number} = \frac{G \times A}{100}$$

where G = specific gravity at 60°F on the API scale
A = aniline point in °F, which is the lowest temperature at which equal parts by volume of freshly distilled aniline and the fuel oil are fully miscible

Calculated carbon aromaticity index (CCAI)

The ignition quality of residual fuels is more difficult to predict than distillate fuels because they comprise blends of many different components but the ignition quality of such fuels may be ranked by determining the Calculated Carbon Aromaticity Index from density and viscosity measurements. It should be noted, however, that the ignition performance of residual fuels is mainly related to engine design and operational factors. Formulae and nomograms for CCAI determination are published by major fuel suppliers and enginebuilders.

Conradson carbon value

This is the measure of the percentage of carbon residue after evaporation of the fuel in a closed space under control. The Conradson or coke value is a measure of the carbon-forming propensity and thus an indication of the tendency to deposit carbon on fuel injection nozzles. The Ramsbottom method has largely replaced the Conradson method of carbon residue testing, but it gives roughly the same results.

Ash content

The ash content is a measure of inorganic impurities in the fuel. Typically, these are sand, nickel, aluminium, silicon, sodium and vanadium. The most troublesome are sodium and vanadium which form a mixture of sodium sulphate and vanadium pentoxide, which melt and adhere to engine components, particularly exhaust valves.

Sulphur content

This has no influence on combustion but high sulphur levels can be dangerous because of acid formation, mentioned earlier in this chapter. In recent years there has been a tendency to equate sulphur content with cylinder liner wear, but opinions differ on this matter. (See also chapter on Emissions).

Water content

This is the amount of water in a given sample of the oil and is usually determined by centrifuging or distillation.

Cloud point

The cloud point of an oil is the temperature at which crystallization of paraffin wax begins to be observed when the oil is being cooled down.

Pour point

This is the lowest temperature at which an oil remains fluid and thus is important to know for onboard handling purposes. An alternative is the solidifying point or the highest temperature at which the oil remains solid. It usually lies some 3°C below the pour point. According to one major enginebuilder, the lowest admissible temperature of the fuel should be about 5–10°C above the pour point to secure easy pumping.

Flash point

The flash point is defined as the lowest temperature at which an oil gives off combustible vapours, or the point at which air/oil vapour mixture can be ignited by a flame or spark.

Specific gravity

This is normally expressed in kg/m^3 or g/cm^3 at 15°C. As the density of the fuel depends upon the density of the individual components, fuels can have identical densities but widely varying individual component densities. Apart from being an indicator of the 'heaviness' of a fuel, when measured by a hydrometer the specific gravity can be used to calculate the quantity of fuel by weight in a tank of given dimensions.

Typical values for standard fuels are given in Table 4.2.

Table 4.2 Typical value for standard fuels

Fuel	SG (g/cm³)	Flash point (°C)	Lower CV (kJ/kg)
Gas oil	0.82–0.86	65–85	44 000–45 000
Diesel oil	0.85	65	44 000
Heavy fuel	0.9–0.99	65	40 000–42 000
(200 secs Redwood No. 1–3500 secs Redwood No. 1)			

The following is the fuel oil specification of a leading low speed diesel engine manufacturer. The properties are considered the worst in each case that can be burnt in the particular engine.

Density at 15°C	kg/m³	max. 991
Kin. viscosity		
at 50°C	mm²/s (cSt)	max. 700
at 100°C	mm²/s (cSt)	max. 55
Carbon residue		
(CCR)	m/m (%)	max. 22
Sulphur	m/m (%)	max. 5.0
Ash content	m/m (%)	max. 0.2
Vanadium	mg/kg (ppm)	max. 600
Sodium	mg/kg (ppm)	max. 100
Aluminium	mg/kg (ppm)	max. 30
Silicon	mg/kg (ppm)	max. 50
Sediment (SHF)	m/m (%)	max. 0.10
Water content	v/v (%)	max. 1.0
Flash point	°C	min. 60
Pour point	°C	max. 30

Combustion equations

Table 4.3 gives a list of the symbols designating the atoms and molecules of the elements and compounds in liquid fuels. For the atomic weights given the suffix denotes the number of atoms accepted as constituting one molecule of the substance.

Table 4.4 summarizes the equations of the chemical reactions taking place during combustion. Nitrogen compounds are ignored. The weight of oxygen needed is evaluated, from the appropriate equation, for every element in the fuel, and the results are added together. The chemically correct quantity of air required to burn exactly the fuel to

Table 4.3 Elements and compounds in liquid fuels and products of combustion

Name	Nature	Atomic symbol	Atomic weight	Molecular symbol	Molecular weight
Carbon	Element	C	12.00	C	12.00
Hydrogen	Element	H	1.008	H_2	2.016
Oxygen	Element	O	16.00	O_2	32.00
Nitrogen	Element	N	14.01	N_2	28.02
Sulphur	Element	S	32.07	S_2	64.14
Carbon monoxide	Compound	—	—	CO	28.00
Carbon dioxide	Compound	—	—	CO_2	44.00
Sulphur dioxide	Compound	—	—	SO_2	64.07
Sulphur trioxide	Compound	—	—	SO_3	80.07
Sulphurous acid	Compound	—	—	H_2SO_3	82.086
Sulphuric acid	Compound	—	—	H_2SO_4	98.086
Water	Compound	—	—	H_2O	18.016

produce carbon dioxide and water vapour and unburnt nitrogen is approximately 15:1, termed the Stoichiometric Ratio. The atmosphere may be assumed to contain 23.15 per cent of oxygen by weight and 20.96 per cent by volume. At standard atmospheric pressure, 1 kg of air occupies a volume of 0.83 m^3 at 20°C and 0.77 m^3 at 0°C. An average diesel fuel should comprise in percentages, carbon 86–87; hydrogen 11.0–13.5; sulphur 0.5–2; oxygen and nitrogen 0.5–1.0. The greatest energy output in the combustion of fuels is from the hydrogen content, and fuels having a high hydrogen:carbon ratio generally liberate greater quantities of heat. Typical examples are given in Table 4.7.

Table 4.4 Combustion equations

Nature of reaction	Equation	Gravimetric meaning
Carbon burned to carbon dioxide	$C + O_2 = CO_2$	12 kg C + 32 kg O = 44 kg CO_2
Carbon burned to carbon monoxide	$2C + O_2 = 2(CO)$	24 kg C + 32 kg O = 56 kg CO
Carbon monoxide burned to carbon dioxide	$2(CO) + O_2 = 2(CO_2)$	56 kg CO + 32 kg O = 88 kg CO_2
Hydrogen oxidised to steam	$2H_2 + O_2 = 2(H_2O)$	4 kg H + 32 kg O = 36 kg steam (or water)
Sulphur burned to sulphur dioxide	$S_2 + 2O_2 = 2(SO_2)$	64 kg S + 64 kg O = 128 kg SO_2
Sulphur dioxide burned to sulphurous acid	$SO_2 + H_2O = H_2SO_3$	64 kg SO_2 + 18 kg H_2O = 82 kg H_2SO_3
Sulphur dioxide burned to sulphur trioxide	$O_2 + 2(SO_2) = 2(SO_3)$	32 kg O + 128 kg SO_2 = 160 kg SO_3
Sulphur trioxide and water to form sulphuric acid	$SO_3 + H_2O = H_2SO_4$	80 kg SO_3 + 18 kg H_2O = 98 kg H_2SO_4

Table 4.5 Volumetric meaning of equations

Nature of reaction	Volumetric result
Carbon burned to carbon dioxide	1 vol C + 1 vol O = 1 vol CO_2
Carbon burned to carbon monoxide	2 vols C + 1 vol O = 2 vols CO
Carbon monoxide burned to carbon dioxide	2 vols CO + 1 vol O = 2 vols CO_2
Hydrogen oxidised to steam	2 vols H + 1 vol O = 2 vols H_2O (steam not water)
Sulphur burned to sulphur dioxide	1 vol S + 2 vols O = 2 vols SO_2
Sulphur dioxide burned to sulphurous acid	—
Sulphur dioxide burned to sulphur trioxide	—
Sulphur trioxide and water to form sulphuric acid	—

Table 4.6 Heat evolved by combustion

Nature of reaction	Thermo-chemical equation	Heat evolved per kg of element burned kcal/kg
Carbon burned to carbon dioxide	$C + O_2 = CO_2 + 96\,900$	8 080
Carbon burned to carbon monoxide	$2C + O_2 = 2(CO) + 58\,900$	2 450
Carbon monoxide burned to carbon dioxide	$2(CO) + O_2 = 2(CO_2) + 315\,000$	5 630 (per kg CO burned)
Hydrogen oxidised to steam	$2H_2 + O_2 = 2(H_2O) + 136\,000$	33 900 (a) 29 000 (b)
Sulphur burned to sulphur dioxide	$S_2 + 2O_2 = 2(SO_2) + 144\,000$	2 260
Sulphur dioxide burned to sulphurous acid	—	—
Sulphur dioxide burned to sulphur trioxide	—	—
Sulphur trioxide and water to form sulphuric acid	—	—

Notes (a) Value if the latent heat of the steam formed is to be *included*
      (b) Value if the latent heat of the steam formed is to be *excluded*

Table 4.7

Fuel	H:C ratio	Calorific value kJ/kg
Methane	4:1	55 500
Ethane	3:1	51 900
Propane	2.7:1	50 400
Kerosene	1.9:1	43 300
Heavy fuel	1.5:1	42 500
Benzene	1:1	42 300
Coal	0.8:1	33 800

FUEL OIL TREATMENT

Low, medium and high speed diesel engines are designed to burn marine diesel oil (ISO 8217, class DMB or BS 6843, class DMB) while low speed and many medium speed engines can also accept commercially available heavy fuel oil (HFO) with a viscosity up to 700 cSt (7000 sec Redwood No. 1).

In order to ensure effective and sufficient cleaning of the HFO (removing water and solid contaminants) the fuel oil specific gravity at 15°C should be below 0.991. Higher densities—up to 1.010—can be accepted if modern centrifugal separators are installed, such as the systems available from Alfa-Laval (Alcap), Westfalia (Unitrol) and Mitsubishi (E-Hidens II).

Alfa-Laval's Alcap system comprises an FOPX separator, a WT 200 water transducer and ancillary equipment including an EPC-400 control unit (Figure 4.2). Changes in water content are constantly monitored by the transducer which is connected to the clean oil outlet of the separator and linked to the control unit. Water, separated sludge and solid particles accumulate in the sludge space at the separator bowl periphery. When separated sludge or water force the water towards the disc stack minute traces of water start to escape with the cleaned oil and are instantly detected by the transducer in the clean oil outlet. The control unit reacts by triggering a sludge discharge or by allowing water to drain off through a separate drain valve, thereby re-establishing optimum separation efficiency.

A number of benefits are cited for the Alcap system over conventional fuel cleaning systems. Continuous optimum separation efficiency is

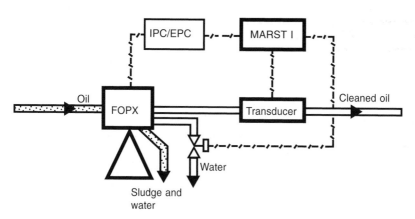

Figure 4.2 Schematic diagram of Alfa Laval's Alcap fuel treatment system based on its FOPX centrifugal separator

achieved since water and sludge in the bowl are never permitted to enter the disc stack. Only one Alcap FOPX separator is required, a single-stage unit achieving the same and often higher separation efficiency than a traditional two-stage configuration featuring two separators (a purifier followed by a clarifier).

Manual cleaning separators are not recommended for attended or unattended machinery spaces. Centrifugal separators should be selfcleaning, either with total discharge or with partial discharge. The nominal capacity of the separator should be as recommended by the supplier. Normally, two separators are installed for HFO treatment, each with adequate capacity to comply with the fuel throughput recommendation.

The schematic representation of a heavy fuel oil treatment system approved by Wärtsilä is shown in Figure 4.3, the company highlighting the following points in designing a particular installation:

Gravitational settling of water and sediment in modern fuel oils is an extremely slow process due to the small density difference between the oil and the sediment. To achieve the best settling results, the surface area of the settling tank should be as large as possible, with low depth, because the settling process is a function of the fuel surface area of the tank, the viscosity and density difference.

The function of the settling tank is to separate the sludge and water contained in the fuel oil, to act as a buffer tank and to provide a suitable constant fuel temperature of 50–70°C. This can be improved by installing two settling tanks for alternate daily use.

Wärtsilä also advises the use of 'new generation' separators without gravity disc to treat heavy fuels of up to 700 cSt/50°C viscosity and facilitate continuous and unattended operation. When using conventional gravity disc separators, two such sets working in series (purifier–clarifier) are dictated. The clarifier enhances the separation result and acts as a safety measure in case the purifier is not properly adjusted.

The effective separator throughput should accord with the maximum consumption of the diesel engine plus a margin of about 18 per cent to ensure that separated fuel oil flows back from the daily tank to the settling tank. The separators should be in continuous operation from port to port. To maintain a constant flow through the separators individual positive displacement-type pumps operating at constant capacity should be installed. The separation temperature is to be controlled within +/–2°C by a preheater.

The fuel treatment system can be improved further by an automatic filtration unit installed to act as a monitoring/safety device in the event of separator malfunction. Such a filtration unit is particularly

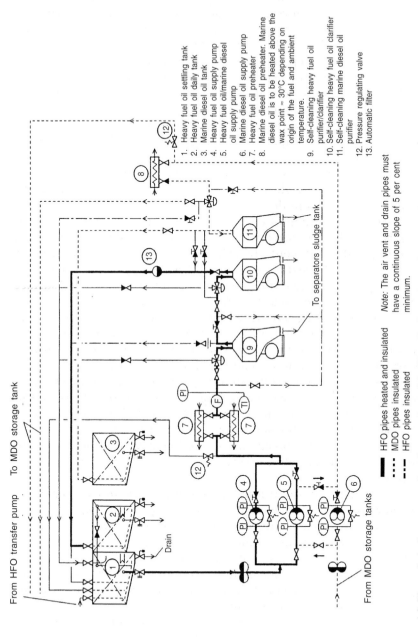

1. Heavy fuel oil settling tank
2. Heavy fuel oil daily tank
3. Marine diesel oil tank
4. Heavy fuel oil supply pump
5. Heavy fuel oil/marine diesel oil supply pump
6. Marine diesel oil supply pump
7. Heavy fuel oil preheater
8. Marine diesel oil preheater. Marine diesel oil is to be heated above the wax point ≈ 30°C depending on origin of the fuel and ambient temperature.
9. Self-cleaning heavy fuel oil purifier/clarifier
10. Self-cleaning heavy fuel oil clarifier
11. Self-cleaning marine diesel oil purifier
12. Pressure regulating valve
13. Automatic filter

HFO pipes heated and insulated
MDO pipes insulated
HFO pipes insulated

Note: The air vent and drain pipes must have a continuous slope of 5 per cent minimum.

Figure 4.3 Heavy fuel oil treatment plant layout approved by Wärtsilä for its low speed engines

recommended when conventional separators with gravity discs are deployed. Separators can also be supported by homogenisers (see section below) in cases where possible asphaltene sludge precipitation due to unstable or incompatible fuels is anticipated or where the size of abrasive particles has to be reduced.

The fuel oil heater may be of the shell and tube or plate heat exchanger type, with electricity, steam or thermal oil as the heating medium. The required heating temperature for different oil viscosities is derived from a fuel oil heating chart, Figure 4.4, based on information from oil suppliers with respect to typical marine fuels with a viscosity index of 70–80. Since the viscosity after the heater is the controlled parameter the heating temperature may vary, depending on the viscosity and viscosity index of the fuel. The viscosity meter setting, reflecting the desired fuel injection viscosity recommended for an engine by the enginebuilder, is typically 10–15 cSt.

To maintain a correct and constant viscosity of the fuel oil at the inlet to the main engine the heater steam supply should be automatically controlled, usually based on a pneumatic or electronic control system.

Alfa Laval believes that two-stage pressurized systems are preferable for fuel conditioning (feeder and booster) systems, suggesting that single-stage systems can be difficult to control and suffer from cavitation and gasification problems associated with high fuel temperatures (120–150°C). Its own two-stage system features a 4 bar low pressure section, while the high pressure side ranges between 6–16 bar, depending on the requirements of the enginebuilder. On the LP side are two pumps (one standby), an auto-filter with back-up manual bypass filter, and a flow transmitter. There is also a mixing tank, where fresh fuel is mixed with hot fuel returning from the engine. Oil from the day tanks is fed to the system under pressure via LP supply pumps to eliminate any gasification and cavitation problems.

From the mixing tank the fuel enters the HP section, where the flow rate is maintained at a multiple of the actual fuel consumption rate so as to prevent fuel starvation at the injectors. The HP section incorporates circulation pumps, heaters and a viscosity transducer. The transducer compares viscosity against a value set by the enginebuilder and sends a signal to the controller allowing it to alter the fuel temperature by adjusting the flow of heating medium (steam or thermal oil) to the heater. 'Self-cure' functions include pump changeover to standby in the event of failure or too low pressure, and change from viscosity to temperature control (or vice versa).

Separator developments

The main aim of centrifugal separators is to remove fuel contaminants—notably water, catalyst fines, abrasive grit and sodium from sea water—that can cause excessive wear when burned in the engine. Continuing development pursues even higher separation efficiencies to maximize engine protection, and more efficient discharge systems that reduce overall discharge volumes, decrease the amount of water used and hence cut the amount of material collected in the sludge tank.

When partial discharge type separators replaced total discharge designs the actual oil consumed at each discharge fell dramatically. In the partial discharge process, however, larger volumes of water are needed for displacing the oil interface and for the operating system. This quantity of water is a major contributor to the overall sludge volume as it too must be processed. Alfa Laval's Separation Unit exploits a new discharge process called CentriShoot, which combines several design improvements to help reduce the sludge consumption volumes associated with separators: by at least 30 per cent and up to 50 per cent compared with previous models.

First, the bowl volume and therefore its contents are much lower than equivalent separators, which means that at each discharge the contents discharged will be significantly less. Secondly, the frequency of discharge has been extended by around four times compared with partial discharging separators. A new design of discharge slide replaces the traditional sliding bowl bottom; the new component is a form of flexible plate that is secured at the centre, the slide outer edge flexing approximately 1–2 mm at each discharge to allow the sludge to be evacuated through the sludge ports.

Installed as standard in the Separation Unit is a REMIND software package which allows the operator to install the program discs on a laptop computer. Connected to the control cabinet, the system can then review and store the alarm history and processing parameters in the computer. The data can be used later to check processing conditions for trouble shooting.

Westfalia Separator's C-generation family of separators exploits Hydrostop and Softstream systems. Hydrostop features special discharge ports and a separator bowl architecture allowing more efficient sludge ejections at full operating speed; this, in turn, extends desludging intervals and reduces maintenance costs. Softstream allows liquids to enter the bowl in a 'super calm' state, thus improving separator efficiency, increasing flow rates and reducing component wear. Typically, says Westfalia, a C-generation separator will reduce sludge volumes by up to 50 per cent during fuel and lube oil purification. The separators can also be specified with Unitrol, a self-thinking system for automatically handling fuel and lube oils of varying quality and density in an unmanned engineroom.

Refinements introduced by Mitsubishi Kakoki Kaisha (MKK) in recent years have sought improved overall performance and capacity, easier operation and reduced maintenance from the Japanese designer's Selfjector Future (SJ-F) series of separators. More compact and lighter units have also been targeted, Figure 4.5.

MKK's Hidens system is offered for treating oils with densities of up to 1.01 g/cm^3 at 15°C, the installation comprising the separator, an automatic control panel, water detector, water detector controller and discharge detector. The water content of the oil is continuously monitored and when a pre-set maximum level is reached a total discharge function is automatically actuated. Among the options that can be specified is a sludge discharge control system that automatically sets the most suitable discharge interval based on feedback from an oil inlet sensor continually measuring sludge concentration in the dirty oil; differences in bunker quality can also be detected at the first stage. A monitoring and diagnostic system can supervise the operating conditions of up to six separators simultaneously.

A homogenizer may be installed to support the separators and filters of a fuel treatment system. The shearing action breaks down particles in the fuel oil to sub-micron sizes and finely distributes any water present as small droplets. A considerable reduction in sludge amounts and improved combustion are claimed. Some homogenizer systems are designed to produce a stable fuel/water emulsion with water amounts up to 50 per cent and higher, contributing to a reduction of NOx emissions in the exhaust gas.

There is some disagreement, however, over the positioning of a homogenizer in the fuel treatment system (before or after the separator). Alfa Laval argues that installing such a unit *prior* to the separator reduces the ability of the treatment system to remove those particles that can cause damage to the fuel injection system and the main engine, and should therefore be avoided.

The role of filtration systems has been strengthened by higher fuel injection pressures and the wider use of common rail fuel systems. Solids remaining after previous treatment stages are removed from the fuel circuit by manual cleaning in a simplex filter or by an automatic backflushing filter. Common rail fuel systems call for stricter filtration regimes, typically dictating 5-micron filtration to protect injection components from extensive wear. Such fine filter elements either translate to a shorter lifetime or larger filter sizes, and this must be compensated by centrifugal separator equipment which is not dimensioned at the maximum flow rate to reach its best possible solid removal performance.

GOOD BUNKERING PRACTICE

Marine fuel oil is the residue after the crude oil has been processed at the refinery to produce diverse high value products, such as gasolines, jet fuels, diesel and petrochemical feedstocks.

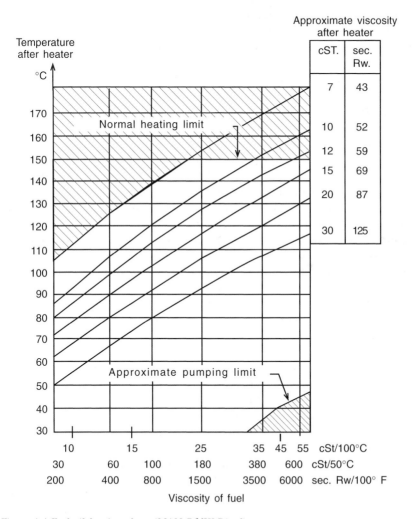

Figure 4.4 Fuel oil heating chart (MAN B&W Diesel)

Refineries traditionally created these products by a relatively simple distillation process. But to satisfy increased demand additional processes—for example, thermal and catalytic cracking—were introduced to use residues and heavier distillates as feedstocks. The volume of the final residue available for bunker fuel was thus reduced, concentrating the impurities remaining.

Between the refinery and the delivery of the bunkers onboard, via any intermediary storage that might be required, are many opportunities for the fuel to become contaminated: sometimes so badly that it may damage the ship's machinery and associated systems. Especially threatening is the presence in the fuel of catfines resulting from the catalytic cracking stage at the refinery in which the oil is heated over aluminium silicate catalyst, a substance harder than metal.

Immediate and longer term risks to ship and machinery are associated with poor fuel quality, including fire and explosion, loss of main engine power and/or auxiliary power black-out, and difficulty in fulfilling charter party clauses with regard to ship speed, engine power and specific fuel consumption.

Little legislation exists to govern bunker fuel but the parameters of the international standard ISO 8217 have been widened in recent years. Since most risk associated with bunkers impacts on the ship and owner (even though the fuel is bought by and belongs to any charterer) the best protection is for all parties to know exactly what fuel is needed for the vessel, and then to test that the delivered bunkers meet these requirements. Most problems and disputes can be traced back either to the ordering or sampling of the fuel. The traditional way of ordering bunkers has been by quoting IFO grades, which only refer to viscosity at 50°C, but this can be technically inaccurate. It is more important in the first instance to make sure that the fuel is suitable for the machinery and fuel system in which it will be used. The best starting point is the fuel specification provided by the enginebuilder. It makes commercial sense for ship operators to follow a recognized specification such as ISO 8217. This will help to secure a charter party and should help to secure economically priced bunkers worldwide. The charter party clause should clearly state what fuel is required and may also include other aspects, such as segregation, testing and the services of a surveyor for quantity determination. The charter party clause should also be kept as simple as possible, Lloyd's Register recommending the following wording as an example: 'Charterer to provide all fuel oil and diesel required, in accordance with ISO 8217:1987, Grades RM . . . and DM . . . , as updated from time to time'. Ship operators are encouraged to stipulate that the bunker supplied is 'fit for its intended use' on the delivery note.

Among the key parameters of a fuel specification are:

- A maximum viscosity: this is required to be within the enginebuilder's specified limit; it is also a guide to the storage and handling properties of the fuel.
- A maximum density or specific gravity: dictated by the ship's fuel treatment plant.
- A minimum flash point: dictated by safety regulations for the storage of fuel at sea.
- A maximum pour point: this limits the amount of heating required to store and handle the fuel.
- A maximum water content.
- A maximum sulphur content: may be necessary to satisfy regional environmental regulations but is also desirable to limit the corrosive products of combustion and to ensure that the lubricants used in the engine can cope with these products.

Bunker fuel short delivery and quality disputes can be extremely contentious. In the event of a dispute the conduct of measurement and sampling techniques will usually come under close scrutiny, and it will be necessary for the ship's interests to demonstrate that proper procedures were carried out during the bunkering.

Diligent participation of both ship and supplier during measurement and sampling is very important in preventing or resolving disputes, and the importance of maintaining proper records and documentation before, during and after bunkering cannot be overstated.

It is essential that prior to bunkering the chief engineer and any attending surveyor should first check the bunker receipt to ensure that the fuel on the barge complies with that ordered. The tanks of the receiving ship and the bunker barge should then be carefully measured to ascertain the exact quantities on both vessels. If no surveyor is in attendance, the task should be performed by either the chief or first engineer assisted by an engineroom rating. The engineer should preferably use his own sounding tape, thermometers, hydrometers and temperature/volume correction tables. Water-finding paste should be used on the measuring tape or sounding rod to check for free water. If found, the quantity of free water should be deducted from the volume of fluids in the tank to arrive at the correct oil volume.

On the receiving vessel, when sufficient empty tanks are available, many chief engineers prefer to segregate fuels from different stemmings: both to ease the calculation of the quantity loaded and to eliminate as far as possible any problems with incompatibilities of fuels produced from different crude feedstocks.

Once the measuring has taken place the quantity onboard the barge should be calculated using its calibration tables. Since the supply of fuel is often under Customs control bunker tankers have officially approved calibration tables adorned with a variety of stamps verifying their authenticity. The receiving ship's representatives must therefore ensure that the tables offered to them are suitably approved and declined if they are at all suspect. Poor quality, unapproved tables may be encountered and even cases when no tables are available. If no tables can be provided measures should be taken to obtain them. If no tables can be found the fuel on the barge should be rejected, even if there is a delivery meter onboard. The veracity of fuel meters should be regarded as suspect even when meter correction factors and calibration certificates are present.

Transfer can commence provided that the quantity in the barge tanks is the same as that on the bunker delivery receipt. Agreement must be reached on the rate of delivery and signals for the starting, stopping and slowing down of delivery. In Japanese ports an interpreter is often provided to liaise between the supplier and the receiver.

It is now mandatory for vessels to have clearly laid down procedures for bunkering. This is an extension of US Coastguard directives which required all tonnage visiting US ports to display diagrams of their fuel loading, transfer and storage systems. For bunkering, these rules required that a senior engineer officer be nominated as 'in charge' and the ranks of his assistants to be given. Before bunkering, a detailed plan was to be prepared showing how much of each type of oil was to be loaded or transferred, which tanks were to be used, and the anticipated finishing ullages. A deck officer was to be nominated to maintain the ship's mooring and other arrangements during bunkering. All deck scuppers were to be sealed and sawdust or other absorbent material provided. Receptacles for spilled oil and oil-contaminated sawdust were to be at hand. Fire extinguishers and hoses were to be deployed nearby and the fire main was to be pressurized. The method of communication between the various participants was to be given, with a back-up mode provided.

After all necessary pollution prevention precautions have been taken the valves of the receiving ship's bunkering lines can be set and delivery started. When the integrity of the delivery hose and its connection to the ship's system have been proven the rate of delivery can be slowly increased to the agreed level.

If possible, during the delivery, a set of three samples of the fuel should be taken using a small bore valve fitted to the ship's loading connection and retained for any necessary future analysis. Proprietary sampling devices which can be bolted to the loading flange are available.

The best of these has a small bore pipe directed against the flow which leads outside to three branches, each fitted with a valve. It is therefore possible to obtain three continuous drip samples during loading to give an 'average' of the fuel quality.

The bunker barge normally provides samples of the fuel delivered and the officer in charge of the operation should always attend to observe the samples being taken and sealed. It is customary for the ship's representative to sign the labels for the sample cans. Both barge and receiving vessel then retain samples. The ship's samples should be retained for at least six months.

When the transfer has been completed and the hose blown through and disconnected the tanks of both vessels should be re-measured to enable the quantity transferred to be calculated. Any discrepancy between the ship and barge figures then becomes a matter for negotiation or protest, depending on the magnitude involved.

Vigilance on the part of personnel at the time of stemming fuel, and during its subsequent correct storage and centrifuge treatment, are essential in preventing or mitigating future problems.

On a motor ship at least two centrifugal separators are usually installed for removing water and other impurities from fuel. Under normal conditions only one of these units operates as a separator on a continuous basis. When it is known, however, that the fuel to be treated contains excessive particulate matter two centrifuges can be operated in series. The first acts as a separator to remove water and particulate matter, and the second as a clarifier to remove only particulate matter. In all cases, the throughput of the centrifuges should be as low as possible to achieve the greatest dwell time in the bowl and hence maximum separation efficiency.

Bunker quality testing

Onboard test kits are available to allow ship's staff to carry out a range of tests during or after bunkering or during storage (if fuel stratification is suspected), for checking purifier or clarifier efficiency, for determining the cause of excessive sludging at the separators, if water ingress into the fuel is suspected, or when transferring bunkers to a new tank. The most comprehensive kits can detect catalytic fines and the presence of microbes, and test comparative viscosity, compatibility/stability, density, insolubles content, pour point, salt water, sludge or wax, and water content.

Fuel testing services—such as those offered worldwide by specialist divisions of Lloyd's Register, Det Norske Veritas and the American Bureau of Shipping—offer a swift analysis and advice on the handling and use of bunkers based on samples supplied. Under ISO 8217, even if a fuel meets all the usual specification criteria, it should still not contain harmful chemicals or chemical waste that would cause machinery damage or pose a health hazard to personnel.

Any ship operator doubting the need to test bunkers should be reminded that some 3-5 per cent of stems are critically off-spec. Among the diverse contaminants (mixed inadvertently with the fuel) and adulterants (added wilfully) found over the years have been acids, polypropylene, organic chlorides and hydrogen sulphite. A number of ships have suffered serious machinery damage in recent years from contaminated or adulterated fuels, the culprit components identified as:

- Petroleum-based solvents, with the potential to destroy lubricant oil film and cause damage to the rubbing parts of fuel pumps and engine cylinders.
- Organic chlorides, which not only damage machinery but are also hazardous to ship staff.
- Methyl esters, while environmentally safe and friendly, are nevertheless solvents which can remove the oil film and destroy lubrication between rubbing parts.

The damage potential is very high. A very high wear rate can be expected between fuel pump plungers and barrels, and also between piston rings and cylinder liners. Additionally, the solvents are hazardous to personnel and exposure, even at low ppm levels, is not recommended. If they are mixed with bunker fuels there is a real risk of engineroom staff coming into contact with the solvent fumes, particularly in the purifier room. In the absence of any method of removing such contaminants from onboard, it is better to offload the bunkers.

Det Norske Veritas cites the case of a fully loaded VLCC which bunkered 3000 tonnes of heavy fuel oil. Shortly after, the tanker suffered main and auxiliary engine breakdown and a full black-out in the Malacca Strait, resulting from sticking fuel pumps and clogged filters. An investigation revealed: fuel apparently within the ISO specification but quite 'aggressive'; solvent-like organic contaminants (detected by sophisticated additional analyses); considerable risk potential; and at least 10 other ships affected with severe problems. The rapid identification of the problem prevented major accidents.

In a contaminated marine diesel oil incident, the fuel injection system failed due to corrosion, paint was stripped off pumps and engine components, rubber seals and gaskets dissolved, and crew members suffered skin irritation.

As an example of the potential impact of catalyst fines on engines, Det Norske Veritas highlights the case of 2400 tonnes of heavy fuel oil bunkered in the Middle East. Catfines in the fuel caused extreme wear on the piston rings, piston rods and cylinder liners during the subsequent voyage to France. A fuel analysis showed an Al+Si content of 91 ppm compared with the specification maximum of 80 ppm. The damage to the engine cost US$500 000 to rectify and the offhire time a similar amount.

Low sulphur fuels

Environmental concerns over damage to ecosystems resulting from sulphur emissions (see the chapter on Emissions) have stimulated the creation of SOx Emission Control Areas (SECAs). An IMO framework proposed a global fuel sulphur content cap of 4.5 per cent and a 1.5 per cent sulphur limit in certain SECAs, the latter posing commercial and technical challenges for marine fuel producers, bunker suppliers, ship operators and machinery designers.

Low sulphur fuel oil is generally produced by refining low sulphur crudes. Refineries may need to sweeten their crude slates or perhaps just tap off certain sweeter grades from their internal production runs in order to yield the desired low sulphur fuels. Suppliers have the challenge of storing and marketing additional grades of fuel. Enginebuilders and lube oil formulators have to match the total base number (see Lubricating Oils section below) of cylinder and system lubricants to the new low sulphur fuel grades. Summarizing the impact of a low sulphur fuel regime on shipowners and onboard staff, ExxonMobil Marine Fuels highlights:

- More has to be paid for the fuel burned in SECAs.
- Procedures need to be established for masters and chief engineers to know when and where they must burn specific fuels.
- Better records must be kept, showing exactly what fuel is burned when and where, and providing proof to authorities.
- Fuel must be tested on bunkering to ensure that it meets sulphur content limits.
- Shipowners must ensure that operation teams and bunker buyers are aware of the fleet's needs for low sulphur fuel in specific areas; and, when chartering out a ship, the owner will need to negotiate a modification to the charter party to ensure that the charterer is required to supply low sulphur fuel if trading to SECAs.
- Owners will need to consider modifying their existing tonnage to permit better segregation of bunkers; and shipbuilders and designers will have to create ships with much better means of

segregating fuels and for switching from one fuel grade to another. (Some newbuildings are already specified with separate low sulphur fuel tanks, and this is likely to become a standard design feature.)

No modifications to engines are dictated, however, and for ships trading exclusively in SECAs it may be possible to use lube oil in the main engine with a lower total base number and hence a lower cost. In addition, heavy fuel oils with a lower sulphur content have a slightly higher energy value, underwriting a slightly improved specific fuel consumption. Engine and exhaust system components will wear less and last longer since sulphuric acid-based corrosion is reduced; and scrubbing exhaust gases for inert gas generation is made simpler and cheaper.

The above section includes information from Lloyd's Register's Fuel Oil Bunker Analysis Service (FOBAS), the International Bunker Industry Association (IBIA), Shell Marine, ExxonMobil Marine Fuels, Tramp Oil and Salters Maritime Consultants.

LUBRICATING OILS

Significantly improved thermal efficiency, fuel economy and ability to burn poor quality bunkers have resulted from the intensive development of low speed crosshead and high/medium speed trunk piston engine designs in the past 20 years. Further progress in performance and enhanced lifetimes can be expected from the exploitation of higher firing pressures and better combustion characteristics.

Such advances pose a continuing challenge to engine lubricant formulators, however: higher pressures and temperatures make it more difficult to provide an oil film in the critical zones, and longer strokes (another design trend) lead to spreadability problems. The success of new generations of engines will therefore continue to depend on the quality of their lubricants.

Lubricants are complex blends of base oils and chemical additives. Base oils are derived from the refining of specific crudes, and their properties—such as stability at high temperature and viscosity—depend on the nature of the crude and the production process. Additives either improve the basic properties of oils or underwrite new properties. Marine lubricants exploit many types of additive:

- Detergent and dispersant additives are used to clean engines. Detergents are most effective in the hot areas (pistons, for example) where they destroy carbon deposits or prevent their formation; dispersants help to maintain particles (such as combustion residues) in suspension in the oil until they are eliminated via filters and centrifuges.

- Anti-wear and high pressure additives contribute to the maintenance of satisfactory lubrication under the most severe wear conditions.
- Anti-oxidation additives retard damage due to thermal and oxidation phenomena at high lubricant pressures and hence extend the life of the lubricant.

Other specific roles are performed by anti-foaming, anti-corrosion, anti-rust and anti-freeze agents.

Environmental legislation has also impacted on the composition of lubricants. A reduction in the sulphur content of fuels, for example, influences Total Base Number (TBN) levels and calls for a new balance of detergent and dispersant additives. Measures taken to reduce NOx emissions from engines—such as delayed fuel timing, the injection of water into the cylinder, emulsified fuel and selective catalytic reduction (SCR) systems based on ammonia—also dictate attention to lubricant detergency, resistance to hydrolysis and compatibility with catalysts.

All additives are created from a number of ingredients performing diverse functions. One part of the chemical molecule of 'overbased' detergents, for example, cleans the engine and another neutralizes acids produced by combustion. Additives can interact with each other either positively (synergy) or negatively (antagonism). The art of the lubricant formulator is to select the best base oils and optimize the proportions of the various additives for the specific duty.

Whether or not the engine is a two-stroke or a four-stroke design, operating on distillate or residual fuel, the function of the cylinder lubricant is the same:

- To assist in providing a gas seal between the piston rings and cylinder liner.
- To eliminate or minimize metal-to-metal contact between piston rings, piston and liner.
- To act as a carrier fluid for the functional alkaline additive systems, particularly that which neutralizes the corrosive acids generated during the combustion process.
- To provide a medium by which combustion deposits can be transported away from the piston ring pack to keep rings free in grooves.
- To minimize deposit build-up on all piston and liner surfaces.

Formulating an 'ideal' lubricant to perform all these functions for all engines on the market is difficult and will always be subject to experience and judgement. First, there are varying requirements depending on engine type (single or dual level lubrication, oil injection position in the liner and oil feed rate applied), operating condition (power and rotational speed set points) and fuel quality used. Secondly, in many cases the requirements are contradictory; for example, the high viscosity necessary to yield good load-carrying performance will adversely affect spreadability.

Cylinder oils

The cylinder conditions in the latest designs of low speed crosshead engines impose a higher maximum pressure, higher mean effective pressure and higher cycle temperature than previous generations. In addition, the combined effect of a longer piston stroke and lower engine rotational speed exposes the cylinder oil film to higher temperatures for a longer period of time.

The diversity of engine designs, operating conditions and worldwide bunker fuel qualities means there can never be a single truly optimum cylinder oil grade. In certain special circumstances other grades may perform better or at least more cost-effectively.

Modern crosshead engine cylinder oils are formulated from good quality base oils with an additive package which performs a variety of functions within the cylinder. The product must possess adequate viscosity at high working temperatures, yet be sufficiently fluid to spread rapidly over the working surfaces; it must have good metal wetting properties and form an effective seal between rings and liners; and it must burn cleanly and leave a deposit which is as soft as possible.

The results of shipboard trials and other investigations by a leading lubricating oil company indicated that a 70 BN/SAE 50 cylinder oil, formulated with the right technology, is the most cost-effective way of providing cylinder lubrication for the majority of low speed engines.

Alternative cylinder lubricants can be considered for special applications or to counter specific problems, or if the ship involved is engaged in a trading pattern where the sulphur level of the bunkers is consistently high or consistently low.

Shell introduced a higher TBN cylinder lubricant to complement its Alexia Oil 50 product. The Shell Alexia X Oil is specially blended at 100 TBN for more highly rated low speed engines consistently burning residual fuels with a sulphur content greater than 3.5 per cent.

Castrol Marine's Cyltech 80 TBN crosshead engine cylinder lubricant has reportedly proven its ability in service consistently to achieve significant reductions in both liner and piston ring wear. Lubricant costs have also been reduced through lower consumption secured by the combination of a high TBN and effective anti-wear properties.

Cyltech 80AW was introduced in 2002 to provide stronger protection against wear and scuffing through enhanced resistance to oil breakdown at high temperatures. Extra protection is also promised from the anti-wear (AW) properties if oil breakdown occurs under extreme conditions, resulting in metal-to-metal contact. Apart from reducing wear rates under normal and extreme operating conditions and increasing safety margins against scuffing, the lubricant is said to foster operation at minimum oil feed rates, thus reducing annual lubrication costs. Excellent cleanliness in piston ring zones and extra protection against corrosive wear are also reported.

Cyltech 70, introduced at the same time and replacing Castrol Marine's S/DZ 70 lubricant, is described as a superior 70 BN cylinder oil, benefiting from the same resistance to breakdown at elevated temperatures as Cyltech 80AW. It is offered as a 70 BN solution to operators not requiring the higher protection of Cyltech 80AW and where a higher base number is less critical.

Investigations in the 1990s by Castrol Marine included shipboard trials with 130 TBN and 30 TBN products to monitor the performance of cylinder lubricants under differing conditions. The 130 TBN lubricant was evaluated for three months in the Sulzer 6RLB90 low speed engine of a containership which operated at 85–90 per cent maximum continuous rating. The aims of the trial were to increase Castrol's experience with higher TBN cylinder lubricants and improve its knowledge of the mechanisms of liner corrosive wear. The 30 TBN cylinder lubricant was evaluated in a two-stroke engined vessel operating in the Baltic and burning low sulphur 180 cSt fuel. The main aim was to determine an optimum between acid neutralization (corrosive wear) and detergency (piston cleanliness).

FAMM (Chevron/Texaco) notes that most commercial cylinder lubricants are 70 TBN/SAE 50 grade but warns the operator to be aware that the different products have a different chemical composition. The latter affects the corrosive wear protection of the liner. The group's R&D laboratory conducted Bolnes engine tests on a series of different 70 TBN/SAE 50 cylinder lubricants, with the focus on measuring corrosive wear (the most dominant wear factor in low speed engines). Differences of up to 100 per cent in the maximum liner wear were found. FAMM advises that corrosive wear can be adequately minimized by a number of measures:

- Keeping the liner temperature above the sulphuric acid dew point.

- Maintaining proper water shedding of the intake air, especially under tropical conditions; the water shedding operation of the charge air should be regularly inspected.
- Applying a sufficiently high oil feed rate to guarantee a satisfactory oil film thickness, minimizing adhesive wear and providing good alkalinity retention.
- Using a cylinder lubricant with superior corrosive wear protection.

Specialists stress that lubricant feed rate is critical and should be carefully balanced between cylinders. Any downward departure from the enginebuilder's recommended feed rate is potentially dangerous, while an excessive rate is generally not cost-effective.

BP Marine's improved Energol CLO 50M 70BN cylinder lubricant promises cost-effectiveness with the highest performance levels. Lower operational costs result from reduced wear and cleaner operation, underwriting a higher margin of safety for low oil feed rates, extended time between overhauls and better control against scuffing. A new generation cylinder oil from ExxonMobil, Mobilgard 570, is claimed to secure maximum protection from adhesive and corrosive wear at the higher operating temperatures and pressures imposed by the latest two-stroke engines. Such conditions reduce the viscosity of the lubricant and increase the loads it must withstand. In addition, longer piston strokes have extended the amount of surface to be protected and the amount of time the lubricant must resist the severe cylinder temperatures and corrosive sulphur acids. New additives with a greater thermal stability and an optimum viscosity (21 cSt at 100°C) foster lubricant distribution and film retention. Superior ring and liner protection is reportedly demonstrated by Mobilgard 570 at the 70 TBN alkalinity level, along with cleanliness under sustained operation with fuel sulphur levels from less than 0.5 up to 5 per cent.

Types of cylinder wear

Cylinder lubricants are an essential component in the control of liner and piston ring wear in low speed crosshead engines. High wear rates can be caused by corrosive, abrasive and adhesive wear, the conditions prevailing in the combustion space being the key influence on the type of wear experienced. In older engine designs, Castrol Marine explains, the wear was predominantly corrosive but in the latest engine designs it is now largely adhesive wear.

Traditionally, cylinder oils had high alkalinity to provide protection against corrosive wear: 70 BN was the industry norm, with higher BN oils available to secure even greater protection. The increase in engine performance, and in particular the rise in cylinder liner temperature, effectively engineered out high corrosive wear. Engines designed and built in the late 1980s and early 1990s had liner temperatures sufficiently high enough to prevent corrosive wear; and the use of a 70 BN lubricant was often the most economical means of providing cylinder lubrication and protection.

Once liner temperatures are raised above 250°C the performance of standard cylinder oils becomes marginal in providing protection against increased liner and piston ring wear. Under these conditions, the type of wear is largely adhesive and the performance of cylinder oils has to be assessed by their ability to minimize such wear. (Corrosive wear, however, is a problem that may not have been completely eliminated. Theory, as well as data published on test engines with very high peak pressures, points to an increased corrosion potential as more of the sulphur is converted to sulphur trioxide. If the engine charge air contains enough water, then its combination with sulphur trioxide leads to the formation of sulphuric acid in sufficient quantities to tip the wear balance towards corrosion.)

Adhesive wear, sometimes referred to as 'scuffing', occurs when the surface asperities on the liner and piston ring have metal-to-metal contact. When contact takes place local welding and metal deformation result, with the amount of contact determining the rate of wear. The rougher the surfaces, the greater the likelihood of adhesive wear occurring (which is why it is usually experienced during the engine running-in period).

Basic separation of the contact surfaces is provided by the lubricant film, and the higher the viscosity the thicker the oil film. Because of the high liner temperatures experienced in modern engines the oil viscosity on the liner surface can be as low as 2 cSt near top centre, the resultant film thickness being one micron or less. Even under these conditions adhesive wear will not take place, provided the surfaces are not rough. The roughness of the liner and ring surfaces has a significant influence on the load at which adhesive wear starts; and the rougher the surfaces, the less the load-carrying ability. Rougher surfaces also require thicker oil films (or higher viscosity oils) to carry the same load.

The term 'boundary lubrication' describes the condition where the peaks of one rough surface (asperities) come into contact with those of another rough surface, and the lubricant film between the surfaces is not thick enough to prevent metal-to-metal contact. Full separation is only achieved when the ratio of film thickness over the mean roughness value (or mean height of asperities) is greater than three. A typical surface finish for the cylinder liner of a modern engine in service is around 0.5 microns, while the oil film thickness of three times this value (1.5 microns)—required to prevent asperity contact—may not be established until some distance down the stroke from top dead centre. The conditions for adhesive wear are therefore present at around top dead centre.

Control of the onset of adhesive wear calls for an adequate oil film thickness which can be enhanced by chemical films from lubricant additives. Such films have lower shear strengths than the metal surface and hence are removed, partially or fully, during sliding. Additives that are effective at reducing friction chemically react with the metal surfaces to form relatively thin planar layers of low shear strength. Replacement of these films during the out-of-contact time of the surfaces is therefore essential to the control of wear.

Additive effectiveness is strongly temperature dependent: as the metal surface temperatures increase, the chemical bond between additive and surface is broken and the additive de-absorps from the metal surface. Only additive systems that absorp at elevated temperatures (250°C) are effective in providing the chemical film that can give added protection against the onset of adhesive wear. Additionally, the calcium detergents used to provide the alkalinity of the cylinder oil have the effect of lowering the coefficient of friction. The long chain natural acid present in such detergents is slowly released during neutralization and migrates to the metal surface to form a friction-modified surface. Consequently, higher base number cylinder oils will provide higher load-carrying performance before the onset of adhesive wear, and then provide some anti-wear performance to reduce the amount of adhesive wear.

Chemically-active surface components give further protection against the onset of adhesive wear by providing a low shear film between the surface asperities. In terms of wear protection, therefore, this addition is equivalent to the use of a higher viscosity cylinder oil: a 19 cSt oil of 80 BN and containing high temperature anti-wear additives is equivalent to a 23 cSt standard 70 BN oil.

Once the conditions for adhesive wear are established then measurable and rapid wear will take place. If the conditions are particularly severe the rate of wear can accelerate, leading to severe damage and seizure. When the metal surface asperities come into contact local temperatures are generated that are much higher than the bulk temperature of the oil or the metal surface temperatures. The local temperatures (or flash temperatures) for cylinder lubrication are typically between 400 and 500°C.

The only way to control the rate of adhesive wear is to limit the local welding and deformation of the asperities, and this can be achieved by using high temperature high performance additives.

Special tests were conducted on a modified test engine in which the liner temperatures were raised to almost 300°C. Under these conditions, severe adhesive wear took place within 100 hours with a standard 70 BN cylinder oil. Using an 80 BN oil incorporating high temperature anti-wear additives, however, minimized the amount of adhesive wear. Other tests showed that this combination of high BN and anti-wear can reduce the amount of adhesive wear by up to 50 per cent.

Severe adhesive wear will increase the surface roughness on the rings and liners, with the result that the load to the onset of adhesive wear is reduced. For example, if an engine which experiences adhesive wear at a certain operating load is shut down (perhaps to examine the wear) then the increased roughness of the surfaces means that when the engine is run again adhesive wear will start at a lower load. If the conditions that resulted in adhesive wear are now removed (by operating at a much lower load) then it may be possible that the surface damage to liners and rings may be polished out to give acceptable running surfaces.

Anti-wear performance summary

- Viscosity has a major influence on the load at which adhesive wear starts; higher viscosity index (VI) cylinder oils provide more protection than lower VI oils.
- The rougher the surfaces, the lower the onset threshold for adhesive wear or the higher the viscosity required to prevent adhesive wear.
- The load at which adhesive wear starts is also influenced by the anti-wear performance of the alkaline detergent and the use of high temperature anti-wear additives.
- Only a standard BN cylinder oil with a viscosity of 23 cSt at 100°C will provide similar protection against the onset of adhesive wear as an 80 BN oil with high temperature anti-wear performance at 19 cSt at 100°C.
- The practical limit on the viscosity of cylinder oils (monogrades) for reasons of pumpability, handling and distribution over the liner surface is 21 cSt at 100°C with a VI of 100.
- The load at the onset of adhesive wear is 30 per cent higher from an 80 BN cylinder oil with high temperature anti-wear performance compared with a 70 BN oil.

- True anti-wear performance is a measure of the ability to reduce the rate of wear once the conditions for adhesive wear are established. A reduction of up to 50 per cent can be achieved by using an 80 BN cylinder oil with high temperature anti-wear performance compared with a standard 70 BN oil.

Medium speed engine system oils

Lubricating medium speed trunk piston engines is considered much more of an exact science than lubricating low speed crosshead engine cylinders. Sulphurous byproducts from the combustion of high sulphur fuel are countered by increasing the amount of over-based additive in the formulation to give an oil of higher alkalinity, as measured by TBN. But this is not quite as simple as merely adding alkaline additives. Research is carried out to establish a balanced blend of the many functionally different additives: for example, anti-oxidants, anti-wear agents, corrosion inhibitors, detergents, dispersants, anti-foam agents and pour-point depressants.

The performance of a lubricating oil in a marine medium speed engine tends to be measured in terms of its ability to maintain an acceptable level of TBN in service. TBN retention is influenced by factors such as fuel sulphur content, combustion quality, lubricating oil make-up, piston ring blow-by and water contamination. With regard to the oil itself, however, retention is directly related to additive balance and the quality of additives used.

Although marine lubricants will continue to rely largely on the types of additives that have ensured success so far, formulators suggest that more sophisticated and effective versions can be expected. The search for more powerful synergisms between additives of different species will intensify, and lube oils that can outlast the life of an engine could be possible in the future.

Liner lacquering

In recent years a number of highly rated medium speed engines of certain types have experienced problems with lacquer formation on cylinder liners. This build-up of lacquer in the honing grooves can result in a smoothed or glazed liner surface that leads to a significant increase in the rate of oil consumption. If left unchecked, lacquering can also be accompanied by hard carbonaceous deposits which lead to scoring or polishing of the liners and increased engine operating costs.

A number of common factors link examples of engines in which liner lacquering has been found: large variations in load (for example,

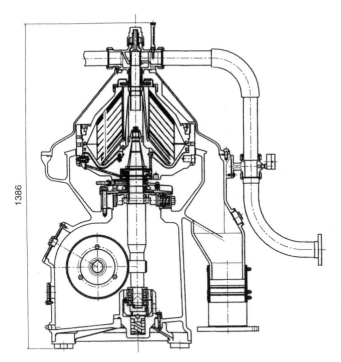

1386

Figure 4.5 Mitsubishi Selfjector SJ-FH design centrifugal separator

frequent and long periods of idle followed by full load operation); high mean effective pressure medium speed designs; and low sulphur, mainly distillate, fuels. Lacquering can typically occur in the propulsion engines of offshore supply vessels and shortsea ships. Lacquer characteristically contains both organic and inorganic (metal salts) material, the colour varies from very light amber to dark brown, and there is uneven distribution over the liner.

Modern, highly rated engines exploit much higher fuel injection pressures and shorter injection times. The total injection, ignition and combustion process has to take place in a few degrees of crank angle under conditions in which, besides complete combustion, thermal cracking of fuel components can be expected. If these cracked components reach the relatively cool cylinder liner surface they will condense, concentrate and start the formation of resinous lacquer by a process of polymerization and evaporation of the light ends. See Fig. 4.6.

Fuel quality has not suffered a step change in recent years that would account for the incidences of liner lacquering problems. However, the complexity of refining processes has increased during the past

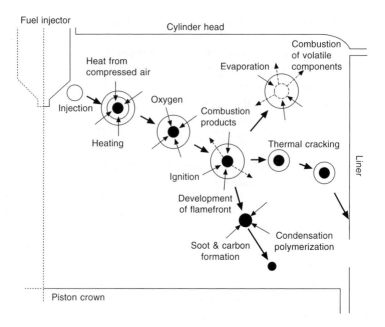

Figure 4.6 Formation of resinous lacquer on the cylinder liner (Shell Marine)

decade and therefore the complexity of components in both distillate and residual fuels has also increased. Experiments have shown that different gas oils produce different distributions of deposits between pistons and liners under conditions known to promote lacquer formation. While the fuel clearly influences liner lacquer formation, however, the link to the sources of gas oil is not clear.

An insight into how fuel contributes to liner lacquering was derived from service experience with a ship running on residual fuel rather than distillate fuel. The ship bunkered a highly cracked, high sulphur residual fuel before each outward voyage leg and a straight-run, low sulphur residual fuel before each return leg on a regular sailing schedule. This practice always resulted in an increase in sump oil top-up and building of liner lacquer during the return voyage. Although distillate fuels do not show the same sensitivity as this example, it does show that the fuel composition is a factor in lacquer formation.

Evidence from liner lacquering in the field as well as in laboratory engines suggests the following mechanism for lacquer formation, characterized by the following steps:

- Fuel droplets injected at high pressure begin to burn rapidly.
- Rapid burning leads to incomplete combustion of some larger fuel droplets but intense heating of droplet cores.

- Thermal cracking of some fuel components occurs within larger droplets.
- Some droplets impinge on the cylinder liner where thermal cracking reactions continue to form unsaturated reactive hydrocarbon products.
- Condensation and polymerization of unsaturated products on the liner forms resinous organic polymeric deposit.
- Sticky resinous deposit is spread over the liner by piston ring movement.
- Lubricant, including metal-containing additives, becomes entrained in the deposit.
- Continual exposure of the deposit to combustion, evaporation of light ends and rubbing by piston movement forms hardened lacquer deposit.
- Build-up of lacquer fills honing grooves and leads to increased rate of oil consumption.

Current understanding suggests that any fuel will contribute to liner lacquer formation under engine conditions that promote incomplete combustion and thermal breakdown of the fuel. However, the combination of particular fuel compositions combined with such engine conditions is the principal cause of liner lacquer formation in medium speed engines.

The lubricant can affect lacquer formation. Research by Shell Marine has shown that a lubricant formulation must have the correct balance of detergency, dispersency, alkalinity and anti-oxidancy characteristics to prevent lacquering from occurring. Lacquering will occur most frequently when the engine is running under very high load conditions. Good maintenance is essential, and the following are dictated to prevent lacquer formation: injection timing should be correct; injectors should not be worn; and liner and inlet air temperatures should not be too low. Shell's investigations showed that high performance lubricants can contribute to minimizing lacquering and thus reducing lube oil consumption.

Black sludge formation

A problem in medium speed engines may arise from the contamination of lube oil by the combustion blow-by gases and/or by direct raw fuel dilution. Higher fuel pump pressures in modern engines have contributed to increased fuel leakage: analyses of used lube oils from engines running on heavy fuel oil have indicated an average HFO contamination of 2 per cent; in some cases, levels of up to 15 per cent were detected.

Most heavy fuel oils supplied today emerge from cracking installations. The cracked asphaltenes, an inherent part of such bunkers, do not dissolve in engine lubricants but instead coagulate and form floating asphalt particles, two to five microns in size. These particles are very sticky and form black deposits on all metal surfaces of the engine, including the cambox and the crankcase. The admixture of residual fuels and some lubricants may lead to coagulation of asphaltenic particles causing sludge formation in engines, clogging the oil scraper ring and increasing piston ring groove deposits. Sludge may overload the lube oil separators and the filters may become clogged.

This problem posed by so-called black sludge or black paint—a sticky, viscous mixture of fuel and combustion-derived deposits— stimulated leading lube oil suppliers to develop new system lubricants able to disperse high concentrations of cracked asphaltenes. Products were developed to meet the challenge posed by engine blackening, undercrown deposits, piston head corrosion, purifier heater fouling, increased lube oil consumption, base number depletion, oil scraper ring clogging and increased piston deposits. Additives prevent the contaminants causing black sludge from depositing in the engine, the lube oil holding the impurities in suspension so that they can be removed by centrifugal separation. The lubricants are claimed to disperse the highest level of cracked asphaltenes, resulting in cleaner engines (especially excellent piston cleanliness), reduced lube oil purifier cleaning intervals and filter consumption, and the near elimination of deposit formation in purifier heaters.

Synthetic lubricants

The vast majority of marine lubricants are still formulated with mineral bases but the development of synthetic lubricants targets enhanced machine running safety and durability and the extension of oil change intervals. Such products are particularly valued for harsh duties and are applied widely in aviation, high speed trains and Formula One racing cars.

Synthetic lubricants are made from synthetic base oils derived from the chemical reaction of one or more products, supplemented by special additives incorporated to boost performance properties. These base oils may be formulated from synthetic hydrocarbons, organic esters, polyalkylene glycols, phosphatic esters, silicates, silicones and fluorosilicones, polyphenyl esters and fluorocarbons.

Marine industry applications currently embrace reciprocating and rotary auxiliary machinery, such as turbochargers, air compressors, refrigeration compressors, LPG/LNG compressors and centrifugal separators. Some high speed ferry engines also exploit synthetic lubricants.

Exxon Mobil's SHC (Synthesized Hydrocarbon) lubricants, formulated from pure chemical lubricant base stocks and special additives, promise the following advantages over mineral oil products:

- Wax-free, underwriting low temperature performance.
- Relatively small viscosity changes with temperature, and completely shear stable.
- Excellent oxidation resistance, providing superior wear protection.
- Significantly extended service life.
- Ability to handle greater loads, higher speeds and more extreme temperatures.

These benefits reportedly translate into reduced overall operating costs and lower machinery downtimes.

Low chlorine, zinc-free Mobilgard SHC 1 is formulated as a crankcase oil for distillate-fuelled high and medium speed engines in installations subject to low temperature and/or frequent starts, rapid loading after start-up, and abrupt shutdown after high speed operation. The lubricant requires no changes in seals, hoses, paints or filters; no flushing is necessary and the oil is fully compatible with hydrocarbon products. Extensive heavy duty field tests with MTU high speed engines in fast ferries are said to have demonstrated the ability to reduce engine wear, extend draining intervals and lower fuel consumption. Mobilgard SHC 1 can also be used in gearboxes.

Cylinder lubricants and low sulphur fuels

SOx Emission Control Areas (SECAs)—see under the section on Fuels given above—impact on cylinder lubrication in low speed engines since the fuel dictated to meet emission controls is generally limited to a maximum 1.5 per cent sulphur content. Concern that contemporary cylinder oils (whether 70 BN or lower) might not be the best choice for service with low sulphur fuels was addressed by MAN B&W Diesel and BP Marine in answering these questions:

Should a newbuilding be specified with a dual storage and piping system for cylinder oils?
Shipowners are advised to seriously consider future operational flexibility for switching in and out of low sulphur fuel. To facilitate optimization in various scenarios it is sensible to plan ships with dual (or multiple) storage tanks to allow more than one cylinder oil grade to be carried.

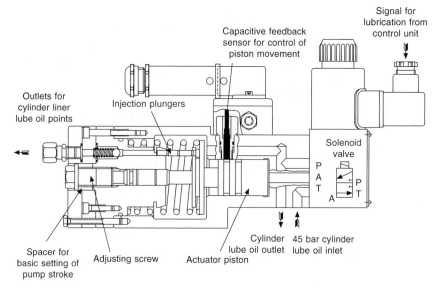

Figure 4.7 Alpha cylinder lubricator unit (MAN B&W Diesel)

The actual tank capacities may be open to debate but it is useful to start allowing for segregation.

What should be done when changing over to low sulphur fuel for short durations on random bunkers?
If regular inspection shows that 70 BN oil is performing satisfactorily, then consider remaining on that lubricant; as a caution, however, avoid running-in on low sulphur fuel with 70 BN oil, if possible.

What should be done when switching regularly between low and high sulphur fuels on a routine trading route?
As above, but if there are two grades of cylinder oil onboard (70 BN and a suitable lower BN grade) then switch accordingly if that is convenient and if monitoring confirms that yields the best results. This could be the policy when SECAs become widespread.

What should be done when on prolonged operation with low sulphur fuel?
Run on low BN oil and closely monitor to ensure it gives the best results.

CYLINDER LUBRICANT FEED RATES

Cylinder oil feed rates should be set with a reasonable margin of safety on the recommendations of the enginebuilder/designer. A high

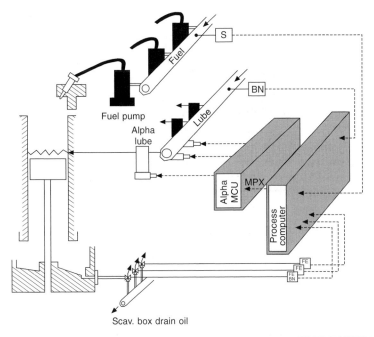

Figure 4.8 Arrangement of Alpha electronic cylinder lubricant system (MAN B&W Diesel)

oil feed rate, far greater than 150 per cent nominal, is seldom required other than for a short period during running-in or when special circumstances justify it. There have been reports from both MAN B&W and Wärtsilä (Sulzer) engines that over-lubrication can in fact be harmful in certain adverse conditions.

Unless there are abnormal circumstances, MAN B&W Diesel advises, it is pointless to pursue prolonged operation with an excess of 2 g/bhph (2.72 g/kWh), especially in conjunction with low sulphur fuel, 70 BN cylinder oil and an engine series well designed against corrosive wear and with no need for excessive BN. Even for a highly detergent oil, over-lubrication can lead to excessive calcium carbonate deposits in the remote top edge of the hot piston crown land. (The BN of all commercial cylinder oils today is derived largely from calcium carbonate.)

In adverse conditions it is sometimes not the oil detergency that cannot cope but rather there is an enormous generation of calcium carbonate deposit. Influenced by MAN B&W Diesel, BP Marine recognized the potential problem of a high feed rate and excessive calcium carbonate deposition during running-in and in low sulphur fuel operation. Its launch of the cylinder lubricant CL-DX 405 addressed such conditions with a reduced BN (40) and reduced calcium carbonate, while maintaining very high detergency.

Enginebuilders are also concerned that an engine well designed against corrosive wear can become sensitive to excessive feed rate on 70 BN oil when burning fuel with a normal sulphur content. This is because if the BN is not used, the unused BN can result in hard calcium carbonate deposits (used BN results in a softer calcium sulphate). A high cylinder oil feed rate, Wärtsilä adds, can lead to a disturbed pressure drop across the ring pack and negatively affect piston running. Excessive cylinder oil also tends to increase the formation of deposits on the top land of a piston. A solid crust is formed, which is softer or harder depending on the additives used in the oil, the type of fuel and its sulphur content, as well as its combustion characteristics. An excessive feed rate can, in fact, also lead to the destruction of a scuffing liner within hours.

Another reason for caution over excessive feed rates is that modern engines are significantly more powerful, burn more fuel and use more cylinder oil per cylinder than before. The cylinder oil ash level could be more than five times higher than in engines of the 1970s and 1980s.

'Intelligent' cylinder lubrication

Demands placed on cylinder oil in two-stroke engines vary widely and depend on the operational conditions. The cylinder of a large bore engine is typically fed with just one gram of lubricant during each stroke, and this small amount must be spread correctly to fulfil all requirements. Studies of the relationship between wear and lube oil dosage reveal that interaction between operational factors—such as load variations, fuel and lube oil qualities, and atmospheric humidity—exert a strong influence on the wear rates of piston rings, ring grooves and cylinder liners. It has also become clear that over-lubrication has a considerable negative effect on the tribological conditions of liners and rings.

Traditional cylinder lubrication systems, whereby the lube oil is injected at a fixed rate proportional to the engine rev/min or mean effective pressure and perhaps adjusted occasionally based on a scavenge port inspection, may yield a reasonable average condition. Such a condition, however, will result from long periods of excessive lubrication and shorter periods of oil starvation because the sum of influencing factors will change on a daily or sometimes even on an hourly basis. Furthermore, there is the risk of ruining the cylinder condition if unfavourable operational factors are not detected and properly counteracted by the engine operators.

Operational conditions beyond the norm are unavoidable, and many shipowners simply over-lubricate their engines under the false assumption that they will thus always be 'on the safe side'. But over-lubrication is not only expensive; it may even be counterproductive in promoting scuffing through excessive carbon deposits and/or 'bore polished' running surfaces. A wear study has proved that the optimal basic setting of cylinder lubricators should be proportional to the engine load and the fuel oil sulphur content. Feed rate control proportional to the load is one of two standard options of MAN B&W Diesel's Alpha Lubricator system; the other is to control lubrication in proportion to the mean effective pressure. Lubricator control in relation to the fuel oil sulphur content may either be carried out automatically—based on a feed-forward signal from the fuel inlet line—or manually, based on the sulphur content from the bunker receipt or fuel oil analysis data.

The electronically-controlled Alpha Lubricator was developed to inject the oil into the cylinder directly on the piston ring pack at the exact time that the effect is optimal. The system features a number of injectors that inject a specific amount of lubricant into the cylinder every four (five, six etc.) revolutions of the engine. The lubricator has a small piston for each quill in the cylinder liner, with power for injecting the oil provided from system pressure generated by a pump station (Figures 4.7 and 4.8).

The properties of cylinder oil scraped from the liner wall reflect the chemical environment in the cylinders as well as the physical condition of rings and liner; and there is a direct relationship between some of the key parameters in the scrape-down oil and the actual cylinder condition. A lubrication algorithm—based on scrape-down analysis data, cylinder oil dosage, engine load and cylinder wear rate—can thus be created. Automatic optimization of lube oil dosage and cylinder lubrication efficiency is facilitated by on-line monitoring of the scrape-down oil composition from each cylinder, feeding the results into a computer (along with the above algorithm), and sending signals to each Alpha Lubricator. Corrosive wear control can be based on either feed rate control or control of the cylinder oil base number; the latter approach calls for two or more lube oil tanks or blending facilities onboard.

Alpha Lubricators—allowing significantly reduced feed rates over mechanical lubricators—can be specified for all MAN B&W MC and MC-C low speed engines as well as retrofitted to engines in service. Large bore engines are equipped with two such lubricators for each cylinder, while smaller bore engines have one unit. Further reductions in cylinder oil consumption are promised from a so-called 'sulphur handle' which, in conjunction with the Alpha Lubricator system, feeds the lubricant in proportion to the amount of sulphur entering the cylinder (Alpha Adaptive Cylinder Oil Control). The load- or kW-dependent rate is applied rather than the rev/min- or mean effective pressure-dependent rate of other lubricator systems.

Two basic requirements have to be addressed: the cylinder oil dosage must not be lower than the minimum needed for lubrication; and the additive amount supplied through the BN (the alkalinity throughput) must be only sufficient for neutralization and for keeping the piston ring pack clean. As the second criterion usually over-rides the first, the following determine the control: the cylinder oil dosage shall be proportional to the engine load (that is, the amount of fuel entering the cylinders), and proportional to the sulphur percentage in the fuel. A minimum cylinder oil dosage is set to address the other duties of the lubricant, such as securing sufficient oil film detergency.

LUBRICANT TESTING

Lubricants are a valuable indicator of the overall functioning and wear of machinery. Under normal engine operating conditions the deterioration of a lubricant takes place slowly, but severe conditions and engine malfunctions promote faster degradation. Monitoring lubricants in service and analysing the results determine not only their condition but that of the engine and auxiliary machinery they serve. All the major marine lubricant suppliers offer used lube oil testing services to customers, with results and advice transmitted by e-mail, fax or post to owner and ship from specialist laboratories. The range of analyses undertaken on engine lubricant samples submitted typically embraces checks on:

- *Water content*: even in small quantities, water is always undesirable in a lubricant since it can act as a contaminant. A check on the water content will indicate defects in the engine, purifier and other auxiliary machinery. Where water is present, testing determines whether it is fresh or sea water.
- *Insoluble products*: lubricants in an engine are contaminated by a number of insoluble products. Monitoring their presence provides a very good indication of the engine's operating condition or the effectiveness of the purifier or filter; it also safeguards against breakdowns since a sudden increase in the insoluble products present indicates a malfunction.
- *Viscosity*: a measure of the ability of a lubricant to flow, viscosity

cannot be used in isolation to assess a lubricant's condition but it yields useful information in conjunction with other determining factors, such as the level of oxidation or contamination by fuel, water or other elements.

- *Flash point*: the temperature at which the application of a flame will ignite the vapours produced by an oil sample under standard conditions. Below a certain value there is a high risk of very serious incidents in service, including crankcase explosion. The flash point must therefore be monitored very carefully.

- *Base number (BN)* or alkalinity reserve: during the combustion process the sulphur contained in the fuel can produce acidic products that can damage the engine; this acid must be neutralized to avoid extensive corrosive wear of the cylinder liner or piston rings. Neutralization is the role of specific additives in the lubricant, the amount and type of which are measured by the BN which represents the alkalinity reserve. Close monitoring of the BN in service is essential to ensure that the lubricant still has sufficient alkalinity reserve to perform correctly.

- *Wear metal elements*: monitoring changes in metal elements (measured in ppm) provides vital information on how patterns of wear develop inside the machinery. Regular monitoring detects any sudden increase in a wear element, such as iron or copper, which would indicate that excessive wear was occurring.

The correct interpretation of results dictates analyses on a regular basis so that curves can be plotted and extrapolated to indicate trends. Any measurement that deviates significantly from the curves must be investigated further; adopting a historical approach to analyses can minimize the risk of machinery malfunction and breakdown. When machinery manufacturers offer no specific recommendations, Lub Marine suggests the following as average sampling periods:

> Main engine...............................3/6 months
> Diesel genset............................6/12 months
> Hydraulic system..........................12 months
> Turbocharger................................6 months
> Gearing.......................................12 months
> Air and gas compressors..................12 months

MICROBIAL CONTAMINATION OF FUELS AND LUBRICANTS

Shipboard microbial fouling and corrosion have become more commonplace and will probably continue to increase due to fuel oil formulation and handling trends, polluted harbour waters and regulatory pressures.

Naval architects and shipbuilders are advised to design systems that are less conducive to harbouring micro-organisms and more readily sampled and decontaminated if problems are experienced. Comprehensive onboard test kits are available for detecting and quantifying microbial contamination and for monitoring the success of anti-microbial measures.

Microbial contamination can manifest itself in the infection of fuel, lubricants, cooling water and bilge and ballast waters. The results can be: engine damage (pistons and cylinder liners may wear faster, and filters and fuel injection nozzles become plugged, sometimes within a few hours); fuel tank pitting corrosion and accelerated localized pitting; and general wastage corrosion of bilge pipes, tank bottoms and hulls. In some cases, internal and external microbial attack has led to rapid ship structural perforation. Of special concern is the possibility that certain strains of bacteria may endanger the health of the crew, as would the incorrect handling and application of hazardous chemical biocides in remedial treatment operations.

Various kinds of microbes are encountered: bacteria; moulds; and fungi and yeasts. Bacteria come in two types: aerobic, which use oxygen to oxidize their food; and anaerobic, which cannot tolerate oxygen. One anaerobic strain, termed sulphate reducing bacteria (SRB), is particularly virulent and produces corrosive sulphide.

Bacteria produce biosurfactants which break down the oil/water interface, promoting emulsions that are detectable when the fuel is tested for water separation. If the infection is severe and longstanding the aerobic bacteria consume all the oxygen and problems with SRB may result.

Microbes need water and a food source to survive, ingredients provided by the hydrocarbons and chemical additives in oil. A temperature between 15°C and 40°C is also required. Warm enginerooms offer an ideal breeding ground. Laid-up tonnage or ships in intermittent service are most vulnerable to attack because microbes dislike movement. Dormant fuel systems may be aggravated by condensation and water leakage.

Various strategies are available for reducing the problems but solutions must be safe and environmentally acceptable. The details needed to build these strategies into full working protocols will depend on individual circumstances, the time available and access to anti-microbial agents.

There are many different microbiological problems which need to be met with specific tailored solutions but there are certain common principles of good practice which can be implemented:

- Physical prevention: preventing ingress of inoculating microbes, particularly those already adapted to growth in relevant environments. Avoid spreading contamination from passing clean fluids through contaminated pipes and filters and into dirty tanks. Minimize the conditions which encourage microbial growth and water.
- Physical decontamination: settling, heat, filtration and centrifugal procedures all aid decontamination, the choice of process depending on equipment and time constraints.
- Chemical prevention: protecting against minor contamination, coupled with good housekeeping to prevent rather than cure infection.
- Chemical decontamination: a wide range of chemical biocides is available. Only a few are appropriate to each specific application (there is no universal eradication fluid). All are toxic and must only be used with due regard to health and safety and environmental impact.

Microbial problems in the shipping industry were originally mainly confined to distillate fuels and lubricants but areas of attack have recently spread to residual fuels and bilge, ballast and potable waters. Specialists cite the following contributory factors:

- Lower levels of shipboard manning and less experienced personnel have undermined stringent housekeeping regimes.
- Adverse trading conditions have led to ships being laid up or deployed intermittently, fostering long and undisturbed incubation periods for opportunist micro-organisms.
- Marine pollution legislation under the Marpol 73/78 regulation which restricts the pumping of bilges has led to water laying stagnant for longer periods.
- Environmental restrictions on the use of toxic biocidal chemicals within bilges and fuels exacerbate the problems associated with microbial contamination.
- Lack of knowledge of the factors which cause microbial contamination and accurate diagnosis of the operational problems being experienced.
- Shortcomings in the design of tanks, bilges and pipe systems whose configuration should foster effective water draining and subsequent treatment.

There will always be an initial source of shipboard fuel contamination, such as a shore tank, dirty pipeline, road tanker or polluted tank wash water. Airborne contamination via tank breathers is less likely. Once infestation has occurred, particular circumstances may encourage microbial proliferation onboard. Further inoculation is then immaterial as the key aggravating factors are water and warmth.

Water accumulates in tanks when there is no drain or water scavenge system, where the drain is not at the lowest point in the tank, and where draining procedures are not enforced. Tanks in the engineroom or other warm locations and tanks receiving recirculated distillate fuel from injectors are ideal for incubating microbes. Double bottom tanks, due to their lower temperatures, are less prone to microbial proliferation.

The following visual symptoms of fuel microbial contamination are given as a guide:

- Aggregation of microbes into a biomass, observed as discolouration, turbidity and fouling.
- Biosurfactants produced by bacteria promote stable water hazes and encourage particulate dispersion.
- Purifiers and coalescers, which rely on a clean fuel/water interface, may malfunction.
- Tank pitting.

Operational indications of fuel microbial contamination are:

- Bacterial polymers may completely plug filters and orifices within a few hours.
- Filters, pumps and injectors foul and fail.
- Non-uniform fuel flow and variations in combustion may accelerate piston ring and cylinder liner wear rates, and affect camshaft torque.

When heavily infected fuel is used all or some of these phenomena will confront the ship's engineers within a few hours. Initially, they will probably be faced with filter plugging, fuel starvation, injector fouling and malfunctioning of any coalescer in use.

The extent of the contamination must first be established by the following procedure:

1. Clear glass sample bottles rinsed with boiling water should be used to take drain or bottom samples from service, header and storage tanks.
2. Microbial contamination in a sample will be apparent as a haze in the fuel from the presence of sludge. This sludge readily disperses in the fuel when the sample is swirled; at this time

sticky 'cling film' flakes may be seen adhering to the wall of the bottle. The water will be turbid and there may be some bottom sludge; if this is black, SRB are present and they will be an added corrosion hazard.

3. The engineer can now evaluate the various fuel locations and instigate an emergency strategy by using the cleanest fuel. If only heavily contaminated fuel is available it should first be allowed to settle for as long as possible and then drawn off from the top to a clean tank, preferably via a filter, purifier, centrifuge or coalescer. Using a biocide at this stage (see section below) is usually not advisable due to a tendency to block filters with the dislodged biofilms.

4. At the earliest opportunity, drain or bottom samples should be taken and the supplier's retained sample forwarded for microbiological examination. This will identify, by sophisticated 'fingerprinting', whether or not the bunkers supplied were contaminated.

The standard methods for curing heavily contaminated fuels involve the use of biocides which may be toxic and incur disposal problems. Careful handling is necessary. Once a biocide has been added, the dead biomass must be allowed to settle before drainage and filtration. The careful monitoring of filters is also required to prevent blockage by the dead microbes.

In expert hands, biocides eliminate the microbial menace but using an incorrect biocide can exacerbate the problem rather than effect a remedy. The choice of biocide is therefore important. Prevention is better than cure and the source of the infections should be identified to prevent their recurrence.

LUBE OIL CONTAMINATION

Lubricant operating temperatures of crankcase oils in wet engines are normally sufficient to control and prevent infestation. Any problems are usually associated with ships which have shut down their lubricant system, allowing temperatures to drop and water to accumulate.

There are many thousands of types of microbes but only a few are able to grow in lubricating oils at the elevated operational temperatures in use. Provided the plant treatment equipment is functional, the purifier heater temperature is maintained and the water content kept at a minimum, the lubricating oil system will be both environmentally and nutrient deficient. Such conditions will prevent microbes from establishing themselves. Should heavy contamination of the system occur, however, this self-regulating mechanism will be unable to prevent microbial proliferation (Figure 4.9). Fortunately, the symptoms of microbial contamination will not occur immediately, allowing engineers to implement remedial physical and/or chemical decontamination programmes.

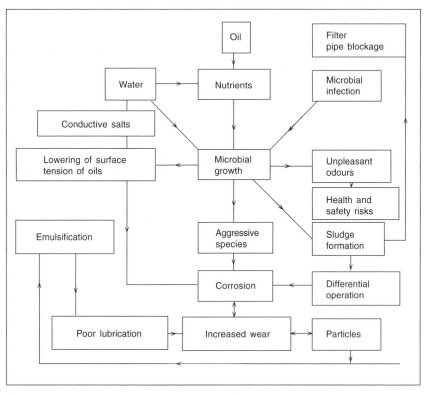

Figure 4.9 Interactive problems associated with microbial corrosion (Lloyd's Register)

The visual symptoms of microbial contamination of lubricants are:

- Slimy appearance of the oil, the slime tending to cling to the crankcase doors.
- Rust films.
- Honey-coloured films on the journals, later associated with corrosion pitting.
- Black stains on whitemetal bearings, pins and journals.
- Brown or grey/black deposits on metallic parts.
- Corrosion of the purifier bowl and newly machined surfaces.

- Sludge accumulation in the crankcase and excessive sludge at the purifier discharge.
- Paint stripping in the crankcase.

Operational symptoms of microbial contamination in lubricants are:

- Additive depletion.
- Rancid or sulphitic smells.
- Increase in oil acidity or sudden loss of alkalinity.
- Stable water content in the oil which is not resolved by the purifier.
- Filter plugging in heavy weather.
- Persistent demulsification problems.
- Reduction of heat transfer in coolers.

Microbial growth occurs in the water associated with the lubricant, and the phenomenon is therefore characteristic of crankcase oils in wet engines, particularly those with water-cooled pistons. Lubricant infection can be identified by slimy film formation on the crankcase doors. There may also be a rancid odour and whitemetal parts may be stained black. As the problem progresses, filters choke up, organic acids are formed and the oils tend to emulsify.

If sulphate reducing bacteria (SRB) are present—a problem particularly associated with laid-up tonnage—copious pitting of ferrous and non-ferrous metals may occur (Figure 4.10).

Since micro-organisms feed upon the additives within the oil the lubricity of the oil may be impaired and its viscosity altered; and there is a greater resultant acidity and increased potential for emulsification and corrosion. Hydrogen sulphide may also be produced as a byproduct. Serious corrosion problems will result within weeks of the initial contamination if these factors occur at the same time.

Sources of contamination within the lubricant are the fuel, cooling water and seawater. Cooling water in particular has featured as a common contaminant of crankcase oil at engine operating temperatures, since the use of chromates as corrosion inhibitors is banned. The use of chromates also acted as an effective anti-microbial biocide.

Prevention is better than cure and it is known that microbial growth is retarded by extreme alkaline conditions. To date, there have been no microbial problems reported with medium speed engines during operation when using highly alkaline lubricants of BN 12 to 40. At elevated temperatures lubricating oils tend to be self-sterilising. Unfortunately, however, there still remains the possibility of infestation from contaminated bilges and tanktops which may inadvertently leak into the oil system.

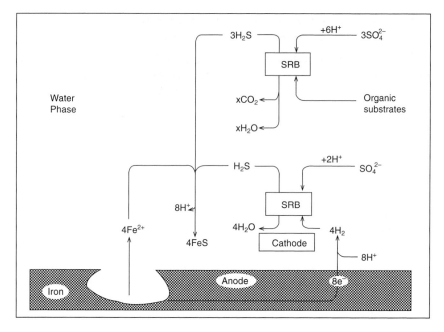

Figure 4.10 Anaerobic corrosion by sulphate reducing bacteria (SRB)

Since the majority of operational microbial problems in lubricating oils arise due to infection from cooling water, seawater from cooler leaks and overflows of contaminated bilges, the effectiveness and efficiency of plant treatment are important remedial measures.

The following procedures are recommended to avoid microbiological problems:

- Ensure that the water content of crankcase oil does not rise above 0.5 per cent by weight.
- Check that purifier suction is as near to the bottom of the sump as possible.
- Maintain a minimum oil temperature after the purifier heater of 70°C or higher for at least 20 seconds and/or 80°C for 10 seconds.
- Circulate the volume of oil in the sump via the purifier at least once every eight to ten hours.
- Regularly check that coolant corrosion inhibitor concentrations are at the manufacturer's recommended values.
- Monitor the microbial population of the cooling water and prevent water leaks into the oil system.
- Test for microbial contamination of the oil system and monitor whether the oil after the purifier is sterile.

- Prevent ingress of contaminated bilge water.
- Inspect storage tanks and regularly check for water.

(The above information is based on advice from Lloyd's Register's Fluid Analytical Consultancy Services (FACS) department which has considerable theoretical and practical expertise in assessing and resolving shipboard microbial contamination problems with fuel and lube oils.)

5 Performance

An important parameter for a marine diesel engine is the rating figure, usually stated as bhp or kW per cylinder at a given rev/min.

Although enginebuilders talk of continuous service rating (csr) and maximum continuous rating (mcr), as well as overload ratings, the rating which concerns a shipowner most is the maximum output guaranteed by the enginebuilder at which the engine will operate continuously day in and day out. It is most important that an engine be sold for operation at its true maximum rating and that a correctly sized engine be installed in the ship in the first place; an under-rated main engine, or more particularly an auxiliary, will inevitably be operated at its limits most of the time. It is wrong for a ship to be at the mercy of two or three undersized and thus over-rated auxiliary engines, or a main engine that needs to operate at its maximum continuous output to maintain the desired service speed.

Prudent shipowners usually insist that the engines be capable of maintaining the desired service speed fully loaded, when developing not more than 80 per cent (or some other percentage) of their rated brake horsepower. However, such a stipulation can leave the full-rated power undefined and therefore does not necessarily ensure a satisfactory moderate continuous rating, hence the appearance of continuous service rating and maximum continuous rating. The former is the moderate in-service figure, the latter is the enginebuilder's set point of mean pressures and revolutions which the engines can carry continuously.

Normally a ship will run sea trials to meet the contract trials speed (at a sufficient margin above the required service speed) and the continuous service rating should be applied when the vessel is in service. It is not unknown for shipowners to then stipulate that the upper power level of the engines in service should be somewhere between 85–100 per cent of the service speed output, which could be as much as 20 per cent less than the engine maker's guaranteed maximum continuous rating.

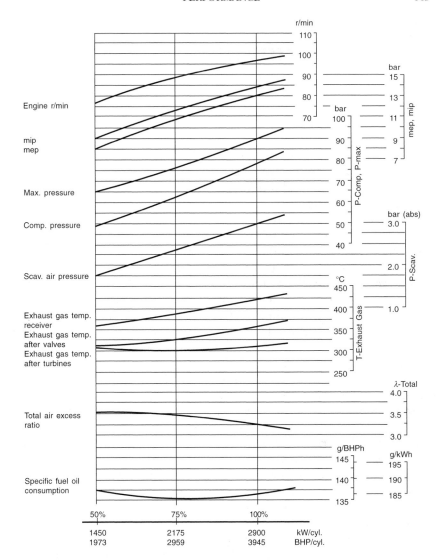

Figure 5.1 *Typical performance curves for a two-stroke engine*

MAXIMUM RATING

The practical maximum output of a diesel engine may be said to have
been reached when one or more of the following factors operate:

1. The maximum percentage of fuel possible is being burned
   effectively in the cylinder volume available (combustion must be

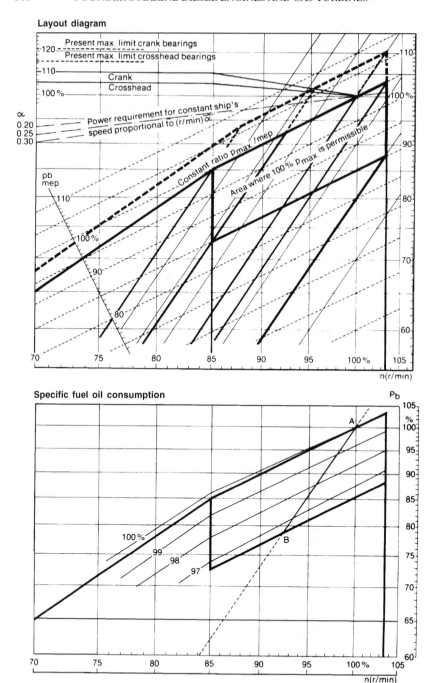

Figure 5.2 Layout diagrams showing maximum and economy ratings and corresponding fuel consumptions

completed fully at the earliest possible moment during the working stroke).

2. The stresses in the component parts of the engine generally, for the mechanical and thermal conditions prevailing, have attained the highest safe level for continuous working.
3. The piston speed and thus revolutions per minute cannot safely be increased.

For a given cylinder volume, it is possible for one design of engine effectively to burn considerably more fuel than one of another design. This may be the result of more effective scavenging, higher pressure turbocharging, by a more suitable combustion chamber space and design, and by a more satisfactory method of fuel injection. Similarly, the endurance limit of the materials of cylinders, pistons and other parts may be much higher for one engine than for another; this may be achieved by the adoption of more suitable materials, by better design of shapes, thicknesses, etc., more satisfactory cooling and so on. A good example of the latter is the bore cooling arrangements now commonly adopted for piston crowns, cylinder liner collars and cylinder covers in way of the combustion chamber.

The piston speed is limited by the acceleration stresses in the materials, the speed of combustion and the scavenging efficiency: that is, the ability of the cylinder to become completely free of its exhaust gases in the short time of one part cycle. Within limits, so far as combustion is concerned, it is possible sometimes to increase the speed of an engine if the mean pressure is reduced. This may be of importance for auxiliary engines built to match a given alternator speed.

For each type of engine, therefore, there is a top limit beyond which the engine should not be run continuously. It is not easy to determine this maximum continuous rating; in fact, it can only be satisfactorily established by exhaustive tests for each size and type of engine, depending on the state of development of the engine at the time.

If a cylinder is overloaded by attempting to burn too much fuel, combustion may continue to the end of the working stroke and perhaps also until after exhaust has begun. Besides suffering an efficiency loss, the engine will become overheated and piston seizures or cracking of engine parts may result; or, at least, sticking piston rings will be experienced, as well as dirty and sticking fuel valves.

EXHAUST TEMPERATURES

The temperature of the engine exhaust gases can be a limiting factor for the maximum output of an engine. An exhaust-temperature graph plotted with mean indicated pressures as abscissae and exhaust temperatures as ordinates will generally indicate when the economical combustion limit, and sometimes when the safe working limit, of an engine has been attained. The economical limit is reached shortly after the exhaust temperature begins to curve upwards from what was, previously, almost a straight line.

Very often the safe continuous working load is also reached at the same time, as the designer naturally strives to make all the parts of an engine equally suitable for withstanding the respective thermal and mechanical stresses to which they are subjected.

When comparing different engine types, however, exhaust temperature cannot be taken as proportionate to mean indicated pressure. Sometimes it is said and generally thought that engine power is limited by exhaust temperature. What is really meant is that torque is so limited and exhaust temperature is a function of torque and not of power. The exhaust temperature is influenced by the lead and dimensions of the exhaust piping. The more easily the exhaust gases can flow away, the lower their temperature, and vice versa.

DERATING

An option available to reduce the specific fuel consumption of diesel engines is derated or so-called 'economy' ratings. This means operation of an engine at its normal maximum cylinder pressure for the design continuous service rating, but at lower mean effective pressure and shaft speed.

By altering the fuel injection timing to adjust the mean pressure/maximum pressure relationship the result is a worthwhile saving in fuel consumption. The horsepower required for a particular speed by a given ship is calculated by the naval architect and, once the chosen engine is coupled to a fixed pitch propeller, the relationship between engine horsepower, propeller revolutions and ship speed is set according to the fixed propeller curve. A move from one point on the curve to another is simply a matter of giving more or less fuel to the engine.

Derating is the setting of engine performance to maximum cylinder pressures at lower than normal shaft speeds, at a point lower down the propeller curve. For an existing ship and without changing the propeller this will result in a lower ship speed, but in practice when it is applied to newbuildings, the derated engine horsepower is that which will drive the ship at a given speed with the propeller optimized to absorb this horsepower at a lower than normal shaft speed.

Savings in specific fuel consumption by fitting a derated engine can be as much as 5 g/bhph. However, should it be required at some later date to operate the engine at its full output potential (normally about 15–20 per cent above the derated value) the ship would require a new propeller to suit both higher revolutions per minute and greater absorbed horsepower. The injection timing would also have to be reset.

MEAN EFFECTIVE PRESSURES

The term brake mean effective pressure (bmep) is widely quoted by enginebuilders, and is useful for industrial and marine auxiliary diesel engines that are not fitted with a mechanical indicator gear. However, the term has no useful meaning for shipboard propulsion engines. It is artificial and superfluous as it is derived from measurements taken by a dynamometer (or brake), which are then used in the calculation of mechanical efficiency. Aboard ship, where formerly the indicator and now pressure transducers producing PV diagrams on an oscilloscope, are the means of recording cylinder pressures, mean indicated pressure (mip) is the term used, particularly in the calculation of indicated horsepower.

Many ships now have permanently mounted torsionmeters. By using the indicator to calculate mean indicated pressure and thus indicated horsepower, and the torsionmeter to calculate shaft horsepower from torque readings and shaft revolutions, the performance of the engine both mechanically and thermally in the cylinders can be readily determined.

Instruments such as pressure transducers, indicators, tachometers and pressure gauges (many of which are of the electronic digital or analogue type of high reliability) allow the ship's engineer to assess accurately the performance of the engine at any time.

The values of brake horsepower, mean indicated pressure and revolutions per minute are, of course, capable of mutual variation within reasonable limits, the horsepower developed per cylinder being the product of mean indicated (or effective) pressure, the revolutions per minute and the cylinder constant (based on bore and stroke). The actual maximum values for horsepower and revolutions to be used in practice are those quoted by the enginebuilder for the given continuous service rating.

PROPELLER SLIP

The slip of the propeller is normally recorded aboard ship as a useful pointer to overall results. While it may be correct to state that the amount of apparent slip is no indication of propulsive efficiency in a new ship design, particularly as a good design may have a relatively high propeller slip, the daily variation in slip (based on ship distance travelled compared with the product of propeller pitch and revolutions turned by the engine over a given period of time) can be symptomatic of changes in the relationship of propulsive power and ship speed; and slip, therefore, as an entity, is a useful parameter. The effects on ship speed 'over the ground' by ocean currents is sometimes considerable.

For example, a following current may be as much as 2.5 per cent and heavy weather ahead may have an effect of more than twice this amount.

PROPELLER LAW

An enginebuilder is at liberty to make the engine mean pressure and revolutions what he will, within the practical and experimental limits of the engine design. It is only after the maximum horsepower and revolutions are decided and the engine has been coupled to a propeller that the propeller law operates in its effect upon horsepower, mean pressure and revolutions.

$$shp \text{ varies as } V^3$$
$$shp \text{ varies as } R^3$$
$$T \text{ varies as } R^2$$
$$P \text{ varies as } R^2$$

where shp = aggregate shaft horsepower of engine, metric or imperial;
$\quad\quad V$ = speed of ship in knots
$\quad\quad R$ = revolutions per minute
$\quad\quad T$ = torque, in kg metres or lbft = Pr
$\quad\quad P$ = brake mean pressure kgf/cm^2 or lbf/in^2
$\quad\quad r$ = radius of crank, metres or feet

If propeller slip is assumed to be constant:

$$shp = KV^3$$

where K = constant from shp and R for a set of conditions.

But R is proportional to V, for constant slip,

$$\therefore \quad shp = K_1 R^3$$

where K_1 = constant from shp and R for a set of conditions.

But $shp = \dfrac{p \times A \times c \times r \times 2\pi \times R}{33\,000}$

$$= K_1 R^3 \text{ (imperial)}$$

when A = aggregate area of pistons, cm^2 or in^2

$\quad c = 0.5$ for two-stroke, 0.25 for four-stroke engines:

or $\quad T = PAcr = \dfrac{33\,000}{2\pi \times R} \times K_1 R^3 = K_2 R_2 \text{ (imperial)}$

or

$$T = PAcr = \dfrac{4500}{2\pi \times R} \times K_1 R^3 = K_2 R_2 \text{ (metric)}$$

i.e. $\quad T = K_2 R^2$

where K_2 = constant, determinable from T and R for a set of conditions.

$$PAcr = T \text{ or } \frac{T}{Acr} = \frac{K_2}{Acr} \times R^2$$

i.e. $\quad P = K_3 R^2$

where K_3 = constant determinable from P and R for a set of conditions.

The propeller law index is not always 3, nor is it always constant over the full range of speeds for a ship. It could be as much as 4 for short high speed vessels but 3 is normally satisfactory for all ordinary calculations. The index for R, when related to the mean pressure P, is one number less than that of the index for V.

Propeller law is most useful for enginebuilders at the testbeds where engine loads can be applied with the dynamometer according to the load and revolutions calculated from the law, thus matching conditions to be found on board the ship when actually driving a propeller.

FUEL COEFFICIENT

An easy yardstick to apply when measuring machinery performance is the fuel coefficient:

$$C = \frac{D^{2/3} \times V^3}{F}$$

where C = fuel coefficient
  D = displacement of ship in tons
  V = speed in knots
  F = fuel burnt per 24 hours in tons

This method of comparison is applicable only if ships are similar, are run at approximately corresponding speeds, operate under the same conditions, and burn the same quality of fuel. The ship's displacement in relation to draught is obtained from a scale provided by the shipbuilders.

ADMIRALTY CONSTANT

$$C = \frac{D^{2/3} \times V^3}{shp}$$

where C = Admiralty constant, dependent on ship form, hull finish and other factors.

If C is known for a ship the approximate *shp* can be calculated for given ship conditions of speed and displacement.

APPARENT PROPELLER SLIP

$$\text{Apparent slip, per cent} = \left(\frac{P \times R - 101.33 \times V}{P \times R} \right) \times 100$$

where P = propeller pitch in ft
  R = speed of ship in knots
101.33 is one knot in ft/min

The true propeller slip is the slip relative to the wake stream, which is something very different. The engineer, however, is normally interested in the apparent slip.

PROPELLER PERFORMANCE

Many variables affect the performance of a ship's machinery at sea so the only practical basis for a contract to build to a specification and acceptance by the owner is a sea trial where everything is under the builder's control. The margin between the trial trip power and sea service requirements of speed and loading must ensure that the machinery is of ample capacity. One important variable on the ship's performance is that of the propeller efficiency.

Propellers are designed for the best combinations of blade area, diameter, pitch, number of blades, etc, and are matched to a given horsepower and speed of propulsion engine; and in fact each propeller is specifically designed for the particular ship. It is important that the engine should be able to provide heavy torque when required, which implies an ample number of cylinders with ability to carry high mean pressures. However, when a propeller reaches its limit of thrust capacity under head winds an increase in revolutions can be to no avail.

In tank tests with models for powering experiments the following particulars are given:

Quasi-propulsive coefficient (Q)

$$Q = \frac{\text{model resistance} \times \text{speed}}{2\pi \times \text{torque} \times \text{rev/min}} = \frac{\text{work got out per min}}{\text{work put in per min}}$$

Total shaft horsepower at propeller (EHP)

$$\text{EHP} = \frac{ehp \times p}{Q}$$

where ehp = effective horsepower for model as determined by tank testing;

       p = increase for appendages and air resistance equal to 10–12 per cent of the naked model EHP, for smooth water conditions.

The shaft horsepower at the propeller for smooth sea trials is about 10 per cent more than in tank tests. The additional power, compared with sea trials, for sea service is about 11–12 per cent more for the South Atlantic and 20–25 per cent more for the North Atlantic. This is due to the normal weather conditions in these areas. The size of the ship affects these allowances: a small ship needs a greater margin. By way of example: 15 per cent margin over trial conditions equals 26.5 per cent over tank tests.

The shaft horsepower measured by torsionmeter abaft the thrust block exceeds the EHP by the power lost in friction at the sterntube and plummer blocks and can be as much as 5–6 per cent.

The brake horsepower (bhp) exceeds the torsionmeter measured shp by the frictional power lost at the thrust block. Bhp can only be calculated onboard ship by multiplying the recorded indicated horsepower by the mechanical efficiency stated by the enginebuilder.

Required bhp for engine = EHP + sea margin + hp lost at sterntube and plummer blocks + hp lost at thrust block.

Typical values for the quasi-propulsive coefficient (Q) are: tanker 0.67–0.72; slow cargo vessel 0.72–0.75; fast cargo liner 0.70–0.73; ferry 0.58–0.62; passenger ship 0.65–0.70.

POWER BUILD-UP

Figures 5.3 and 5.4 are typical diagrams showing the propulsion power data for a twin-screw vessel. In Figure 5.3 curve A is the EHP at the trial draught; B is the EHP corrected to the contract draught; C is the shp at trial draught on the Firth of Clyde; D is the shp corrected to the contract draught; E is the power service curve from voyage results; F shows the relation between speed of the ship and engine revolutions on trials; G is the service result. The full rated power of the propelling engines is 18 000 bhp; the continuous service rating is 15 000 bhp.

In Figure 5.3 curves A to D show shp as ordinates and the speed of the ship as abscissae. In Figure 5.4 powers are shown as ordinates, revolutions as abscissae.

Figure 5.5 shows the relationship between revolutions, power and brake mean pressure for the conditions summarized in Figures 5.3 and 5.4. A fair line drawn through the observed points for the whole range shows the shp to increase approximately as the cube of the revolutions and the square of the bmep. For the range 95–109 rev/min, the index increases to 3.5 for the power and 2.5 for the bmep. Between 120 and 109 rev/min a more closely drawn curve shows the index to rise to 3.8 and the bmep to 2.8. In Figure 5.5 ordinates and abscissae are plotted to a logarithmic base, thus reducing the power/revolution and the pressure/revolution curves to straight lines, for simplicity.

TRAILING AND LOCKING OF PROPELLER

In Figure 5.6 there are shown the normal speed/power curves for a twin-screw motor vessel on the measured mile and in service.

The effect upon the speed and power of the ship when one of the propellers is trailed, by 'free-wheeling', is indicated in the diagram. The effect of one of the propellers being locked is also shown.

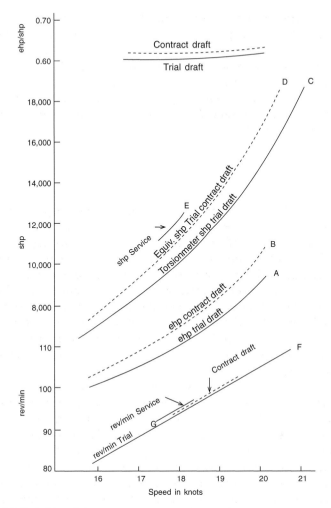

Figure 5.3 Propulsion data

Figure 5.7 shows the speed/power curves for a four-screw motorship:

1. When all propellers are working.
2. When the vessel is propelled only by the two centre screws, the outer screws being locked.
3. When the ship is being propelled only by the two wing screws, the two inner screws being locked.

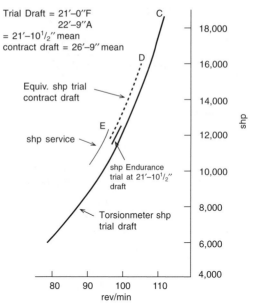

Trial Draft = 21'–0"F
             22'–9"A
= 21'–10^1/$_2$" mean
contract draft = 26'–9" mean

Equiv. shp trial
contract draft

shp service

shp Endurance
trial at 21'–10^1/$_2$"
draft

Torsionmeter shp
trial draft

Figure 5.4 Propulsion data

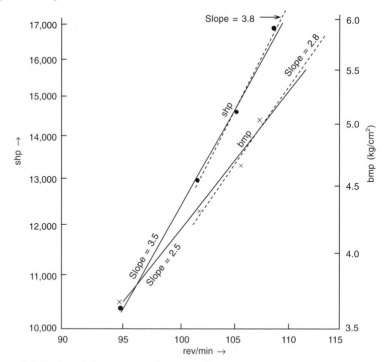

Figure 5.5 Engine trials: power, revolutions and mean pressure

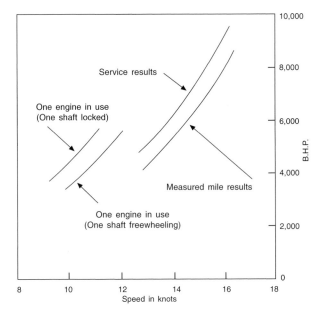

Figure 5.6 Speed/power curves, twin-screw vessel

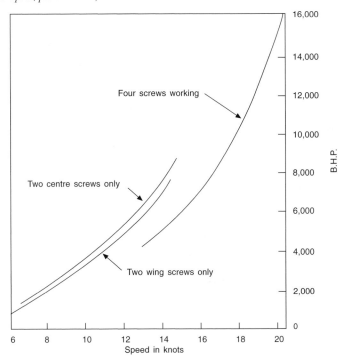

Figure 5.7 Speed/power curves, quadruple-screw ship

ASTERN RUNNING

Figure 5.8 summarizes a series of tests made on the trials of a twin-screw passenger vessel, 716 ft long, 83 ft 6 in beam, trial draught 21 ft forward, 26 ft aft, 26 000 tons displacement.

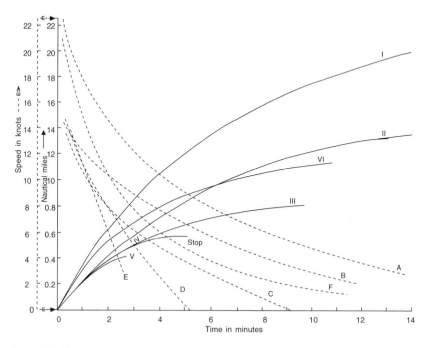

Figure 5.8 Ship stopping trials

As plotted in Figure 5.8, tests I to VI show distances and times, the speed of approach being as stated at column 2 in Table 5.1. The dotted curves show reductions of speed and times.

The dotted curves A to F respectively correspond to curves I to VI. In test I, after the ship had travelled, over the ground, a distance of two nautical miles (1 nm = 6080 ft) the test was terminated and the next test begun.

In Table 5.2 a typical assortment of observed facts related to engine stopping and astern running is given. Where two or three sets of readings are given, these are for different vessels and/or different engine sizes.

Trials made with a cargo liner showed that the ship was brought to rest from 20 knots in 65 seconds. Another cargo liner, travelling at 16

Table 5.1 Ship stopping trials

1 Test No.	2 Ahead speed of approach knots (rev/min)	3 Propellers	4 Propellers stopped (min)		5 Ship stopped (min)	6 Distance travelled (nautical miles)
			P.	S.		
I	23.0 (119)	Trailing; unlocked	16.4	14.9	—	2.0
II	14.5 (75)	Trailing; unlocked	12.5	12.1	17.0	1.4
III	13.5 (75)	Trailing; locked	1.5	1.5	13.0	0.8
IV	15.0 (75)	Ahead running checked; no additional astern power	1.3	1.3	5.2	0.5
V	14.8 (75)	Engine stopped; astern as quickly as possible	0.7	0.8	3.1	0.4
VI	22.4 (116)	Trailing; unlocked	1.5	1.6	15.0	1.1

Table 5.2 Engine reversing and ship stopping

Ship	Engine type	Ahead (rev/ min)	Time for engine stopping (sec)	Engine moving astern (sec)	Astern running		Ship stopped	
					rev/min	sec	min	sec
Large passenger	D.A. 2C. (twin)	65	—	30	—	—	4	2
Small fast passenger	S.A. 2C. tr. (twin)	217 195 220	53 35 45	63.5 45 59	160 160 170	72.5 60 81	2 1 2	15 54 5
Passenger	Diesel- electric (twin)	92	110	—	80	225	5	0
Cargo	S.A. 2C. (twin)	112	127.5	136	100	141	—	
Cargo	D.A. 2C. (single)	90 88 116	24 12 35	26 14 40	90 82 95	115 35 50	2 3 —	26 25
Cargo	S.A. 2C. (single)	95 110 116	31 12 10	33 21 55	75 110 110	45 53 110	— 3 4	21 6

knots, was brought to a stop in a similar period. The engine, running full power ahead, was brought to 80 rev/min astern in 32 seconds, and had settled down steadily at full astern revolutions in 50 seconds.

A shipbuilder will think of ship speed in terms of the trial performance in fair weather but to the shipowner ship speed is inevitably related to scheduled performance on a particular trade route. Sea trials are invariably run with the deadweight limited to fuel, fresh water and ballast. Because of the difference between loaded and trials draught, the hull resistance may be 25–30 per cent greater for the same speed. This has a consequential effect upon the relation between engine torque and power, and in the reaction on propeller efficiency.

Adverse weather, marine growth and machinery deterioration necessitate a further power allowance, if the service speed is to be maintained. The mean wear and tear of the engine may result in a reduction of output by 10–15 per cent or a loss of speed by up to one knot may be experienced.

When selecting a propulsion engine for a given ship a suitable power allowance for all factors such as weather, fouling, wear and tear, as well as the need to maintain the service speed at around 85 per cent of the maximum continuous rating, should all be taken into consideration. C.T.W.

6 Engine and plant selection

Choosing a propulsion engine or engines and the most suitable plant configuration for a given newbuilding or retrofit project is not a simple decision. It dictates careful study of the machinery options available and the operating profile of the ship.

In the past the shipowner or designer had the straight choice of a direct-coupled low speed two-stroke engine driving a fixed pitch propeller or a geared medium speed four-stroke engine driving either a fixed or controllable pitch (CP) propeller. Today, ships are entering service with direct-coupled (and sometimes geared) two-stroke engines driving fixed or CP propellers, geared four-stroke engines or high/medium speed diesel–electric propulsion plants. Diverse diesel–mechanical and diesel–electric configurations can be considered (Figures 6.1 and 6.2).

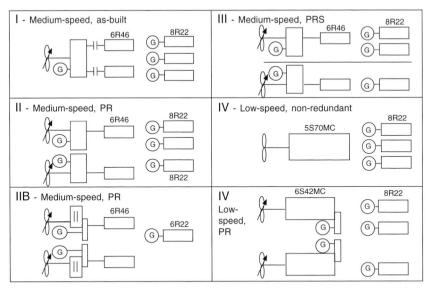

Figure 6.1 Low and medium speed diesel-based propulsion machinery options for a 91 000 dwt tanker and associated auxiliary power generating (G) source

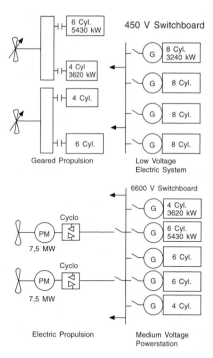

Figure 6.2 Schematics of geared medium speed diesel propulsion plant with gensets (top) and diesel–electric plant serving propulsion and auxiliary power demands (below)

Low speed engines are dominant in the mainstream deepsea tanker, bulk carrier and containership sectors while medium speed engines are favoured for smaller cargo ships, ferries, cruise liners, RoRo freight carriers and diverse specialist tonnage such as icebreakers, offshore support and research vessels. The overlap territory has nevertheless become more blurred in recent years: mini-bore low speed engines can target shortsea and even river/coastal vessels while new generations of high powered large bore medium speed designs contest the traditional deepsea arenas of low speed engines.

Many shipowners may remain loyal to a particular type or make and model of engine for various reasons, such as reliable past operating experience, crew familiarity, spares inventories and good service support. But new or refined engine designs continue to proliferate as enginebuilders seek to maintain or increase market share or to target a new sector. In some cases an owner who has not invested in newbuildings for several years may have to consider a number of models and plant configurations which are unfamiliar.

Much decision-making on the engine focuses on cost considerations—

not just the initial cost but the type of fuel which can be reliably burned, maintenance costs, desirable manning levels and availability/ price of spares. The tendency now is to assess the total life cycle costs rather than the purchase price of the main engine. Operating costs over, say, 20 years may vary significantly between different types and makes of engine, and the selected plant configuration in which the engine has to function. Key factors influencing the choice of engine may be summarized as:

- Capability to burn heavy fuel of poor quality without detrimental impact on the engine components and hence maintenance/ spares costs.
- The maintenance workload: the number of cylinders, valves, liners, rings and bearings requiring periodic attention in relation to the number of crew carried (bearing in mind that lower manning levels and less experienced personnel are now more common).
- Suitability for unattended operation by exploiting automated controls and monitoring systems.
- Propulsive efficiency: the ability of the engine or propeller shaft to be turned at a low enough speed to drive the largest diameter (and hence most efficient) propeller.
- Size and weight of the propulsion machinery.
- Cost of the engine.

The size of the machinery space is largely governed by the size of the main engine which may undermine the cargo-carrying capacity of the ship. The available headroom is also important in some ships, notably ferries with vehicle decks, and insufficient headroom and surrounding free space may make it difficult or impossible for some engines to be installed or overhauled.

DIESEL–MECHANICAL DRIVES

The direct drive of a fixed pitch propeller by a low speed two-stroke engine remains the most popular propulsion mode for deepsea cargo ships. At one time a slight loss of propulsive efficiency was accepted for the sake of simplicity but the introduction of long stroke and, more recently, super- and ultra-long stroke crosshead engines has reduced such losses. For a large ship a direct-coupled speed of, say, 110 rev/min is not necessarily the most suitable since a larger diameter propeller turning at speeds as low as 60 rev/min is more efficient than one of a smaller diameter absorbing the same horsepower at 110 rev/min. The longer stroke engines now available develop their rated outputs at speeds ranging from as low as 55 rev/min (very large bore models) up to around 250 rev/min for the smallest bore models. It is now possible to specify a direct-drive engine/propeller combination which will yield close to the optimum propulsive efficiency for a given ship design.

Large bore low speed engines develop high specific outputs, allowing the power level required by many ship types to be delivered from a small number of cylinders. Operators prefer an engine with the fewest possible cylinders, as long as problems with vibration and balance are not suffered. Fewer cylinders obviously influence the size of the engine and the machinery space, the maintenance workload, and the amount of spares which need to be held in stock. In most deepsea ships the height restriction on machinery is less of a problem than length, a larger bore engine with fewer cylinders therefore underwriting a shorter engineroom and more space for cargo. Larger bore engines also generally return a better specific fuel consumption than smaller engines and offer a greater tolerance to heavy fuels of poor quality.

A direct-coupled propulsion engine cannot operate unaided since it requires service pumps for cooling and lubrication, and fuel/lube oil handling and treatment systems. These ancillaries need electrical power which is usually provided by generators driven by medium or high speed diesel engines. Many genset enginebuilders can now offer designs capable of burning the same heavy fuel grade as the main engine as well as marine diesel oil or blended fuel (heavy fuel and distillate fuel mixed in various proportions, usually 70:30) either bunkered as an intermediate fuel or blended onboard. 'Unifuel' installations—featuring main and auxiliary engines arranged to burn the same bunkers—are now common.

Auxiliary power generation

The cost of auxiliary power generation can weigh heavily in the choice of main machinery. Developments have sought to maximize the exploitation of waste heat recovery to supplement electricity supplies at sea, to facilitate the use of alternators driven by the main engine via speed-increasing gearing or mounted directly in the shaftline, and to power other machinery from the main engine.

Gear-based constant frequency generator drives allow a shaft alternator to be driven by a low speed engine in a fixed pitch propeller installation, with full alternator output available between 70 per cent and 104 per cent of propeller speed. A variety of space-saving arrangements are possible with the alternator located alongside or at either end of a main engine equipped with compact integral power take-off gear. Alternatively, a thyristor frequency converter system can be specified to serve an alternator with a variable main engine shaft speed input in a fixed pitch or CP propeller installation.

The economic attraction of the main engine-driven generator for electrical power supplies at sea is that it exploits the high thermal efficiency, low specific fuel consumption and low grade fuel-burning capability of the ship's diesel prime mover. Other advantages are that the auxiliary diesel gensets can be shut down, yielding benefits from reduced running hours in terms of lower fuel and lubricating oil consumptions, maintenance demands and spares costs.

System options for electricity generation have been extended by the arrival of power turbines which, fed with exhaust gas surplus to the needs of modern high efficiency turbochargers, can be arranged to drive alternators in conjunction with the main engine or independently.

These small gas turbines are also in service in integrated systems linking steam turbo-alternators, shaft alternators and diesel gensets; the various power sources—applied singly or in combination—promise optimum economic electricity production for any ship operating mode. Some surplus electrical output can also be tapped to support the propulsive effort via a shaft alternator switched to function as a propulsion motor.

Such a plant is exploited in a class of large low speed engine-powered containerships with significant reefer capacity whose overall electrical load profile is substantial and variable. Crucial to its effectiveness is a computer-controlled energy management system which co-ordinates the respective contributions of the various power sources to achieve the most economical mode for a given load demand.

Integrated energy-saving generating plants have been developed over the years by the major Japanese shipbuilding groups for application to large tankers and bulk carriers. The systems typically exploit waste heat (from low speed main engine exhaust gas, scavenge air and cooling water) to serve a steam turbo-alternator, air conditioning plant, heaters and distillers. System refinements were stimulated by the diminishing amount of energy available from the exhaust gas of low speed engines, in terms of both temperature and volumes, with the progressive rise in thermal efficiencies. The ability of the conventional waste heat boiler/turbo-alternator set to meet electrical demands at sea was compromised, any shortfall having to be plugged by supplementary oil firing of an auxiliary boiler or by running a diesel genset and/or shaft alternator. The new integrated systems, some also incorporating power gas turbines, maximize the exploitation of the waste heat available in ships whose operating profiles and revenues can justify the added expense and complexity.

Geared drives

The most common form of indirect drive of a propeller features one or more medium speed four-stroke engines connected through clutches and couplings to a reduction gearbox to drive either a fixed pitch or CP propeller (Figures 6.3 and 6.4). The CP propeller eliminates the need for a direct-reversing engine while the gearing allows a suitable propeller speed to be selected. There is inevitably a loss of efficiency in the transmission but in most cases this would be cancelled out by the improvement in propulsive efficiency when making a comparison of direct-coupled and indirect drive engines of the same horsepower. The additional cost of the transmission can also be offset by the lower cost of the four-stroke engine since two-stroke designs, larger and heavier, cost more. The following generic merits are cited for geared multi-medium or high speed engine installations:

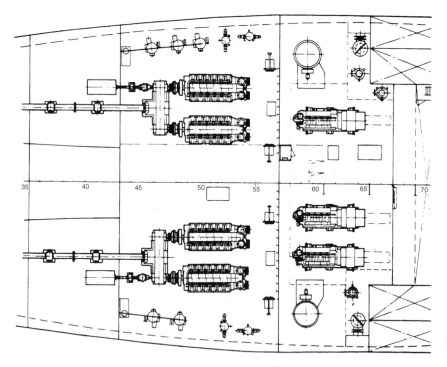

Figure 6.3 A typical multi-engine geared twin-screw installation

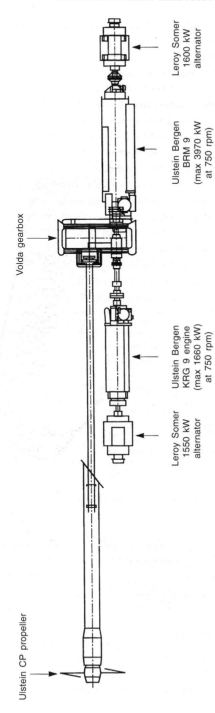

Leroy Somer
1600 kW
alternator

Ulstein Bergen
BRM 9
(max 3970 kW
at 750 rpm)

Volda gearbox

Ulstein Bergen
KRG 9 engine
(max 1660 kW)
at 750 rpm)

Leroy Somer
1550 kW
alternator

Ulstein CP propeller

Figure 6.4 One of the twin medium speed geared propulsion/auxiliary power lines of a Norwegian coastal ferry

- Ships with more than one main engine benefit from enhanced availability through redundancy: in the event of one engine breaking down another can maintain navigation. The number of engines engaged can also be varied to secure the most economical mode for a given speed or deployment profile. Thus, when a ship is running light, partially loaded or slow steaming, one engine can be deployed at its normal (high efficiency) rating and some or all of the others shut down. In contrast, in similar operational circumstances, a single direct-coupled engine might have to be run for long periods at reduced output with lower efficiency.
- The ability to vary the number of engines deployed allows an engine to be serviced at sea, easing maintenance planning. This flexibility is particularly valued in an era when port turnround times are minimized. Engines can also be overhauled in port without the worry of authorities demanding an unscheduled shift of berth or sudden departure.
- By modifying the number of engines per ship and the cylinder numbers per engine to suit individual power requirements the propulsion plant for a fleet can be standardized on a single engine model, with consequent savings in spares costs and inventories, and benefits in crew familiarity. The concept can also be extended to the auxiliary power plant through 'uniform machinery' installations embracing main and genset engines of the same model.
- Compact machinery spaces with low headrooms can be created, characteristics particularly valued for RoRo ferries.

Designers of marine gearing, clutches and couplings have to satisfy varying and sometimes conflicting demands for operational flexibility, reliability, low noise and compactness from transmission systems.

Advances in design, materials and controls have contributed to innovative solutions for versatile single- and multi-engine propulsion installations featuring power take-offs for alternator drives and power take-ins to boost propulsive effort. In many propulsion installations a gearbox is expected to: determine the propeller speed and direction of rotation, and provide a reversing capability; provide a geometric coupling that can connect and separate the flow of power between the engine and propeller shaft or waterjet drive; and absorb the thrust from the propeller.

An impressive flexibility of operating modes can be arranged from geared multi-engine propulsion plants which are, in practice, overwhelmingly based on four-stroke machinery (a number of geared two-stroke engine installations have been specified, however, for special purpose tonnage such as offshore shuttle tankers).

Father-and-son layouts

Flexibility is enhanced by the adoption of a so-called 'father-and-son' (or 'mother-and-daughter') configuration: a partnership of similar four-stroke engine models, but with different cylinder numbers, coupled to a common reduction gearbox to drive a propeller shaft. Father-and-son pairs have been specified to power large twin-screw cruise vessels. An example is provided by the 1995-built P&O liner *Oriana* whose 40 000 kW propulsion plant is based on two nine-cylinder and two six-cylinder MAN B&W L58/64 medium speed engines. Each father-and-son pair drives a highly skewed Lips CP propeller via a Renk Tacke low noise gearbox which also serves a 4200 kW shaft alternator/propulsion motor.

Propulsion can be effected either by: the father-and-son engines together; the father engines alone or the son engines alone; and with or without the shaft alternators operating as propulsion motors (fed with electrical power by the diesel gensets).

DIESEL–ELECTRIC DRIVE

An increasingly popular form of indirect drive is diesel–electric propulsion based on multi-medium speed main gensets. New generations of AC/AC drive systems exploiting cycloconverter or synchroconverter technology have widened the potential of electric propulsion after years of confinement to specialist niches, such as icebreakers, research vessels and cablelayers.

The diesel–electric mode is now firmly entrenched in large cruise ships and North Sea shuttle tanker propulsion, references have been established in shortsea and deepsea chemical carriers, as well as in Baltic RoRo passenger/freight ferries, and project proposals argue the merits for certain classes of containership. Dual-fuel diesel-electric propulsion has also penetrated the offshore supply vessel and LNG carrier sectors.

A number of specialist groups contest the high powered electric drive market, including ABB Marine, Alstom, STN Atlas and Siemens which highlight system capability to cope with sudden load changes, deliver smooth and accurate ship speed control, and foster low noise and vibration. Applications have been stimulated by continuing advances in power generation systems, AC drive technology and power electronics.

Two key AC technologies have emerged to supersede traditional DC electric drives: the Cycloconverter system and the Load Commutated Inverter (LCI) or Synchrodrive solution. Both are widely used in electric propulsion but, although exploiting the same basic Graetz bridge prevalent in DC drive systems, they have different electrical characteristics. Development efforts have also focused on a further AC system variant, the voltage source Pulse Width Modulation (PWM) drive.

A variable-speed AC drive system comprises a propulsion motor (Figure 6.5) and a frequency converter, with the motor speed controlled by the converter altering the input frequency to the motor.

Figure 6.5 One of two 15 000 kW STN Atlas propulsion motors installed on the cruise liner Costa Victoria

Both Cyclo and PWM drives are based on advanced AC variable speed technology that matches or exceeds the performance characteristics of conventional DC drives. Each features motor speed control with maximum torque available from zero to full speed in either direction, facilitating operation with simple and robust fixed pitch (rather than controllable pitch) propellers. Smooth control underwrites operation at very low speeds, and vector control yields a rapid response to enhance plant and ship safety. Integrated full electric propulsion (IFEP) is now standard for cruise liners and specified for a widening range of other ship types, in which the economic and operational benefits of the central power station concept can be exploited.

Electric propulsion requires motors to drive the propellers and gensets to supply the power. It seems somewhat illogical to use electric generators, switchgear and motors between prime movers and propeller when direct-coupling or geared transmission to the shaft may be all that is necessary. There are obviously sound reasons for some installations to justify the complication and extra cost of electric propulsion, Lloyd's Register citing:

Flexibility of layout: the advantage of electric transmission is that the prime movers and their associated generators are not constrained to have any particular relationship with the load—a cable run is a very versatile medium. In a propulsion system it is therefore possible to install the main diesel gensets and their support services in locations remote from the propeller shaft (Figure 6.6). Diesel gensets have even been installed in containers on deck to provide propulsive power, and other vessels equipped with a 10 000 kW generator mounted in a block at the stern above the RoRo deck.

Another example of the flexibility facilitated by an electric propulsion system is a semi-submersible rig installation whose gensets are mounted on the main deck and propulsion motors in the pontoons. In cruise ships, ferries and tankers the layout flexibility allows the naval architect to create compact machinery installations releasing extra revenue-earning space for passenger accommodation/amenities and cargo.

Opportunities to design and build tankers more cost-effectively are also offered. The optimized location of the main machinery elements allows the ship's overall length to be reduced and steel costs correspondingly lowered for a given cargo capacity; alternatively, the length of the cargo tank section can be extended within given hull dimensions.

Diesel–electric solutions facilitate the modularization and delivery of factory-tested turnkey packages to the building yard, with the complete main gensets mounted on common bedplates ready for coupling to their support systems. The compactness of the machinery outfit fosters shorter runs for the cabling and ancillary systems, and the engine casing and exhaust gas piping are also shortened.

It is difficult for a diesel–electric plant to match the fuel economy of a single direct-drive low speed main engine which is allowed to operate at its optimum load for the long transoceanic leg of a tanker's voyage. But some types of tanker (oil products and chemical tonnage, as well as North Sea shuttle carriers) are deployed in varied service

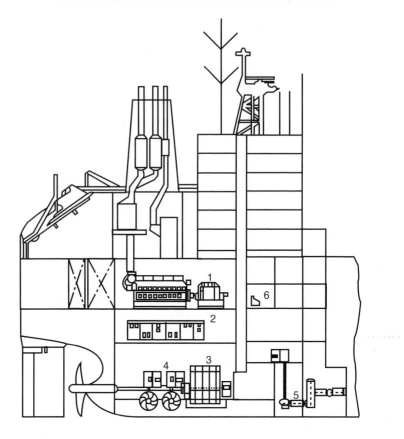

Figure 6.6 Machinery arrangement in a diesel–electric tanker
  1. *Diesel gensets*
  2. *Main switchboard and cycloconverters*
  3. *Propulsion motor*
  4. *Stern thrusters*
  5. *Cargo pumps*
  6. *Engine control room*

profiles and spend a considerable time at part loads: for example, in restricted waters, during transit in ballast and manoeuvring.

A diesel–electric tanker can exploit the abundance of electrical power to drive a low noise cargo pumping outfit, perhaps reducing time in port; and (particularly in the case of a dynamically positioned shuttle carrier) to serve powerful bow and stern thrusters during sensitive manoeuvring at the loading buoy (Figure 6.7).

Load diversity: certain types of tonnage have a requirement for substantial

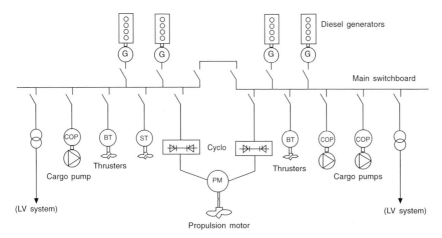

Figure 6.7 Diesel–electric 'power station' meeting the load demands of propulsion, cargo pumping, thrusters and hotel services in a tanker

amounts of power for ship services when the demands of the propulsion system are low: for example, tankers and any other ship with a significant cargo discharging load. A large auxiliary power generating plant for cruise ships and passenger ferries is dictated by the hotel services load (air conditioning, heating and lighting) and the heavy demands of transverse thrusters during manoeuvring. The overall electrical demand may be 30–40 per cent of the installed propulsion power and considerable standby capacity is also called for to secure system redundancy for safety. These factors have helped to promote the popularity of the medium speed diesel–electric 'power station' concept for meeting all propulsion, manoeuvring and hotel energy demands in large passenger ships.

Economical part load running: this is best achieved when there is a central power station feeding propulsion and ship services. A typical medium speed diesel–electric installation features four main gensets (although plants with up to nine sets are in service) and, with parallel operation of all the sets, it is easy to match the available generating capacity to the load demand. In a four-engine plant, for example, increasing the number of sets in operation from two that are fully loaded to three partially loaded will result in three sets operating at 67 per cent load: not ideal but not a serious operating condition.

It is not necessary to operate gensets at part load to provide the spare capacity for the sudden loss of a set: propulsion load reduction may be available instantaneously and in most ships a short-term reduction in propulsive power does not represent a hazard. The propulsion plant controls continuously monitor generating capability, any generator overload immediately resulting in the controls adjusting the input to the propulsion motors. During manoeuvring, propulsion plant requirements are below system capacity and the failure of one generator is unlikely to present a hazardous situation.

Ease of control: the widespread use of CP propellers means that control facilities once readily associated with electric drives are no longer able to command the same premium. It is worth noting, however, that electric drives are capable of meeting the most exacting demands with regard to dynamic performance, exceeding by a very wide margin anything that is required of a propulsion system.

Low noise: an electric motor can provide a drive with very low vibration characteristics, a quality valued for warships, research vessels and cruise ships where, for different reasons, a low noise signature is required. In the case of warships and research vessels it is noise into the water which is the critical factor; in cruise ships it is structure-borne noise and vibration to passenger spaces that are desirably minimized. Electric transmission allows the prime movers to be mounted in such a way that the vibration transmitted to the seatings, and hence to the water or passenger accommodation, is minimized. The use of double resilient mounting may be necessary for very low noise installations.

Environmental protection and ship safety: controls to curb noxious exhaust gas emissions—tightening nationally, regionally and globally—also favour the specification of electric transmission since the constant speed running of diesel prime movers optimized to the load is conducive to lower NOx emission levels. An increasing focus on higher ship safety through redundancy of propulsion plant elements is another positive factor. A twin-propeller installation is not the only method of achieving redundancy. Improving the redundancy of a single-screw ship can also be secured by separating two or more propulsion motors in different compartments and coupling them to a reduction gear located in a separate compartment. The input shafts are led through a watertight bulkhead.

Podded propulsors

Significant economic and technical benefits in ship design, construction and operation are promised by electric podded propulsors, whose appeal has extended since the early 1990s from icebreakers and cruise liners to offshore vessels, ferries and tankers.

A podded propulsor (or pod) incorporates its electric drive motor in a hydrodynamically-optimized submerged housing which can be fully rotated with the propeller(s) to secure 360-degree azimuthing and thrusting capability (Figure 6.8).

Figure 6.8 Three 14 000 kW Azipod propulsors power Royal Caribbean Cruises' Voyager-class liners

The motor is directly coupled to the fixed pitch propeller(s) mounted at either or both ends of the pod. Pusher and tractor or tandem versions can be specified, with input powers from 400 kW to 30 000 kW supplied by a shipboard generating plant. The merits cited for pods over traditional diesel-electric installations with shaftlines highlight:

- Space within the hull otherwise reserved for conventional propulsion motors can be released and exploited for other purposes.
- More creative freedom for ship designers since the propulsors and the prime movers require no direct physical connection.
- Steering capability is significantly better than with any conventional rudder system; stern thruster(s) can be eliminated, along with rudder(s) and shaftline(s).

- Excellent reversing capability and steering during astern navigation, and enhanced crash stop performance.
- Low noise and vibration characteristics associated with electric drive systems are enhanced by the motor's underwater location; and hull excitations induced by the propeller are very low thanks to operation in an excellent wake field.
- Propulsor unit deliveries can be late in the shipbuilding process, reducing 'dead time' investment costs; savings in overall weight and ship construction hours are also promised.

Combined systems

A combination of geared diesel and diesel–electric drive is exploited in some offshore support vessels whose deployment profile embraces two main roles: to transport materials from shore bases to rigs and platforms, and to standby at the offshore structures for cargo and anchor handling, rescue and other support operations. In the first role the propulsive power required is that necessary to maintain the free running service speed. In the second role little power is required for main propulsion but sufficient electrical power is vital to serve winches, thrusters and cargo handling gear.

The solution for these dual roles is to arrange for the main propellers to be driven by medium speed engines through a reduction gear which is also configured to drive powerful shaft alternators. By using one shaft alternator as a motor and the other to generate electrical power it is possible to secure a diesel–electric main propulsion system of low horsepower and at the same time provide sufficient power for the thrusters. Such a twin-screw plant could simultaneously have one main engine driving the alternator and the propeller while the other main engine is shut down and its associated propeller driven by the shaft alternator in propulsion motor mode.

A further option is to fit dedicated propulsion motors as well as the shaft alternators, yielding greater plant flexibility and fuel saving potential but at the expense of extra cost and complexity. For the above power plants to operate effectively CP main and thruster propellers are essential.

Combined medium or high speed diesel engine and gas turbine (CODAG) systems are now common for large fast ferry propulsion, while some large diesel-electric cruise liners feature medium speed engines and gas turbines driving generators in a CODLAG or CODEG arrangement.

7 Pressure charging

A naturally aspirated engine draws air of the same density as the ambient atmosphere. Since this air density determines the maximum weight of fuel that can be effectively burned per working stroke in the cylinder, it also determines the maximum power that can be developed by the engine. Increasing the density of the charge air by applying a suitable compressor between the air intake and the cylinder increases the weight of air induced per working stroke, thereby allowing a greater weight of fuel to be burned with a consequent rise in specific power output.

This boost in charge air density is accomplished in most modern diesel engine types by exhaust gas turbocharging, in which a turbine wheel driven by exhaust gases from the engine is rigidly coupled to a centrifugal type air compressor. This is a self-governing process which does not require an external governor.

The power expended in driving the compressor has an important influence on the operating efficiency of the engine. It is relatively uneconomical to drive the compressor direct from the engine by chain or gear drive because some of the additional power is thereby absorbed and there is an increase in specific fuel consumption for the extra power obtained. About 35 per cent of the total heat energy in the fuel is wasted to the exhaust gases, so by using the energy in these gases to drive the compressor an increase in power is obtained in proportion to the increase in the charge air density.

The turbocharger comprises a gas turbine driven by the engine exhaust gases mounted on the same spindle as a blower, with the power generated in the turbine equal to that required by the compressor.

There are a number of advantages of pressure charging by means of an exhaust gas turboblower system:

- A substantial increase in engine power output for any stated size and piston speed, or conversely a substantial reduction in engine dimensions and weight for any stated horsepower.
- An appreciable reduction in the specific fuel consumption rate at all engine loads.

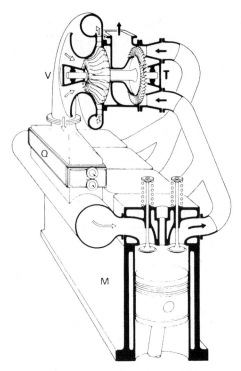

Figure 7.1 Basic arrangement of exhaust turbocharging. M. Diesel engine; Q. Charge air cooler; T. Turbocharger exhaust turbine; V. Turbocharger compressor

- A reduction in initial engine cost.
- Increased reliability and reduced maintenance costs, resulting from less exacting conditions in the cylinders.
- Cleaner emissions (see section below).
- Enhanced engine operating flexibility.

Larger two-stroke engines may be equipped with up to four turbochargers, each serving between three and five cylinders.

FOUR-STROKE ENGINES

Exhaust gas turbocharged single-acting four-stroke marine engines can deliver three times or more power than naturally aspirated engines of the same speed and dimensions. Even higher power output ratios are achieved on some engine types employing two-stage turbocharging where turbochargers are arranged in series. At one time almost all four-stroke engines operated on the pulse system, though constant pressure turbocharging has since become more common as it provides greater fuel economy while considerably simplifying the arrangement of exhaust piping.

In matching the turboblower to the engine, a free air quantity in excess of the swept volume is required to allow for the increased density of the charge air and to provide sufficient air for through-scavenging of the cylinders after combustion. For example, an engine with a full load bmep of 10.4 bar would need about 100 per cent of excess free air, about 60 per cent of which is retained in the cylinders, with the remaining 40 per cent being used for scavenging. Modern engines carry bmeps up to 28 bar in some cases, requiring greater proportions of excess air which is made possible by the latest designs of turbocharger with pressure ratios as high as 5:1.

To ensure adequate scavenging and cooling of the cylinders, a valve overlap of approximately 140° is normal with, in a typical case, the air inlet valve opening at 80° before top dead centre and closing at 40° after bottom dead centre. The exhaust value opens at 50° before bottom dead centre and closes at 60° after top dead centre.

For a low-rated engine, with a bmep of 10.4 bar compared with 5.5 bar for a naturally aspirated engine, a boost pressure of some 0.5 bar is required, corresponding to a compressor pressure ratio of 1.5:1. As average pressure ratios today are between 2.5 and 4, considerable boost pressures are being carried on modern high bmep medium speed four-stroke engines.

Optimum values of power output and specific fuel consumption can be achieved only by utilization of the high energy engine exhaust pulses. The engine exhaust system should be so designed that it is impossible for gases from one cylinder to contaminate the charge air in another cylinder, either by blowing back through the exhaust valve or by interfering with the discharge of gases from the cylinder. During the period of valve overlap it follows that the exhaust pressure must be less than air charging pressure to ensure effective scavenging of the cylinder to remove residual gases and cooling purposes. It has been found in practice that if the period between discharge of successive cylinders into a common manifold is less than about 240°, then interference will take place between the scavenging of one cylinder and the exhaust of the next. This means that engines with more than three cylinders must have more than one turbocharger or, as is more common, separate exhaust gas passages leading to the turbine nozzles.

The exhaust manifold system should be as small as possible in terms of pipe length and bores. The shorter the pipe length, the less likelihood there is of pulse reflections occurring during the scavenge period. The smaller the pipe bore, the greater the preservation of exhaust pulse energy, though too small a bore may increase the frictional losses of the high velocity exhaust gas to more than offset the increased pulse energy. Sharp bends or sudden changes in pipe cross-sectional area should be avoided wherever possible.

TWO-STROKE ENGINES

Compared with four-stroke engines, the application of pressure charging to two-stroke engines is more complicated because, until a certain level of speed and power is reached, the turboblower is not self-supporting.

At low engine loads there is insufficient energy in the exhaust gases to drive the turboblower at the speed required for the necessary air-mass flow. In addition, the small piston movement during the through scavenge period does nothing to assist the flow of air, as in the four-stroke engine. Accordingly, starting is made very difficult and off-load running can be very inefficient; below certain loads it may even be impossible. A solution was found by having mechanical scavenge pumps driven from the engine arranged to operate in series with the turboblowers. Standard on modern engines, however, are electrically driven auxiliary blowers.

Two-stroke engine turbocharging is achieved by two distinct methods, respectively termed the 'constant pressure' and 'pulse' systems. It is the constant pressure system that is now used by all low speed two-stroke engines.

For constant pressure operation, all cylinders exhaust into a common receiver which tends to dampen-out all the gas pulses to maintain an almost constant pressure. The advantage of this system is that it eliminates complicated multiple exhaust pipe arrangements and leads to higher turbine efficiencies and hence lower specific fuel consumptions. An additional advantage is that the lack of restriction, within reasonable limits, on exhaust pipe length permits greater flexibility in positioning the turboblower relative to the engine. Typical positions are at either or both ends of the engine, at one side above the air manifold or on a flat adjacent to the engine.

The main disadvantage of the constant pressure system is the poor performance at part load conditions and, owing to the relatively large exhaust manifold, the system is insensitive to changes in engine operating conditions. The resultant delay in turboblower acceleration, or deceleration, results in poor combustion during transition periods.

For operation under the pulse system the acceptable minimum firing order separation for cylinders exhausting to a common manifold is about 120°. The sudden drop in manifold pressure, which follows each successive exhaust pulse, results in a greater pressure differential across the cylinder during the scavenge period than is obtained with the constant pressure system. This is a factor which makes for better scavenging.

Figure 7.2 shows the variation in rotational speed of a turboblower during each working cycle of the engine, i.e. one revolution. This diagram clearly illustrates the reality of the impulses given to the turbine wheel. The fluctuation in speed is about 5 per cent. Each blower is coupled to two cylinders, with crank spacing 135°, 225°. It is now standard practice to fit charge air coolers to turbocharged two-stroke engines, the coolers being located between the turbochargers and the cylinders (Figure 7.3).

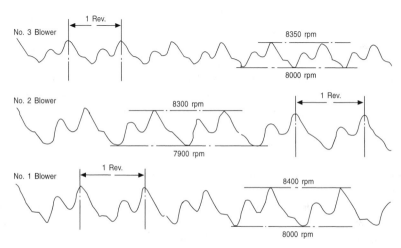

Figure 7.2 Cyclic variations in turbocharger revolutions

CHARGE AIR COOLING

The increased weight or density of air introduced into the cylinder by pressure charging enables a greater weight of fuel to be burned, and this in turn brings about an increase in power output. The increase in air density is, however, fractionally offset by the increase of air temperature resulting from adiabatic compression in the turboblower, the amount of which is dependent on compressor efficiency.

This reduction of air density due to increased temperature implies a loss of potential power for a stated amount of pressure charging. For example, at a charge air pressure of, say, 0.35 bar, the temperature rise is of the order of 33°C—equivalent to a 10 per cent reduction in

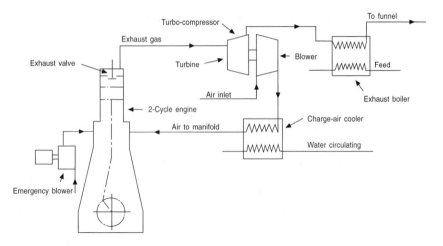

Figure 7.3 Exhaust gas turbocharging system with charge air cooling

charge air density. As the amount of pressure charging is increased, the effect of turboblower temperature rise becomes more pronounced. Thus, for a charge air pressure of 0.7 bar, the temperature rise is some 60°C, which is equivalent to a reduction of 17 per cent in the charge air density.

Much of this potential loss can be recovered by the use of charge air coolers. For moderate amounts of pressure charging, cooling of the charge air is not worthwhile, but for two-stroke engines especially it is an advantage to fit charge air coolers, which are standard on all makes of two-strokes and most medium speed four-stroke engines.

Charge air cooling has a double effect on engine performance. By increasing the charge air density it thereby increases the weight of air flowing into the cylinders, and by lowering the air temperature it reduces the maximum cylinder pressure, the exhaust temperature and the engine thermal loading. The increased power is obtained without loss—and, in fact, with an improvement in fuel economy. It is important that charge air coolers should be designed for low pressure drop on the air side; otherwise, to obtain the required air pressure the turboblower speed must be increased.

The most common type of cooler is the water-cooled design with finned tubes in a casing carrying seawater over which the air passes. To ensure satisfactory effectiveness and a minimum pressure drop on the charge air side and on the water side, the coolers are designed for air speeds of around 11 m/s and water speeds in the tubes of 0.75 m/s. Charge air cooler effectiveness is defined as the ratio of charge air temperature drop to available temperature drop between air inlet temperature and cooling water inlet temperature. This ratio is approximately 0.8.

SCAVENGING

It is essential that each cylinder should be adequately scavenged of gas before a fresh charge of air is compressed, otherwise the fresh charge is contaminated by residual exhaust gases from the previous cycle. Further, the cycle temperature will be unnecessarily high if the air charge is heated by mixing with residual gases and by contact with hot cylinders and pistons.

In the exhaust turbocharged engine the necessary scavenging is obtained by providing a satisfactory pressure difference between the air manifold and the exhaust manifold. The air flow through the cylinder during the overlap period has a valuable cooling effect; it helps to increase the volumetric efficiency and to ensure a low cycle temperature. Also, the relatively cooler exhaust allows a higher engine output to be obtained before the exhaust temperature imposes a limitation on the satisfactory operation of the turbine blades.

In two-stroke engines the exhaust/scavenge overlap is necessarily limited by the engine design characteristics. In Figure 7.4 a comparison of the exhaust and scavenge events for poppet valve engines and opposed piston engines is given. In the poppet valve engine the camshaft lost motion coupling enables the exhaust pre-opening angle to be 52° ahead and astern. In the opposed piston engine the exhaust pre-opening angle is only 34° ahead and 20° astern. Against this, however, the rate of port opening in opposed-piston engines is quicker than in

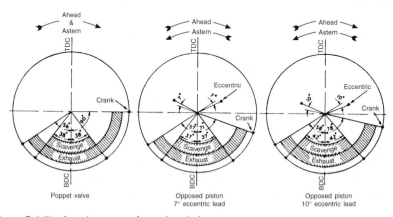

Figure 7.4 Single-acting two-stroke engine timing

poppet valve engines. It should be noted, however, that opposed-piston low speed engines are no longer in production and poppet valves are used in the majority of new designs.

In the four-stroke engine the substantial increase in power per cylinder obtained from turbocharging is achieved without increase of cylinder temperature. In the two-stroke engine the augmented cylinder loading—if any—is without significance.

A strange fact is that the engine exhaust gas raises the blower air to a pressure level greater than the mean pressure of the exhaust gas itself. This is because of the utilization of the kinetic energy of the exhaust gas leaving the cylinder, and the energy of the heat drop as the gas passes through the turbine. In Figure 7.5 the blower pressure gradient A exceeds the turbine pressure gradient G by the amount of the scavenge gradient. The design of the engine exhaust pipe system can have an important influence on the performance of turboblowers.

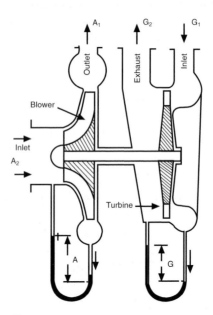

Figure 7.5 Scavenge gradient

The test results of turbocharged engines of both two- and four-stroke types will show there is an increase in temperature of the exhaust gas between the cylinder exhaust branch and the turbine inlet branch, the rise being sometimes as much as 95°C. The reason for this apparent anomaly is that the kinetic energy of the hot gas leaving the cylinder is converted, in part, into additional heat energy as it adiabatically compresses the column of gas ahead of it until, at the turbine inlet, the temperature exceeds that at the cylinder branch. At the turbine some of the heat energy is converted into horsepower, lowering the gas temperature somewhat, with the gas passing out of the turbine to atmosphere or to a waste heat recovery boiler for further conversion to energy. Though the last thing to emerge from the exhaust parts or valves is a slug of cold scavenge air which can have a cooling effect on the recording thermometer, adiabatic compression can still be accepted as the chief cause of the temperature rise.

MATCHING OF TURBOBLOWERS

The correct matching of a turboblower to an engine is extremely important. With correct matching, the engine operating point should be close to optimum efficiency, as shown by the blower characteristic curve (Figure 7.6). In this diagram the ordinates indicate pressure ratio; the abscissae show blower capacity.

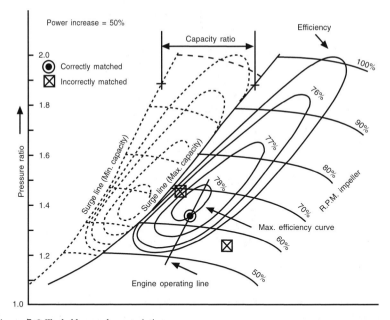

Figure 7.6 Turboblower characteristics

During the initial shop trials of a new engine, with the turbocharger recommended by the manufacturer, test data are recorded and analysed.

If the turboblower is correctly matched, nothing more needs to be done. Should the matching be incorrect, however, the turboblower will supply charge air at either too low or too high a pressure, or surging may occur at the blower. Mis-matching can usually be corrected by a change of turbine capacity and/or blower diffuser.

Compressors can be designed which maintain a high efficiency at a constant pressure ratio over a wide range of air mass flows by providing alternate forms of diffuser for any one design of impeller. This range can be further extended by the use of different impeller designs, each with its own set of diffusers, within a given frame size of turboblower.

BLOWER SURGE

Too low an air-mass flow at a given speed, or pressure ratio, will cause the blower to surge, while too high a mass flow causes the blower to choke, resulting in loss of pressure ratio and efficiency at a given speed. The blower impeller, as it rotates, accelerates the air flow through the impeller, and the air leaves the blower with a velocity that is convertible into a pressure at the diffuser. If, for any reason, the rate of air flow decreases, then its velocity at the blower discharge will also decrease; thus there will come a time when the air pressure that has been generated in the turboblower will fall below the delivery pressure. There will then occur a sudden breakdown of air delivery, followed immediately by a backward wave of air through the blower which will continue until the delivery resistance has decreased sufficiently for air discharge to be resumed. 'Surging' is the periodic breakdown of air delivery.

In the lower speed ranges surging is manifested variously as humming, snorting and howling. If its incidence is limited to spells of short duration it may be harmless and bearable. In the higher speed ranges, however, prolonged surging may cause damage to the blower, as well as being most annoying to engineroom personnel. Close attention to the surging limit is always necessary in the design and arrangement of blowers.

If one cylinder of a two-stroke engine should stop firing, or is cut out for mechanical reasons, when the engine is running above say 40–50 per cent of engine load, it is possible that the turboblower affected may begin to surge. This is easily recognized from the repeated changing in the pitch of the blower noise. In these circumstances the engine revolutions should be reduced until the surging stops or until firing can be resumed in all cylinders.

TURBOCHARGING SYSTEMS

Four basic turbocharging arrangements are now used on marine diesel engines (Figure 7.7):

CONSTANT PRESSURE METHOD—All the exhaust pipes of the cylinders of an engine end in a large common gas manifold to reduce the pressure pulses to a minimum with a given loading. The turbine can be built with all the gas being admitted through one inlet and therefore a high degree of efficiency is reached. For efficient operation the pressure generated by the compressor should always be slightly higher than that of the exhaust gas after the cylinder.

IMPULSE TURBOCHARGING METHOD—The exhaust gas flows in pulsating form into the pipes leading to the turbine. From there it flows out in a continuous stream. The gas pulses from the separate cylinders are each fed to a corresponding nozzle ring segment of the expansion turbine. By overlapping the opening times of inlet and exhaust valves after the pulse decay, efficient scavenging of the cylinders is possible.

PULSE CONVERTER METHOD—Interactive interference limits the ways in which the normal impulse system can be connected to groups of cylinder exhaust pipes. However, with a pulse converter such cylinder groups can be connected to a common ejector. This prevents return flows and has the effect of smoothing out the separate impulses. It also improves turbine admission, increases efficiency and does not mechanically load the blading as much as normal impulse turbocharging.

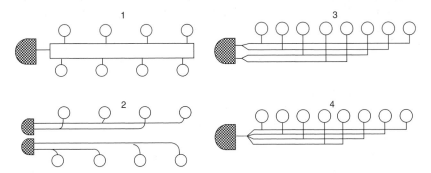

Figure 7.7 Exhaust turbocharging methods: 1. Constant pressure; 2. Impulse; 3. Pulse converter; 4. Multi-pulse system

MULTI-PULSE METHOD—This is a further development of the pulse converter. In this case, a number of exhaust pipes feed into a common pulse converter together with a number of nozzles and a mixing pipe. With this form of construction the pressure waves are fed through with practically no reflection because the turbine nozzle area is larger. The multi-pulse method makes a noticeable performance increase possible, in comparison with the normal impulse turbocharging method.

TURBOCHARGER CONSTRUCTION

An example of turbocharger construction is provided by ABB Turbo Systems' (Brown Boveri) VTR design which remained unchanged for many years (Figure 7.8).

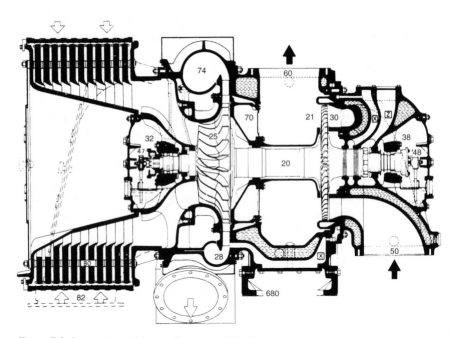

Figure 7.8 Cross-section of Brown Boveri type VTR501 turbocharger

The diesel engine exhaust gases enter through the water-cooled gas inlet casing (50), expand in the nozzle ring (30) and supply energy to the shaft (20) by flowing through the blading (21). The gases exhaust to the open air through the gas outlet casing (60), which is also water cooled, and the exhaust piping. The charge air enters the compressor through an inlet stub (82) or through the silencer filter (80). It is then compressed in the inducer and the impeller (25), flows through the diffuser (28) and is fed to the engine via the pressure stubs on the compressor casing (74).

Air and gas spaces are separated by the heat insulating bulkhead (70). In order to prevent exhaust gases from flowing into the balance channel (2) and the turbine side reservoir, barrier air is fed from the compressor to the turbine rotor labyrinth seal via channel X.

The rotor (20) has easily accessible bearings (32, 38) at both ends, which are supported in the casing with vibration damping spring elements. Either roller or plain bearings are used but for the most common construction using roller bearings a closed loop lubrication system with an oil pump directly driven from the rotor is used (47, 48). These pumps are fitted with oil centrifuges to separate out the dirt in the lubricating oil. The bearing covers are each fitted with an oil filter, an oil drain opening and an oil gauge glass. On models with plain bearings, where the quantity of oil required is large, these are fed from the main engine lubricating oil system.

A key feature of this VTR design was the modular construction to match a wide range of diesel engine types. The separate modules of the turbocharger, as shown in Figure 7.8, are: the silencer filter (80); the air inlet casing (82); the compressor housing (74); the gas outlet casing (60); the outlet casing feet (680); and the gas inlet casing (50). The fixing screws are placed so that the radial position of all the separate casings can be arranged in any position relative to the other.

Compressor wheels for MAN B&W turbochargers are almost exclusively milled from an aluminium forged blank. Cast aluminium alloys were once used as well but superseded as mechanical strength requirements increased. Titanium alloys may also be exploited for applications imposing high air outlet temperatures at high pressure ratios. Radial turbine wheels are purchased as casting blanks of a nickel base alloy. 'Built' rotors are used only for axial turbine wheels (the cast or forged individual blades are retained in the disk by a fir-tree foot). For smaller diameter ranges, progress in casting technology has yielded axial turbine wheel castings incorporating the blades (integral wheels).

TURBOCHARGER PERFORMANCE AND DEVELOPMENTS

A typical pressure–volume curve for a turbocharger is shown in Figure 7.9, with reference to the operating characteristics of a two-stroke diesel engine. The operating curve runs almost parallel to the surge

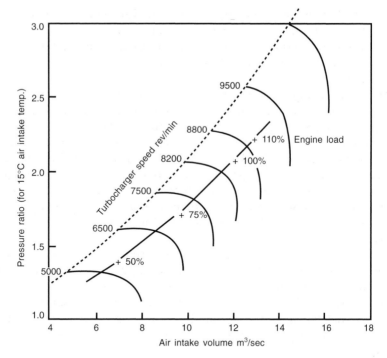

Figure 7.9 Typical pressure–volume curve

line. Every value of the engine output corresponds to a point on the curve, and this point, in turn, corresponds to a particular turbocharger speed which is derived automatically. There is, consequently, no need for any system of turbocharger speed control.

If, for example, at a specific turbocharger speed the charge air pressure is lower than normal the conclusion may be drawn that the compressor is contaminated. By spraying a certain amount of water into the compressor the deposit can be removed, provided it has not already become too hard. A special duct for water injection is arranged on VTR turbochargers.

The operating curve and the behaviour of the turbocharger may vary, depending on whether constant pressure or pulse turbocharging is employed and also on how any volumetric (i.e. mechanically driven) compressors and turbochargers are connected together. By observing changes in behaviour it is usually possible to deduce causes and prescribe the measures necessary to rectify the problems.

Turbocharger design and charging system refinements continue to underwrite advances in diesel engine fuel economy and specific output. Turbocharger developments also contribute to improved engine reliability (lower gas and component temperatures) and operational flexibility (better part-load performance, for example). Rising engine mean effective pressure ratings call for an increase in charge air pressures, with turbocharger efficiencies remaining at least at the same level.

While satisfying increasing demands for higher efficiencies and charge air pressures, turbocharger designers must also address the requirement for compact configurations, reliability and minimal maintenance: qualities fostered by designs with fewer components. All servicing and overhaul work should ideally match the intervals allowed for the major engine components. In addition, safe engine operation dictates a maximum width of the compressor map in order to avoid compressor surging (liable to occur due to changes in operating conditions influenced by hull fouling which has the same effect as speed reduced at constant torque).

Demand for higher charge air pressures necessitates higher circumferential velocities of the compressor wheel and raises the sound level; this in turn has promoted the development of more effective intake silencers for turbochargers. Turbocharger development also seeks to raise the specific flow rate for a given frame size to achieve as compact a unit as possible for cost efficiency, space saving and ease of installation.

Turbocharger designers have pursued higher pressure ratios, overall thermodynamic efficiencies and specific volumetric flow rates, as well as lower noise levels, from more simple and compact modular configurations with uncooled casings, inboard plain bearings lubricated directly from the engine oil circuit and significantly fewer parts than earlier generations. Refinements have sought easier servicing and enhanced reliability and durability.

TURBOCHARGING AND EMISSIONS

A high turbocharger efficiency contributes to reduced carbon dioxide emissions by improving the engine efficiency. The same turbocharger, combined with a 'smart' turbocharging system able to guarantee an optimum air/fuel ratio under all conditions, would contribute even more to the control of soot emissions.

The role turbocharging can play in reducing NOx emissions is not easily recognised, explains ABB Turbo Systems. This is because NOx is produced during combustion at very high temperatures—something which can hardly be influenced by changes in the turbocharging system since the flame temperature depends on local conditions in the cylinder and not on the global mean air/fuel ratio. Nevertheless, an improvement can be achieved through a joint development programme involving both the turbocharging system and the engine, aimed at reducing the temperatures of the working cycle in the cylinders.

The first idea was turbocooling, in which the charge air is cooled in a process that makes use of a special turbocharger. If the pre-compressed air is further compressed in a second-stage compressor, then cooled and expanded through a turbine, very low temperatures can be obtained at the cylinder inlet. First evaluations revealed that the available turbocharger efficiencies for this process were not high enough for reasonable engine efficiencies, ABB Turbo Systems reports.

The Miller cycle promises much better results. The idea is similar to that on which turbocooling is based. The charge air is compressed to a pressure higher than that needed for the engine cycle, but filling of the cylinders is reduced by suitable timing of the inlet valve. Thus, the expansion of the air and the consequent cooling take place in the cylinders and not in a turbine. The Miller cycle was initially used to increase the power density of some engines (see Niigata engines). Reducing the temperature of the charge allows the power of a given engine to be increased without making any major changes to the cylinder unit. When the temperature is lower at the beginning of the cycle the air density is increased without a change in pressure (the mechanical limit of the engine is shifted to a higher power). At the same time, the thermal load limit shifts due to the lower mean temperatures of the cycle.

Promising results were obtained on an engine in which the Miller cycle was used to reduce the cycle temperatures at constant power for a reduction in NOx formation during combustion: a 10 per cent reduction at full load was achieved, while fuel consumption was improved by around 1 per cent. This was mainly due to the fact that with the Miller cycle—at the same cylinder pressure level—the heat losses are reduced due to the air/fuel ratio being slightly higher, and the temperatures lower.

The significantly higher boost pressure for the same engine output with the same air/fuel ratio has mitigated against widespread use of the Miller cycle. If the maximum achievable boost pressure by the turbocharger is too low in relation to the desired mean effective pressure for the engine, the Miller cycle results in a significant derating. Even when the achievable boost pressure is high enough, the fuel consumption benefits of the Miller cycle are partially neutralized when the efficiency of the compressor and turbine of the turbocharger decreases too rapidly at high compression ratios. The practical implementation of the Miller cycle requires a turbocharger capable of achieving very high compressor pressure ratios, and offering excellent efficiency at high pressure ratios.

See Niigata engines for further information on the Miller cycle.

TURBOCHARGER DESIGNERS

ABB Turbo Systems

Switzerland-based ABB Turbo Systems, through Brown Boveri (BBC), has strong links with the pioneering days of turbocharger development by Alfred Büchi and marine engine applications of turbocharging. Büchi applied for his first patent in 1905 but it was not until after World War 1 (during which the first turbochargers were applied to aero engines) that turbochargers entered seagoing service. The 1925-commissioned liners *Preussen* and *Hansestadt Danzig* were the first ships with exhaust gas turbocharged machinery, the output of the MAN 10-cylinder four-stroke engines being raised from 1250 kW (naturally aspirated) to 1840 kW by BBC turbochargers.

BBC's launch of the VTR..0 series in 1944 marked the end of an era in which each turbocharger was custom built for the application and made available a range of volume-produced standard turbochargers for serving engines with power outputs from 370 kW to 14 700 kW. Until then, turbochargers had been applied commercially to four-stroke engines but it now became possible to turbocharge two-stroke engines equipped with engine-driven scavenging air pumps. In order to eliminate the scavenging pumps and reduce fuel consumption, however, the turbocharging system had to be a pulse system. This feature was successfully introduced in 1952 on a B&W engine powering the Danish tanker *Dorthe Maersk*, the first ship to be equipped with a turbocharged two-stroke engine. The two VTR 630 turbochargers boosted engine output by some 35 per cent to 5520 kW.

The use of VTR-type turbochargers on two- and four-stroke engines proliferated rapidly from the 1950s, the series subsequently benefiting from continual design refinements to raise overall efficiencies and compressor pressure ratios (Figure 7.10). A turbocharger efficiency factor of 74.7 per cent was achieved by a VTR 714E model in 1989 (contrasting with the 50–55 per cent efficiency of BBC turbochargers in the early 1950s); and a VTR 304P model attained a pressure ratio of 5:1 in 1991. Apart from performance improvements, R&D has sought enhanced capability to serve engines burning low grade fuels and

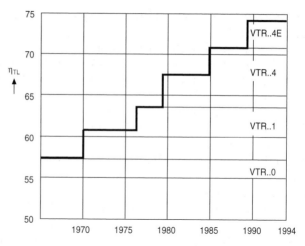

Figure 7.10 The total efficiency of successive ABB VTR turbocharger generations has increased significantly

operating under severe loading conditions, with reliability and minimal maintenance.

ABB Turbo Systems' current programme embraces the following series:

- VTR series: designed for operation with engines delivering from around 700 kW up to 18 500 kW per turbocharger.
- VTC series: particularly favoured where compactness is required, serving engines with outputs from 1000 kW to 3200 kW per turbocharger.
- RR series: mainly applied to high speed engines and smaller medium speed engines with ratings between 500 kW and 1800 kW per turbocharger.
- TPS series: a new generation small turbocharger, detailed in the next section.
- TPL series: a new generation large turbocharger, detailed in subsequent sections.
- NTC power turbine: exploits superfluous exhaust gas energy diverted from the manifold to raise the fuel efficiency of the engine (see *Turbo Compound Systems* below).

VTR..4 series turbochargers are fitted with single-stage axial turbines and radial compressors. The combination of various turbine blade heights with a large number of guide vane variants (nozzle rings)—as well as different compressor wheel widths and a variety of corresponding diffusers—fosters optimum matching of engines and turbocharger operating characteristics.

External bearings result in lower bearing forces and allow the use of self-lubricating anti-friction bearings. Sleeve bearings with external lubrication (engine lubrication) are available as an alternative for larger turbocharger types. Any imbalance of the rotor shaft caused by contamination or foreign particles is absorbed by damping spring assemblies in the bearings, the lower bearing forces associated with external bearings being an advantage in this context. External bearings are also accessible for servicing and can be removed without dismantling the compressor wheel; and the rotor shafts can be removed without dismantling the gas pipes.

ABB Turbo Systems recommends the use of self-lubricating anti-friction bearings whose reduced friction promotes a higher mechanical efficiency and other advantages when starting up the engine and during manoeuvring. Another benefit cited is the ability of this type of bearing to withstand short term oil failure. Oil centrifuges in the closed lubrication system of the anti-friction bearing separate out any dirt particles, and ensure constant lubrication with pure lube oil, even in an inclined position. The turbine oil used has a positive effect on service life. Sleeve bearings with external lubrication are available but these normally involve substantially higher costs, particularly for large turbochargers. Separate oil filtering systems and a standby tank for emergency lubrication must be provided.

The gas inlet, gas outlet and compressor casings of the VTR turbocharger are split vertically and bolted together. The three casings and the supports can be rotated with respect to one another in increments of 15 or 30 degrees to give the engine designer freedom in mounting the turbocharger to the engine. A large selection of gas inlet casing variants is available with different numbers and arrangements of inlets. Flexibility in the matching of the turbocharger to various charging systems, engine types (vee or in-line) and numbers of cylinders is thereby fostered.

Uncooled casings are available as an alternative to the water-cooled casing on VTR..4 turbochargers for applications where heat recovery is paramount (Figure 7.11). All the gas ducts are entirely uncooled and not in contact with cooling water at any point, making the greatest possible amount of exhaust heat available for further exploitation (steam generation for electricity production or ship's services, for example).

Functional reliability is fostered by arranging for the bearing housing on the turbine end to be cooled with a small amount of water, helping to keep the lubricating oil temperature low. The simultaneous cooling

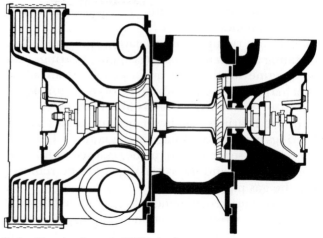

Section through water-cooled VTR turbocharger

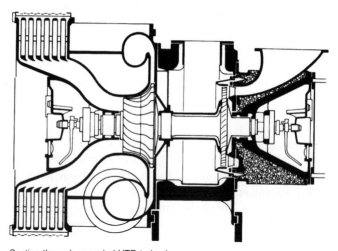

Section through uncooled VTR turbocharger

Figure 7.11 Water-cooled and uncooled ABB VTR turbochargers

of the jacket of the gas outlet casing makes it possible for the temperature
of the entire surface of the casing to be kept within the limits stipulated
by classification societies for the prevention of fire and protection
against accidental contact.

Air to be compressed is drawn in either through air suction branches
or filter silencers. An integral part of every silencer is a filter which

intercepts coarse dirt particles from the intake air and counteracts compressor contamination. The filter can be cleaned during operation without having to dismantle the silencer. The compressor noise is reduced in line with international regulations by felt-covered, shaped plates.

Addressing the demands of low speed two-stroke engine designers, ABB Turbo Systems introduced the VTR..4D turbocharger series with a compressor pressure ratio of up to 4. Three model sizes (VTR 454D, VTR 564D and VTR 714D) cover engine outputs from 5000 kW to 70 000 kW. Efficiency levels for contemporary low speed engines can be met by the established VTR..4 and VTR..4E turbochargers (Figures 7.12 and 7.13). The significantly higher efficiency of the VTR..4E series allows surplus exhaust energy to be exploited in a power gas turbine connected mechanically to the engine crankshaft or a generator (see the section on Turbo Compound Systems below). But neither series is able to meet fully the demands of low speed engines running with mean effective pressures of up to 20 bar, requiring compressor pressure ratios of 3.9 at full load and turbocharger efficiencies of 67 per cent and higher. The VTR..4 does not reach the required efficiency level, and the aluminium alloy compressor wheel of the VTR..4E limits the achievable pressure ratio. The VTR..P turbocharger, developed for four-stroke engines (see section below), does not represent a solution either, although it offers an efficiency improvement at higher pressure ratios compared with the VTR..4.

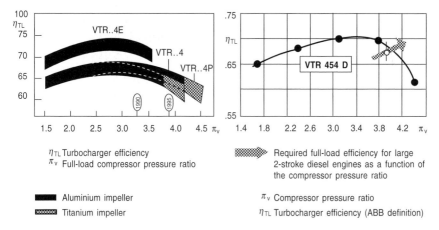

η_{TL} Turbocharger efficiency
π_V Full-load compressor pressure ratio

Required full-load efficiency for large 2-stroke diesel engines as a function of the compressor pressure ratio

■■■ Aluminium impeller
▨▨▨ Titanium impeller

π_V Compressor pressure ratio
η_{TL} Turbocharger efficiency (ABB definition)

Figure 7.12 (a) Turbocharger efficiency as a function of the compressor pressure ratio for today's ABB turbochargers (valid for applications on large two-stroke diesel engines). (b) Turbocharger efficiency as a function of the compressor pressure ratio for the VTR 454D along a typical two-stroke diesel engine operating line

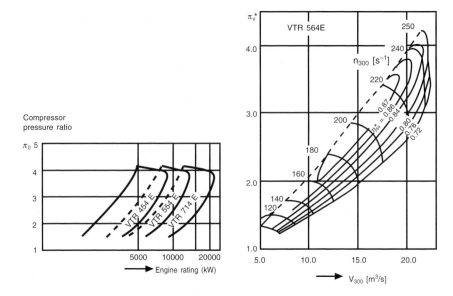

Figure 7.13 (a) Range of application of VTR..4E turbochargers and (b) the performance map of a VTR 564E model

Development goals for the VTR..4D series were defined as: a compressor pressure ratio of 4 while improving efficiency; the highest level of reliability despite an increase in mechanical load; and the retention of key elements of the proven VTR design and its external dimensions, and hence facility for full interchangeability.

A turbocharger combining the pressure ratio of the VTR..4/4P and the high efficiency of the VTR..4E was targeted. The thermodynamic potential of existing compressors and turbine designs led to a strategy for developing a new series optimally blending well-proven components and matching the fitting dimensions of the VTR..4E. The high pressure compressor of the VTR..4P turbocharger was selected for thermodynamic and economic reasons, the designer citing its:

- High efficiency level for the whole range of competitive volume flow rate.
- Greater margin for further increases in mean effective pressure and overload margin.
- Use of aluminium as the impeller material instead of the significantly more expensive titanium; a full load pressure ratio of 4.2 is underwritten.

Two compressors of different diameters are available to cover a competitive volume range at the required high efficiency level within the same overall turbocharger envelope.

The turbine of the VTR..4E turbocharger was chosen for the VTR..4D series since it offers efficiency advances of up to around 3.5 points over the VTR..4/4P turbine. This is due to better recovery of kinetic energy at the turbine wheel outlet, resulting from an optimally adapted outlet diffuser in combination with a new enlarged exhaust gas casing. The latest anti-friction bearing technology was tapped to benefit the VTR..4D series, particularly on the turbine side. Improved mechanical reliability is promised from a fully machined steel cage with surface treatment, an optimized oil feed for better lubrication, optimized smaller rollers for less slip and a larger cage guiding area.

Like other ABB turbochargers, the VTR..4D is designed to ensure that in the worst case scenario—bursting of the impeller or turbine— all the parts remain within the casings. Casing material was changed from grey (laminar) cast iron to ductile (nodular) cast iron and the turbine enclosed in a protective shroud. Burst tests on the compressor and turbine confirmed the containment capability.

ABB Turbo Systems' VTR..4P series addresses the needs of enginebuilders seeking to raise the output of medium speed designs. Such engines may run with meps of up to 24–25 bar, requiring a turbocharger compressor ratio of 3.8–4.0 and an overall turbocharger efficiency exceeding 60 per cent. Meps of up to 28 bar, required by some four-stroke engine designs, dictate a compressor ratio of around 4.2–4.5 while maintaining or even improving the overall turbocharger efficiency. The VTR..4P series offers ratios up to 5:1 from a single-stage turbocharging system whose simplicity is favoured over more complex two-stage arrangements.

Advanced flow calculations, stress analyses, CAD/CAM methods and laboratory tests resulted in a one-piece compressor wheel with backward-curving main and splitter blades discharging through an airfoil diffuser. An aluminium alloy impeller will suit most needs but titanium can be specified for extreme applications. The turbine disc is of a high grade nickel alloy, with new configurations applied for the blade roots and damping wires.

The VTR..4P programme was evolved to cover the demands of engines with outputs ranging from around 1000 kW to 10 000 kW. The high performance units are designed to be physically interchangeable with VTR..4 models for convenient retrofitting as well as for application to new engines. The turbocharger can be flanged directly onto existing air and exhaust gas ducting to improve engine performance by increasing the boost pressure. The pressure-optimized series complemented the established efficiency-optimized VTR..4E series in the ABB Turbo Systems' portfolio.

TPS series

Advances in small medium speed and large high speed diesel engine technology called for more compact turbochargers with increased pressure ratios at higher overall efficiency levels. ABB Turbo Systems addressed the market in the mid-1990s with the TPS series, designed to serve four-stroke engines with outputs from 500 kW to 3200 kW per turbocharger (Figure 7.14).

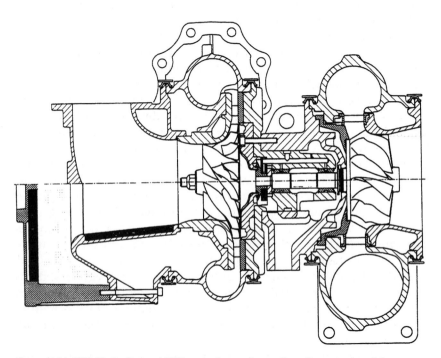

Figure 7.14 ABB Turbo Systems' TPS turbocharger for small medium speed and large high speed engines

To ensure the full range of pressure ratios required by such engines, the TPS..D/E turbocharger is available with two completely different compressor stages. One stage is designed for pressure ratios up to 4.2 (TPS..D) and the other for pressure ratios up to 4.7 (TPS..E). Thanks to mechanically optimized single-piece designs incorporating bore-less compressor wheels, pressure ratios up to 4.5 can be reached for a wide range of applications using aluminium alloy as the compressor material. Both compressors feature a single-piece aluminium alloy wheel with splitter-bladed impeller and backswept blades for high efficiency and a wide compressor map. Peak efficiencies of more than 84 per cent are obtainable.

Experience with the RR-type turbocharger was tapped by the designers in creating the TPS series. The resulting configuration is modular and offers a high level of flexibility with a minimum of sub-assemblies. A significantly reduced number of parts is used in comparison with the RR type, itself a simple design. A modular configuration fosters the easy integration or fitting of options, such as a jet assist system. Special attention was paid during design and development to ease of maintenance and simple mounting of the turbocharger on the engine.

Criteria for turbine development included the ability to operate at gas temperatures of up to 750°C and yield a long lifetime on heavy fuel operation and under extreme load conditions. The requirement for a large volumetric range had to be met with a minimum number of parts. Considerable aerodynamic potential was offered by the turbine of the RR..1 turbocharger which had also demonstrated excellent reliability and the highest efficiencies. Operation at pressure ratios far above today's levels was possible with the turbine, for which a nozzle ring was developed to satisfy the goals for the TPS turbocharger. A new scroll casing and nozzle ring were designed using proven inverse calculation methods that ensure losses are kept very low and turbine blading is not too strongly excited. Thermodynamic measurements showed that the turbine efficiency of the new design was as high as that of its predecessor. Blade vibration excitations are even lower.

Previously, stress distribution for the turbine wheel was optimized by trial and error: a geometry was defined and the resulting stresses calculated, the geometry then being varied until the stresses lay at an acceptable level. Using the latest computer-aided tools for optimizing structures, the stress level was reduced by around 30 per cent over the earlier version.

Turbocharger bearings must deliver a good load-carrying capability with respect to both static and dynamic forces, good stability and minimal mechanical losses from a cost-effective design fostering modest maintenance. The TPS turbocharger bearings are lubricated directly from the engine lube oil system.

An increase in turbocharger pressure ratio means a rise in the load exerted on the thrust bearing. A turbine wheel with a high back wall was chosen to counteract the main thrust of the compressor as far as possible. Extensive calculations and temperature measurements were performed on the thrust bearing, both in steady state operation and with the compressor in surge mode, to optimize its dimensions and to achieve the lowest possible bearing losses without compromising on load-carrying capability.

The key design factor for the radial plain bearings is their stability characteristic. Considered in terms of this criterion, ABB Turbo Systems notes, a squeezed film damped multi-lobe plain bearing would be the best choice. But such a bearing has higher losses than a bearing with freely rotating bushes. The latter, on the other hand, exhibits an inadequate stability characteristic. A new bearing design with rotating hydraulically braked multi-lobe bearings was therefore developed to blend the respective merits of the two types.

By providing a tangential lube oil supply it is possible to reduce the speed at which the floating bushes rotate by around 20 per cent compared with freely rotating bushes. This measure, the designer says, has ensured a stability which is effectively as good as with the squeezed film damped multi-lobe plain bearings; and mechanical losses can be reduced by more than 20 per cent.

Development of the TPS..-F33 series, introduced in 2001, focused on increasing the pressure ratio (up to 4.7 with a compressor wheel made of aluminium alloy) and the flow capacity, while retaining high efficiency, reliability, long lifetime and ease of maintenance (Figure 7.15).

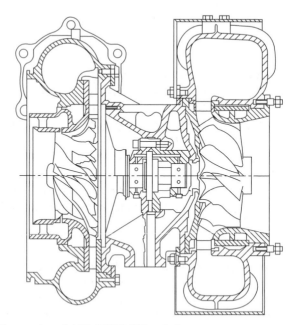

Figure 7.15 Cross-section of ABB TPS57-F33 turbocharger

The new series comprises four different sizes, with the same outline dimensions as the established TPS..D/E range and hence representing an increased power density. The interchangeability offers good solutions for upgrading engines. Among the options are variable turbine nozzle geometry and an air recirculation system (a flow recirculating device in the compressor inlet that substantially enhances the surge margin and thereby broadens the compressor map). The recirculating flow is driven by pressure differences between the downstream and the upstream slots in the compressor casing.

The rise in brake mean effective pressure ratings of engines in recent years has dictated a considerable increase in the power density of turbocharger rotors. Burst tests are performed to ensure that even in the worst case scenario (for example, a fire in the exhaust system)—in which the compressor or turbine wheel can burst—all parts remain within the turbocharger casings and do not endanger personnel. The turbine casing design, which includes inner and outer burst protection rings, seeks optimum containment.

TPL series

A completely new generation turbocharger from ABB Turbo Systems—the TPL series—was launched in 1996 to meet the demands of the highest output low and medium speed engines, the designer having set the following development goals:

- High compressor pressure ratio for increased engine power output.
- High turbocharger efficiency levels for reduced engine fuel consumption and lower exhaust gas temperatures.
- High specific flow capacity, resulting in a compact design and low weight.
- High reliability, long lifetime and extended periods between overhauls.
- Easy maintenance, even for machines operated under adverse conditions (such as those serving engines burning low quality heavy fuel oils).

The uncooled design features an overall number of components reduced by a factor of six compared with the VTR series, contributing to extended times between overhauls (see Figure 7.16 a, b, c).

A modular configuration provides the TPL series with the flexibility to meet the turbocharging requirements of two-stroke and four-stroke engines. Two-stroke engines with outputs from 5000 kW to 25 000 kW

Figure 7.16 (a) Cutaway of ABB TPL65 turbocharger

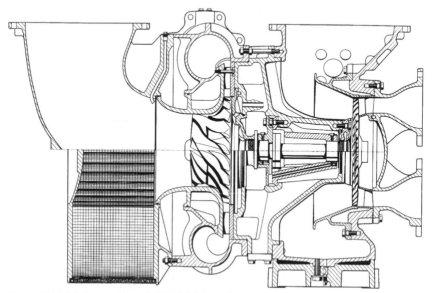

Figure 7.16 (b) Cross-section of ABB TPL65 turbocharger

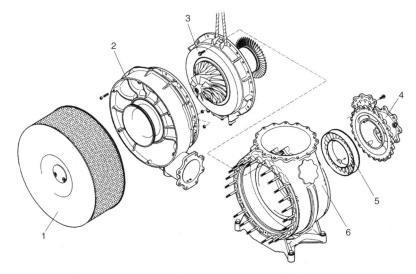

Figure 7.16 (c) Modular design concept of ABB TPL turbocharger: 1. Filter-silencer; 2. Air volute; 3. Rotor block; 4. Gas inlet casing; 5. Nozzle ring; 6. Gas outlet casing

per turbocharger are served by the TPL-B variant, introduced in 1999, while the original TPL-A model addresses demands from four-stroke engines delivering 1250 kW to 18 000 kW per turbocharger.

Despite differences in their thermodynamics and design, the TPL-A and TPL-B series share established features: for example, the basic modular TPL concept and the central bearing housing with the complete bearing assembly. All TPL turbochargers consist of a central cartridge unit with the rotor assembly. The casings on the compressor side and on the turbine side are connected by flanges; and all casings are uncooled. Complete dismantling of the turbocharger, including the nozzle ring, is possible from the 'cold side', leaving the exhaust pipes in place and untouched.

All TPL turbochargers feature the same bearing technology, with plain bearings selected to secure a long lifetime. The non-rotating, radial bearing bushes are suspended in a squeeze film damper, which increases the stability of the rotor while reducing the dynamic forces and hence the load on the radial bearings. The main thrust bearing comprises a free-floating disk between the rotating shaft and the stationary casing. This bearing concept halves the speed gradient in the axial gap so that the losses and the risk of wear are significantly reduced; thanks to this technology, the thrust bearing is also very tolerant to inclinations of the rotor.

Since two-stroke engines usually have a lubrication system driven by electric pumps rather than by the engine itself, an emergency oil supply has to be provided to ensure that the turbocharger runs out safely in the event of a black-out. The TPL-B turbocharger can be optionally equipped with an integrated emergency oil tank mounted on top of the bearing casing. Since this system is based on gravity it does not require any auxiliary support.

Thermodynamic improvements from the TPL-B series derive from a new axial turbine family (the TV10) which was especially developed for turbocharging systems for two-stroke engines. Five turbine wheels and over 20 nozzle rings can be specified for each TPL-B model, these options allowing the turbocharger characteristic to be adapted to the requirements of the application within a wide range.

Two new radial compressor stages were developed for the TPL family. The TPL-B compressor features a single-piece aluminium alloy wheel with splitter-bladed impeller design and backswept blades for high efficiency and a wide compressor map. Peak efficiencies exceeding 87 per cent are reportedly obtainable. Enlarged compressor diameters further increased the volume flow, thereby fostering optimized matching of the turbochargers to the engine application.

The TPL 91B model, which supplemented the TPL 73, 77, 80 and 85B models, was introduced as the world's largest turbocharger, yielding a volume flow rate above 50 m^3/s and a compressor pressure ratio up to 4.5. Co-developed by ABB and its Japanese licensee IHI, this model is primed to boost even more powerful two-stroke engines than those currently on the market.

ABB planned to extend variable turbine geometry to more of its turbocharger designs, the system having been successfully engine-tested with a TPL65 model before entering service.

Napier

UK-based Napier Turbochargers, now part of the Siemens group, offers its Napier 7-series and 8-series turbochargers for application to two- and four-stroke engines developing up to 11 000 kW per turbocharger. Napier 7-series turbochargers are suitable for engines requiring a single turbocharger installation from 1500 kW to 6500 kW. Developed from the company's earlier generation, the 297 and 357 models deliver a 5:1 pressure ratio and an overall efficiency exceeding 70 per cent.

Development of the 8-series was stimulated by the need to offer turbochargers capable of yielding pressure ratios in excess of 5:1 and serving engines with outputs up to 11 000 kW. The new axial-turbine series (Figure 7.17) was headed by the NA298 and NA358 models with a turbocharger efficiency exceeding 70 per cent and a 5.5 pressure ratio promoting lower engine fuel consumption and thermal loading. Enhanced levels of component durability are also claimed, while the air-cooled design contributes to a reduced overall weight. Inboard bearings were applied for the first time on a Napier turbocharger of this size, the hydrodynamic units reportedly securing stable rotor dynamic performance and underwriting a long life. The new bearing arrangement facilitates the use of axial and radial exhaust gas flow into the turbocharger, extending the options for the enginebuilder.

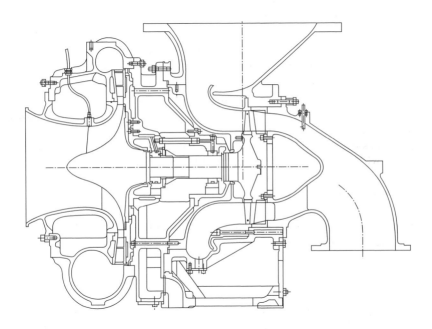

Figure 7.17 Cross-section of Napier 458 turbocharger for medium speed engines

Two separate compressor wheel designs are offered to provide differing characteristics for optimized efficiency in diverse applications. Other features include a circular exhaust flange (eliminating the need for transition pieces) and a mounting flange incorporating both the lube oil inlet and outlet. Another concept pioneered by Napier—

cartridge construction—benefits serviceability of the 7-series and 8-series, this configuration allowing operators to remove and repair the rotor and bearings through a simple procedure without disturbing the exhaust connections. Good access is also provided to key functional areas, such as the nozzle ring.

Titanium compressors, offering enhanced aerodynamic potential and mechanical properties, were initially planned to replace the aluminium components but design refinements enabled the required performance to be attained with aluminium.

Smaller engine requirements are addressed by the Napier 047, 057 and 067 all-radial air-cooled turbochargers in the 7-series, which are suitable for engines developing outputs from 500 kW to around 1700 kW and yield pressure ratios up to 5. Typical applications include the Wärtsilä 20 engine and other *circa*-200 mm bore genset engines. The 047 model features inboard bearings and is uncooled. In common with other turbochargers in the series, the bearings are designed to run on the engine's lube oil system and to yield an extremely long life, even when operating on contaminated lube oil. The 047 model incorporates 50 per cent fewer parts than its predecessor, the Napier CO45. Cost savings result from using a reduced number of components; and maintenance is simplified, with no special tooling required or clearances to set. Simple clamps are used to hold the casings together and provide total indexability of connecting flange positions.

Earlier axial-flow turbine Napier 457 and 557 models, which serve engines from 4000 kW to 10 000 kW output, featured long-life outboard plain hydrodynamic bearings fed directly from the engine lube oil system, and water-cooled casings.

High efficiency at all pressure ratios was sought from the 5- and 7-series through advanced compressor and turbine aerodynamic designs. The compressor stage features a one-piece aluminium alloy wheel manufactured on five-axes milling machines (Figure 7.18). Vanes with sweepback and rake are specified to secure high performance, and a divergent circular arc (DCA) diffuser design yields maximum pressure recovery from a compact configuration. A high efficiency fabricated nozzle with profiled blades serves the turbine section of the small and large axial-flow turbochargers.

Napier turbochargers exploit the HF bearing design to secure a long service life, even when operating with the poorest quality engine lubricating oil: a bearing life exceeding 20 000 hours is reportedly common.

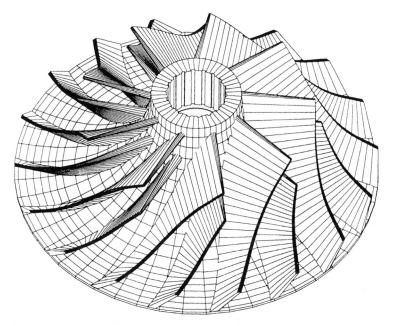

Figure 7.18 A 3D CAD visualization of the Napier 457 turbocharger impeller, showing the one-piece intervaned swept-back design with tip rake

Napier turbocharger series performance data

	7-series	8-series
Pressure ratio capability	5:1	5.5:1
Efficiency @ 5:1	65%	70%
Design point pressure ratio	4:1	4.5:1
Design point efficiency	70%	75%
Efficiency @ 2.5:1	65%	70%

MAN B&W

A full range of turbochargers for two-stroke and four-stroke engines is offered by MAN B&W Diesel of Germany, the current programme embracing the NR/R series, the NR/S series, the NA/S/T9 series, and the new generation TCR and TCA series. In addition, the PT/PTG power turbine series is available for turbo compound systems. The NA/S and NR/S series were progressively introduced to the market in the1990s, their uncooled casings and inboard plain bearing features subsequently being adopted by other turbocharger manufacturers.

The radial-flow turbine NR/S series was designed for: safe control of the charge air pressure ratio up to 4.5; adequate extension of the application range; a higher efficiency level; improved reliability in all operating ranges; unrestricted heavy fuel capability; and reduced maintenance demands. Seven models—the NR12/S, NR14/S, NR17/S, NR20/S, NR24/S, NR29/S and NR34/S—form the programme which overlaps with the smallest of the established axial-flow turbine NA/S series. The totally water-free series (Figure 7.19) was a further development of the established NR/R turbocharger, featuring a revised compressor and turbine wheel. MAN B&W's proven system of internal plain bearings was retained, the axial thrust now being absorbed by a separate thrust bearing. The five models in the radial-flow turbine NR/R series serve engine outputs from around 400 kW to 4400 kW per turbocharger.

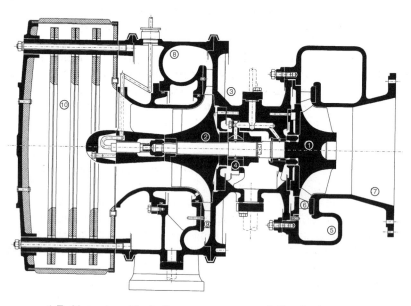

1 Turbine rotor with shaft	6 Nozzle ring
2 Compressor wheel	7 Outlet diffusor
3 Bearing casing	8 Compressor casing
4 Bearing bush	9 Diffusor
5 Gas inlet casing	10 Silencer

Figure 7.19 MAN B&W NR/S radial-flow turbocharger

The development targets for MAN B&W's axial-flow turbine NA/S and NA/T9 series turbochargers listed the following performance criteria:

- Medium speed four-stroke engines: a compressor ratio in excess of 4; and a turbocharger efficiency in excess of 60 per cent.
- Low speed two-stroke engines: a compressor ratio in excess of 3.6; a turbocharger efficiency of at least 64 per cent (without an associated power turbine system); and a turbocharger efficiency of at least 69 per cent (with a power turbine system).
- A pressure ratio of up to 4.5 for the NA/S and 4 for the NA/T9.

MAN B&W carried out extensive theoretical and experimental investigations on the key individual turbocharger components: the compressor, turbine and bearings. Apart from higher efficiencies from the turbine and compressor, the goals included the attainment of high vibratory stability for the bladings and a smooth running performance at low bearing temperatures.

A one-piece radial compressor wheel with continuously backswept blades was selected to secure an uprated pressure ratio and flow rate, and improved vibratory stability. The wheel carries ten main blades and ten splitter blades. The introduction of splitter blades reportedly yielded a marked increase in efficiency for pressure ratios greater than 3.5; and an adequate safety margin against surging in the part-load range was maintained. Various coatings, such as Keplacoat and Alumite, have been investigated to protect the compressor wheels against corrosive attack undermining the fatigue strength under reversed bending stresses.

A few percentage points improvement in efficiency was also gained by the development of a new turbine blade and by optimizing the outlet diffuser. The investment-cast turbine blade offers a significant cost reduction over a forged blade, while retaining the material properties and the fatigue strength under reversed bending stresses. The resistance of the blading to vibrations and 'exceptionally smooth' running at pressure ratios up to 4.5 were promoted by the incorporation of a floating bush in the turbine bearing system.

The S-type turbochargers are characterized by uncooled turbine casings and bearing housings, dispensing with the previous double-walled heavy casings and cooling connections. The water-cooled bearing casing was retained, however, for the NA/T9 turbocharger. Tests with an NA57/T9 turbocharger serving a MAN B&W 6L60MC low speed engine achieved an efficiency of around 67.5 per cent at the 100 per cent load point; the charge air pressure ratio was around 3.6. The maximum efficiency measured at near-70 per cent engine load was around 69 per cent. A high efficiency level over the remaining load range was also logged.

The NA/S series embraces the NA29, NA34, NA40 and NA48 models, and the NA/T9 series the NA57 and NA70 models, together covering engine output requirements from 1650 kW to 24 500 kW. An efficiency of 63 per cent was reached by an NA34/S model serving a MAN B&W 6L40/54 medium speed engine at its full load rating of 4320 kW, corresponding to a charge air pressure ratio of 4.

TCA Series

In 1999/2000 MAN B&W decided to develop a completely new turbocharger generation to replace the NA series, resulting in the launch in 2002 of the TCA (TurboCharger Axial) series with radial-flow compressor wheel and axial-flow turbine blading. The designers pursued the following goals: high specific flow rates; high efficiencies; low noise emissions; ease of maintenance; ease of engine mounting; and high reliability with a long service life.

The TCA series was headed into production by the TCA77 model, Figure 7.20, with additional frame sizes from the TCA33 to the TCA99 planned to follow progressively and cover the requirements of the highest powered two-stroke and four-stroke engines (see table). Most are available in three versions:

- Two-stroke version—pressure ratio up to 4.2, primarily for two-stroke engines.
- Four-stroke version—pressure ratio up to 4.7, for both four-stroke and two-stroke engines.
- High pressure version—pressure ratio up to 5.2, for future four-stroke engines with very high mean effective pressures.

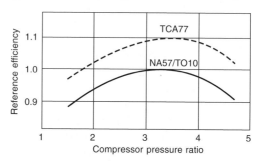

Figure 7.20 Efficiency map of the prototype MAN B&W TCA77 turbocharger plotted against the maximum efficiency of the earlier NA57 design

A modular design allows the turbocharger to cover as broad a range of applications as possible. Air is sucked in either via an intake air silencer, a 90-degree intake casing or an intake manifold; exhaust gas flows into the turbine via either an axial- or radial-flow casing. The design features the separate casing-base concept proven in the NA series. The compressor volute is available with either one or two opposing thrust pieces: the single-piece variant has advantages for two-stroke and four-stroke in-line cylinder engines, while the two-piece volute offers advantages when attaching a single turbocharger to V-type four-stroke engines.

Both the compressor wheel and the turbine can be adjusted to engine requirements by choosing from a range of meridians and blading configurations. Diffusers and nozzle rings that are very finely stepped in their mass-flow areas allow the turbochargers to be fine-tuned to the engine. A nozzle ring capable of adjustment during operation is available optionally for maximum variability.

A minimal number of external connections were sought. On the engine side and in the plant system itself the only additional connections—apart from the air and exhaust gas connections—are an oil inlet and drain pipe, together with a breather connection for the bracket. No cooling water or lock-air connections, additional oil overflow piping or external lube oil supply are required. A special emergency operation facility is integrated into the turbocharger to allow continued operation under emergency conditions. All the hot casing components are equipped with a new temperature- and noise-reducing casing with a sheet metal cover, significantly reducing the risk of engineroom personnel coming into contact with excessively hot surfaces.

In order to limit bearing losses the shaft diameter of the rotating assembly was reduced, paying special attention to rotor dynamics so that, despite the reduction, it was possible to increase rotor stability compared with the NA series turbochargers. The rotating assembly is mounted in the bearing casing on two floating-sleeve radial bearings. Since the floating-sleeve design significantly reduces the rate of shear in the bearing gaps, in comparison with the fixed-sleeve type, both bearing losses and bearing wear are reduced.

The thrust bearing is equipped with a floating centre plate which acts in the same way as the floating-sleeve radial bearings. The thrust bearing is located outside the radial bearing on the compressor side, facilitating removal without having to dismantle the radial bearing and rotating assembly. The fact that the thrust bearing is subject to far greater stresses than radial bearings (because of the axial thrust of the turbine and compressor) makes this a significant feature for ease of maintenance, MAN B&W notes.

Bearing points are supplied with oil by a ring channel arranged in the inside of the bearing casing. Oil feed, either from the right or left side of the turbocharger, lube oil distribution to all the bearing points, and the supply from the emergency lube oil service tank are via this ring channel.

No bearing case is water cooled. The heat input from the compressor and turbine is dissipated in the lube oil flung off the shaft of the rotating assembly. The oil mist thus generated can drop down the walls of the generously-dimensioned interior of the bearing casing, thereby evenly absorbing the heat to be dissipated. The bearing casing has its own air vent, ensuring that the leakage air which the compressor inevitably forces into the casing through the shaft seal of the rotating assembly does not increase crankcase pressure in the engine but dissipates it directly.

All airflow components were optimized with regard to flow configuration and stress reduction using 3D CFD and FEM analyses. The result was a turbine with wide-chord blades arranged in a fir-tree root in the turbine disc. A characteristic feature of such blades is their very high chord-to-height ratio, this creating a compact, very stiff and hard-wearing blade. The turbine blades can be of varying angles and lengths. With the aid of advanced design tools, it is now possible to dispense with lacing wire to dampen exhaust-generated vibrations, even in four-stroke engine applications. Apart from improving the blade profile, this has significantly boosted efficiency, MAN B&W explains.

As in the NA series, the turbine shaft is friction welded to the turbine disc. The disc (which in the case of the TCA77 turbocharger must absorb a centrifugal force per blade equivalent to the weight of a fully loaded truck and trailer) is made from a forged steel alloy. The nozzle rings, with a new blade profile design, and a carefully matched turbine outlet diffuser, also contribute to efficiency gains.

The compressor wheel is operated at circumferential speeds well in excess of 500 m/s, generating considerable centrifugal forces. To withstand these forces, the wheel is made of a high-strength aluminium alloy. For applications where the components will come into contact with corrosive media a special corrosion-resistant coating can be applied to the compressor wheel.

A new design of compressor volute and new nozzle ring designs contribute to optimized turbocharger matching and high efficiency, while a new compressor mounting system was developed to ease removal. The compressor wheel is deformed by the powerful centrifugal forces acting on it; this tends to cause it to expand across its diameter, inevitably provoking a shortening in axial direction and causing the bore to widen. The new design ensures good centring of the wheel on both the inlet side as well as to the labyrinth ring in the plant system itself. A tie bolt compensates for the axial shortening; and a tightening system allows a small torque wrench to produce the required tiebolt tensioning. The excellent centring makes it possible to remove and replace the compressor wheel without the need to rebalance the rotating assembly.

Noise emissions in a turbocharger occur primarily at the compressor wheel. Any measures aimed at reducing noise therefore concentrate on the cold area. Noise is generated by a combination of pressure fluctuation in way of the compressor wheel inlet and the interaction between the wheel and the diffuser. Inlet side noise emission was reduced on the TCA series by optimizing the fluid mechanics of the compressor wheel, seeking the best compromise between high compressor efficiency and noise emission. In addition, the silencer was designed with radially-arranged curve plate elements, dampening noise emissions to a high degree while keeping intake pressure losses low.

In order to reduce noise radiation in the engine's charge air manifold the number of blades in the diffuser was matched to the number of blades in the compressor; furthermore, the compressor volute was provided with a noise-reducing casing. All these measures together significantly reduced the level of noise emissions compared with the NA series, and comfortably achieved the goal of attaining emissions below the 105 dB(A) at 1 m distance level demanded by seafarer associations and classification societies.

In the event of the worst possible form of turbocharger damage—the rupture of a compressor wheel or one or more of turbine blades—steps must be taken to ensure that no fragments can escape from the turbocharger casings and cause injury. All the anti-burst measures used on the TCA series are based on the idea that the forces occurring cannot be absorbed by solid, rigid elements. The strategy instead is to employ flexible elements to transform the kinetic energy generated by the individual ruptured components into friction heat, thereby preventing the casing components from failing as a result of local overload.

The compressor volute, bearing casing and turbine gas outlet casing, for example, are held together by 24 very strong tiebolts. The compressor volute is of very rigid design, while the insert piece directly surrounding the compressor wheel is made flexible by means of long tiebolts. When the compressor wheel bursts, the compressor vanes are crushed against the insert piece. The remaining components of the hub body become jammed between the insert piece and the bearing casing. The insert piece is dragged towards the silencer, using up a major part of the kinetic energy in the component fragments. The fragments then become lodged between the compressor volute and the bearing casing; the remaining kinetic energy in the fragments is completely dissipated.

On the turbine side, a controlled failure point was targeted at the connection between the turbine blade and the disc. Burst protection is therefore easier on the turbine side than on the compressor side since, in the worst-case scenario, one or more turbine blades can fracture when, for example, one of the engine's exhaust valves fractures. The components surrounding the turbine impeller are designed to be able to absorb the kinetic energy of several broken blades.

MAN B&W TCA turbocharger series/Engine output per turbocharger in kW

	Two-stroke engines Ie = 8.5*		Four-stroke engines Ie = 6.5*
TCA33		2800–4300 kW	
TCA44		4100–6200 kW	
TCA55	4400–8200 kW		5800–10 800 kW
TCA66	6200–11 600 kW		8200–15 200 kW
TCA77	8800–16 400 kW		11 500–21 500 kW
TCA88	12 400–23 300 kW		16 300–30 300 kW
TCA99	18 700–31 300 kW		

* Specific air consumption in kg/kWh

TCR Series

Complementing the TCA series, a new programme of radial flow TCR turbochargers is offered in six sizes—from TCR12 to TCR22 models—covering engine outputs from 390 kW to 5000 kW with pressure ratios up to 5.2. The completely new design features a turbine with CFD-optimized rotor blades, nozzle ring and inlet and outlet casings; a variable nozzle ring is available optionally. Also CFD-optimized for increased efficiency are the compressor wheel, diffusor ring and compressor volute, with internal flow recirculation extending the surge margin. The high performance plain bearings are optimized for minimized mechanical losses.

Mitsubishi

Successive generations of MET turbochargers introduced since 1965 by Mitsubishi have featured non-water-cooled gas inlet and outlet casings (Figure 7.21). Demand for higher pressure ratios and greater efficiency was addressed by the Japanese designer with the development of the MET-SD series, an evolution of the MET-SC programme. The main features of the earlier series were retained but the new models exploited a more efficient compressor wheel incorporating 11 full and 11 splitter blades arranged alternately.

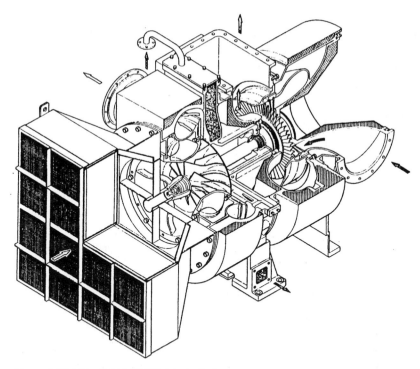

Figure 7.21 Mitsubishi MET-SD type turbocharger

The MET-SD turbocharger shadowed the efficiency of the MET-SC equivalent in the low speed range but yielded 4–5 per cent more in the high speed range when the pressure ratio exceeds 3.5. The air, gas and lube oil pipe installations were unchanged, as were the mounting arrangements. The five models in the MET-SD programme individually cover the turbocharging demands of engines with outputs ranging from 1800 kW to 18 000 kW at a pressure ratio of 3.5. The air flow ratings from the smallest (MET33SD) to the largest model (MET83SD) at that pressure ratio vary from 3.5 to 37.5 m^3/s.

The MET turbocharger rotor is supported by an internal bearing arrangement based on a long-life plain bearing lubricated from the main engine system. The large exhaust gas inlet casing is of two-piece construction comprising a detachable inner casing fitted with the turbine nozzle assembly and a fixed outer casing containing the rotor shaft fitted with the turbine wheel. The inner casing can be readily unbolted and pulled from the outer casing, complete with the nozzle assembly, thus exposing the turbine wheel and nozzle assembly for inspection or servicing. If necessary, the turbine wheel and rotor can be removed from the outer casing.

Mitsubishi targeted pressure ratios of 4 and beyond with the subsequent MET-SE/SEII series, which shares the same main characteristics of earlier MET generations:

- Completely non-cooled (no sulphuric acid corrosion of the casing).
- Inboard bearing arrangement (ease of release).
- Sliding bearings (long life).

In addition, as with previous models, an inner/outer double-walled structure was adopted for the gas inlet casing for ease of release on the turbine side for maintenance checks; and a lubricant header tank is incorporated in the structure for effective lubrication of the turbocharger rotor following emergency shutdown of the engine. MET-SE turbochargers also exploit wide chord-type blades for improved performance in the high pressure ratio region.

The seven models in the MET-SE/SEII programme—embracing 33SE to 90SE types—deliver air flow rates from 3.1 to 53 m³/s at a pressure ratio of 3.6, satisfying engines with outputs ranging from 1300 kW to 24 100 kW per turbocharger, (Figure 7.22.) Total turbine efficiency is enhanced by a compressor impeller featuring 3D profile blading to maximize aerodynamic performance, and turbine wheel blading that requires no damping wire.

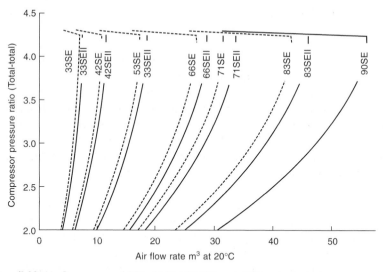

Figure 7.22 Air flow ranges of Mitsubishi MET-SE and SEII series turbochargers

Introduced in 2001 to boost two-stroke engines powering large containerships, the MET90SE design has an air flow range some 30 per cent higher than that of the 83SE (formerly the highest capacity model) but is only 10 per cent larger. The higher capacity from a modest increase in dimensions was achieved by proportionally enlarging the rotor shaft and developing a new compressor impeller wheel and turbine blades. Expanding the capacity of such wheels and blades without compromising reliability and performance is a key challenge for turbocharger designers, Mitsubishi notes.

Different compressor impeller materials—such as titanium and cast steel—have been investigated as an alternative to the forged aluminium traditionally used in MET turbochargers. A pressure ratio of 4 with an aluminium impeller was nevertheless achieved thanks to an impeller air cooling system and a redesigned compressor blade. By reducing the impeller backward curvature angle from 35 degrees to 25 degrees the pressure ratio could be raised without the need to increase the impeller peripheral speed.

Mitsubishi successfully developed a cast stainless steel compressor impeller in preference to an aluminium alloy component for its MET-SH turbocharger. Raising the compressor pressure ratio to the higher levels increasingly demanded by the market dictates attention to compatibility between the impeller strength and design and the aerodynamic design. Mitsubishi cites, respectively, the problems of higher impeller peripheral speed, elevated impeller blade temperature and increased impeller blade surface load; and the problem of optimizing the aerodynamic characteristics of the higher speed air flow in the impeller.

Limitations in resolving all these problems are presented by an impeller of aluminium alloy, a material which Mitsubishi has traditionally used for MET turbochargers because of its high specific strength and excellent machinability. Running the aluminium alloy impeller at temperatures above the ageing treatment level (190–200°C), however, threatens a reduction in material strength and should be avoided. Depending on the suction air temperature, the impeller blade could reach this temperature level when operating at a pressure ratio of around 3.7.

An aluminium alloy forging also experiences progressive creep deformation at temperatures of around 160°C and above, calling for consideration in design to preventing the stress from becoming excessive at impeller blade areas subject to high temperature.

The stainless steel impeller casting for the SH series turbocharger delivered improved high temperature strength and made it possible to raise the maximum working pressure ratio to 4 and higher, without reducing the impeller peripheral speed. Improved material strength allowed the impeller blade to be smaller in thickness than its aluminium alloy counterpart, despite the higher pressure ratio and increased peripheral speed. This feature in turn made it possible to increase the number of blades while ensuring the necessary throat area, hence improving the impeller's aerodynamic characteristics. The stainless steel impeller of the MET42SH turbocharger, the first production model in the MET-SH series, features 15 solid-formed full and splitter blades.

A higher density impeller casting called for special consideration of the rotor stability. The rotor shaft of the SH series turbocharger is larger in diameter than that of the SD series, with the compressor side journal bearing located outside the thrust bearing to increase the distance between the two journal bearings. Tests showed that, despite the higher mass of the stainless steel impeller, the 3rd critical speed for the rotor is around 50 per cent higher than the maximum allowable speed, underwriting rotor security.

A turbocharger total efficiency of 68 per cent was logged by the MET42SH version fitted with impeller blading having a large backward angle and operating at a pressure ratio of 4.3 on a diesel genset.

KBB Turbochargers

A subsidiary of the Ogepar group, the German specialist KBB Turbochargers (formerly Kompressorenbau Bannewitz) offers radial- and axial-flow turbine turbochargers to serve four-stroke and two-stroke engines developing from 300 kW to 8000 kW. The designs are characterized by compactness, inboard plain bearings, and both water-cooled and uncooled bearing housings.

The HPR series embraces four models (HPR 3000, 4000, 5000 and 6000) with radial turbines covering an engine power range from 500 kW to 3000 kW with a maximum pressure ratio of 5 and overall efficiencies ranging from 63 to 68 per cent. The six-model radial-turbine R series addresses engines with outputs from 300 kW to 2800 kW with maximum pressure ratios of 4 or 4.5 and overall efficiencies from 62 to 66 per cent.

An optional jet-assist system available for the HPR and R-series accelerates the charger rotor on the compressor side by injecting compressed air into the area of the impeller for a limited time span. Such a system partly overcomes the disadvantages of single-stage turbocharging in the partial load region, and reportedly yields a clear improvement in the start-up behaviour of the engine.

The axial-turbine M series features four models (M 40, 50, 60 and 70) serving engines with outputs from 900 kW to 8000 kW with a maximum pressure ratio of 4 and overall efficiencies ranging from 65 to 70 per cent (Figure 7.23).

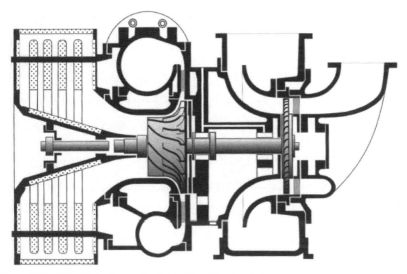

Figure 7.23 KBB Turbochargers' axial-turbine M-series design

Both axial and radial turbine series share some common features, such as: internal sliding journal bearings; lubrication from the engine oil system; a bearing life up to 20 000 hours; uncooled turbine casings; and a variety of intake casings and silencer outlines to facilitate installation on the engine. An option for the R-3 series, variable radial turbine (VRT) geometry, delivers: an extended operating range with high torque at low engine speed; improved acceleration; reduced smoke; and lower fuel consumption.

Turbomeca (Hispano-Suiza)

SNECMA group member Hispano-Suiza of France teamed up with the USA's Allied Signal Aerospace in the 1990s to develop the NGT (New Generation Turbocharger) series for high and medium speed engines in the 1000 kW to 9000 kW power range. In 2002, SNECMA transferred its turbocharger activities to another group member, Turbomeca.

The NGT family embraces four overlapping members: the HS 4800, HS 5800/HS 5800 P, HS 5900 and HS 6800. The following measures of performance are cited for the HS 4800 and 5800: compressor isentropic efficiency up to 86 per cent in usable zones; overall efficiency 73–75 per cent on engine; and continuous pressure ratios of 4.5 (aluminium compressor) and 5 (titanium compressor) (Figure 7.24).

Compactness and light weight resulted from exploiting the internal thrust and journal bearing concept and applying the principle of split

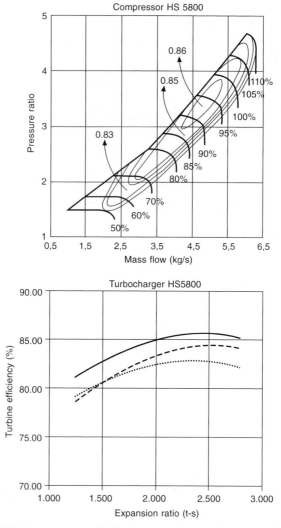

Figure 7.24 Performance maps for Turbomeca HS5800 turbocharger

(radial + axial) diffusion at the compressor impeller outlet. The compressor stage features a high strength forged alloy impeller with swept back blades (single piece), light alloy cast casing, and over 20 combinations of cuts and diffuser stagger angles. A compressor internal recirculation (CIR) option optimizes efficiency at part loads and enhances the surge margin above 20 per cent.

The turbine stage is specified with: chromium nickel alloy for the nozzle rings; high resistant refractory alloy optimized for insensitivity to overheating (an anti-vibration device is fitted for pulse turbocharging systems); 28 combinations of cuts and nozzle ring area; and optimized diffusion exhaust contributing to the achievement of a high stage efficiency. Full compatibility with heavy fuel operation was addressed in the design.

A maintenance-friendly design was sought from the total elimination of water cooling, and an optimized configuration of the internal thermal flows simplifies connector pipe installation and servicing. Completely interchangeable components are also used; all rotor parts, for example, are balanced separately and may be changed on site as required without total dismantling and overall balancing. Among the options offered are an air filter silencer; titanium compressor; jet assistance for instantaneous load increase; speed sensor fitting without dismantling; and a facility for cleaning and washing the turbine and compressor stages during operation.

TURBOCHARGER RETROFITS

The benefits of turbocharger retrofits are promoted by manufacturers, arguing that new generation models—with significantly higher overall efficiencies than their predecessors—can underwrite a worthwhile return on investment. Retrofitting older engines with new high efficiency turbochargers can be economically justified when one or more of the following conditions prevail, according to ABB Turbo Systems:

- The engine suffers from high exhaust temperatures; air manifold pressure does not rise to its normal level; and full load can no longer be attained.
- Need to reduce fuel costs and/or exhaust emissions.
- Need to increase engine operating safety for an extended period.
- Engine performance requires optimizing for slow steaming or another new operating profile.
- Unsatisfactory lifetimes are experienced with key turbocharger elements due to erosion or corrosion; older generation turbocharger spares from some manufacturers are difficult to obtain and/or are costly; and a long delivery time is quoted for replacement turbochargers.

The performance benefits from a retrofit (Figure 7.25) are reportedly realized in lower exhaust temperatures (down by around 50–80°C) and a lower fuel consumption (reduced by 3–5 per cent). The uncooled gas casings of new turbochargers also foster a longer lifetime for critical parts subject to wear from erosion and corrosion. Furthermore, ABB

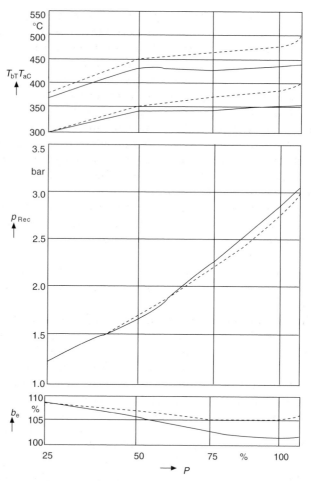

Figure 7.25 Turbocharger retrofit benefits exhaust-gas temperatures and specific fuel consumption of an ABB VTR..1 turbocharger exhibiting low efficiency (dashed lines) compared with a replacement VTR..4 turbocharger operating with a markedly higher efficiency (continuous lines)

Turbo Systems notes, the ball and roller anti-friction bearings normally fitted to its VTR..4 type turbochargers deliver better starting characteristics, improved part load performance and faster acceleration. Enhanced reliability is also promised by modern turbochargers.

TURBO COMPOUND SYSTEMS

The high efficiency of some modern large turbochargers is comfortably in excess of that required to pressure charge engines, allowing some of the exhaust gas to be diverted to drive a power turbine. A number of these turbo compound installations are in service, mainly in conjunction with MAN B&W and Sulzer low speed engines, the investment particularly worthwhile in an era of high fuel prices.

The power gas turbine is arranged either to supplement directly the propulsive effort of the engine via power take-in gearing to the crankshaft or to drive a generator via power take-out/power take-in (PTO/PTI) gearing. Sulzer's Efficiency-Booster system—based on an ABB Turbo Systems concept—promises fuel savings of up to 5.5 g/kWh, raising the maximum engine efficiency to 54 per cent with virtually no effect on the exhaust gas energy available for conventional waste heat recovery (Figure 7.26).

Apart from yielding worthwhile fuel savings throughout the engine load range, the Efficiency-Booster system can be an attractive alternative to engine derating for an equivalent brake specific fuel consumption. Compared with a derated engine, the system may allow the saving of one cylinder (depending on the type of engine and its power/speed rating), with consequent benefits in both a lower first cost and a shorter engineroom.

The Efficiency-Booster is described as a straightforward on/off system which can be exploited above 40–50 per cent engine load without undermining engine reliability. Exhaust energy is recovered simply by diverting a proportion of the exhaust gas from the exhaust manifold to a standard ABB turbocharger-type power turbine arranged in parallel with the engine turbochargers. The power turbine—forming a compact module with its integrated epicyclic gear—is mechanically coupled to the Sulzer RTA low speed engine's integral PTO gear. The recovered energy is thus directly fed to the engine crankshaft, allowing the engine to run at a correspondingly reduced load to deliver the same total power to the propeller shaft—but with a correspondingly lower fuel consumption. The result depends on the engine's contract-maximum continuous rating since fuel savings increase with brake mean effective pressure.

The simple mechanical arrangement, based on standard components, underwrites a low first cost and offers a high transmission efficiency. It also allows the Efficiency-Booster to be shut off at engine part-load and when manoeuvring.

Similar benefits are reported for MAN B&W's Turbo Compound System/Power Take In (TCS/PTI) arrangement which can be fitted to its MC low speed engines with bore sizes from 600 mm to 900 mm (Figure 7.27). The system is not applicable to smaller models because of their size and the lower efficiency of the smaller turbochargers used. MAN B&W and ABB power turbines are used, respectively, in

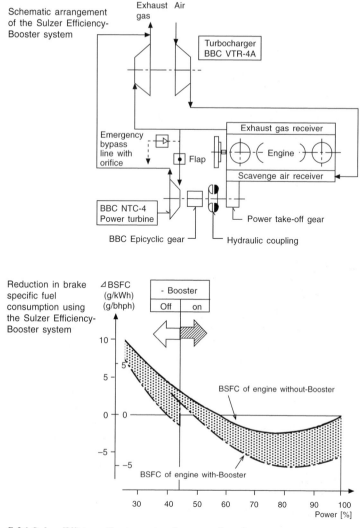

Figure 7.26 Sulzer Efficiency-Booster system layout and performance

association with the MAN B&W and ABB turbochargers serving the engine. At engine loads above 50 per cent of the optimized power, the power turbine is connected to the crankshaft via a hydraulic clutch and flexible coupling.

The amounts of exhaust gas actually available for the TCS depends on the turbocharger efficiency and turbocharger matching chosen for the individual case. TCS power falls off at lower loads owing to the smaller gas amounts: at 50 per cent of the optimized engine power

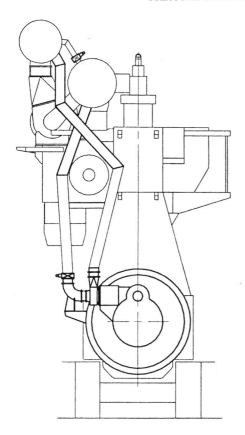

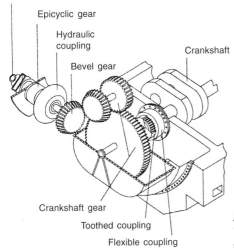

Power turbine

Epicyclic gear

Hydraulic
coupling

Bevel gear

Crankshaft

Crankshaft gear

Toothed coupling

Flexible coupling

*Figure 7.27 Arrangement of MAN B&W
TCS/PTI system on main engine and the
power turbine drive to the crankshaft*

the output of the TCS/PTI unit is around 25 per cent, returning negligible savings in the specific fuel consumption of the engine. The maximum power turbine output obtainable is 4.2 per cent of the engine's optimized power, the yield depending on the turbocharger selected.

MAN B&W's Power Take Off/Power Take In system combines a PTO serving a generator and a TCS/PTI unit (power turbine coupled to the crankshaft gear). The combined benefits of both systems are achieved without incurring the full first cost associated with each since some of the most costly elements (the crankshaft gear, for example) are common.

Engines equipped with high efficiency turbochargers can operate without auxiliary blowers at lower loads (down to around 25 per cent) than engines with standard turbochargers (around 35 per cent).

See Chapter 30 for Sequential Turbocharging (MTU and SEMT-Pielstick) and Two-stage Turbocharging systems (Niigata and Paxman); Chapter 27 for Wärtsilä SPEX system; and Chapter 11 for Mitsubishi Two-stage Turbocharging system.

8 Fuel injection

The emphasis in this chapter is on medium speed engines with camshaft-actuated individual jerk pumps for each cylinder. Higher speed engines are also covered, including those which use camshaft pumps (or block pumps): those in which all the jerk pump elements are grouped into one or more complete units, each equipped with a common camshaft. This section does not apply (except in a general way) to low speed two-stroke engines whose systems are described later under individual makes.

INJECTION AND COMBUSTION

The essence of a diesel engine is the introduction of finely atomized fuel into the air compressed in the cylinder during the piston's inward stroke. It is, of course, the heat generated by this compression, which is normally nearly adiabatic, that is crucial in achieving ignition. Although the pressure in the cylinder at this point is likely to be anything up to 200 bar, the fuel pressure at the atomizer will be of the order of 1300–1800 bar.

There is a body of evidence to suggest that high injection pressure at full load confers advantages in terms of fuel economy, and also in the ability to digest inferior fuel. Most modern medium speed engines attain 1200–1800 bar in the injection high-pressure pipe. Some recent engine designs achieve as much as 2300 bar when pumping heavy fuel. For reasons of available technology, the earliest diesel engines had to use compressed air to achieve atomization of the fuel as it entered the cylinder (air blast injection), and while airless (or solid) injection delivered a significant reduction in parasitic loads it also presented considerable problems in the need for high precision manufacture, and the containment of very high and complex stresses.

The very high standard of reliability and lifetime now attained by modern fuel injection systems, notwithstanding their basic simplicity, reflects a considerable achievement in R&D by fuel injection equipment manufacturers.

In the early days of airless injection many ingenious varieties of combustion chamber were used, sometimes mainly to reduce noise or smoke, or to ease starting; but often in part to reduce, or to use modest, injection and combustion pressures. A growing emphasis on economy and specific output, coupled with materials development and advances in calculation methods allowing greater loads to be carried safely, has left the direct injection principle dominant in modern medium speed and high speed engine practice.

Direct injection is what it says it is: the fuel is delivered directly into a single combustion chamber formed in the cylinder space (Figure 8.1), atomization being achieved as the fuel issues from small drillings in the nozzle tip.

For complete combustion of the fuel to take place, every droplet of fuel must be exposed to the correct proportion of air to achieve complete oxidation, or to an excess of air. In the direct injection engine the fuel/air mixing is achieved by the energy in the fuel spray propelling the droplets into the hot, dense air. Additional mixing may be achieved by the orderly movement of the air in the combustion chamber, which is called 'air swirl'. Naturally aspirated engines usually have a degree of swirl and an injection pressure of around 800 bar. Highly turbocharged engines with four-valve heads have virtually no swirl, but have an injection pressure of 1200–1800 bar to provide the mixing energy. Where indirect injection is exploited, some high speed engines retain a pre-chamber in the cylinder head into which fuel is injected as a relatively coarse spray at low pressure, sometimes using a single hole. Combustion is initiated in the pre-chamber, the burning gases issuing through the throat of the chamber to act on the piston (Figure 8.2).

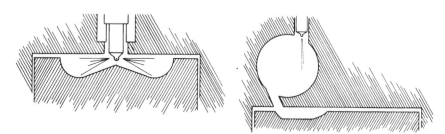

Figure 8.1 Cross-section of direct injection combustion chamber

Figure 8.2 Indirect injection: the Ricardo pre-chamber

Fuel/air mixing is achieved by a very high air velocity in the chamber, the air movement scouring the walls of the chamber and promoting good heat transfer. Thus the wall can be very hot-requiring heat resistant materials—but it can also absorb too much heat from the air in the initial compression strokes during starting and prevent ignition. It is these heat losses that lead to poor starting and inferior economy. Further forms of assistance, such as glow plugs, have therefore sometimes been necessary to achieve starting when ambient pressures are low. The throttling loss entailed by the restricting throat also imposes an additional fuel consumption penalty.

One engine designer, SEMT-Pielstick, achieved an ingenious combination of the two systems by dividing the pre-chamber between cylinder head and piston crown. At TDC a stud on the piston enters the pre-chamber to provide a restricted outlet. On the expansion stroke the restriction is automatically removed and fuel economy comparable with normal direct injection engines is attainable (Figure 8.3).

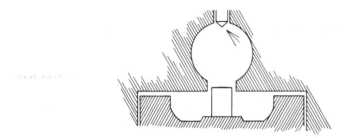

Figure 8.3 The variable geometry combustion chamber of SEMT-Pielstick

Direct injection, too, has variants which reflect the fact that, despite considerable expenditure on research into its mechanism, the detail of how combustion develops after ignition is achieved is still largely empirical. The essentials are:

1. That some at least of the fuel injected is atomized sufficiently finely to initiate combustion. Ignition cannot take place until a droplet of fuel has reached the temperature for spontaneous ignition. Since heat is taken up as a function of surface area (proportional to the square of the diameter) and the quantity of heat needed to achieve a temperature rise is a function of volume (varying as the cube of the diameter) only a small number of fine droplets are needed to initiate combustion. High speed photography of combustion does indeed show that ignition takes place in a random manner near the injector tip and usually outside the main core of the spray.

2. That the fuel should mix with the air in order to burn. Since

most of the air in a roughly cylindrical space is, for geometrical reasons, near the periphery most of the fuel must penetrate there, and this is easier with large droplets; hence, also, the use of wide core angles and multiple spray holes.

3. That under no circumstances must fuel reach the liner walls or it will contaminate the lubricating oil. An advantage of combustion spaces formed in piston crowns is that the walls of the chamber form a safe target at which spray may be directed. This type of combustion chamber in the piston has the further advantage that, during the piston descent, air above the piston periphery is drawn into the combustion process in a progressive manner.

4. The injection period should be reasonably short and must end sharply. Dribble and secondary injection are frequent causes of smoke, and also of lubricating oil becoming diluted with fuel or loaded with insoluble residues.

Dribble is the condition where fuel continues to emerge from the nozzle at pressures too low to atomize properly; it is caused by bad seating faces or slow closure.

Secondary injection is what happens when the pressure wave caused by the end of the main injection is reflected back to the pump and then again to the injector, reaching it with sufficient pressure to reopen the injector at a relatively late stage in combustion. Any unburned or partly burned fuel may find its way onto the cylinder walls and be drawn down by the piston rings to the sump.

INJECTOR

Working backwards from the desired result to the means to achieve it, the injector has to snap open when the timed high pressure wave from the pump travelling along the high pressure pipe has reached the injector needle valve. Needle lift is limited by the gap between its upper shoulder and the main body of the holder. Needle lift is opposed by a spring, set to keep the needle seated until the 'blow-off pressure' or 'release pressure' of the injector is reached by the fuel as the pressure wave arrives from the pump. This pressure is chosen by the enginebuilder to ensure that there is no tendency for the needle to reopen as the closing pressure waves are reflected back and forth along the high pressure pipe from the pump. The setting also has some effect on injection delay and the quality of injection; it is usually chosen to be between 200–300 bar.

The needle valve is invariably provided with an outer diameter on which the fuel pressure acts to overcome the spring pressure and cause the initial lift. This brings into play the central diameter of the needle so that it snaps open to the lift permitted.

The needle and seat cones are usually ground to a differential angle so that contact is made at the larger diameter (Figure 8.4). When the needle is only slightly open the greater restriction to flow is at the outer rather than the inner diameter of the seat. This ensures that as the needle lifts there is a sudden change of pressurized area and the needle force against the spring changes correspondingly, giving a very rapid lift to fully open; and conversely when closing. (This is why it is bad practice to lap the needle to its seat.)

Control of the temperature of the injector, particularly of the sensitive region round the needle seat and the sac, is very important. This is especially so when heavy fuels are used, both because the fuels themselves have to be heated and because they tend to burn with a more luminous flame that increases the heat input to all the metal surfaces. If the tip temperature rises too high, the fuel remaining in the sac after injection boils and spits from the spray holes: it is not properly burned and forms a carbon deposit around each spray hole. Such carbon formation, which builds up fairly rapidly around the boundary of the sprays, causes a rapid deterioration in the quality of combustion.

With smaller engines up to about 300 mm bore, it is usually sufficient to rely on the passage and recirculation of the pumped fuel itself to achieve the necessary cooling. The injector clearances may, in fact, be eased slightly to promote such circulation. This back leakage, normally about 1 per cent of the pumped fuel, is led away to waste, or via a

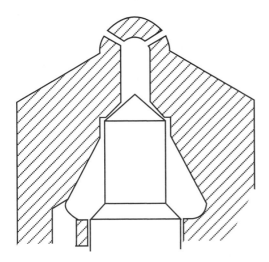

Figure 8.4 Detail of needle and seat showing differential angle (exaggerated)

monitoring tank to the fuel tank. For larger engines it is usually necessary
to provide separate circuits specifically for a coolant flow, either of
treated water, light fuel or lubricating oil.

All of these considerations lead to a relatively bulky injector and
thus to tricky compromises with the available space within the cylinder
head on single piston engines, where space is also needed to permit
the largest possible valve areas, cages and cooling. Nonetheless, there
are performance advantages in highly rated engines in reducing the
inertia of the components directly attached to the needle of the injector;
and this leads fuel injection equipment manufacturers to try to find
space for the injector spring at the bottom rather than at the top of
the injector, as in Figure 8.5.

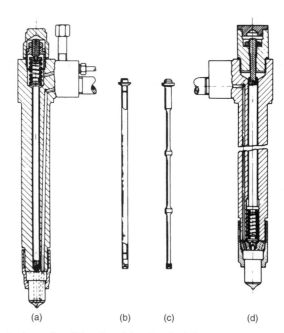

(a)   (b)   (c)   (d)

Figure 8.5 Comparison of traditional and low inertia injectors:
(a) conventional injector
(b) and (c) lightweight spindles
(d) low inertia injector (Delphi Bryce)

FUEL LINE

Figure 8.6 shows a typical fuel line pressure diagram and also the
needle lift diagram for an engine having a camshaft speed of 300 rev/
min, both at high load (Figure 8.6(a)) and at low load (Figure 8.6(b)).

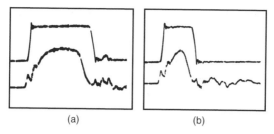

(a)   (b)

Figure 8.6 Fuel line pressure (lower line) and needle lift diagrams
(a) At high load
(b) At low load (Delphi Bryce)

Figure 8.6(a) shows a very brief partial reopening of the nozzle, or secondary injection, after the main opening period. In Figure 8.6 the breaks in each trace are made to indicate intervals of 3 cam degrees.

The fuel line has to preserve as much as possible of the rate of pressure rise, the maximum pressure and the brevity of injection that the fuel pump has been designed to achieve. To do this, the fuel line must have the lowest practical volume and the maximum possible stiffness against dilation. There is also a limit to the minimum bore because of pressure loss considerations.

The only real degree of freedom left to the designer is to make the fuel line as short as possible. Even so, a given molecule of fuel may experience two or three injection pulses on its way from the pump chamber to the engine cylinder in a medium speed engine running at full load. This rises to 30 or more at idling.

If the fuel pump is mounted at a greater distance from the injector the elastic deformation of the longer pipe and compression of the larger volume of fuel will dissipate more and more of the pressure developed at the fuel pump and prolong the period of injector needle lift compared with the pump delivery. In all medium speed engines this compromise on injection characteristics, and its adverse effect on fuel economy (and heat flux in combustion chamber components), is considered unacceptable, and each cylinder has an individual fuel pump with a high pressure fuel line as short as possible.

The smaller the engine, the more manufacturers (in fact all manufacturers of automotive sizes) opt for the production convenience of a camshaft pump (i.e. the pump elements serving all or several cylinders are mounted in one housing with their own camshaft) and accept any compromise in performance. In such engines a given molecule of fuel may endure 20 or more injection pulses while migrating along the high pressure pipe at full load. In such cases the elastic characteristics of the HP pipe (and of the trapped fuel) come to equal the pumping element as a major influence on the qualities and duration of injection.

Naturally, all injection pipes must have identical lengths to equalize cylinder behaviour. On a 150 kW auxiliary engine of approximately 150 mm bore running at 1200 rev/min, the effect of a 600 mm difference in pipe length, due to running pipes each of different length directly from pump to the cylinder it served, was 8-degrees in effective injection timing, and consequently about 25 bar difference in firing pressure.

Camshaft or block pumps are, of course, much cheaper and one maker (Deutz) adopted for a higher speed engine the ingenious solution of grouping up to three or four pump elements in one housing, each housing being mounted close to the cylinders concerned.

The HP fuel lines are made of very high quality precision drawn seamless tube almost totally free from internal or external blemishes. They must, moreover, be free of installation stress or they risk fracture at the end connections, and they must be adequately clipped, preferably with a damping sleeve, if there is any risk of vibration.

Fuel line failures due to pressure are not as frequent as they once were, thanks to improvements in the production of seamless tube. Nevertheless, current regulations insist that in marine use fuel lines are sleeved or enclosed to ensure that in the event of a fracture escaping fuel is channelled to a tank where an alarm may be fitted. This is to avoid the fire hazard of fuel finding a surface hot enough to ignite it, or of accumulating somewhere where it could be otherwise accidentally ignited. Most designers now also try to ensure that leaking fuel cannot reach the lubricating oil circuit via the rocker gear return drainage or by other routes.

Several manufacturers shorten and segregate the HP line by leading it to the injector through, rather than over, the cylinder head. Some others incorporate the pump and the injector in a single component, avoiding the high pressure pipe altogether (see section below).

PUMP

The final item for consideration is the fuel injection pump, which is the heart of the fuel injection system. Essentially, the pumping element is a robust sleeve or barrel which envelops a close-fitting plunger. Each is produced to a very fine surface finish of 1–2 micro-inches and to give a clearance of approximately 7–10 µm (microns) depending on the plunger diameter.

In its simplest form (Figure 8.7) the barrel has a supply port at one side and a spill port at the other. (In early practice these ports were

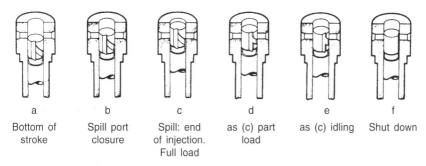

a	b	c	d	e	f
Bottom of stroke	Spill port closure	Spill: end of injection. Full load	as (c) part load	as (c) idling	Shut down

Figure 8.7 Principle of operation of the fuel injection pump (Delphi Bryce)

combined.) The plunger is actuated by the fuel cam through a roller follower. Before, and as the plunger starts to rise (Figure 8.7(a)), the chamber is free to fill with fuel and to spill back into the gallery in the pump housing outside until the rising top edge of the plunger closes the supply and spill ports (Figure 8.7(b)). Thereafter the fuel is pressurized and displaced through the delivery valve towards the injectors. The initial travel ensures that the plunger is rising fast when displacement commences, so that the pressure rises sharply to the desired injection pressure.

When the plunger has risen far enough, a relieved area on it uncovers the spill port, the pressure collapses, and injection ceases (Figure 8.7(c)). The relief on the plunger has a helical top (control) edge so that rotation of the plunger by means of the control rod varies the lift of the plunger during which the spill port is closed, and therefore the fuel quantity injected and the load carried by the engine, as shown in Figures 8.7(c, d, e). At minimum setting the helical edge joins the top of the plunger and if the plunger is rotated so that this point, or the groove beyond it, coincides with the spill port, the latter is never closed. In that case no fuel is pumped and that cylinder cuts out (Figure 8.7(f)). The area of the plunger between the top and the control edge is termed 'the land'.

There are, of course, many complications of this simple principle. The first is that a delivery valve, essentially a non-return valve, is needed at the pump outlet to keep the fuel line full, as shown in Figure 8.8. The delivery valve usually has an unloading collar to allow for the elasticity of the pipe and fuel, and to help eliminate the pressure pulsations which cause secondary injection. Some types of delivery valve, however, sense and unload the line pressure directly.

The second cause of complication is that, when the spill port opens and a pressure of anything up to 1600 bar is released, cavitation and/or erosion is likely to occur, affecting both the housing directly opposite

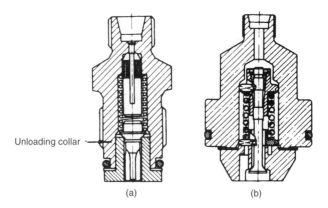

Unloading collar

(a)   (b)

Figure 8.8 Delivery valves (a) volume unloading (b) direct pressure unloading (Delphi Bryce Ltd)

the port, and the plunger land which is still exposed to the port at the instant of release. Modern designs seek to eliminate these effects by using respectively a sacrificial (or in some cases a specially hardened) plug in the housing wall, and by special shaping of the ports.

Another complication is the filling process. When the pump spills, further plunger lift displaces fuel through the spill port. As it falls on the descending flank of the cam, the opposite effect applies until the falling control edge closes the spill port. Further fall, under the action of the plunger return spring, has to expand the fuel trapped above the plunger until the top edge uncovers the supply port again. This it can only do by expanding trapped air and fuel vapour, and by further vaporization, so that a distinct vacuum exists, and this is exploited to refill the pump.

There is also the effect of dilation due to pressure, the need for lubrication, and the need to prevent fuel from migrating into spaces where it could mingle with crankcase oils.

Early pumps could get by without undue complications: the fuel itself provided adequate plunger lubrication, and fuel which had leaked away in order to do this was simply rejected to a tank which was emptied to waste.

Today, grooves are provided on the plunger or in the barrel wall. The first groove collects leaking fuel. Where there are two grooves the second will spread lubrication, but there is sometimes an additional groove between them to collect any further migration either of fuel or lubricant. Figure 8.9 shows a section of a modern heavy duty fuel pump incorporating most of the features described above.

The pump is actuated by a follower which derives its motion from

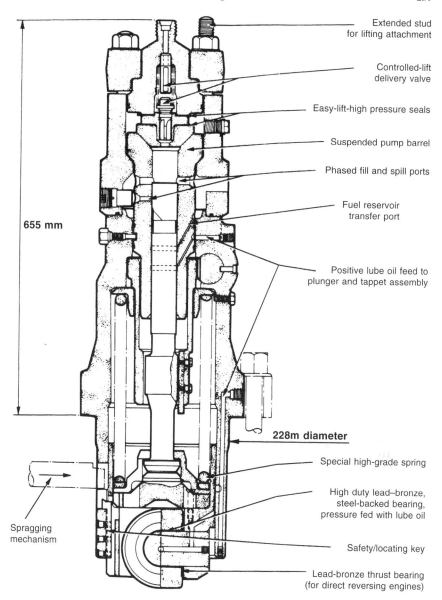

Extended stud
for lifting attachment

Controlled-lift
delivery valve

Easy-lift-high pressure seals

Suspended pump barrel

Phased fill and spill ports

Fuel reservoir
transfer port

Positive lube oil feed to
plunger and tappet assembly

655 mm

228m diameter

Special high-grade spring

High duty lead–bronze,
steel-backed bearing,
pressure fed with lube oil

Safety/locating key

Lead-bronze thrust bearing
(for direct reversing engines)

Spragging
mechanism

Figure 8.9 Cross-section of Bryce FFOAR type flange-mounted unit pump suitable for four-stroke engines with crankshaft speeds up to 450 rev/min (Delphi Bryce)

the cam. Block pumps incorporate all these components in a common
housing for several cylinders, while individual pumps used on some
larger engines incorporate the roller. This not only makes servicing

more straightforward, it also ensures unified design and manufacturing responsibility.

The roller is a very highly stressed component, as is the cam, and close attention to the case hardening of the cam and to the roller transverse profile is essential to control Hertzian (contact) stresses. The roller is sometimes barrelled very slightly, and it must bear on the cam track parallel to the roller axis under all conditions. On the flanks of the cam this is a function of slight freedom for the follower assembly to rotate, while on the base circle and the top dwell of the cam it depends on precise geometry in the assembly.

The roller pin requires positive lubrication, and the cam track must at least receive a definite spray. Lubrication channels have to be provided from the engine system through the pump housing to the roller surface, its bearings, and to the lower part of the plunger.

Timing

The desired timing takes into account all the delays and dynamic effects between the cam and the moment of ignition. The desired timing is that which ensures that combustion starts and continues while the cylinder pressure generated can press on the piston and crank with the greatest mechanical advantage, and is complete before the exhaust valve opens at 110–130 degrees ATDC. The enginebuilder always specifies this from his development work, but it usually means that the only settable criterion, the moment of spill port closure (SPC), is about 20–25 degrees BTDC.

In earlier practice this was done by barring (or manually turning over) the engine with the delivery valve on the appropriate cylinder removed, and with the fuel circulating pump on (or more accurately with a separate small gravity head), and the plunger rotated to a working position. When fuel ceased to flow, the spill port had closed. Sometimes this criterion was used to mark a line on a window in the pump housing through which the plunger guide could be seen, to coincide with a reference line on the guide.

The accuracy of this method is, however, disappointing. There are differences between static and dynamic timing which vary from pump to pump and many makers now rely on the accuracy of manufacture, and specify a jig setting based on the geometry of the pump and the SPC point to set the cam rise or follower rise at the required flywheel (crank) position.

Some makers allow users to make further adjustments of the timing according to the cylinder pressure readings but this places a premium on the accuracy, not only of the calibration of instruments, but of the method itself. It should be borne in mind that neither the instruments nor the method are as accurate as the pump manufacture, by an order of magnitude.

If the cam is fixed it is usually possible, within stated limits, to adjust the height of the tappet in relation to the cam. Raising it has the effect that the plunger cuts off the supply spill port (and starts injection) sooner for a given cam position, and vice versa (Figure 8.10). The limit is set at one extreme by the need for there to be 'top clearance' at the top of the follower lift (i.e. of the plunger stroke) between the top of the plunger and the underside of the delivery valve, in order to avoid damage. Too small a clearance also means that injection is commencing too near the base of the cam, and therefore at too low a velocity to give the required injection characteristics. On the other hand, too large a clearance (to compensate for an over-advanced cam by lowering the plunger) risks interfering with the proper spill function which terminates injection.

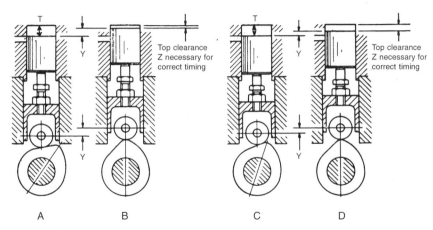

A   B   C   D

Figure 8.10 Adjusting timing by top clearance. Configuration of cam, tappet and pump at commencement of injection (A and C); and maximum cam and follower lift (B and D). In each case the pump is set correctly for the required injection point. A shows the case for an over-retarded cam, C that for an over-advanced cam within permitted limits of top clearances. In either case Y (the residual lift) + Z = the timing dimension of the pump, T (GEC Ltd)

The cam with its follower may be designed to give a constant velocity characteristic throughout the injection stroke. Recent designs with a high initial velocity aim to give a constant pressure characteristic, but many manufacturers find that the performance advantage or stress advantage over the constant velocity design is too small to warrant the greater production cost.

Uniformity

Bearing in mind the random way in which ignition occurs during each injection, and perhaps also because the filling process of the pump chamber is largely dependent on suction, it is not surprising that successive combustion cycles in the same cylinder at the same steady load are not identical. A draw card taken slowly to show a succession of cycles will show this visibly, and a maximum pressure indicator will show by feel, or otherwise, that maximum pressures vary from cycle to cycle, perhaps by 5 bar or more. (Any governor corrections will have an effect, too, but progressively over several cycles.) These variations are a fact of life, and not worth worrying about unless they are 20 bar or more. It does mean that if indicator cards are being taken as a check on cylinder performance, several cycles per cylinder should be taken to ensure that a properly representative diagram is obtained.

Differences will also occur between cylinders for reasons of component tolerance (unless compensated adequately by the setting method), and because of the dynamic behaviour of the camshaft, which deflects in bending due to cam loading and torsionally due to the interaction of the cam loadings of adjacent cylinders. These will characteristically fall into a pattern for a given engine range and type.

While enginebuilders should (and usually do) allow for variations in cylinder behaviour to occur, it is obviously good practice to minimize these to achieve maximum economy with optimum stresses. However, it would be wise to bear in mind that it is inherently very difficult to make pressure and temperature measurements of even the average behaviour of combustion gases in the cylinders.

To achieve, without laboratory instrumentation techniques, an accuracy which comes within an order of magnitude of the standard of accuracy to which fuel injection components are manufactured and set is impossible. This is particularly true of exhaust temperature as a measure of injection quantity, notwithstanding that this practice has been generally accepted by many years of use. Fuel injection settings on engines should not therefore be adjusted without good reason, and then normally only on a clear indication that a change (not indicative of a fault) has occurred.

Fuel quantity imbalance is far from being the only cause of a variation in the pattern of exhaust temperatures, nor timing of firing pressure. Consider, for instance, variations in compression ratio, charge air mass, scavenge ratio, not to mention calibration errors and the accuracy of the instruments used.

Fuel injection pumps (and injectors) have their own idiosyncrasies.

The relationship between output and control rod position (the calibration) is not absolutely identical in different examples of pumps made to the same detail design. The output of a family of similar pumps can be made sensibly identical at only one output, namely at the chosen balance point. At this output a stop, or cap, on the control rod is adjusted to fit a setting gauge (or, in older designs, a pointer to a scale), so that it can be reproduced when setting the pumps to the governor linkage on the engine.

Obviously, the calibration tolerances will widen at outputs away from the balance point, which is usually chosen to be near the full load fuelling rate. As the tolerances will be physically widest at idling, and will also constitute a relatively larger proportion of the small idling quantity (perhaps 35 per cent), the engine may well not fire equally on all cylinders at idling, but this sort of difference is not sufficient to cause any cylinder to cut out unless another fault is present.

When a pump is overhauled it is advisable that its calibration should be checked, and the balance point reset. This ideally entails the use of a suitable heavy duty test rig, but in the largest sizes of pump it is sometimes more practical to rely on dimensional settings. D.A.W.

RECENT DEVELOPMENTS AND TRENDS

Fuel injection systems have a significant influence on the combustion process and hence a key role to play in improving engine fuel consumption and reducing noxious exhaust emissions. The following characteristics of an injection system are desirable in achieving these goals:

- Injection pressures during the whole process should be above 1000–1200 bar for a good spray formation and air–fuel mixture; a tendency in practice to 1600–1800 bar and higher is noted.
- Total nozzle area should be as small as possible in relation to cylinder diameter for good combustion, particularly at part load.
- Total injection duration should be 20 degrees of crank angle or less for achieving a minimum burning time in order to exploit retarded combustion for reduced NOx emissions without loss in efficiency. A high compression ratio is desirable.
- High pressures at the beginning of injection promote reduced ignition delay, while increased mass flow can result in an overcompensation and increased pressure gradients. Consequently, rate shaping is necessary in some cases, particularly with high speed engines.

- High aromaticity fuels cause increased ignition delay in some cases. Pre-injection with high injection pressures is necessary and can achieve non-sensitivity to fuel quality.
- Electronically-controlled adjustment of injection timing should be applied for optimised NOx emissions at all loads, speeds and other parameters.
- The load from the torque of the injection equipment on the camshaft and/or the gear train should be as low as possible in order to prevent unwanted additional stresses and noise.
- For safety reasons, even a total breakdown of electrical and other power supplies should not result in the engine stopping.

To understand fully the complex hydraulic events during the fuel injection process, the Swiss-based specialist Duap says it is essential to appreciate the function of all the elements (pump, pressure pipe, fuel valve and nozzle) forming the injection system.

It is commonly believed by non-specialists that the plunger within the pump pushes the fuel upwards like a pillar, thus effecting the immediate injection of fuel into the combustion chamber. The reality is, however, that the plunger moves with such a high speed that the fuel likely to be conducted is highly compressed locally (elasticity of fuel). The compressed fuel now generates a pressure wave which runs through the pipe and valve, causing the nozzle to open and inject. The pressure wave of the fuel is forced through the system at a speed of around 1300 m/s. In the same way that sound in the air is reflected by houses and hills, the wave is reflected between the fuel valve and the pump. It may therefore easily run back and forth between those components several times before the nozzle is actually forced to open and inject.

This process, Duap warns, underlines the importance of maintaining the injection system in an excellent condition and ensuring that the fuel is properly treated and free of dirt. If, for example, the nozzle spray holes are partially blocked by extraneous elements or carbon particles the pressure wave may not be sufficiently reduced within the system. This eventually results in the destruction of the fuel pump cam or other vital parts of the injection system upon the next stroke.

The task of the injection system is to feed fuel consecutively to each cylinder within a very short period of time (0.004–0.010s, depending on the engine rev/min). It is also essential that the same amount of fuel is delivered with each stroke: deviations in the quantity supplied to different cylinders will adversely affect the performance of the engine and may result in crankshaft damage due to resonance.

All the key elements of the system must be manufactured to very small tolerances; the clearance between plunger and barrel, for example, is not larger than 4–16 microns (depending on the plunger diameter). Such a high precision clearance also dictates an adequate surface quality (within 0.2–0.5 microns), a finish which can only be achieved by careful grinding and lapping.

Before the fuel is pushed into the pressure pipe (linking the pump and fuel valve) it has to pass the pressure valve, which has several tasks. The first task is to separate hydraulically the pump from the pressure pipe after the fuel has passed the valve; the second is to smooth the pressure wave running back and forth within the pipe: this calming down is necessary to secure the proper closing of the fuel valve without having additional and uncontrolled injection; and the third is to maintain a certain pressure within the pressure pipe for the next injection stroke (rest pressure). The fulfilment of these tasks can only be guaranteed by using pressure valves manufactured to very tight tolerances (bearing in mind that the valves play an important role in ensuring that identical amounts of fuel are delivered to each engine cylinder).

Another highly underestimated component of the fuel injection system is the pressure pipe. Despite its small size and apparent simplicity, Duap notes, the pipe has to endure pressure waves of up to 1800 bar; even the smallest mark may thus lead to fracture.

The last link in the injection system chain is the fuel valve (nozzle holder assembly and nozzle) which is designed to inject and atomize the fuel into the combustion chamber when the pressure wave has reached a pre-determined strength.

UNIT INJECTOR VERSUS PUMP/PIPE/INJECTOR

A unit injector is more likely to be considered for engines today than in the past but the choice between unit injector (Figure 8.11) and pump/pipe/injector (Figure 8.12) systems can be complex. The UK-based specialist Delphi Bryce (formerly Lucas Bryce) cites the following merits and demerits of the two configurations:

Unit injector positive features: hydraulic volumetric efficiency (smaller plunger); longer nozzle life; fewer components; no high pressure pipe; and fast end of injection. More complex cylinder head designs and higher cylinder head loads are negative features.

Pump/pipe/injector system positive features: less space required by injector; facilitates use of residual fuels and water injection; simple pushrod mechanism; generally easier servicing; and higher low load operating

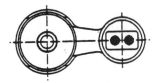

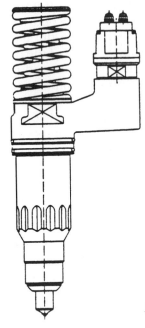

Figure 8.11 Outline of electronic unit injector (Delphi Bryce)

pressure. Large trapped fuel volume and parasitic power loss are negative features.

Unit injectors

The initial concern with the application of unit injectors is usually cylinder head installation and the drive. Increased space demands and the need to provide fuel passages (and cooling passages in the case of residual fuel operation) can conflict with cylinder head air and exhaust passages and cooling requirements. Such issues can be resolved through close co-operation at the design stage.

A unit injector has a low trapped fuel volume, which yields several major benefits:

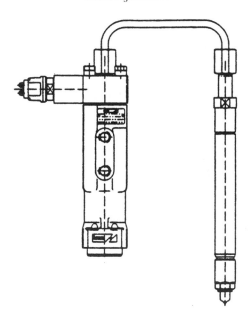

Figure 8.12 Pump/pipe/injector system (Delphi Bryce)

- Pumping efficiency is high, allowing smaller plungers to be used and hence fostering reduced camshaft loading; this can be important as a means of minimizing camshaft diameters.
- Very responsive to drive input, thus reducing hydraulic delays. The injection rate diagram more closely follows, in both time and shape, the cam profile, making it easier to optimize injection system performance requirements.
- Fast end of injection with limited high pressure wave effects; this eliminates the need for a delivery valve and, experience has shown, leads to enhanced durability, particularly of the nozzle.

The fuel system drive must be sufficiently stiff to counteract the high operating loads or the hydraulic performance benefits can be lost. A typical unit injector drive is illustrated (Figure 8.13). The responsiveness of unit injectors can lead to greater pressure turndown at reduced speed and load than exhibited by pump/pipe/injector systems. This can be addressed by cam design selection. The cam design can be tailored to achieve maximum plunger velocity when delivering low fuel quantities: Figure 8.14 shows the effect of using a falling rate cam on the injection pressure.

Unit injectors offer many advantages, particularly as operating pressures increase towards 2000 bar, but they have installation drawbacks.

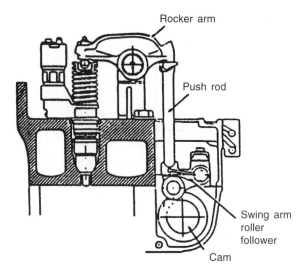

Figure 8.13 Typical unit injector drive (Delphi Bryce)

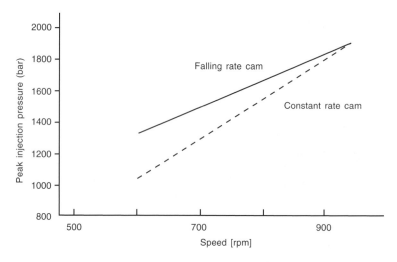

Figure 8.14 Effect of cam design on the injection pressure (Delphi Bryce)

In practice, unit injectors are limited to engines of under 300 mm bore. In larger engines, Delphi Bryce suggests, they are bulky, difficult to handle and their withdrawal height from the cylinder head can pose a problem in some applications.

Pump/pipe/injector systems

Engine designs with bore sizes larger than 300 mm can benefit from the advantages of pump/pipe/injector systems. The primary requirement for installation is to minimize the trapped high pressure fuel volume. This is usually achieved by having a high camshaft and the high pressure connection to the injector as low as possible through the cylinder head. The resulting low trapped fuel volume maximizes hydraulic efficiency and responsiveness.

Pressure wave effects in the high pressure passages will still be sufficient to require a delivery valve to eliminate secondary injection and cavitation erosion. Pressure unloading delivery valves have been used for many years and, with experience, have become reliable. The need to create the fastest possible pressure collapse at the end of injection and a trapped residual pressure, and to ensure durability of the valve components, led to a valve design with a series of tailoring features (Figure 8.15). Valve opening pressures, an unloading feature on the pro flow valve and a control orifice can all be used to tune hydraulic performance.

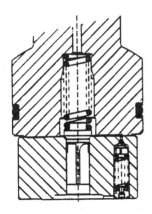

Figure 8.15 Typical pressure unloading delivery valve design (Delphi Bryce)

Pump/pipe/injector systems have hydraulic disadvantages but offer flexibility. Elements and valves can be changed relatively easily during servicing and to offer uprating opportunities. Pump/pipe/injector systems also facilitate water injection and high pressure gas injection which are of increasing interest to engine designers.

The nozzle is the interface between the injection system and the combustion process: good nozzle performance is essential for good combustion. The detailed design of the nozzle seat, sac and spray hole entry is vital in minimizing pressure flow losses and maximizing energy in the spray (Figure 8.16). This ensures minimum droplet size and an

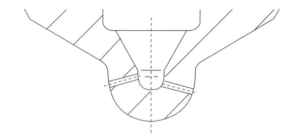

Figure 8.16 Typical nozzle 'sac end' design; low volume for reduced emissions (Delphi Bryce)

even spray distribution and penetration. Abrasive flow machining techniques are typically exploited to secure rounded edges and polished surfaces to improve discharge coefficients.

Major durability and reliability concerns for injection equipment have been cavitation erosion at high pressure spill and element seizure. Cavitation erosion is being addressed by improved delivery valve designs and pressure decay techniques at the barrel port/fuel gallery. To promote seizure resistance at the increased plunger velocities required to generate higher pressures it is necessary to improve element geometries and control surface textures. The use of surface coatings is also becoming widespread.

Programs are available to design fuel cams and analyse unit injector rocker arrangements. The results can be used to predict injection pressure, period and rate. Accurate system modelling is a considerable aid in reducing application development time and cost; a wide range of variables can be analysed with confidence.

ELECTRONIC FUEL INJECTION

It was inevitable that electronic fuel injection technology, successfully introduced for high speed diesel engines, would eventually be applied to benefit medium speed designs.

Lucas Bryce (now Delphi Bryce) first demonstrated electronic fuel injection in the late 1970s, reportedly showing that improved fuel consumption and emissions could be achieved. The market was not ready, however: the technology was new and there was then no environmental legislation to provide the incentive to improve engine emissions performance. The introduction of electronic fuel injection in the heavy truck market to meet stringent emission controls gave that impetus and the investment to advance the technology.

The Lucas Bryce system comprised a single-cylinder, electronically-controlled plunger pump which supplies fuel to the injector via a high pressure pipe and is driven directly from the engine camshaft; alternatively, an electronically-controlled unit injector may be used (the choice depends on the engine design). Fuel control is achieved by solenoid valves operated from the electronic control unit which Delphi Bryce calls the smart drive unit (SDU). Lucas Electronics also developed an engine governor, called a control and protection card (CPC); in some cases, however, the governor is supplied by specialist manufacturers or the enginebuilder.

With larger volume fuel flows to handle, Lucas Bryce's main effort was to achieve a suitable hydraulic performance while using the solenoid/valve capsule developed for the truck engine market by its then sister company Lucas EUI Systems. This approach avoided the potentially prohibitive cost required for a special production line for a low volume solenoid assembly.

Fuel is introduced via a port into the space above the pump plunger. As the plunger rises beyond the port fuel is displaced through the open solenoid and past the secondary valve seat. The solenoid is energized, the valve closes and pressure builds up above the secondary valve. The pressure drops across the secondary valve orifice, then closes it rapidly. With the spill passages closed, fuel is forced through the delivery valve to the injector. At the computed time for end of injection the solenoid is de-energized and pressure collapses above the secondary valve, causing it to open and spill fuel. The injector nozzle then closes to finish injection.

As well as limiting harmful engine emissions, Delphi Bryce claims its electronic fuel injection system offers enhanced reliability, diagnostic capability, optimized timing and fuel control for all load and speed conditions, including transient operation.

Electronically-controlled fuel injection equipment is playing a key role in meeting exhaust emissions legislation. In isolation, however, electronic systems do not provide a total solution. Other important complementary developments are: higher injection rates; higher injection pressures; balanced systems; initial rate control; optimized spray pattern; flow-controlled nozzles; and low sac volume nozzles.

The move to electronic control of fuelling brings advantages in accuracy of delivery and timing and underwrites sophisticated scheduling of those parameters over the full range of engine operating conditions. Developments are expected in many areas, Delphi Bryce citing:

- Pilot injection/rate shaping and control.

- Skip cylinder firing and reduced cylinder operation under low loads.
- Individual cylinder balancing by end-of-line test or adaptive software compensating for in-service wear.
- Dual-fuel engine management systems.
- Variable fuel quality compensation.
- Combustion diagnostics.
- Advanced closed loop controls with 'smart sensors'.
- Water injection controls.
- Engine health monitoring/failure prediction for preventative maintenance.
- Data collection for engine life history monitoring.

Electronic fuel control offers the opportunity to cut-out fuel to individual cylinders, thereby keeping cylinders equally warm and preventing other problems which can occur on unfired cylinders, such as lubricating oil 'souping' in the exhaust passages. Cylinder cut-out is also applied on some engines to secure more stable idling.

Generally, Delphi Bryce explains, if a transducer can be fitted to the engine to measure a feature which can be beneficial to enhancing engine control then it can be incorporated into an electronic control/ engine management system. An obvious example is cylinder pressure, whereby fuel delivery can be continually adjusted within agreed maximum limits throughout the operating period to the next overhaul to compensate for cylinder wear and injection system wear.

A schematic of a typical electronically-controlled fuel injection system (Figure 8.17) shows the main transducers.

COMMON RAIL INJECTION SYSTEMS

In mechanical fuel injection systems the injection pressure is a function of engine speed and load. At low load, therefore, the pressure drops and the result is very large fuel droplets which survive as such until they contact the combustion space surfaces. Common rail fuel injection technology offers the possibility to maintain a high injection pressure (and hence small droplets) at all loads down to idling (Figure 8.18).

Common rail systems existed in the early 1900s in mechanically-activated form but the recent re-emergence of the concept in four-stroke and two-stroke engine applications reflected advances in reliable and cost-effective electronic controls, as well as developments in materials and manufacturing technology underwriting systems capable of handling pressures of 1500 bar and higher. The following advantages

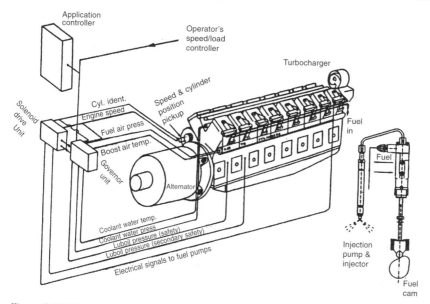

Figure 8.17 Schematic of typical electronically controlled fuel injection system showing the main transducers (Delphi Bryce)

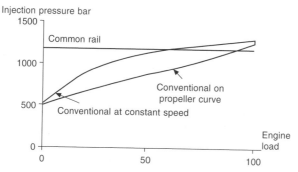

Figure 8.18 In contrast to conventional systems, common rail fuel injection allows a high injection pressure and small fuel droplets to be maintained down to idling (Wärtsilä)

are cited for common rail systems over conventional mechanical and electronically-controlled jerk pumps:

- Any injection pressure can be used, at any engine load or speed operation point, with the injection rate control close to the nozzle; this makes it possible to improve performance over the full engine load range.
- Injection timing can be varied during engine running to optimize

engine performance without the timing range being limited by cam working lengths; this is beneficial for emissions optimization as well.

- Camshaft peak torque is smaller; this allows a smaller camshaft to be used and makes possible greater engine power density and smaller engine outlines.
- The common rail pumping principle gives higher efficiency (no helix spill) and the potential to reduce mechanical drive losses, underwriting lower fuel consumption.
- The design is simpler, allowing the elimination of the one pump-per-cylinder principle, the fuel rack control shaft and the mechanical governor; this results in cost reductions for the overall fuel injection system.

Among the potential operational benefits are: lower fuel consumption; lower NOx emissions; no visible smoke at any load (and the possibility to start the engine without visible smoke); load cycling without smoke; and lower maintenance costs.

MTU COMMON RAIL SYSTEM FOR HIGH SPEED ENGINES

Development work on common rail fuel injection systems by MTU from 1990 benefited the German designer's Series 4000 high speed model (see Chapter 30) which was introduced in 1996 as the first production engine in its class to feature such a system (Figure 8.19). The group's Stuttgart-based specialist subsidiary L'Orange Einspritzsysteme was responsible for developing system components.

Research and development progressed from a pilot project on single-cylinder engines and an eight-cylinder demonstrator engine. The positive results encouraged MTU to develop the Series 4000 engine from the outset with a common rail injection system so that its benefits could be realised not just in terms of injection but in the overall design. The system embraces the following key elements:

- The injectors in the cylinder heads.
- The high pressure pump, driven by a gear system.
- The fuel rail (fuel supply line running along the engine length).
- The fuel lines between the fuel rail and the injectors.

One fuel rail is provided for the injectors and their respective solenoid valves on each bank of cylinders. The fuel pressure is generated by a high pressure pump driven via a gear train at the end of the crankcase.

The electronic control system controls the amount of fuel delivered

Figure 8.19 A common rail fuel system was an innovation on the MTU Series 4000 high speed engine

to the injectors by means of the solenoid valves, and the injection pressure is also adjusted to the optimum level according to the engine power demand. Separate injection pumps for each cylinder are eliminated and hence the need for a complicated drive system for traditional pumps running off the camshafts (which, due to high mechanical stresses, calls for reinforced gearing systems or even a second system of gears on larger engines). Fewer components naturally foster greater reliability.

Another benefit cited for the common rail system is its flexible capability across the engine power band, and the fact that it is able to deliver the same injection pressure at all engine speeds from rated speed down to idling. In the case of the Series 4000 engine that pressure is around 1200 bar, corresponding to 1600–1800 bar on a conventional system.

Success is reflected in the fuel consumption (below 195 g/kWh) and emission characteristics of the engine. Future fuel injection systems must be distinguished by extremely flexible quantity control, commencement of injection timing and injection rate shaping: capabilities achieved only with the aid of electronic control systems.

Conventional injection systems with mechanical actuation include in-line pumps, unit pumps with short high pressure (HP) fuel lines and unit injectors. A cam controls the injection pressure and timing while the fuel volume is determined by the fuel rack position. For future engines with high injection pressures, however, the in-line pump system can be ignored because it would be hydraulically too 'soft' due to the long HP lines, MTU asserts.

A comparison has been made between unit pump and unit injector systems, assuming the unit injector drive adopts the typical camshaft/push rod/rocker arm principle. Using simulation calculations, the relative behaviour of the two systems was investigated for a specified mean injection pressure of 1150 bar in the injector sac. This time-averaged sac pressure is a determining factor in fuel mixture preparation, whereas the frequently used maximum injection pressure is less meaningful, MTU explains. The pressure in a unit pump has been found to be lower than in a unit injector but, because of the dynamic pressure increase in the HP fuel line, the same mean injection pressure of 1150 bar is achieved with less stress in the unit pump. With the unit injector, the maximum sac pressure was 1670 bar (some 60 bar higher than in the unit pump). To generate 1150 bar the unit injector needed 3.5 kW, some 6 per cent more power. During the ignition delay period 12.5 per cent of the cycle-related amount of fuel was injected by the unit pump compared with 9.8 per cent by unit injector. The former is, therefore, overall the stiffer system. Translating the pressure differential at the nozzle orifice and the volume flow into the mechanical energy absorbed, the result was a higher efficiency of 28 per cent for the unit pump compared with 26 per cent for the unit injector.

From the hydraulic aspect, MTU reports, the unit pump offers benefits in that there is no transfer of mechanical forces from the push rod drive to the cylinder head, and less space is required for the fuel injector (yielding better design possibilities for inlet and exhaust systems). With conventional systems, the volume of fuel injected is controlled by the fuel rack; and matching the individual cylinders dictates appropriate engineering effort. The effort is increased significantly if injection timing is effected mechanically.

The engineering complexity involved in enabling fuel injection and timing to be freely selected can be reduced considerably by using a solenoid valve to effect time-orientated control of fuel quantity. To produce minimum fuel injection quantity extremely short shift periods must be possible to ensure good engine speed control. Activation of the individual solenoid valves and other prime functions, such as engine speed control and fuel injection limitation, is executed by a microprocessor-controlled engine control unit (ECU). Optional adjustment of individual cylinder fuel injection calibration and injection timing is thus possible with the injection period being newly specified and realized for each injection phase. Individual cylinder cut-out control is only a question of the software incorporated in the ECU.

With cam-controlled injection systems, the injection pressure is dependent on the pump speed and the amount of fuel injected. For engines with high mean effective pressures in the lower speed and low load ranges, this characteristic is detrimental to the atomization process as the injection pressure drops rapidly.

Adjusting the injection timing also influences the in-system pressure build-up. For example, if timing is advanced the solenoid valve closes earlier, fuel compression starts at a lower speed and thus leads to lower injection pressures which, in turn, undermines mixture preparation.

To achieve higher injection pressures, extremely steep cam configurations are required. As a result, high torque peaks are induced into the camshaft which dictates a compensating degree of engineering effort on the dimensioning of the camshaft and gear train, and may even call for a vibration damper.

While the solenoid valve-controlled system has a number of advantages, MTU argues, it retains the disadvantages of the conventional systems; and in the search for a new flexible injection system it represents only half a step. A full step is only achieved when the pressure generation, fuel quantity and injection timing functions can be varied independently by exploiting the common rail injection system (CRIS).

In the CRIS configuration for a V16-cylinder engine (Figure 8.20) the HP pump delivers fuel to the rail which is common to all cylinders.

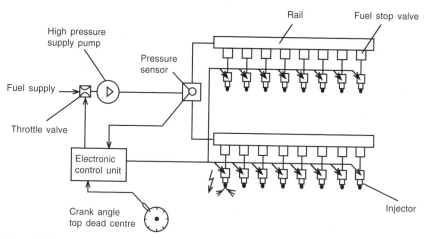

Figure 8.20 Common rail injection system arranged for a V16-cylinder high speed engine (MTU)

Each injector is actuated in sequence by the ECU as a function of the crankshaft angle. The injector opens when energized and closes when de-energized. The amount of fuel injected per cycle is determined by the time differential and the in-system pressure. The actual in-system pressure is transmitted to the control unit via a pressure sensor, and the rail pressure is regulated by the ECU via the actuator in the fuel supply to the HP pump.

In such a system the injector incorporates several functions. The nozzle needle is relieved by a solenoid valve and thus opened by the fuel pressure. The amount of fuel injected during the ignition delay period is regulated by the nozzle opening speed. After the control valve is de-energized an additional hydraulic valve is activated to secure rapid closure of the nozzle needle and, therefore, a minimum smoke index. This servo-assisted injector allows the opening and closing characteristics to be adjusted individually and effected very precisely. It is capable of extremely high reaction speeds for controlling minimum fuel quantities during idle operation or pilot injection, MTU claims.

Compared with a conventional injection system, the pumping force is considerably lower: pressure generation is accomplished by a multi-cylinder, radial piston pump driven by an eccentric cam. Pressure control is realized by restricting the supply flow. Locating the HP pump on the crankcase presents no problems, while the deletion of the fuel injection control cams from the camshaft allows that component to be dimensioned accordingly.

The rail system is required to supply all injectors with fuel at an identical pressure, and its design is based on criteria such as minimum fuel injection quantity deviations between cylinders, faster pressure build-up after start and minimum pressure loss from the rail to the nozzle sac. Simulation exercises have shown that pressure fluctuations can be very low.

The ECU determines the engine speed and calculates the amount of fuel to be injected, based on the difference between actual and preset engine speeds. The individual injectors are energized as a function of crankshaft angle and firing order.

Control of the common rail in-system pressure is also carried out by the ECU. The advantages inherent in the solenoid valve-controlled conventional injection system (regarding speed control due to the improved actuator dynamic) can also be applied to the same extent to CRIS: at the moment of each injection the governor-generated data can be immediately processed. Pressure regulation is highly effective, says MTU. Load shedding (required, for example, when a waterjet propulsion unit emerges from the sea during rough conditions) can be effected from full to zero load in just 10 ms. While this leads to a rapid rise in engine speed the governor reduces fuel injection to zero which, in turn, causes a pressure increase in the rail system as the individual injectors become inactive. The pressure regulator reacts and restricts the HP pumps to a maximum of 1330 bar (a pressure excess of 130 bar over the specified value of 1200 bar). Additionally, the ECU assumes the control and monitoring functions that are standard for MTU engines, including sequential turbocharging control.

With common rail injection the complete system is permanently subjected to extremely high pressures. In the event of failure of a single injector the engine is protected in that the injector is cut off from the fuel supply by the fuel stop valve: no fuel can enter the combustion chamber. The engine can, however, still be operated at reduced (get-you-home) power.

A mechanical pressure relief valve is actuated if, due to a pressure regulating system malfunction, the pressure rises to an unacceptable level. Leaks in the system are identified by the pressure monitoring system.

During tests, MTU reports, the unit pump system displayed the best characteristics due to the high specified injection pressures. Greater flexibility, improved dynamics, reduced design effort and, as a result, lower costs are offered by the unit pump system with solenoid valve control. Only a common rail injection system, however, makes it possible to achieve injection rate shaping over the complete operating range and thus to underwrite the specifications for future diesel engines, MTU asserts.

The high pressure pump for the Series 4000 fuel system was changed from a four- or eight-cylinder radial piston pump to a four-cylinder in-line unit to increase its performance and achieve a higher delivery rate for the V20-cylinder engine model.

MTU Series 8000 engine CR system

MTU subsequently applied an electronic common rail system to the group's largest-ever high speed engine, the 265 mm bore Series 8000, introduced in 2000 (see Chapter 30), whose V20-cylinder version develops up to 9000 kW at 1150 rev/min. As on the Series 4000 engine, the system enables infinite variation of injection timing, volume and pressure over the entire engine performance curve. The supply of pressurized fuel to the injectors, however, is achieved by different means. On the Series 4000 engine there is a rigid fuel accumulator (common rail) for each row of cylinders, while the Series 8000 engine has a separate accumulator for each cylinder located in the vicinity of the injector and with a capacity of 0.4 litres.

This arrangement facilitates a smaller overall engine width and optimum progression of combustion, and reduces the interplay between the injectors and the pressure fluctuations in the fuel pipes. Only two high pressure pumps are required to generate the substantial pressure and the necessary flow volume for the entire engine.

The L'Orange system for the Series 8000 engine is based on two 3-cylinder high pressure fuel pumps producing a maximum rail pressure of 1800 bar. These pumps are integrated in the gear drive at the front side of the engine and driven by the crankshaft with a gear ratio of 1.65 at speeds up to 1800 rev/min. The high pressure part of the injection system, with the functions of high pressure generation and accumulation as well as fuel metering, comprises:

- Two high pressure fuel pumps with integrated pump accumulators.
- Up to 20 accumulators, each equipped with a flow limiting valve.
- High pressure fuel lines.
- Up to 20 electro-magnetically controlled injectors.

Double-walled high pressure fuel lines distribute the fuel from the pumps along the engine to each cylinder. An accumulator serving each cylinder acts as a storage space to maintain an almost constant fuel pressure during injection, despite the relatively small cross-section in the fuel supply lines. The compressibility of the fuel ensures an almost constant injection pressure during injection. The intervals between injection are used to replace the injected fuel mass in the accumulators and build up the fuel pressure again. The accumulator incorporates a so-called fuel-stop valve which interrupts the fuel supply to the relevant injector in cases of excessive fuel flow rates (for example, caused by leakages or continuous injection). The engine management system will then adjust the power output of the engine accordingly.

Rounded-off spray hole intake geometries on the electronically-controlled injector ensure an optimally sprayed injection of fuel into the combustion chamber. The start of injection and the quantity injected is accurately controlled to the microsecond by the engine control system in line with the engine operating condition.

The system fuel pressure is available at the nozzle seat as well as in the control unit above the plunger for needle lift control (the so-called connecting rod). Compared with the usual two-cycle systems, this offers the advantage of a faster operating response. The different sized pressurized surfaces of connecting rod and needle result in a hydraulic closing force. When the solenoid is energized by an electrical signal the armature is lifted. This opens an accurately machined cross-section of the outlet throttle in the control unit, causing a rapid drop in the hydraulic pressure on the connecting rod. The nozzle needle lifts off the seat and starts the injection process, which is stopped when the solenoid is de-energized. The closing flank of the injection characteristics can be shaped very rapidly through an intermediate valve after the control unit; it opens additional supply cross-sections during closing of the solenoid.

Fuel is injected in the precise quantity and at exactly the moment required for the conditions under which the engine is operating. The pump's intake flow throttle ensures that only the precise amount of fuel required is pressurized, resulting in a very high pump efficiency and consequently low fuel consumption.

The advantages of the common rail injection system are fully exploited in conjunction with the engine control system. The flexibility of the injection system with respect to injection pressure and timing allows the option to access and optimize the engine operating parameters throughout the performance map (addressing fuel consumption and emissions, for example).

For each operating condition of the engine the engine control system analyses the sensor information and computes them into a setting for system fuel pressure, injection start and timing. The computed data takes into account engine speed, power demand and transient operating conditions, such as engine warm-up after cold starting. The scope also embraces surrounding conditions and possible faults in the drive system, with appropriate countermeasures automatically being initiated. In addition, the engine control system monitors the limits to certain operating parameters (maximum pressure in the fuel system, for example) to avoid component damages.

Wärtsilä CR system for medium speed engines

A common rail fuel system developed by Wärtsilä in conjunction with Delphi Bryce for application to the group's heavy fuel-burning medium speed engines is modular, with common components designed to serve engines with cylinder numbers from 4 to 18. Development was stimulated by increasing market pressure, especially from passenger tonnage, for no visible smoke emissions under any circumstances.

The Wärtsilä system (Figure 8.21) is based on a pressurizing pump and a fuel accumulator for every two cylinders, the series of accumulators interconnected by small-bore piping. Connections from every accumulator feed the injectors of two cylinders (in engines with an odd number of cylinders an extra accumulator is installed with a connection to one cylinder.)

Functionally, the main reason for choosing such a system is that by splitting up the fuel volumes in several accumulators the risk of pressure

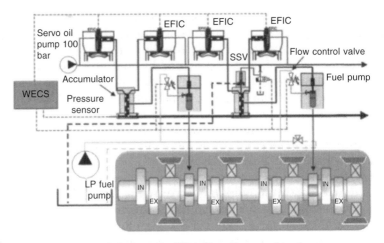

Figure 8.21 Common rail fuel system for Wärtsilä medium speed engines

waves in the common rail is avoided. From a safety viewpoint, an advantage is that high pressure fuel exists only in the same area as in a conventional engine (in the hot box). A third merit cited is economic, the system using a camshaft that is present anyway to drive the valves.

The computer-controlled fuel injection system opens up possibilities to optimize engine performance thanks to the freedom to choose key parameters: injection timing; injection pressure independent of engine speed; maximum injection quantity dependent on charge air pressure and rotational speed; optimized injection during starting; and optimized injection timing at load pick-up.

Each common rail fuel pump is driven by a two-lobe cam, whose pumping profile can be considerably slower than the conventional jerk pump type that it replaces. In a common rail system there is no longer the requirement to pump fuel to the injector directly within the typical range of 6-10 microseconds; this fosters a reduction in mechanical drive noise and drive energy consumed. The interconnected accumulator arrangement reduces the pulsation migration ability along the rail. Additionally, the rail components are rigidly fixed to the engine block, reducing the risks of piping leakage. Each accumulator is connected directly to two injectors, and each pipeline contains a hydraulic flow fuse for safety reasons; each accumulator also contains a pressure sensor.

Wärtsilä's engine control system (WECS) regulates the common rail system, monitoring several system parameters; it also controls fuel pressure in the rail and injection quantities and timings according to specified maps. The solenoids are not responsible for all control;

pressurized engine oil is used to achieve the necessary forces in the injectors and stop solenoids (this oil system is termed 'control oil').

The common rail pump is constructed with a tappet and barrel element integrated in one housing. The use of heavy fuel oil makes it preferable to divide the pump internally; this places the pumping element in a fuel-only environment, which eliminates the risk of lacquer accumulation from the fuel and chemical interaction with the engine oil. The pump can be driven by two- or three-cam lobes. The two-lobe alternative is preferred both for the roller/roller pin oil film duty cycle and because it creates fewer pumping pulses to settle into the rail prior to injections.

The volumetric efficiency of the pump is typically above 80 per cent. This level is generally higher than that of jerk pumps with a plunger spill helix, where the end of the plunger stroke marks the rejection of fuel into the return circuit. Smaller plunger sizes are thus allowed, which in turn significantly reduces the maximum camshaft drive torque for an equivalent pumping pressure target.

Charging of the common rail is controlled by regulating the volume of fuel allowed to enter the pumping chamber; this is effected by a flow control valve installed in either the barrel or the pump housing. The flow control valve itself is rotated by a rotary solenoid, which has a position feedback sensor for fault detection prior to engine start or during engine operation. Another detection device is incorporated in the barrel. A thermocouple sensor is installed close to the pumping chamber to detect delivery valve leakage; if the delivery valve were to leak, fuel from the rail would be re-pumped, causing a temperature increase in the pumping chamber.

Each accumulator, mounted rigidly on the engine block, receives fuel from the high pressure pump, stores it under high pressure and distributes it to the injectors. The accumulator itself is an empty space. Flow fuses located at outlets from the accumulator to the injector limit the quantity of fuel that can be transferred from the accumulator to the injector if the latter malfunctions. The accumulator is also fitted with a device that makes it easy to find a leaking pipe if the leak alarm is activated.

At least one of the accumulators is also equipped with a start-up and safety valve (SSV), which opens the flow path from the accumulator and is controlled by the WECS. The SSV is opened when the engine is in the pre-warming phase or when the engine is to be stopped. Incorporated in the SSV is a mechanical safety valve to protect the system against excessive rail pressure, this valve operating completely independently of the main valve. Connected to the SSV is an empty space in the fuel outlet channel, whose function is to dampen the pressure waves when the SSV is opened.

The fuel injector was designed for efficiency and safe operation, with four main elements: nozzle, shuttle valve, solenoid valve and piston acting on the nozzle needle. The injector is fitted with a conventional nozzle that is linked to the common rail fuel source via the shuttle valve. Two key functions are performed by the shuttle valve: it controls injections and de-pressurizes the injector needle. In its closed (static) position the valve isolates the injector passages and nozzle from the pressure source, so that pressure is not applied to the nozzle seat between injections. The shuttle is opened by applying a hydraulic load using the control oil source. The control oil is switched on and off using a fast-acting solenoid valve (a reliable automotive component produced in high volumes).

Such an approach fosters full flexibility of the start of injection and fuelling quantity, Wärtsilä explains. The piston above the nozzle, together with any pressure on it, acts to hold the nozzle closed; pressure is normally only applied to the piston at the start and end of injections. The piston arrangement has three functions:

- The pressure on the piston at the start of injection slightly delays opening of the nozzle, allowing time for the shuttle to complete its travel—even for the smallest injections. This promotes fuelling consistency because the need for partial shuttle travel at the lowest outputs is avoided.
- At the end of injection the pressure on the piston helps to close the nozzle. The fuel pressure at the nozzle seat at the end of injection is much higher than in conventional systems and thus each injection is terminated very rapidly.
- A safety benefit: if the shuttle becomes stuck in a partially open position the pressure is continuously applied to the piston, this force keeping the nozzle closed in this fault condition.

These design features resulted in an injector with a high mean operating pressure, capable of working up to 2000 bar. Injections start crisply and terminate with the nozzle being closed while injecting a nearly full pressure. The injector was also designed for robust safety in the event of a fault condition:

- Rail pressure is not applied to the nozzle seat between injections; this ensures that the fuel is cut off if the nozzle needle is prevented from fully closing.
- The piston acting on the nozzle needle prevents injection if the shuttle seizes or is prevented from fully closing.

Engine operation on heavy fuel oil is addressed by the following features:

- During the engine's pre-start warm-up phase the pump flow control valves are shaken and opened separately every five seconds. Heated HFO flows through each pump in turn and then onward to the rail and back to the tank via the bypass SSV valve system.
- The fuel injectors do not have HFO present in the solenoid armature area; a servo oil system is used.
- In the fuel injectors the rail pressure does not act directly on the nozzle needle, reducing the risk of needle sticking; and the rail pressure is released from the nozzle needle between injections to improve safety.

The Wärtsilä common rail system was designed for retrofitting to the group's existing engines by simply removing the original conventional injection pumps and replacing them with common rail fuel pumps and accumulators that fit exactly into current engine blocks.

Acknowledgements to Duap, Fiedler Motoren Consulting, Delphi Bryce, MTU, L'Orange and Wärtsilä Corporation.

Low speed engine fuel injection systems, including electronically-controlled configurations, are covered in Chapter 9 (Low Speed Engines introduction), Chapter 10 (MAN B&W), Chapter 11 (Mitsubishi) and Chapter 12 (Sulzer).

9 Low speed engines—introduction

Low speed two-stroke engine designers have invested heavily to maintain their dominance of the mainstream deepsea propulsion sector formed by tankers, bulk carriers and containerships. The long-established supremacy reflects the perceived overall operational economy, simplicity and reliability of single, direct-coupled crosshead engine plants. Other factors are the continual evolution of engine programmes by the designer/licensors in response to or anticipation of changing market requirements, and the extensive network of enginebuilding licensees in key shipbuilding regions. Many of the standard ship designs of the leading yards, particularly in Asia, are based on low speed engines.

The necessary investment in R&D, production and overseas infrastructure dictated to stay competitive, however, took its toll over the decades. Only three low speed engine designer/licensors—MAN B&W Diesel, Mitsubishi and Sulzer (now part of the Wärtsilä Corporation)—survived into the 1990s to contest the international arena.

The roll call of past contenders include names either long forgotten or living on only in other engineering sectors: AEG-Hesselman, Deutsche Werft, Fullagar, Krupp, McIntosh and Seymour, Neptune, Nobel, North British, Polar, Richardsons Westgarth, Still, Tosi, Vickers, Werkspoor and Worthington. The last casualties were Doxford, Götaverken and Stork whose distinctive engines remain at sea in diminishing numbers. The pioneering designs displayed individual flair within generic classifications which offered two- or four-stroke, single- or double-acting, and single- or opposed-piston configurations. The Still concept even combined the Diesel principle with a steam engine: heat in the exhaust gases and cooling water was used to raise steam which was then supplied to the underside of the working piston.

Evolution decreed that the surviving trio of low speed crosshead engine designers should pursue a common basic configuration: two-stroke engines with constant pressure turbocharging and uniflow scavenging via a single hydraulically-operated exhaust valve in the cylinder head. Current programmes embrace mini-to-large bore models with short, long and ultra-long stroke variations to match the propulsive power demands and characteristics of most deepsea (and even some coastal/shortsea) cargo tonnage. Installations can be near-optimized for a given duty from a permutation involving the engine bore size, number of cylinders, selected output rating and running speed. Bore sizes range from 260 mm to 1080 mm, stroke/bore ratios up to 4.2:1, in-line cylinder numbers from four to 14, and rated speeds from around 55 to 250 rev/min. Specific fuel consumptions as low as 154 g/kW h are quoted for the larger bore models whose economy can be enhanced by optional Turbo Compound Systems in which power gas turbines exploit exhaust energy surplus to the requirements of modern high efficiency turbochargers.

Progress in the performance development of low speed engines in the popular circa-600 mm bore class is illustrated in Figure 9.1.

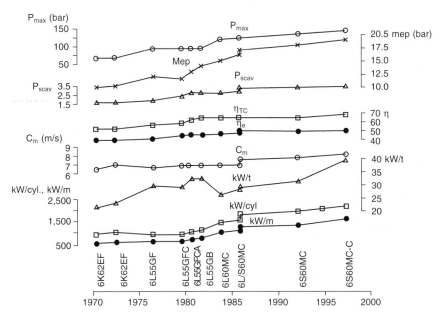

Figure 9.1 Development of key performance parameters for low speed engines (circa 600 mm bore) over a 30-year period (MAN B&W Diesel)

Recent years have seen the addition of intermediate bore sizes to enhance coverage of the power/speed spectrum and further optimize engine selection. Both MAN B&W Diesel and Sulzer also extended their upper power limits in the mid-1990s with the introduction of super-large bore models—respectively of 980 mm and 960 mm bore sizes—dedicated to the propulsion of new generations of 6000 TEU-plus containerships with service speeds of 25 knots or more. The 12-

cylinder version of MAN B&W's current K98MC design, delivering 68 640 kW, highlights the advance in specific output achieved since the 1970s when the equivalent 12-cylinder B&W K98GF model yielded just under 36 800 kW. Large bore models tailored to the demands of new generation VLCC and ULCC propulsion have also been introduced.

Successively larger and faster generations of post-Panamax containerships have driven the development of specific output and upper power limits by low speed engine designers. To ensure that containership designs with capacities up to and exceeding 10 000 TEU can continue to be specified with single engines, both MAN B&W and Wärtsilä extended their respective MC and Sulzer RTA programmes. Wärtsilä raised the specific rating of the Sulzer RTA96C design by 4 per cent to 5720 kW/cylinder at 102 rev/min, and introduced an in-line 14-cylinder model delivering up to 80 080 kW. (The previous power ceiling had been 65 880 kW from a 12-cylinder model.)

MAN B&W responded to the challenge with new 13- and 14-cylinder variants of the K98MC and MC-C series, offering outputs from 74 230 kW to 80 080 kW at 94 or 104 rev/min. These series can also be extended to embrace 15- to 18-cylinder models, if called for, taking the power threshold to just under 103 000 kW. Such an output would reportedly satisfy the propulsive power demands of containerships with capacities up to 18 000 TEU and service speeds of 25–26 knots. In 2003 MAN B&W Diesel opened another route to higher powers: a 1080 mm bore version of the MC engine was announced with a rating of 6950 kW/cylinder at 94 rev/min, the 14-cylinder model thus offering 97 300 kW.

V-cylinder configurations of existing low speed engine designs have also been proposed by MAN B&W Diesel to propel mega-containerships, promising significant savings in weight and length per unit power over traditional in-line cylinder models. These engines would allow the higher number of cylinders to be accommodated within existing machinery room designs (Figure 9.2).

Parallel development by the designer/licensors seeks to refine existing models and lay the groundwork for the creation of new generations of low speed engine. Emphasis in the past has been on optimizing fuel economy and raising specific outputs but reliability, durability and overall economy are now priorities in R&D programmes, operators valuing longer component lifetimes, extended periods between overhauls and easier servicing.

Lower production costs through more simple manufacture and easier installation procedures are also targeted, reflecting the concerns of enginebuilder/licensees and shipyards. More compact and lighter

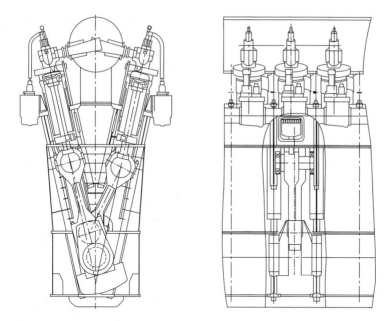

Figure 9.2 V-cylinder versions of larger bore MC/MC-C engines have been proposed by MAN B&W Diesel

weight engines are appreciated by naval architects seeking to maximize cargo space and deadweight capacity within given overall ship dimensions.

In addition, new regulatory challenges—such as noxious exhaust emission and noise controls—must be anticipated and niche market trends addressed if the low speed engine is to retain its traditional territory (for example, the propulsion demands of increasingly larger and faster containerships which might otherwise have to be met by multiple medium speed engines or gas turbines).

A number of features have further improved the cylinder condition and extended the time between overhauls through refinements in piston design and piston ring configurations. A piston cleaning ring incorporated in the top of the cylinder liner now controls ash and carbon deposits on the piston topland, preventing contact between the liner and these deposits which would otherwise remove part of the cylinder lube oil from the liner wall.

Computer software has smoothed the design, development and testing of engine refinements and new concepts but the low speed engine groups also exploit full-scale advanced hardware to evaluate innovations in components and systems.

Sulzer began operating its first Technology Demonstrator in 1990, an advanced two-stroke development and test engine designated the 4RTX54 whose operating parameters well exceeded those of any production engine (Figure 9.3). Until then, the group had used computer-based predictions to try to calculate the next development stage. Extrapolations were applied, sometimes with less than desirable results. The 4RTX54 engine, installed at the Swiss designer's Winterthur headquarters, allowed practical tests with new parameters, components and systems to be carried out instead of just theory and calculations. Operating data gathered in the field could be assessed alongside results derived from the test engine.

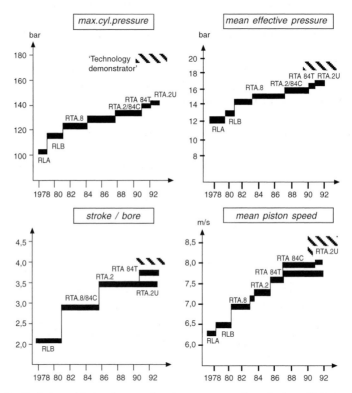

Figure 9.3 Evolution of Sulzer low speed engine parameters from the late 1970s in comparison with those of its RTX 54 Two-stroke Technology Demonstrator engine

The four-cylinder 540 mm bore/2150 mm stroke engine had a stroke/bore ratio approaching 4:1 and could operate with mean effective pressures of up to 20 bar, maximum cylinder pressures up to 180 bar and mean piston speeds up to 8.5 m/sec. Operating without a camshaft—reportedly the first large two-stroke engine to do so—the RTX54 was equipped with combined mechanical, hydraulic, electronic (mechatronics) systems for fuel injection, exhaust valve lift, cylinder lubrication and starting, as well as controllable cooling water flow. The systems underwrote full flexibility in engine settings during test runs.

Sulzer's main objectives from the Technology Demonstrator engine were to explore the potential of thermal efficiency and power concentration; to increase the lifetime and improve the reliability of components; to investigate the merits of microprocessor technology; and to explore improvements in propulsion efficiency. A number of concepts first tested and confirmed on the 4RTX54 engine were subsequently applied to production designs. The upgraded RTA-2U series and RTA84T, RTA84C and RTA96C engines, for example, benefit from a triple-fuel injection valve system in place of two valves. This configuration fosters a more uniform temperature distribution around the main combustion chamber components and lower overall temperatures despite higher loads. Significantly lower exhaust valve and valve seat temperatures are also yielded.

An enhanced piston ring package for the RTA-2U series was also proven under severe running conditions on the 4RTX54 engine. Four rings are now used instead of five, the plasma-coated top ring being thicker than the others and featuring a pre-profiled running face. Excellent wear results are reported. The merits of variable exhaust valve closing (VEC) were also investigated on the research engine whose fully electronic systems offered complete flexibility. Significant fuel savings in the part-load range were realized from the RTA84T 'Tanker' engine which further exploits load-dependent cylinder liner cooling and cylinder lubrication systems refined on the 4RTX54. The 4RTX54 was replaced as a research and testing tool in 1995 by the prototype 4RTA58T engine adapted to serve as Sulzer's next Two-stroke Technology Demonstrator (Figure 9.4).

The widest flexibility in operating modes and the highest degree of reliability are cited by Copenhagen-based MAN B&W Diesel as prime R&D goals underwriting future engine generations, along with:

- Ease of maintenance.
- Production cost reductions.
- Low specific fuel consumption and high plant efficiency over a wide load spectrum.
- High tolerance towards varied heavy fuel qualities.
- Easy installation.
- Continual adjustments to the engine programme in line with the evolving power and speed requirements of the market.

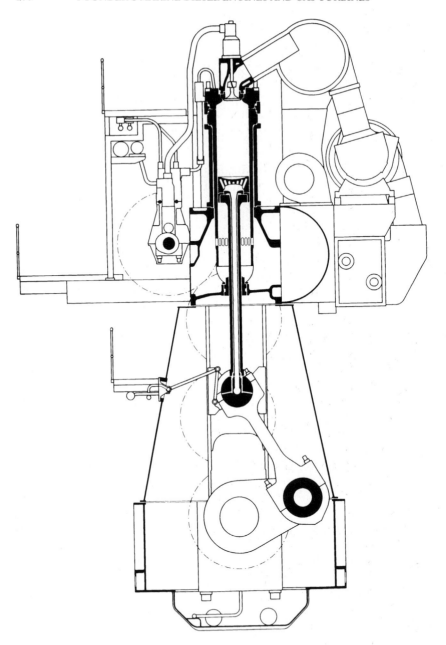

Figure 9.4 The current Sulzer Two-stroke Technology Demonstrator research engine is based on an RTA58T model

- Compliance with emission controls.
- Integrated intelligent electronic systems.

Continuing refinement of MAN B&W Diesel's MC low speed engine programme and the development of intelligent engines (see section below) are supported by an R&D centre adjacent to the group's Teglholmen factory in Copenhagen.

At the heart of the centre is the 4T50MX research engine, an advanced testing facility which exploits an unprecedented 4.4:1 stroke/bore ratio. Although based on the current MC series, the four-cylinder 500 mm bore/2200 mm stroke engine is designed to operate at substantially higher ratings and firing pressures than any production two-stroke engine available today. An output of 7500 kW at 123 rev/min was selected as an initial reference level for carrying out extensive measurements of performance, component temperatures and stresses, combustion and exhaust emission characteristics, and noise and vibration. The key operating parameters at this output equate to 180 bar firing pressure, 21 bar mean effective pressure and 9 m/sec mean piston speed. Considerable potential was reserved for higher ratings in later test running programmes.

A conventional camshaft system was used during the initial testing period of the 4T50MX engine. After reference test-running, however, this was replaced by electronically controlled fuel injection pumps and exhaust valve actuators driven by a hydraulic servo-system (Figure 9.5). The engine is prepared to facilitate extensive tests on primary methods of exhaust emission reduction, anticipating increasingly tougher regional and international controls in the future. Space was allocated in the R&D centre for the installation of a large NOx-reducing selective catalytic reduction (SCR) facility for assessing the dynamics of SCR-equipped engines and catalyst investigations.

The research engine, with its electronically controlled exhaust valve and injection system, has fully lived up to expectations as a development tool for components and systems, MAN B&W Diesel reports. A vast number of possible combinations of injection pattern, valve opening characteristics and other parameters can be permutated. The results from testing intelligent engine concepts are being tapped for adoption as single mechanical units as well as stand-alone systems for application on current engine types. To verify the layout of the present standard mechanical camshaft system, the 4T50MX engine was rebuilt with a conventional mechanical camshaft unit on one cylinder. The results showed that the continuous development of the conventional system seems to have brought it close to the optimum, and the comparison gave no reason for modifying the basic design.

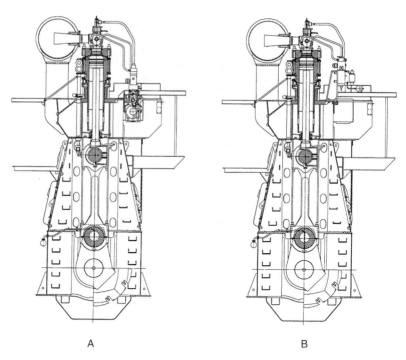

A   B

Figure 9.5 MAN B&W Diesel's 4T50MX low speed research engine arranged with a conventional camshaft (A) and with electronically controlled fuel injection pumps and exhaust value actuating pumps (B)

An example of the degrees of freedom available is shown by a comparison between the general engine performance with the firing pressure kept constant in the upper load range by means of variable injection timing (VIT) and by variable compression ratio (VCR). The latter is obtained by varying the exhaust valve closing time. This functional principle has been transferred to the present exhaust valve operation with the patented system illustrated in the diagram (Figure 9.6). The uppermost figure shows the design of the hydraulic part of the exhaust valve; below is the valve opening diagram. The fully drawn line represents control by the cam while the dotted line shows the delay in closing, thus reducing the compression ratio at high loads so as to maintain a constant compression pressure in the upper load range. The delay is simply obtained by the oil being trapped in the lower chamber; and the valve closing is determined by the opening of the throttle valve which is controlled by the engine load.

Traditionally, the liner cooling system has been arranged to match the maximum continuous rating load. Today, however, it seems

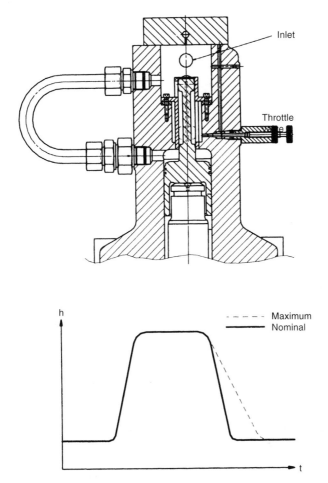

Figure 9.6 Mechanical/hydraulic variable compression ratio (MAN B&W Diesel)

advantageous to control the inside liner surface temperature in relation to the load. Various possibilities for securing load-dependent cylinder liner cooling have therefore been investigated. One system exploits different sets of cooling ducts in the bore-cooled liner, the water supplied to the different sets depending on the engine load. Tests with the system have shown that the optimum liner temperature can be maintained over a very wide load range. The system is considered perfectly feasible but the added complexity has to be carefully weighed against the service advantages.

The fuel valve used on MC engines operates without any external control of its function. The design has worked well for many years but could be challenged by the desire for maintaining an effective performance at very low loads. MAN B&W Diesel has therefore investigated a number of new designs with the basic aim of retaining a simple and reliable fuel valve without external controls. Various solutions were tested on the 4T50MX engine, among them a design whose opening pressure is controlled by the fuel oil injection pressure level (which is a function of the engine load). At low load the opening pressure is controlled by the spring alone. When the injection pressure increases at higher load, this higher pressure adds to the spring force and the opening pressure increases.

Another example of fuel valve development is aimed at reduced emissions. This type incorporates a conventional conical spindle seat as well as a slide valve inside the fuel nozzle, minimizing the sac volume and thus the risk of after-dripping. Significantly lower NOx emissions are reported, as well as reduced smoke and even carbon monoxide, but at the expense of a slightly higher fuel consumption. This type of fuel valve is now included in the options for special low NOx applications of MC engines (Figure 9.7).

The 4T50MX engine was used to test a triple fuel valve-per-cylinder configuration, the measurements mirroring Sulzer's results in yielding reduced temperature levels and a more even temperature distribution than with a two-valve arrangement. The K80MC-C, K90MC/MC-C, S90MC-T and K98MC-C engines were subsequently specified with triple fuel valves to enhance reliability.

Intelligent engines

Both MAN B&W Diesel and New Sulzer Diesel demonstrated 'camshaftless' operation with their research engines, applying electronically controlled fuel injection and exhaust valve actuation systems. Continuing R&D will pave the way for a future generation of highly reliable 'intelligent engines': those which monitor their own condition and adjust parameters for optimum performance in all operating regimes, including fuel-optimized and emissions-optimized modes. An 'intelligent engine-management system' will effectively close the feedback loop by built-in expert knowledge.

Engine performance data will be constantly monitored and compared with defined values in the expert system; if deviations are detected corrective action is automatically taken to restore the situation to normal. A further step would incorporate not only engine optimizing functions but management responsibilities, such as maintenance planning and spare parts control.

To meet the operational flexibility target, MAN B&W Diesel explains,

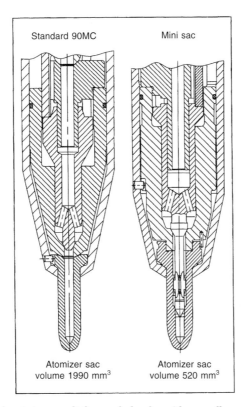

Standard 90MC   Mini sac

Atomizer sac   Atomizer sac
volume 1990 mm³   volume 520 mm³

Figure 9.7 Reduced emissions result from a fuel valve with a smaller sac volume (MAN B&W Diesel). See page 78 for valve with sac volume eliminated

it is necessary to be able to change the timing of the fuel injection and exhaust valve systems while the engine is running. To achieve this objective with cam-driven units would involve a substantial mechanical complexity which would undermine engine reliability. An engine without a traditional camshaft is therefore dictated.

The concept is illustrated in Figure 9.8 whose upper part shows the operating modes which may be selected from the bridge control system or by the intelligent engine's own control system. The centre part shows the brain of the system: the electronic control system which analyses the general engine condition and controls the operation of the engine systems shown in the lower part of the diagram (the fuel injection, exhaust valve, cylinder lube oil and turbocharging systems).

To meet the reliability target it is necessary to have a system which can actively protect the engine from damage due to overload, lack of maintenance and maladjustments. A condition monitoring system must be used to evaluate the general condition of the engine, thus maintaining

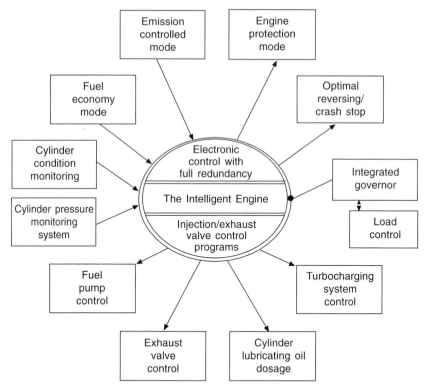

Figure 9.8 Schematic of the Intelligent Engine (MAN B&W Diesel)

its performance and keeping its operating parameters within prescribed limits. The condition monitoring and evaluation system is an on-line system with automatic sampling of all 'normal' engine performance data, supplemented by cylinder pressure measurements. The system will report and actively intervene when performance parameters show unsatisfactory deviations. The cylinder pressure data delivered by the measuring system are used for various calculations:

- The mean indicated pressure is determined as a check on cylinder load distribution as well as total engine output.
- The compression pressure is determined as an indicator of excessive leakage caused by, for example, a burnt exhaust valve or collapsed piston rings (the former condition is usually accompanied by an increased exhaust gas temperature in the cylinder in question).
- The cylinder wall temperature is monitored as an additional indicator of the piston ring condition.

- The firing pressure is determined for injection timing control and for control of mechanical loads.
- The rate of pressure rise (dp/dt) and rate of heat release are determined for combustion quality evaluation as a warning in the event of 'bad fuels' and to indicate any risk of piston ring problems in the event of high dp/dt values.

The cylinder condition monitoring system is intended to detect faults such as blow-by past the piston rings, cylinder liner scuffing and abnormal combustion. The detection of severe anomalies by the integrated systems triggers a changeover to a special operating mode for the engine, the 'engine protection mode'. The control system will contain data for optimum operation in a number of different modes, such as 'fuel economy mode', 'emission controlled mode', 'reversing/ crash stop mode' and various engine protection modes. The load limiter system (load diagram compliance system) aims to prevent any overloading of the engine in conditions such as heavy weather, fouled hull, shallow water, too heavy propeller layout or excessive shaft alternator output. This function will appear as a natural part of future governor specifications.

The fuel injection system is operated without a conventional camshaft, using high pressure hydraulic oil from an engine-driven pump as a power source and an electronically controlled servo system to drive the injection pump plunger. The general concept of the InFI (intelligent fuel injection) system and the InVA (intelligent valve actuation) system for operating the exhaust valves is shown in Figure 9.9. Both systems, when operated in the electronic mode, receive the electronic signals to the control units. In the event of failure of the electronic control system the engine is controlled by a mechanical input supplied by a diminutive camshaft giving full redundancy.

Unlike a conventional, cam-driven pump the InFI pump has a variable stroke and will only pressurize the amount of fuel to be injected at the relevant load. In the electronic mode (that is, operating without a camshaft) the system can perform as a single injection system as well as a pre-injection system with a high degree of freedom to modulate the process in terms of injection rate, timing, duration, pressure, single/double injection, cam profile and so on. Several optimized injection patterns can be stored in the computer and chosen by the control system in order to operate the engine with optimum injection characteristics at several loads: from dead slow to overload as well as for starting, astern running and crash stop. Changeover from one to another of the stored injection characteristics is effected from one injection to the next. The system is able to adjust the injection amount

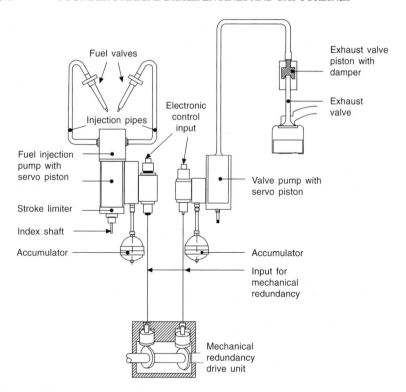

Figure 9.9 Electronically controlled hydraulic systems for fuel injection and exhaust valve operation on MAN B&W Diesel's 4T50MX research engine

and injection timing for each cylinder individually in order to achieve the same load (mean indicated pressure) and the same firing pressure (Pmax) in all cylinders; or, in protection mode, to reduce the load and Pmax on a given single cylinder if the need arises.

The exhaust valve system (InVA) is driven on the same principles as the fuel injection system, exploiting the same high pressure hydraulic oil supply and a similar facility for mechanical redundancy. The need for controlling exhaust valve operation is basically limited to timing the opening and closing of the valve. The control system is thus simpler than that for fuel injection.

Cylinder lubrication is controllable from the condition evaluation system so that the lubricating oil amount can be adjusted to match the engine load. Dosage is increased in line with load changes and if the need is indicated by the cylinder condition monitoring system (in the event of liner scuffing and ring blow-by, for example). Such systems are already available for existing engines.

The turbocharging system control will incorporate control of the scavenge air pressure if a turbocharger with variable turbine nozzle geometry is used, and control of bypass valves, turbocompound system valves and turbocharger cut-off valves if such valves are incorporated in the system. Valves for any selective catalytic reduction (SCR) exhaust gas cleaning system installed will also be controlled.

Operating modes may be selected from the bridge control system or by the system's own control system. The former case applies to the fuel economy modes and the emission-controlled modes (some of which may incorporate the use of an SCR system). The optimum reversing/crash stop modes are selected by the system itself when the bridge control system requests the engine to carry out the corresponding operation. Engine protection mode, in contrast, will be selected by the condition monitoring and evaluation system independently of actual operating modes (when this is not considered to threaten ship safety).

The fruit of MAN B&W Diesel's and Sulzer's R&D is now available commercially, their respective electronically-controlled ME and RT-flex engines being offered alongside the conventional models and increasingly specified for a wide range of tonnage. The designs are detailed in Chapters 10 and 12.

Research and development by Mitsubishi, the third force in low speed engines, has successfully sought weight reduction and enhanced compactness while retaining the performance and reliability demanded by the market. The Japanese designer's current UEC-LS type engines yield a specific power output of around three times that of the original UE series of the mid-1950s. The specific engine weight has been reduced by around 30 per cent over that period and the engine length in relation to power output has been shortened by one-third.

Mitsubishi strengthened its long relationship as a Sulzer licensee in 2002 by forging a joint venture with the Wärtsilä Corporation to develop a new 500 mm bore design to be offered in two versions: a 'mechanical' RTA50C and an 'electronic' RT-flex 50C.

10 MAN B&W low speed engines

MAN B&W Diesel's roots are closely entwined with the early days of the diesel engine through both its German and Danish branches, respectively MAN and Burmeister & Wain. Both groups evolved distinctive low speed two-stroke crosshead engine designs before MAN acquired control of B&W in 1980. MAN subsequently discontinued the development of its own loop-scavenged engines at Augsburg and centred all low speed design and R&D operations in Copenhagen, pursuing the refinement of the MAN B&W uniflow-scavenged programme in the shape of the MC series.

The prototype, an L35MC model, entered service in 1982 and the first large bore example, a six-cylinder L60MC engine, was started on the Christianshavn testbed in September 1983. The full L-MC programme was introduced in 1982 with bore sizes of 350 mm, 500 mm, 600 mm, 700 mm, 800 mm and 900 mm. Refinement of the MC design and the introduction of new bore sizes (260 mm, 420 mm, 460 mm and 980 mm) and stroke options continued through the 1980s/1990s in line with service experience and market trends. The current portfolio embraces 11 bore sizes from 260 mm to 1080 mm and K (short stroke), L (long stroke) and S (super-long stroke) variations, with stroke–bore ratios ranging from 2.44 to 4.2:1 and rated speeds from 56 rev/min to 250 rev/min.

Output demands from around 1100 kW to 97 300 kW are now covered by four- to 14-cylinder in-line models in the MC programme, whose individual rating envelopes are illustrated in Figure 10.1. Additions to the portfolio and key parameter refinements over the years—progressing to Mark 7 versions with mean effective pressures of 19 bar or higher—are shown in Table 10.1. Performance curves are typified by those for an S60MC engine (Figure 10.2).

Electronically-controlled ME versions of the 500 mm to 1080 mm bore models (detailed below) were progressively introduced from 2001 to offer the same outputs as their MC engine equivalents.

Choosing the most appropriate model for a given ship and propulsion duty is eased by MAN B&W's Computerized Engine Application System, which includes an integral speed and power prediction programme

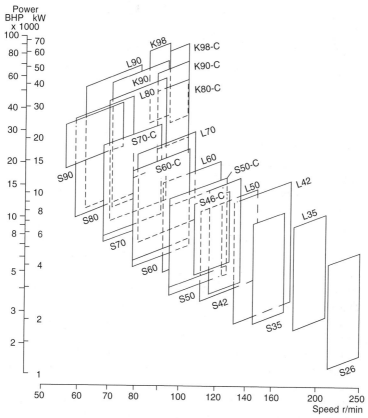

Figure 10.1 MAN B&W Diesel's MC engine programme. A K108MC model has taken the output threshold to 97 300 kW.

for ships and the corresponding main engine selection programme. Optimized solutions (bore size, number of cylinders, speed/power ratings) can be determined on the basis of both technical and economical data input. The system—comprising a number of integrated sub-programmes which are hierarchically arranged—facilitates calculations of such parameters as fuel consumption, utilization of exhaust gas heat and maintenance costs, and allows economic comparisons of the various options.

The layout diagram in Figure 10.3 (in this case for an S60MC engine) shows the area within which there is full freedom to select the combination of engine power (kW) and speed (rev/min) that is the optimum for the projected ship and the expected operating profile. The engine speed (horizontal axis) and engine power (vertical axis)

Table 10.1 Milestones in the evolution of MAN B&W Diesel's MC programme

Year	Development	Mark	Mep (bar)	Mean piston speed (m/sec)	Pmax (bar)
1981	L35MC introduced				
1982	Full L-MC programme	1	15.0	7.2	
1984	L-MC upgraded	2	16.2		
1985	L42MC introduced	2	16.2	7.2	
1986	K-MC introduced	3	16.2	7.6	130
	S-MC introduced	3	17.0	7.8	130
	L-MC upgraded	3	16.2	7.6	130
1987	S26MC introduced		16.8	8.2	
1988	K-MC-C introduced	3	16.2	8.0	130
1991	MC programme updated			8.0	
	K and L-MC	5	18.0	8.0	140
	S-MC	6	18.0		
1992	S26MC and L35MC updated		18.5	8.2	
1993	S35MC and S90MC-T introduced				
	K90MC/MC-C updated	6	18.0	8.0	
1994	S42MC introduced	6	18.5	8.0	
	K98MC-C introduced	6	18.2	8.3	140
1995	K80MC-C upgraded	6	18.0	8.0	
1996	L70MC upgraded	6	18.0	8.2	
	S70MC-C, S60MC-C, S50MC-C	7	19.0	8.5	150
	and S46MC-C introduced	7	19.0	8.3	150
1997	L80MC upgraded	6	18.0	8.0	
	K98MC introduced	6	18.2	8.3	140
1998	S80MC-C, S90MC-C,				
	L90MC-C introduced	7	19.0	8.1	150
	S35MC upgraded	7	19.1	8.1	145
1999	S42MC upgraded	7	19.5	8.0	
2001	L70MC-C introduced	7	19.0	8.5	150
	L60MC-C introduced	7	19.0	8.3	150

are shown in percentage scales. The scales are logarithmic which means that, in this diagram, exponential curves like propeller curves (3rd power), constant mean effective pressure curves (1st power) and constant ship speed curves (0.15 to 0.30 power) are straight lines. The constant ship speed lines, shown at the uppermost part of the diagram, indicate the power required at various propeller speeds in order to keep the same ship speed, provided that, for each ship speed, the optimum propeller diameter is used, taking into consideration the total propulsion efficiency.

An engine's layout diagram is limited by two constant mean effective pressure lines L1–L3 and L2–L4, and by two constant engine speed lines L1–L2 and L3–L4. The L1 point refers to the engine's nominal maximum continuous rating (mcr). Based on the propulsion and

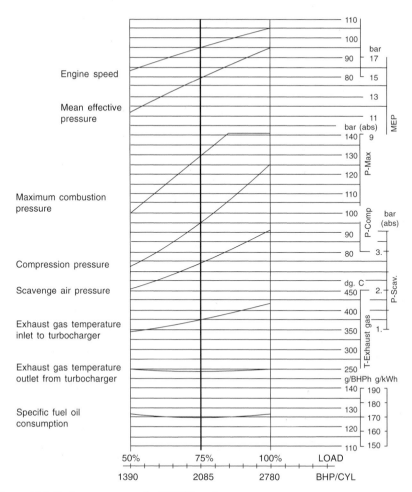

Figure 10.2 Performance curves for S60MC engine

engine running points, the layout diagram of a relevant main engine can be drawn in. The specified mcr point (M) must be inside or on the limitation lines of the layout diagram; if not, the propeller speed has to be changed or another main engine model must be chosen. It is only in special cases that point M may be located to the right of line L1–L2.

The specified mcr is the maximum rating required by the yard or owner for continuous operation of the chosen engine. Point M can be any point within the layout diagram. Once the specified mcr point has been selected, and provided that the shaftline and auxiliary equipment are dimensioned accordingly, that point is now the maximum rating

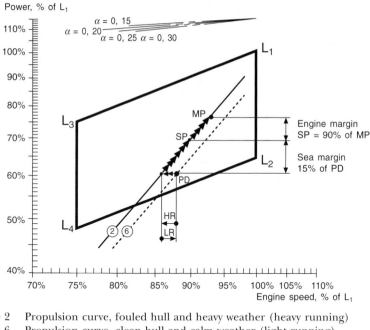

Line 2   Propulsion curve, fouled hull and heavy weather (heavy running)
Line 6   Propulsion curve, clean hull and calm weather (light running)
MP   Specified MCR for propulsion
SP   Continuous service rating for propulsion
PD   Propeller design point
HR   Heavy running
LR   Light running

Figure 10.3 Ship propulsion running points and engine layout for S60MC model

at which an overload of 10 per cent is permissible for one hour in 12 hours. The continuous service rating (S) is the power at which the engine is normally assumed to operate; point S is identical to the service propulsion point (SP) unless a main engine-driven shaft generator is installed.

All MC engines can be delivered to comply with the IMO speed-dependent NOx emission limits for the exhaust gas, measured according to the ISO 8178 test cycles E2/E3 for heavy duty diesel engines. NOx emissions from a given engine will vary according to the engine load and the optimizing power. Specific fuel consumption (sfc) and NOx emission levels are interrelated parameters and an engine offered with both a guaranteed sfc and the IMO NOx limitation will be subject to a tolerance of 5 per cent on the fuel consumption. (*See also the chapter on Exhaust Emissions and Control.*)

MC ENGINE DESIGN FEATURES

All MC series engine models are based on the same design principles and aim for simplicity and reliability, the key elements comprising:

Bedplate: the rigid bedplate for the large bore engines is built up of longitudinal side girders and welded crossgirders with cast steel bearing supports. For the smaller bore engine types the bedplate is of cast iron. It is designed for long, elastic holding-down bolts arranged in a single row and tightened with hydraulic tools. The main bearings are lined with whitemetal and the thrust bearing is incorporated in the aft end of the bedplate. The aft-most crossgirder is therefore designed with ample stiffness to transmit the variable thrust from the thrust collar to the engine seating.

Frame box: this is a single welded unit for the large bore models and a cast iron unit for the smaller types, the design contributing to a very rigid engine structure. The frame box is equipped on the exhaust side with a relief valve and on the camshaft side with a large door for each cylinder providing access to the crankcase components.

Cylinder frame: the cast iron cylinder frames on the top of the frame box make another significant contribution to the rigidity of the overall engine structure. The frames include the scavenge boxes which are dimensioned to ensure that scavenge air is admitted uniformly to the cylinders. Staybolts are tightened hydraulically to connect the bedplate, the frame box and cylinder frames and form a very rigid unit.

Crankshaft: the conventional semi-built, shrink-fitted type crankshaft is provided with a thrust collar. The sprocket rim for the camshaft chain drive is fitted on the outer circumference of the thrust collar in order to reduce the overall length of the engine—except for the high cylinder numbers of the 800 mm bore model and upwards in the programme, for which the chain drive is located between two cylinders. An axial vibration damper is integrated on the free end of the crankshaft.

Connecting rod: in order to limit the height of the engines a relatively short connecting rod, comprising few principal parts, is specified. The large area of the lower half of the crosshead bearing allows the use of whitemetal or tin–aluminium on the small bore engine models. Floating guide shoes mean that most of the alignment work formerly required with crosshead engine pistons can be eliminated. The crankpin bearing for all engine models features thin shells lined with whitemetal.

Cylinder liner: a simple symmetrical design fosters low lubricating oil consumption and low wear rates. The liner is bore cooled on the larger engine models and available in two different configurations—with or without insulation of the cooling water jet pipes—so as to match the cooling intensity closely to the different engine ratings. The joint between liner and cover is located relatively low. This arrangement means that a larger part of the heat-exposed combustion chamber is contained in the steel cylinder cover rather than in the cast iron cylinder liner. Smaller bore engines feature a simple slim-type liner without cooling bores. For both types of liner adequate temperature control of the liner surfaces safeguards against cold corrosion caused by the condensation of sulphuric acid (originating from the sulphur content of heavy fuel) and, at the same time, ensures stable lubrication conditions by preventing excessive temperatures (Figure 10.4).

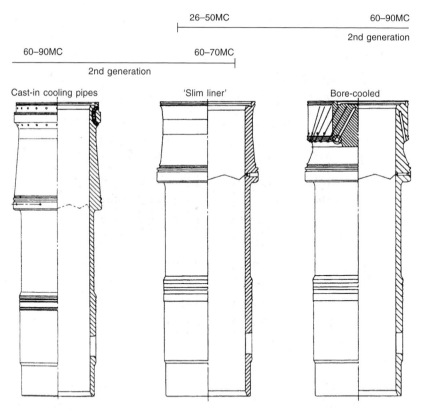

Figure 10.4 Cylinder liner designs

Cylinder cover: a solid steel component provided with bored passages for cooling water, a central bore for the exhaust valve, and bores for fuel valves, safety valve, starting valve and indicator valve.

Piston: an oil-cooled piston crown, made of heat-resistant chrome–molybdenum steel, is rigidly bolted to the piston rod to allow distortion-free transmission of the firing pressure. The piston has four ring grooves which are hard chrome plated on both upper and lower surfaces of the grooves. A cast iron piston skirt (with bronze sliding bands on the large bore engines) is bolted to the underside of the piston crown (Figures 10.5 and 10.6).

S-MC-Mk 5   S50/60/70MC-C and L70MC Mk 6
                                          with high topland

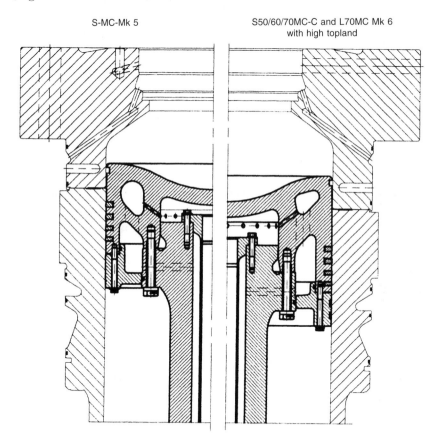

Figure 10.5 Piston/ring pack assembly MC vs MC-C engines

Piston rod: the rod is surface treated to minimize friction in the stuffing box and to allow a higher sealing ring contact pressure. The piston rod stuffing box provides effective sealing between the clean

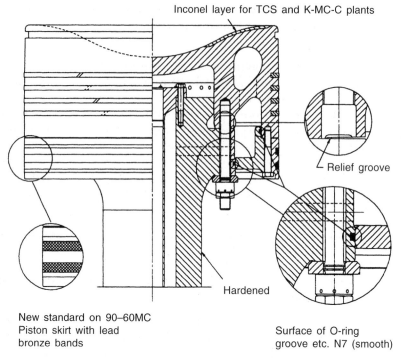

Inconel layer for TCS and K-MC-C plants

Relief groove

Hardened

New standard on 90–60MC
Piston skirt with lead
bronze bands

Surface of O-ring
groove etc. N7 (smooth)

Figure 10.6 Piston assembly

crankcase and the 'combustion area', and has a proven record of very
low amounts of drainage oil.

Camshaft: the fuel injection pumps and the hydraulic exhaust valve
actuators are driven by the camshaft. Cams are shrink-fitted to the
shaft and can be individually adjusted by the high pressure oil method.
Like its predecessors, the MC engine uses a chain drive to operate the
camshaft and thus secures high reliability since a chain is virtually
immune to foreign particles. It also enables the camshaft to be positioned
higher, shortening the hydraulic connections to the fuel valves and
the exhaust valves and, in turn, minimizing timing errors due to elasticity
and pressure fluctuations in the pipe system.

Exhaust valve: hydraulic oil supplied from the actuator opens the
exhaust valve and the closing force is delivered by a 'pneumatic spring'
which leaves the valve spindle free to rotate. The closing of the valve
is damped by an oil cushion on top of the spindle. The rotation force
is provided by exhaust gas acting on vanes fitted to the valve stem.
Extended service life from the valve is underwritten by Nimonic valve

spindles and hardened steel bottom pieces, specified as standard on the large bore engines. The bottom piece features patented 'chamber-in-seat' geometry.

Fuel pump: larger engine models incorporate pumps with variable injection timing for optimizing fuel economy at part load; the start of fuel injection is controlled by altering the pump barrel position via a toothed rack and a servo unit. Individual adjustment can be made on each cylinder. Additionally, collective adjustment of the maximum pressure level of the engine can be carried out to compensate for varying fuel qualities, wear and other factors. Both adjustments can be effected while the engine is running. The pump is provided with a puncture valve which prevents fuel injection during normal stopping and shutdown.

Fuel oil system: the engine is served by a closed pressurized fuel oil system, with the fuel preheated to a maximum of 150°C to ensure a suitable injection viscosity. The fuel injection valves are uncooled. The fuel system is kept warm by the circulation of heated fuel oil, thus allowing pier-to-pier operation on heavy fuel oil.

Reversing mechanism: the engine is reversed by a simple and reliable mechanism which incorporates an angularly displaceable roller in the fuel pump drive of each cylinder. The link connecting the roller guide and the roller is self locking in the Ahead and Astern positions. The link is activated by compressed air which has proved to be a very reliable method since each cylinder is reversed individually. The engine remains manoeuvrable even if one cylinder fails: in such a case the relevant fuel pump is set to the zero index position.

Shaft generators: all MC engines can be arranged to drive shaft generators. The PTO/RCF (Renk constant frequency) system is MAN B&W Diesel's standard configuration for power take-offs from 420 mm bore models and upwards coupled to a fixed pitch propeller. The system is mounted on brackets along the bedplate on the exhaust side of the engine. The generator and its drive are isolated from torsional and axial vibrations in the engine by an elastic coupling and a tooth coupling mounted on a flange on the free end of the crankshaft. The drive comprises a three-wheel gear train. To ensure a constant frequency the planetary gear incorporates a hydraulic speed control arrangement which varies the gearing ratio. The frequency is kept constant down to 70 per cent of the main engine's specified maximum continuous rating speed, corresponding to 30 per cent of the engine's specified mcr power, making the system suitable for fixed pitch propeller installations.

The PTO/GCR (gear constant ratio) system is MAN B&W Diesel's standard solution for power take-off in plants featuring controllable pitch propellers. The system comprises a compact unit with a step-up gear coupled directly to the generator which is located above the elastic coupling.

Power turbines: larger MC engines (from 600 mm bore upwards) can be specified with a turbo compound system (TCS) which exploits exhaust energy surplus to the requirements of a high efficiency turbocharger to drive a power gas turbine (see Chapter 7). For engine loads above 50 per cent of the optimized power the gas turbine is mechanically/hydraulically connected to the crankshaft. Power is fed back to the main engine, thus reducing the total fuel consumption. The standard system, designated TCS/PTI (power take in), is delivered as a complete unit built on the engine. Also offered is a combined PTO/PTI unit embracing a power take off with associated generator and a TCS/PTI unit with a power turbine coupled to the crankshaft gear.

PROGRAMME EXPANSION

New models have entered the MC programme since its launch in response to propulsion market trends at both ends of the power spectrum. A notable addition in 1986—the S26MC series—took the low speed engine deep into small-ship propulsion territory, offering an output per cylinder of 365 kW at 250 rev/min. The nominal power of the 260 mm bore/980 mm stroke design has since been raised to 400 kW at 250 rev/min and the four- to twelve-cylinder programme covers an output range from 1100 kW to 4800 kW. With a stroke–bore ratio of 3.77:1, this rating corresponds to a mean piston speed of 8.2 m/sec and a mean effective pressure of 18.5 bar. A firing pressure of around 170 bar yields a specific fuel consumption of 179 g/kWh at the maximum continuous rating. Reliable operation on poor quality heavy fuel oil with a viscosity of up to 700 cSt/50°C is promised.

The key components of the S26MC engine (Figure 10.7) are based on those well proven in the established larger bore models but with modifications to suit the intended small-ship market. Design amendments since its introduction include an improved axial vibration damper and refinements to the crankshaft. The cylinder cover and fuel injection system have also been upgraded, the auxiliary blower arrangement simplified and the exhaust valve provided with a new closing system with built-in damping.

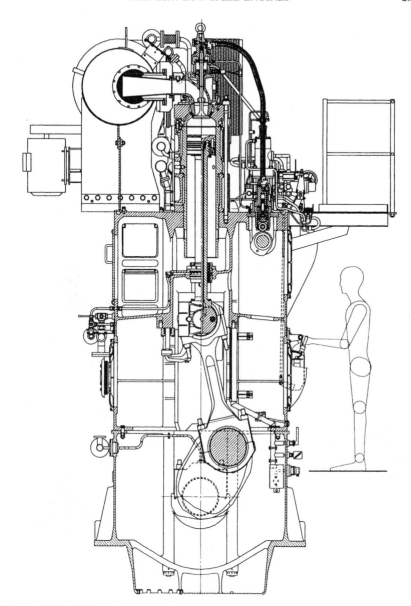

Figure 10.7 S26MC engine cross-section

Similar refinements have benefited the L35MC engine, which pioneered the MC programme. An upgrading in 1992 sought improved performance and reduced maintenance costs, the new maximum power rating of 650 kW/cylinder at 210 rev/min, mean effective pressure of

18.4 bar and mean piston speed of 7.35 m/sec calling for increased diameters of the main and crankpin journals, and reinforcement of the main bearing housing and bedplate. The crosshead bearing, thrust bearing and piston and piston rod were also reinforced, and the cylinder frame, camshaft arrangement and auxiliary blower arrangement simplified.

A longer-stroke S-version of the 350 mm bore design was introduced in 1993, this S35MC model delivering 700 kW/cylinder at 170 rev/min (since increased to 740 kW/cylinder at 173 rev/min.) A number of components were modified in line with the longer stroke (1400 mm compared with 1050 mm) but the main design features were unchanged.

The strong and rigid main structure of the S35MC is essentially the same as the L35MC but the bedplate and framebox are wider and higher due to the longer stroke of the semi-built crankshaft. The connecting rod and piston rod are necessarily longer but the cylinder frame is identical. The main bearings are wider in order to keep the bearing load inside the area ensuring good service results. Most of the components in the S35MC combustion chamber are also similar to those of the L-version. The exhaust valve design was unchanged but an improved sealing system was introduced to reduce further the wear of the valve stem. The cylinder cover is slightly higher to allow for the increased combustion space. Multi-level cylinder lubrication was adopted to ensure sufficient lubrication of the liner. A modified piston pack comprises two high rings in grooves 1 and 2, and two rings of normal height in grooves 3 and 4. All the rings are made from an improved material. The fuel pump plunger diameter was increased to satisfy the larger injection volume per stroke; and improved sealing was introduced to prevent any fuel from leaking into the main lube oil which is common to the crankcase and camshaft. The camshaft diameter was increased to accept the larger torque from the fuel pumps.

The S35MC and the 1994-launched S42MC series were conceived to improve the propulsive efficiency of plants for small-to-medium sized ships by lowering the engine speed, these 350 mm bore and 420 mm bore models, respectively, distinguished by stroke–bore ratios of 4:1 and 4.2:1. The later introduction of the S46MC-C model (also exploiting a 4.2:1 stroke–bore ratio, the highest of any production engine in the world) addressed the needs of smaller tankers and bulk carriers which can also be served by the S42MC and L or S50MC engines. The 460 mm bore model extended the number of ideal combinations of power, speed and number of cylinders for a given project (Table 10.2 and Figure 10.8).

Table 10.2 S46MC-C engine

Bore	460 mm
Stroke	1932 mm
Stroke–bore ratio	4.2:1
Speed	129 rev/min
Mean effective pressure	19 bar
Output/cylinder	1310 kW
Mean piston speed	8.3 m/sec
Cylinders	4–8
Specific fuel consumption (mcr)	174 g/kWh

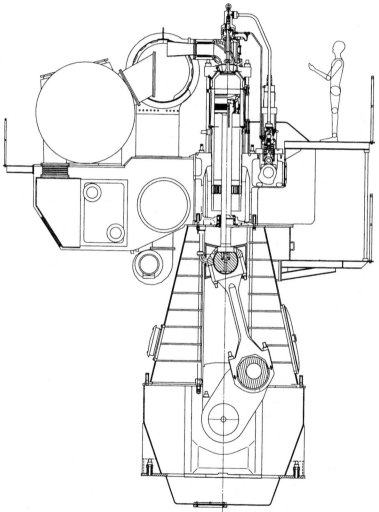

Figure 10.8 S46MC-C engine cross-section

Choice in the mid-bore range (500 mm, 600 mm and 700 mm models) was widened from 1996 by uprated versions supplementing the existing S50MC, S60MC and S70MC engines. These new S50MC-C, S60MC-C and S70MC-C models, which share the same design features as the S46MC-C, are more compact (hence the C-designation) and offer higher outputs than their established equivalents (an L70MC engine is shown in Figure 10.9). Stroke–bore ratios were raised from 3.82 to 4:1 and the increased power ratings correspond to a rise in the mean effective pressure to 19 bar. Supporting the higher ratings are modified turbocharging and scavenge air systems as well as modifications of the combustion chamber configuration and bearings. Installation space savings were achieved by reducing the overall length of the C-engines (by around 1000 mm in the case of the six-cylinder S50MC-C engine) and the overhauling height requirement compared with the original designs. The masses are also lower (by 25 tonnes or 13 per cent for the 6S50MC-C) which yields benefits in reduced vibration excitations. The MC-C engines can be 100 per cent balanced.

VLCC project planners were given a wider choice with the introduction of a longer stroke 900 mm bore model, the S90MC-T design, tailored to the propulsion of large tankers. The layout flexibility enables operators to select maximum continuous speeds between 64 and 75 rev/min in seeking optimum propeller efficiency. The design parameters (Table 10.3) addressed the key factors influencing the selection of VLCC propulsion plant, defined by MAN B&W as: the projected ship speed, the propeller diameter that can be accommodated and engine compactness. A specific fuel consumption of 159 g/kWh is quoted for a derated S90MC-T engine served by high efficiency turbocharger(s).

The dimensions of the S90MC-T engine were required not to exceed those of the 800 mm bore S80MC series which had established numerous references in the 'new generation' VLCC market. MAN B&W suggests that VLCCs with speed requirements of more than 15 knots can take advantage of the S90MC-T engine, either as a full-powered six-cylinder model or as an economy-rated seven-cylinder model. The S80MC engine, in seven-cylinder form, is considered an attractive solution for VLCCs required to operate with speeds up to 15 knots.

Refinements also sought to maintain the competitiveness of the S80MC design. Engine length, and hence engineroom length, represents dead volume and is normally to be minimized. The S80MC therefore benefited from a remodelling of its bedplate and chain drive/thrust bearing to shorten the length by 700 mm. The engine can now be pushed deeper aft into the hull, promoting increased cargo capacity from a given overall ship length or reduced ship length for the same

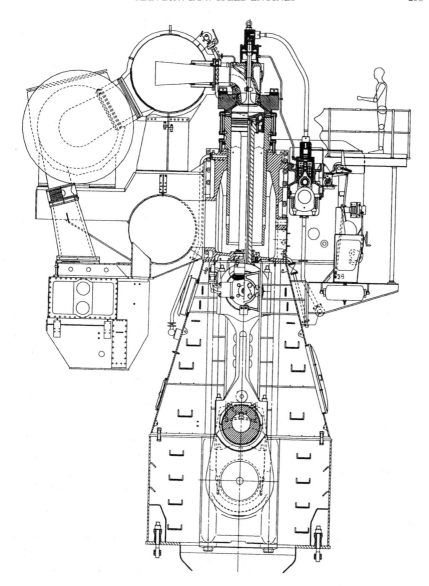

Figure 10.9 L70MC Mk 5 engine cross-section

capacity. The introduction of an optimized fuel system in conjunction with improved timing (already implemented on smaller bore MC engines) raised thermal efficiency equivalent to a reduction in specific fuel consumption (sfc) approaching 3 g/kWh. The sfc can be further improved at part load by specifying a high efficiency turbocharger

Table 10.3 S90MC-T engine

Bore	900 mm
Stroke	3188 mm
Stroke–bore ratio	3.54:1
Nominal speed	75 rev/min
Mean piston speed	7.97 m/sec
Mean effective pressure	18 bar
Output/cylinder	4560 kW
Cylinders	5, 6, 7
Specific fuel consumption (mcr)	166 g/kWh

and an adjustable exhaust gas bypass, an option which can be applied in conjunction with a TCS power turbine system.

The S-MC-C series was expanded by a 900 mm bore design, the first example of which entered service in mid-2000; the first units of another new type, the S80MC-C, were ordered in the same year. The 7S80MC-C and 6S90MC-C models have proved popular for VLCCs and ULCCs requiring higher service speeds than before, their outputs respectively 7 per cent and 15 per cent higher than the 7S80MC Mark 6 engine (Figure 10.10).

Demands for slightly higher outputs from the L60MC and L70MC engines were addressed by adopting S-MC-C design principles for these two types, with a mean effective pressure of 19 bar and a mean piston speed of approximately 8.5 m/s. The output of the L60MC-C engine is 2230 kW/cylinder at 123 rev/min, while the L70MC-C engine delivers 3110 kW at 108 rev/min. Among the targeted applications of the L-MC-C series are feeder containerships.

980 mm bore MC/MC-C engines

The propulsion demands of the largest and fastest longhaul containerships were comfortably met until the mid-1990s by short stroke K90MC and K90MC-C engines (Figure 10.11) with an upper output limit of 54 840 kW. Market interest in a new generation of 6000 TEU-plus ships with service speeds of 25 knots or more stimulated the introduction in 1994 of the K98MC-C series which, in 12-cylinder form, could deliver 68 520 kW (Figure 10.12). This 980 mm bore/ 2400 mm stroke model shared key operating parameters with the 900 mm bore MC/MC-C engines, notably a mean effective pressure of 18.2 bar and a mean piston speed of 8.32 m/sec. A propeller speed requirement of 104 rev/min was selected as the design basis, resulting in a stroke–bore ratio of 2.45 and an output per cylinder of 5710 kW. (A cylinder bore of 980 mm was not new to the marine engine industry:

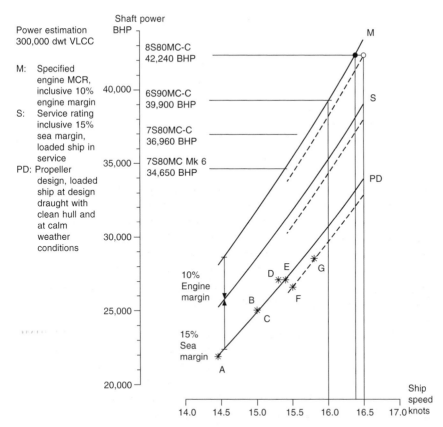

Figure 10.10 Engine selection for VLCC propulsion

in 1966 Burmeister & Wain introduced a K98 design, whose 12-cylinder version developed 33 530 kW; a number of these K98FF and K98GF engines remain in service.)

A longer-stroke K98MC version was introduced in 1997 to complement the K98MC-C engine, with the same maximum continuous rating but at a lower speed of 94 rev/min, and the number of cylinder options for both series was expanded from 9–12 to 6–12. The first K98MC engine, a 7-cylinder model, was tested in mid-1999 (Figure 10.13). The most powerful diesel engines ever commissioned – 12-cylinder K98MC models with a maximum rating of 68 640 kW at 94 rev/min – entered service in 7500 TEU containerships in 2001/2002.

The 12-cylinder K98MC/MC-C models are suitable for powering containerships up to 8000 TEU capacity in single-engine installations. Larger ships can adopt twin-engine plants or specify in-line 13- and 14-cylinder K98MC/MC-C models, now in the standard programme,

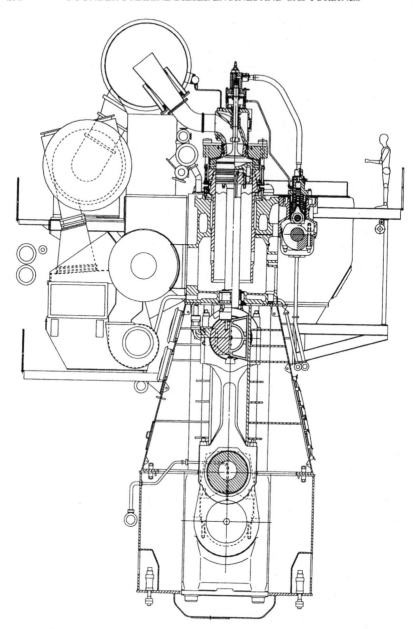

Figure 10.11 K90MC-C engine cross-section

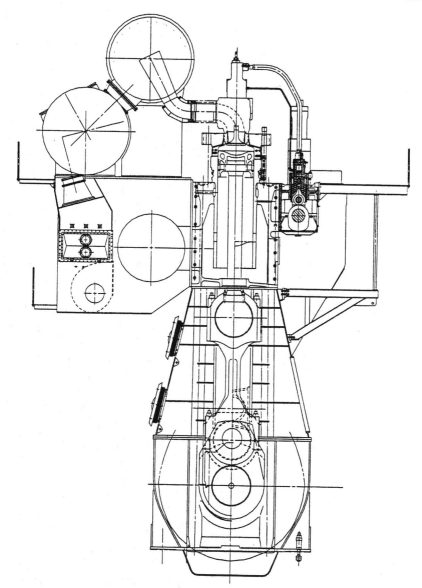

Figure 10.12 K98MC-C engine cross-section

offering outputs from 74 230 kW to 80 080 kW at 94 rev/min or
104 rev/min. The programme also allows for in-line 15- to 18-cylinder
models, taking the power threshold of the K98MC/MC-C series to just
under 103 000 kW. Such an output could satisfy the single-screw

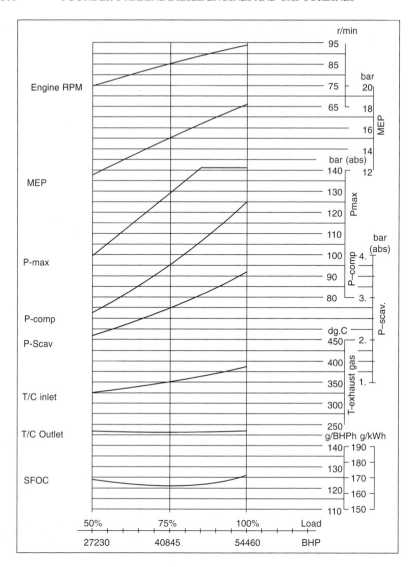

Figure 10.13 Performance curves for 7K98MC engine at ISO reference conditions

propulsive demands of containerships with capacities up to 18 000 TEU and service speeds of 25–26 knots.

Up to the 14K98MC-C model, the usual camshaft chain arrangement is applicable; for larger cylinder numbers, however, two separate chain drives are needed. Two camshafts (one for each 7–8 cylinder line) offer the advantage of providing drives of well proven dimensions for all cylinder numbers. Space for the extra chain drive is readily available because a three-part crankshaft is required to keep the crankshaft elements within the lifting capacities generally available in workshops.

Another route to higher powers for containerships was provided by MAN B&W in 2003 with the release of a 1080 mm bore version of the K-MC/MC-C series developing 6950 kW/cylinder at 94 rev/min. The 14-cylinder model of this, the world's most powerful diesel engine, can thus deliver 97 300 kW.

V-cylinder configurations for large bore MC engines have also been proposed by MAN B&W for mega-containerships (*see Chapter 9, Low speed engines introduction*), promising significant savings in weight and length per unit power over traditional in-line cylinder models. Such engines would reportedly allow the higher number of cylinders to be accommodated within existing machinery room designs.

LARGE BORE ENGINES

The structural design of the K98MC, K98MC-C, S90MC-C and S80MC-C engines is primarily based on that of the compact S-MC-C medium bore models. The main differences between the established K90MC/MC-C engines and the new design are as follows:

Bedplate

All the new large bore engines are designed with thin-shell main bearings of whitemetal. The rigidity of the bearing housing was increased substantially to reduce stress levels and deformations. Furthermore, the change from the traditional stay bolts to the twin stay bolt design, with bolts screwed into the top of the cast bearing part, means that the geometry of the main bearing structure is simplified and an improved casting quality is achieved. Finally, to improve the fatigue strength, a material with upgraded mechanical properties was specified. The use of twin stay bolts, fitted in threads in the top of the bedplate, has almost eliminated deformation of the main bearing housing during bolt tightening. The match between main bearing cap and saddle, secured in the final machining, is thus maintained in the operational condition, yielding a highly beneficial effect on main bearing performance.

A triangular plate design engine frame box was adopted for later K98MC/MC-C engines when this was introduced on the S90MC-C and S80MC-C engines. The new design provides continuous support of the guide bar, thus ensuring uniform deformation of the bar and a more even contact pattern between guide shoe and guide bar that enhances guide shoe performance. In addition, the continuous support contributes to a significant reduction of the stress level in the areas in question. The holes in the supporting plate make it possible to inspect all longitudinal welding seams from the back, and thus ensure that the specified quality is secured.

Main bearing

Tin-aluminium (Sn40Al) has proved very reliable as the bearing material for the smaller engines, the alloy having a much higher fatigue strength at elevated temperatures. The material may be applied to large bore engines to maintain the reliability of the main bearings. Dry-barring on the testbed in some cases caused light seizure of Sn40Al main bearing shells on S-MC-C engines, but the problem was eliminated either by pre-lubrication with grease, high viscosity oil or by PTFE-coating the running surface of the shells.

Engine alignment

Traditionally, bedplate alignment, especially on large tankers, has been performed on the basis of a pre-calculated vertical position of the bearings, as well as of the engine as such, and possibly also involving an inclination of the entire main engine. On completing the pre-calculated alignment procedure, it has been normal practice to check the alignment by measuring the crankshaft deflections. Such checks are normally carried out either in drydock or with the ship afloat alongside at the yard in a very light ballast condition.

Owing to repeated cases of bearing damage, presumably caused by the lack of static loads in the normal operating conditions (ballast and design draught), MAN B&W introduced modified alignment procedures for bedplate and shafting (crankshaft and propulsion shafting) as well as modifying the vertical offset of the main bearing saddles. In the modified bedplate alignment procedure, the importance of the so-called sag of the bedplate is emphasised in order to counteract the hog caused by hull deflections as a result of the loading down of the ship, and partly by deformations due to the heating up of the engine and certain tanks.

Combustion chamber

A reconfigured combustion chamber (Figure 10.14) was developed for the new large bore engines (800 mm bore and above), the key features being:

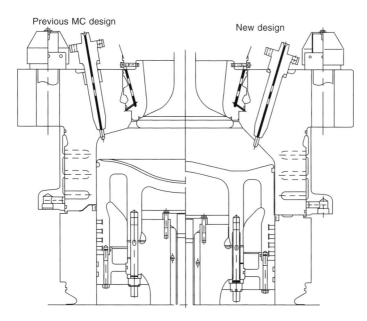

Figure 10.14 New and previous design of MC engine combustion chamber

- *Piston crown with high topland.* In order to protect the piston rings from the thermal load from combustion the height of the piston topland was increased (Figure 10.15). The resulting increased buffer volume between the piston crown and the cylinder wall improves conditions for the rings and allows longer times-between-overhauls. The high topland was first introduced in the mid-1990s, the positive service experience leading to its use for all new engine types.
- *Piston crown with Oros shape.* With increasing engine ratings, the major development challenge with respect to the combustion chamber components is to control the heat load on them. The short-stroke large bore engines have a rather flat combustion chamber because of the relatively smaller compression volume; this makes it more difficult to distribute the injected fuel oil without getting closer to the combustion chamber components. Furthermore, the higher rating means an increased amount of fuel injected per stroke. All this makes it more difficult to control the heat load on the components in short-stroke engines compared with long-stroke engines of the same bore size.

The heat load on the cylinder liner has been reduced by lowering the mating surface between cylinder liner and cylinder cover as much

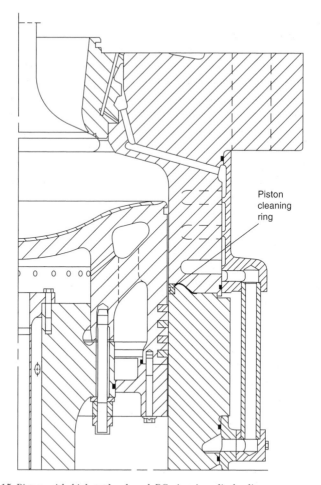

Figure 10.15 Piston with high topland and PC ring in cylinder liner

as possible (to just above the uppermost piston ring at top dead centre). This means that the greater part of the heat load is absorbed by the cylinder cover, which is made of steel and thus more resistant to high heat loads. In addition, the cylinder cover is water cooled, making it relatively easy to control the temperature level.

The piston is cooled by system oil, which means a lower cooling efficiency compared with the cooling of the cylinder cover. Oil cooling of the piston, however, offers a number of advantages. The optimum way of reducing the temperature level on the piston is to reduce the heat load on it, this being secured by redesigning the shape of the combustion chamber, including the piston, to provide more space around the fuel valve nozzles. The new piston shape was termed Oros (Greek for 'small mountain').

The result of the increased distance from the fuel valve nozzles to the piston surface was simulated by CFD analyses, and the optimum shapes of piston crown and cylinder cover determined from these simulations. Tests on several engine types verified the simulations. A significant reduction in temperature was obtained after development tests on K90MC engines with various layouts of fuel oil spray pattern.

Temperature measurements on the piston crown and exhaust valve are shown in Figure 10.16. The reduction in maximum piston temperature was approximately 90°C, this result being attained without impairing the temperature level on the oil side of the piston or the temperature on the exhaust valve.

- *Piston ring pack.* The Controlled Pressure Relief (CPR) top ring with relief grooves is now standard on all MC engines and has proved very effective in protecting the cylinder liner surface as well as the lower piston rings against excessive heat load. The CPR ring has a double lap joint, and an optimum pressure drop across the top piston ring is ensured by relief grooves (Figure 10.17). With increasing mean indicated pressures, the traditional angle-cut ring gap may result in higher thermal load on the cylinder liner; this load is significantly reduced by the CPR ring as no gas will pass through its double lap joint. The relief grooves ensure an almost even distribution of the thermal load from the combustion gases over the circumference of the liner, resulting in a reduced load on the liner as well as on the second piston ring.

Measurements confirmed that the peak temperature on No.2 piston ring was reduced from 300°C in association with an oblique cut top ring to 150°C with the CPR top ring. No.2 ring retains its spring force and times-between-overhauls are considerably extended. Furthermore, the pressure drop across the top piston ring has been optimized with respect to wear on the liner, piston rings and ring grooves. Thanks to the double lap joint, the pressure drop will be almost constant irrespective of the wear on the liner and rings. This contrasts with the traditional angle-cut ring, with which the cylinder condition slowly deteriorates as the liner wears. With the CPR rings, MAN B&W asserts, a continually good cylinder condition and low wear rate can be expected over the whole lifetime of the liner.

Alu-coating of the sliding surface of piston rings was introduced to ensure safe running-in. The aluminium-bronze alloy coating, a type of bearing material, has proved effective in protecting ring and liner

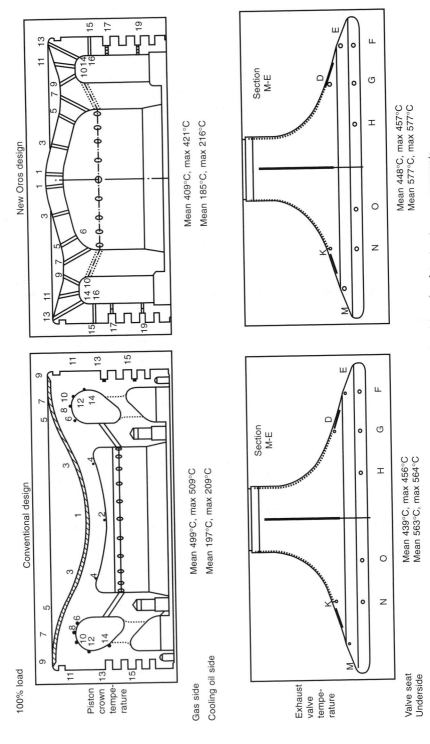

Figure 10.16 New Oros-shaped combustion chamber geometry (right) has contributed to reduced temperature measurements

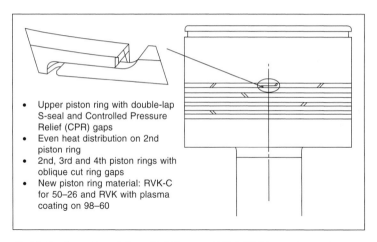

- Upper piston ring with double-lap S-seal and Controlled Pressure Relief (CPR) gaps
- Even heat distribution on 2nd piston ring
- 2nd, 3rd and 4th piston rings with oblique cut ring gaps
- New piston ring material: RVK-C for 50–26 and RVK with plasma coating on 98–60

Figure 10.17 Configuration of Controlled Pressure Relief (CPR) top piston ring

surfaces during the running-in period. Alu-coated rings make it possible to load up a completely new engine on the testbed within only five hours, fostering time and fuel savings; running-in can also be performed with reduced cylinder lubrication. The lifetime of the coating is 1000–2000 hours, depending on such factors as the cylinder oil feed rate and surface roughness of the liner. The technical explanation is that no scuffing occurs with the combination of aluminium bronze/grey cast iron as the wear components in the cylinder. Alu-coated rings allow the normal increase in cylinder oil feed rate after changing rings and/or liners to be dispensed with.

- *Piston ring groove.* In order to extend the interval before reconditioning of the piston ring grooves is required the chromium layer has been increased to 0.5 mm and induction hardening of the grooves before chrome plating introduced. The chromium layer is thus better supported by the base material and the risk of cracking of the brittle chrome reduced, fostering a longer service life.
- *Piston cleaning (PC) ring.* Incorporated in the top of the cylinder liner, the PC-ring has a slightly smaller inner diameter than the liner and hence scrapes off ash and carbon deposits built up on the piston topland (Figure 10.18). Without such a ring, contact between the topland and the liner wall could wipe off the injected cylinder oil, preventing the lubricant from performing its optimized role. In some cases, deposit formation on the topland could cause bore polishing of the liner wall, contributing to deterioration of the cylinder condition. Introducing the PC-ring

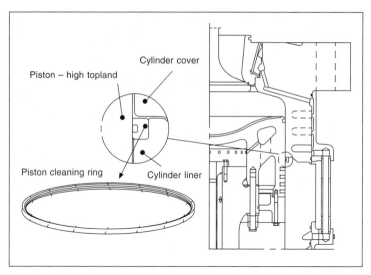

Figure 10.18 Location of piston cleaning (PC) ring in cylinder liner

eliminates contact between deposits on the topland and the liner, promoting an enhanced cylinder condition and lube oil performance.

Cylinder liner

The liners of the large bore engines are bore cooled; the cooling intensity is adjusted to maintain an optimum temperature level and to ensure optimum tribological conditions for the cylinder lube oil. For years MAN B&W Diesel used a wave-cut liner surface, which was modified a few years ago to a semi-honed surface to facilitate running-in of the highly-loaded engines. The original wave-cut had a depth of approximately 0.02 mm. Experience showed that a deeper cut is advantageous in increasing the lifetime of the oil pockets, which, together with the Alu-coated piston rings, leads to very safe running-in. An improved semi-honed wave-cut liner surface was therefore introduced.

Fuel valve

The fuel valve design was changed a few years ago from a conventional type to the mini-sac type. The aim was to reduce the sac volume in the fuel nozzle and curb dripping, thus improving combustion; introducing the mini-sac valves reduced the sac volume to approximately one-third of the original (*see the chapter on Fuel Injection*). To improve combustion even further, however, a new slide-type valve was introduced for all large bore engines, completely eliminating the sac volume (*see the chapter on Exhaust Emissions and Control*). A significant improvement in combustion is accompanied by reduced NOx, smoke and particulate emissions. The reduced particulates also improve the cylinder condition, and the wear rates of cylinder liner, piston rings and ring grooves are also generally lower with slide-type fuel valves.

Slide-type valves were introduced on the K98MC engine from the beginning, their positive effect confirmed on the testbed. With fuel nozzles optimized with respect to heat distribution and specific fuel consumption, the NOx emission values associated with this engine are described as very satisfactory.

Electronic high pressure cylinder lubricator

Cutting cylinder lube oil consumption—which represents a significant annual cost, especially for large bore engines—has been an important R&D target, resulting in the development of a computer-controlled high pressure cylinder lubricator. Development of the electronic Alpha Lubricator started in 1997 and the prototype was installed on a 7S35MC engine in the following year; refinements were carried out on MAN B&W's 4T50MX research engine.

System flexibility makes it possible to choose any number of engine revolutions (four, five, six, etc) between injections of a specific amount of oil into the cylinder. For example, lubrication can be effected every fifth revolution in the compression stroke and every tenth in the expansion stroke if that turns out to be the optimal. The aim is to inject the cylinder oil exactly where and when it is needed: in the piston ring pack as it passes the lube oil quills. Thanks to the high pressure, it is possible to establish an injection period that starts just when the uppermost piston ring is passing the quills and ends exactly when the lowermost ring is passing. Furthermore, injection is in the tangential direction, ensuring optimal distribution of the oil in the complete ring pack and ring grooves.

The Alpha Lubricator features a small piston for each lubrication quill in the cylinder liner. Power for injecting the oil is derived from the system pressure, supplied by a pump station. A conventional common rail system is used on the driving side and a high pressure positive displacement system on the injection side. Equal amounts of oil are supplied to each quill and the highest possible safety margin against clogging of individual quills is secured. The basic oil feed rate can be set by a screw which limits the stroke of the main lubricator piston.

All MC and MC-C engines (new and in service) can benefit from the system; large bore engines are provided with two Alpha Lubricators per cylinder and small bore engines with one. The actual savings from the reduced cylinder oil feed rates achieved depend on the engine size but MAN B&W Diesel has suggested 20 per cent-plus cost reductions as possible. Further reductions are promised by applying the 'Sulphur Handle' to dose oil in proportion to the amount of sulphur entering the cylinder in the fuel; this Alpha ACC (Adaptive Cylinder oil Control) is now standard on Alpha Lubricators. (*See the chapter on Fuels and Lubes for more details of the system.*)

WATER MIST CATCHER

Water droplets entering the cylinder can have a negative effect on the cylinder condition by disturbing the lube oil film on the cylinder liner surface; all two-stroke engines are therefore equipped with a water mist catcher. With increasing pressures and air amounts, however, it has become more difficult to obtain adequate efficiency.

Seeking the optimum method of separating water droplets from the scavenge air, MAN B&W Diesel carried out advanced CFD (Computational Fluid Dynamics) calculations to find the trajectories of the droplets as they leave the cooler. An analysis of the influence of different guide vanes on the air flow made it possible, by calculation, to predict that all droplets larger than 0.1 mm would be separated before the air entered the traditional water mist catcher element.

A new air flow reversing chamber was designed with a pre-catcher guide plate diverting the flow in front of the water mist catcher element and securing 80 per cent water separation before the normal element. In addition, inspection covers were fitted before and after the air cooler element, and improved drains (to prevent the re-entry of water) applied.

MC DESIGN REFINEMENTS

Feedback from MAN B&W MC engines in service has shown average liner wear rates as low as 0.05 mm/1000 hours and cylinder overhauling intervals of 6000–8000 hours. R&D targeting enhanced durability and extended mean times-between-overhauls (TBOs) has focused on:

- *Engine structure:* experience has shown that the jacket cooling water around the lower part of the cylinder frame may in some cases result in a corrosion attack and thus lead to higher wear rates. Cooling water in the lower part of the liner has therefore been omitted, thus increasing the temperature of the liner surface. The modification has been introduced on the latest engine versions (Figure 10.19). Cracks were experienced in the horizontal support for the forward main bearing in the thrust bearing housing. Investigations revealed that, under special service conditions, the design seemed to leave too narrow a margin; improper welding quality was also a contributory factor. While such rib cracks do not imply any risk with respect to the service of the ship in the short term, they should be rectified at the first available

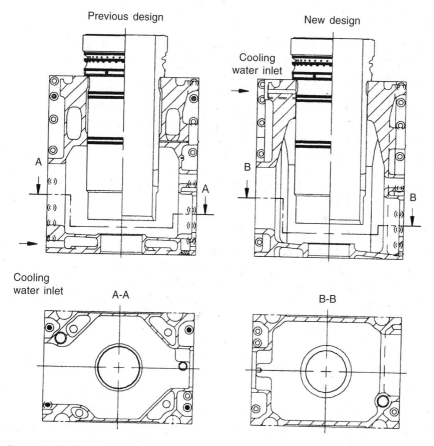

Figure 10.19 Simplified cylinder frame (right) alongside previous design

opportunity. For engines in service, the designer has evolved a rectification method which, in addition to remedying the cracks, also increases the margins. Repairs have been successfully carried out using this procedure. The standard design has also been changed to address the above conditions.

- *Piston rings:* investigation showed that the overhauling interval could be prolonged by introducing approximately 30 per cent higher piston rings in the two uppermost ring grooves. Such rings of a special alloyed grey cast iron were therefore introduced as standard on new engines. A further extension of maintenance intervals can be secured by plasma (ceramic) coating or chrome plating the surface of the uppermost piston ring. These coated rings are available as options. Experience showed that in some cases the normal oblique-cut piston ring gap led to a high heat input to the liner when all cuts coincided. An improved design of S-seal ring (double-lap joint) was developed and tested to achieve a completely tight upper ring and then reduce the pressure drop across this ring by introducing a number of small oblique slots to secure a controlled leakage. The temperature distribution and pressure drop measured across the ring were promising and very clean ring lands resulted. Rings of vermicular iron have also shown promising results. The higher piston topland and higher top rings (the uppermost featuring a pressure-balanced design) were introduced on engines with higher mean effective pressures to enhance reliability and extend times between overhauls (Figure 10.5).

- *Cylinder liner:* factors to be considered when designing cylinder liners include material composition, strength, ductility, heat transfer coefficient and wear properties. The original MC-type liners with cast in pipes unfortunately suffered from cracks but the introduction of the bore-cooled liner solved the problem (Figure 10.20). Such liners have been fitted to all new engines produced in recent years. The change to a bore-cooled liner made of grey cast iron for engines originally supplied with liners of the cast-in cooling pipe design, however, could not be easily effected. Various parallel developments were therefore initiated, based on the use of stronger materials, and improvements to the original design introduced. A modified design of the cast-in pipes, in combination with a tightening-up of the production and quality specifications, has led to considerably improved reliability from this type of liner. The modified liner is the standard spare part supply for older MC engines. The standard specification for new engines is a bore-cooled type made of Tarkalloy C grey cast iron.

Previous design of
cast-in cooling pipes

Changed position and shape
of cast-in cooling pipes

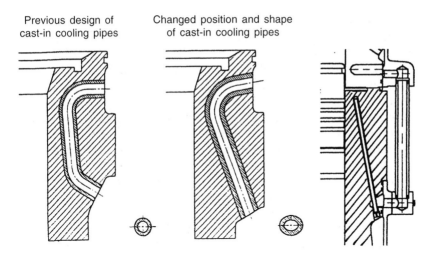

Figure 10.20 MC engine cylinder liner modifications. A current bore-cooled standard is shown (right)

The inside surface temperature of the liner greatly influences the general cylinder condition. Traditionally, the cooling system has been laid out to match the maximum continuous rating load but there is an advantage in controlling the inside liner surface temperature in relation to the load. MAN B&W has investigated and tested different solutions for load-dependent cylinder liner cooling. One system simply adjusted the cooling water flow through the original cooling ducts in the liner but the results were not promising.

Another system features different sets of cooling ducts in the bore-cooled liner, the set deployed depending on the engine load. At nominal power and high loads the inner row of ducts is used to cool the liner, yielding the highest cooling intensity. In the intermediate load range the cooling function is shifted to the next set of ducts which are located further away from the inner surface; this means that the cooling intensity is reduced and the liner surface temperature is kept at the optimum level. At very low loads both rows of cooling ducts are bypassed in order to further reduce the cooling intensity. Tests showed that the optimum liner temperature could be maintained over a very wide load range and that this system was feasible but the added complexity had to be weighed against the service advantages.

The operating condition of cylinder liners and piston rings is to a great extent a function of the temperature along the liner. The upper part is particularly important and a triple-fuel valve configuration (see section below) reduces thermal load while, at the same time, the pressure-balanced piston ring and high topland ensure an appropriate pressure drop across the ring pack and control the temperature regime for the individual piston rings. For monitoring the temperature of the upper part of the liners MAN B&W offers embedded temperature sensors and a recorder.

Alarm and slowdown temperature settings allow the operator to take proper action to restore proper running conditions if, for example, a piston ring or fuel valve is temporarily or permanently out of order. Other features, such as an uncooled cylinder frame, serve to increase slightly the wall temperature on the lower part of the cylinder liner while, at the same time, reducing production costs (Figure 10.19). The rise in wall temperature is aimed at counteracting the tendency towards cold corrosion in the lower part of the liner.

- *Exhaust valves:* a high degree of reliability is claimed for the MC exhaust valve design, and the recommended mean TBOs have generally been achieved. The average time between seat grinding can also be considerably increased for a large number of engines. Examples of more than 25 000 running hours without overhaul have been logged with both conventional and Nimonic-type exhaust valves. The Nimonic spindle is now standard for the 600 mm bore engines and above; and a steel bottom piece with a surface-hardened seat is specified to match the greater hardness of the Nimonic spindle at high temperatures. The combination of spindle and bottom piece fosters a mean TBO of a minimum 14 000 hours.

Some cases of wear of the spindle stem chromium plating may be related to the sealing air arrangement. A new sealing air system was therefore designed, incorporating oil mist and air supply from the exhaust valve's air spring (Figure 10.21). In-service testing gave promising results: completely clean sealing air chambers and virtually no wear of either the spindle stem or the sealing rings. The system is now standard for new engines and can be easily retrofitted to those in service.

Cold corrosion of the exhaust valve housing gas duct led to lower-than-expected lifetimes for a number of valves, particularly those installed in large bore engines. The corrosion attack occurs adjacent to the spindle guide boss and in the duct areas at the cooling water inlet positions (Figure 10.22). The problem has been addressed by new housings designed with thicker gas walls which are now standard fitments for new engines and spares (Figure 10.23). For engines in service, the following repair methods and countermeasures have proved effective in dealing with corrosion attacks in the exhaust valve housing:

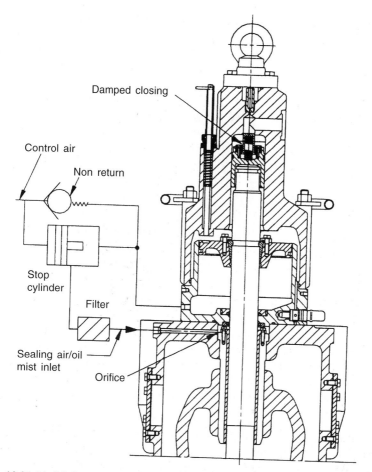

Figure 10.21 Modified exhaust spindle guide sealing air arrangement (schematic)

high velocity sprayed Diamalloy 1005 coating in the gas duct; and repair welding with gas metal arc welding (MIG-type), preferably in conjunction with the sprayed coating.

- *Bearings:* there should be no problems related to production or materials since the manufacture of whitemetal-type main and crankpin bearings is well controlled (Figures 10.24 and 10.25). Some cases of production-related problems were traced to the use of copper- or lead-polluted whitemetal or the lack of proper surface treatment of the steel back before casting the whitemetal. Based on service feedback, the main bearing design has been modified to secure a wider safety margin. The modifications

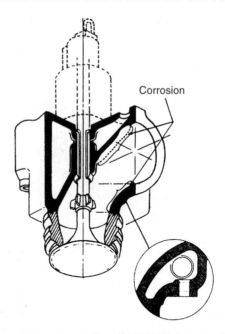

Figure 10.22 Exhaust valve housing, indicating where corrosion problems were encountered and remedied

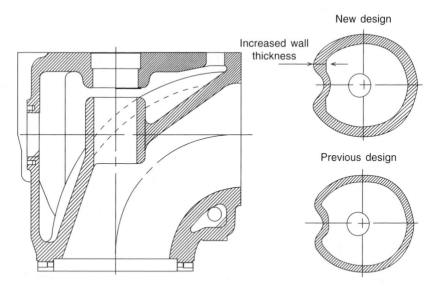

Figure 10.23 An exhaust valve housing with an increased wall thickness serves later MC engines

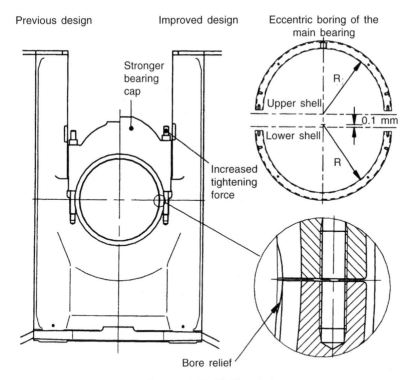

Figure 10.24 Main bearing development (MC Mk V engine)

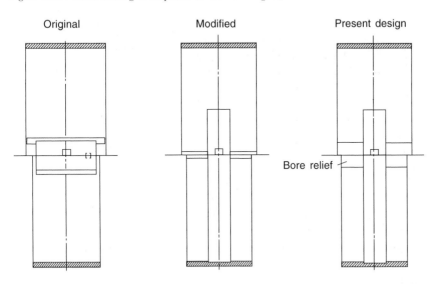

Figure 10.25 Crankpin bearing development

include a larger bore-relief in order to prevent the mating faces from acting as oil scraping edges in case of displacement, and an increase of the side and top clearances to raise the oil flow to the bearing, thus keeping the temperature at a level which sustains the fatigue strength of the whitemetal. The modified bearing design was introduced as the standard for new engines and also made available as spares for engines in service. It is fully interchangeable with existing bearings but the shells must always be replaced in pairs.

- *Fuel injection system:* the fuel valve used on MC engines operates without any external control of its function. The design has worked very well for many years but may be challenged by the desire to maintain effective performance at very low loads as the opening pressure has to be increased with increasing maximum combustion pressure to prevent the blow-back of gases into the fuel system at the end of the injection cycle. At low loads this high opening pressure might lead to irregular injection, speed variations and a risk of fouling of the engine gasways.

Various solutions have been investigated, among them a fuel valve with variable opening pressure (Figure 10.26) which allows low load operation to be improved by reducing the opening pressure in the relevant load range. The opening pressure is controlled by a spring and the actual fuel oil injection pressure. At low load the opening pressure is controlled by the spring alone but, when the injection pressure increases at higher loads, this higher pressure adds to the spring force and the opening pressure increases (a small amount of fuel oil enters the space between the upper spring guide and the slide; the spring that controls the opening pressure is thereby compressed and the spring force increased). Under decreasing load, the leakages will reduce the opening pressure to an appropriate level within the course of a few injections.

Burning of the neck of the fuel valve nozzle on engine types equipped with three fuel valves per cylinder (B&W 90GFCA and 90GB models) has been noted with nozzles made from standard precision-cast Stellite which, in other respects, has yielded excellent service and lifetimes well above 15 000 hours. Several improved production processes and new materials have been investigated. A promising alternative to casting was found in the hot isostatic pressure (HIP) method which delivers a more uniform material composition. Fuel nozzles produced by the process achieved a considerable reduction in the neck-burning rate when tested on the engines concerned.

Research also produced a new patented material composition with

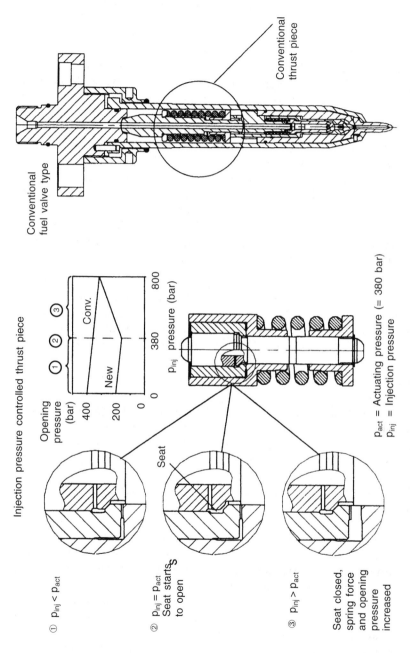

Figure 10.26 Fuel valve with variable opening pressure

the same excellent wear properties as those of Stellite but additionally offering an ability to withstand hot corrosion neck burning. A truly heavy duty fuel valve nozzle is thus available when required.

MAN B&W has addressed the potential for enhancing the reliability of combustion chamber components under increasing mean effective pressure conditions by rearranging the fuel valves in the cylinder cover. Originally, the 90-type large bore engine was provided with three fuel valves per cylinder. A number of tests were made in the early 1980s using only two valves which, with the contemporary mean effective pressures and maximum pressures, showed an advantage in the form of a slightly reduced specific fuel consumption and only a minor increase in the heat load. A two-valve configuration was consequently introduced. Rising mean effective pressures in the 1990s, however, encouraged further testing with three fuel valves per cylinder, the measurements showing a reduced temperature level as well as a more uniform temperature distribution. A three-valve configuration was subsequently introduced on the K80MC-C and K90MC/MC-C models to enhance engine reliability, while the S90MC-T and K98MC-C models were designed from inception with such an arrangement.

Many owners today prefer to have their fuel equipment overhauled by a shoreside workshop but this means that engineroom staff lose familiarity with some vital components, particularly the fuel valves. In a number of cases MAN B&W has found that nozzles with different spray hole sizes and spray patterns have been mixed. Up to four different types have been found on the same engine and even different types on the same cylinder unit. Such mixing may have an adverse effect on the running condition. A problem noted on many ships is incorrect tightening of the fuel valve spring packs—in some cases caused by faulty spring packs. This can lead to malfunctioning of the fuel valve (such as sticking) due to deformation of the components inside the valve. The consequences have been 'jumping' fuel valves and cracked high pressure injection pipes. It is therefore important, MAN B&W Diesel warns, to follow instructions covering the correct tightening of the spring packs.

- *Fuel pumps:* the lifetime of the MC fuel pump plunger and barrel has proved to exceed 50 000 hours. Only a few cases of sticking pump plungers have been reported, and these generally resulted from foreign particles entering the system. The introduction of puncture valves at the top of the pump improved crash stop performance and increased engine shutdown safety. Some problems were experienced with the sealing rings mounted around the plunger. The rings are intended to prevent fuel oil from leaking into the camshaft oil and lowering its flashpoint. Different sealing rings were tested after the introduction of the MC engine, the original ring—with an O-ring as a spring element—having been used on the previous engine type with relatively good results. Under attack from fuel oil, however, the O-rings became hard and lost their tension too quickly, making the sealing ring ineffective. Work on improving the material quality as well as the sealing ring design resulted in spring-loaded lip rings which are now standard.

A new so-called 'umbrella' type fuel oil pump design (Figure 10.27) features a sealing arrangement which eliminates the risk of fuel oil penetrating the camshaft lube oil system (a separate camshaft lube oil system is therefore no longer necessary). The uni-lube oil system introduced as standard allows reductions in installation costs, maintenance and space over the separate systems previously used: tanks, filters, pumps and piping for the camshaft system are eliminated.

- *Chain drives:* the chain drives for the camshaft and moment compensators have functioned well and, if the chains are kept tightened to MAN B&W instructions, the rubber guide bars will hardly ever need replacement. Since the chain tightening procedure on some ships might prove difficult for crews, automatic chain tightening was introduced as an aid to reducing the onboard maintenance workload. The tightening force is applied by a hydraulic piston to which oil is fed via the engine's lubricating/cooling oil system. The hydraulic chain tightener is also available for the front second-order moment compensator drive. Positive service experience encouraged the fitment of the automatic tightener as standard on 50–90 bore MC engines.

- *Installation refinements:* a simplified type of hydraulically-adjustable top bracing introduced for vibration control in 1989 was designed to cope with hull deflections giving rise to excessive stresses in the traditional mechanical type of top bracing. The system features a hydraulic cylinder which is a self-contained and does not need an external pump station. Instead, a pneumatic accumulator is built into the piston. The hydraulic cylinder is pre-tightened by the air pressure in the accumulator. When the engine is started the oil pressure in the cylinder increases because of the influence of the guide force moments, and the hydraulic cylinder adjusts its length to cope with slow hull deflections and thermal influence (Figure 10.28).

The system worked satisfactorily after teething problems were solved

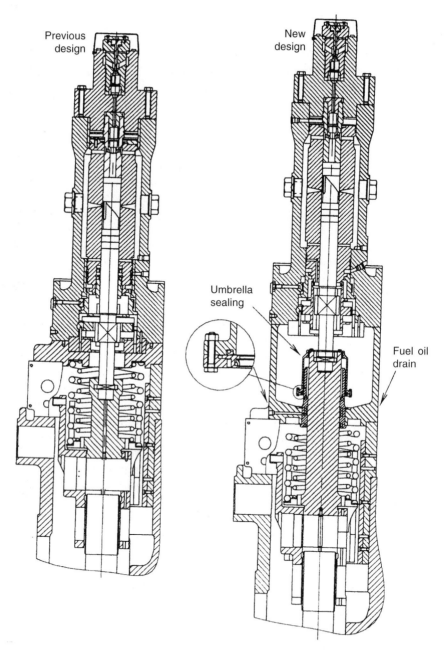

Figure 10.27 Previous type and new umbrella-type of fuel injection pump

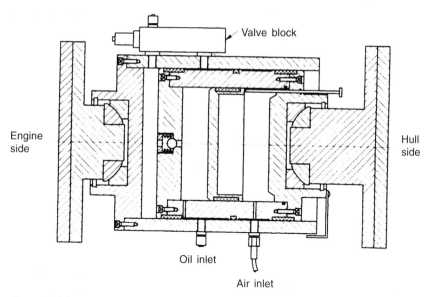

Figure 10.28 Hydraulic top bracing

but a drawback of the hydraulically-adjustable top bracing is its higher
first cost over the traditional mechanical bracing. An improved
mechanical top bracing, incorporating two beams, was designed and
tested in service from early 1997 (Figure 10.29). This is rigid in the
athwartship direction and sufficiently flexible in the longitudinal
direction to adapt to movements between engine and hull.

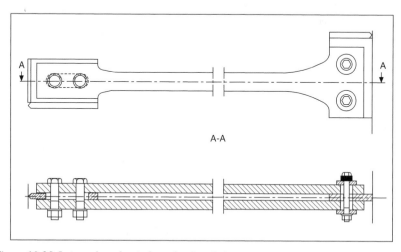

Figure 10.29 Improved mechanical top bracing design

Simplification has also benefited engine seating arrangements. Feedback from shipyards indicated a preference for epoxy supporting chocks; epoxy chocks therefore became the standard in an effort to reduce installation costs, with the previously used cast iron supporting chocks as an option. Normal holding-down bolts are used in association with epoxy chocks but with the tightening force substantially reduced. The lower force is due to the higher coefficient of friction as well as the need to comply with the permissible specific surface pressure of epoxy. The general application of epoxy thus permits the use of thinner bolts which—following successful testing in service—have been introduced as standard. Larger engines (L42MC and upwards) have had one side chock per crossgirder per side. Investigations and experience showed that this number could be halved without any problems. An even simpler solution proposed features a side chock design that allows epoxy to be used as a liner. Installation man-hours are saved since the location of the side chock itself is much less demanding in terms of accuracy (Figure 10.30).

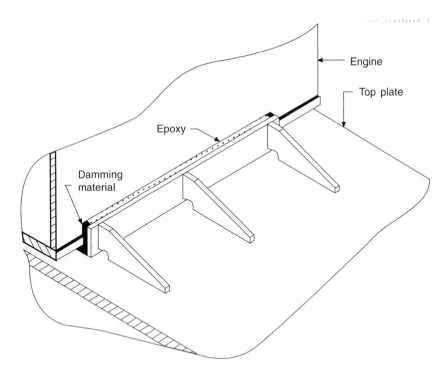

Figure 10.30 Epoxy-type side chock liners

POST-1997 SERVICE EXPERIENCE

A service experience report covering mainly larger bore MC and MC-C engines commissioned after 1997 indicated that a significant increase in reliability and longer times-between-overhauls had resulted from continuous development and updating. The following contributions were highlighted:

Cylinder condition

A number of measures have contributed to positive developments in the cylinder condition in recent years, notably the Oros combustion chamber, high topland pistons, the piston cleaning (PC) ring, controlled pressure relief (CPR) piston rings, alu-coating of piston rings and wave-cut cylinder liners.

With the Oros configuration, the combustion air is concentrated around the fuel nozzles and the distance from the nozzles to the piston top is increased. This results in a lower heat load on the piston top and a basically unchanged heat load on the cylinder cover and exhaust valve. The higher topland and the PC ring have proved very beneficial in avoiding a build-up of lube oil-derived deposits on the topland. Such deposits (via the hard face sponge effect) scrape off and absorb the oil film, leaving the naked liner wall vulnerable to extensive wear and/or scuffing. A high topland piston also means that the mating surfaces between the cylinder liner and the cover are lowered, thus reducing the thermal load on the liner and improving the conditions for lubricating the liner. (This was taken into account before introducing the Oros configuration.)

Cracked cylinder liners are now rarely reported thanks to the successful introduction of countermeasures, such as bore-cooled liners and (for smaller bore engines) slim liners. For engines originally specified with cast-in cooling pipes in the liners, the later design with oval pipes has stopped the occurrence of cracks.

Safe and stable running-in is secured by the semi-honed surface of the liner in conjunction with alu-coated piston rings. A thin layer of alu-coating is worn off the rings during the first 1000–2000 hours, making it possible to reduce the breaking-in and running-in time as well as the cylinder oil feed rate during most of the breaking-in period. This benefits the enginebuilder, who can cut the delivery time of an engine, and the operator, who receives an engine with improved running-in conditions.

The surface of MAN B&W cylinder liners is described as semi-honed. The semi-honing process cuts off the tops of the wave-cut, thus reducing the necessary breaking-in between rings and liner surface, while still retaining circumferential pockets for lube oil. The alu-coated piston rings remove the remaining broken or damaged cementite from the liner surface during the initial wear period (effectively performing a 'free of charge' full honing).

Low and uniform cylinder wear rates and a good liner condition are revealed by inspections of S-MC, S-MC-C, K-MC and K-MC-C engines. While initial (running-in) wear is naturally higher, the wear rate of the largest bore (90 and 98) models is reduced to less than 0.05 mm/1000 hours after around 1500 hours' running, which is considered very satisfactory. Initial wear is part of the running-in of liners and piston rings, and high wear is expected during this period. Experience with the Alpha Lubricator (see above) indicated that there was significant potential for cylinder lube oil reduction while retaining a fully acceptable wear rate and mean time-between-overhauls.

The top ring design of the K98 engine was upgraded to increase the safety margin against breakage in response to a number of failures. The production process at the sub-suppliers was also changed to reduce such incidents. The upgrading involved a number of changes, including relocation of the controlled leakage grooves, reduction of the number of grooves from six to four (the same leakage area is achieved by applying wider grooves) and modified surface machining of the grooves to avoid fine cracks from the outset (Figure 10.31).

Bearings

A decrease in the number of reported main bearing failures has been noted since 1998, when a number of features were introduced to the bearing design, bearing adjustment and engine/shaftline installation.

These major updates included the introduction of the Optimum Lemon Shape (OLS)-type main bearing as an evolution of the Mark 5 bearing type, featuring reduced top and side clearances. Service experience has confirmed the efficiency of the new bearing type, reflected in a significantly reduced number of reported failures. Main bearing damage can sometimes still occur, however: for example, due to poor bonding of the bearing metal. In almost all cases, main bearing damage is initiated from a fatigue crack at the edge of the bearing, the aft edge/manoeuvring side being the most common point of initiation. Geometrical non-conformities, often involved in these cases, further increase the damage frequency as margins established during the design phase are reduced.

Calculations, combining the dynamics of the complete crankshaft with the hydrodynamic and elastic properties of the bearing, have

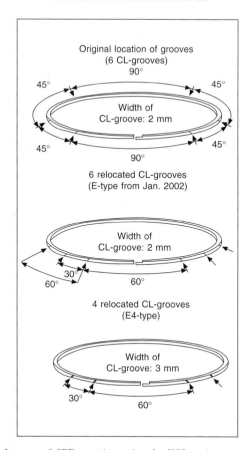

Original location of grooves
(6 CL-grooves)
90°
45°   45°
Width of
CL-groove: 2 mm
45°   45°
90°
6 relocated CL-grooves
(E-type from Jan. 2002)

Width of
CL-groove: 2 mm
30°
60°   60°
4 relocated CL-grooves
(E4-type)

Width of
CL-groove: 3 mm
30°   60°

Figure 10.31 Development of CPR top piston ring for K98 engine

provided detailed information on the mechanisms leading to local loading of the main bearing edges. The calculations have indicated that a slight radial flexibility of the bearing edge will significantly increase the overall minimum oil film thickness. At the same time, the maximum oil film pressure will be reduced.

A bearing design with flexible forward and aft edges of the bearing shell was successfully tested, the flexibility achieved by removing the contact between the shell and bedplate at the end portions of the shell. The unsupported width of the shell is equal to the shell thickness. Apart from the flexible edges, the properties of the bearing are similar to those of the OLS-type and the design provides a larger safety margin in the event of geometrical non-conformities.

A thin shell bearing design has been introduced to the latest engine types. The main bearings of the small and medium bore models (S46MC-

C to S70MC-C) are lined with AlSn40 and provided with a PTFE running-in coating as standard. The bearings of the large bore models are lined with white metal. Few damage incidents to the thin shell main bearings have been reported.

A revised engine installation recommendation—including an updated shaftline alignment procedure and a differentiated bearing height in the aft end of the engine—resolved cases of repeated damage to the aft-end bearings. This was presumably caused by missing static load, particularly in the second aftmost main bearing during normal operating conditions. The new alignment procedure, exploiting pre-calculated bedplate sagging as well as vertical offsets to the main bearing saddles, achieved a significant drop in the number of reported damage incidents to the aft-end bearings.

The crosshead bearings of both MC and MC-C engines generally perform very satisfactorily, but cases of wiping have been observed. Such wiping is of a cosmetic nature but can sometimes cause blockage of the oil-wedges that normally build up the oil film to the 'pads' inside the bearing. Disturbance of this oil film build-up could result in slight fatigue damage just behind the blocked area of the oil-wedge. If observed at an early stage during inspections, however, the problem can be solved by removing the wiped lead from the oil-wedge.

The crankshaft thrust bearing introduced on Mark 5 engines has solved the problem of cracks in the horizontal support plates. By making this plate (which connects the fore and aft thrust bearing supports) in one piece, and shaping it like a calliper, a significantly wider design margin is derived even though engine outputs (and hence propeller thrust) have been increased. No cracks have been reported on engines with this so-called Calliper design thrust bearing (Figure 10.32). The bearing saddles have remained free of cracks, in compliance with pre-calculated stress levels.

Exhaust valves

Nimonic exhaust valve spindles are well accepted now that operators have become acquainted with the long-lasting seat performance, despite dent marks. Nimonic spindles are standard for 50MC and 60MC/MC-C engine models upwards; and Stellite spindles are standard for smaller models.

Corrosion in the valve housing was effectively minimized by introducing the optimized cooling water system; this raises the wall temperature in the housing above the critical level for the formation of acid on the gas side of the duct. The high temperature level dictates a cast iron spindle guide bushing.

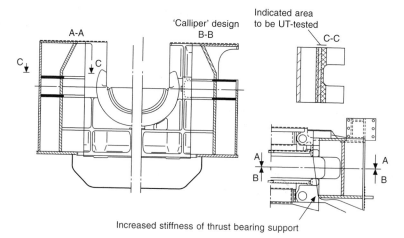

Figure 10.32 Calliper-design thrust bearing

Wear of the previously chrome-plated spindle stem has been effectively reduced by the HVOF-based cermet coating introduced in 1997. Furthermore, wear and corrosion problems at the spindle guide/ spindle/seal area, caused by combustion products, have been minimized over the years by design changes to the sealing air system. Since the stem seal had difficulty in reaching lifetimes similar to those of the valve seats, however, it was decided to replace the sealing air system by a lubricating device tested with good results in long term service.

Reduced wear of the spindle stem (HVOF coating) and of the long spindle guide (grey cast iron) has been achieved but the lifetime of the stem seal itself is still sometimes too short. Tests with oil as the sealing medium instead of air showed very low wear rates on the seals, along with a high cleanliness level on the surfaces of the spindle stem and spindle guide. A system delivering the necessary dosage of only approximately 1 kg/cylinder/day was developed for the medium and large bore engines. It is located in the top of the exhaust valve and fed with oil from the valve's hydraulic system; oil is fed to the spindle guide via a small pipe. The sealing oil is taken from the circulating oil and is therefore part of the necessary minimum oil consumption for keeping the system oil viscosity and base number (BN) level at the prescribed equilibrium.

The latest design of exhaust valve on the small and medium bore engines occasionally suffered from cooling water leaks at the lowermost O-ring between the bottom piece and cylinder cover. Investigations resulted in this O-ring being replaced by a special Teflon seal with spring back-up (U-seal).

Tests showed that the best way of increasing valve seat lifetime was by altering the seat geometry of the bottom piece to the patented W-seat configuration, now standard on all MC/MC-C engines (Figure 10.33). Results are even better when used with the new type of slide fuel valves.

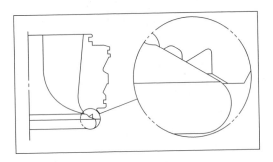

Figure 10.33 W-seat configuration for exhaust valve

Engines featuring the Oros combustion chamber (see section above) leave little distance between the piston top and the underside of the exhaust valve spindle. The usual extended lift system for releasing high hydraulic pressure cannot therefore be applied. A safety valve located in the actuator is used instead. Unfortunately, a few cases of damaged exhaust valves and camshaft sections were experienced due to different external factors, including an insufficient release action of the safety valve. The valve was subsequently redesigned so that, once activated, it implements a special function to keep it open for around 20 seconds. In addition, a disc spring was introduced in the exhaust valve on top of the spindle guide to avoid damage to the air piston in the event of 'over-shoot'/extended lift of the valve spindle.

Fuel injection system

In general, the fuel pumps work well and without difficulties, although a few incidents have been experienced and addressed:

- A combined puncture and suction valve used on the S60MC-C, S70MC-C, S90MC-C and K98MC-C engines originally featured a bellow as a substitute for the conventional sealing rings in order to benefit from a component needing little or no maintenance. The reliability of the bellow was not satisfactory, however, and a new design eliminating this element was introduced for the above engine models.

- Fuel pumps without shock absorbers were introduced on these same engine types as a cost reduction measure but this resulted in annoying (although not damaging) pressure fluctuations in the fuel supply system. Even though these measured fluctuations were within MAN B&W's and the classification societies' design limits, shipowners experienced problems with shipyard-installed equipment such as pumps, filters and pre-heaters. The problems were avoided by re-introducing a shock absorber on the fuel pumps of these engines.
- A new type of gasket between the fuel pump housing and the top cover was introduced to cure leakage experienced on some K98MC/MC-C and S90MC-C engines. The primary sealing is secured with a viton ring, protected against corrosion attack by a steel bushing. A soft iron plate of the same shape as the original seal forms the 'groove' for the square viton ring.
- Fuel pump top cover fractures were experienced on small and medium bore engines, initiated at the position where the inclined drillings for the high pressure pipes intersect with the central bore. The cause of the failure in all cases was related to roundings that did not fulfil the MAN B&W specification. A design change has improved the safety margin against failures, simplified manufacture and made the component less sensitive to minor tolerance deviations (Figure 10.34).

Fractured fuel valve nozzles were found on large bore engines where slide-type valves had been standard for some years. The main reason

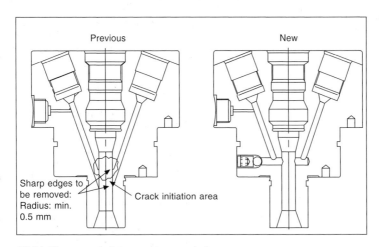

Figure 10.34 Change to fuel pump top cover design

for the cracks was residual stress from machining, but the high temperature of the valve nozzle itself also contributed to fracturing because of the consequential high mean stress. The problem was cured by optimizing the production parameters.

Pressure testing fuel valves

In pressure testing conventional fuel valves, MAN B&W notes, the testing device is only capable of supplying 1–2 per cent of the normal fuel flow rate on the engine, which is not sufficient to ensure proper atomization. If the remaining test items in the procedure are fulfilled, the fuel valve nozzle will work perfectly. As the atomization test can be omitted it is not described in new testing procedures, so verification of a humming sound is no longer possible nor necessary.

Pressure testing procedures for slide fuel valves are quite different from those for conventional valves and—perhaps for that reason—have in some cases been misinterpreted by operators. Slide fuel valves must be disassembled and cleaned before pressure testing, and an atomization test must not be performed on such valves. Cleaning is necessary because the cold and sticky heavy fuel oil, in combination with the very small clearance between the cut-off shaft and the fuel valve nozzle, would significantly restrict movement of the spindle.

An atomization test is not acceptable because the very small needle lift obtained during such a test would result in an unequal pressure distribution on the cut-off shaft, resulting in a relatively hard contact in a small area. This, together with the high frequency oscillations during an atomization test, and the low lubricity of the test oil, would significantly increase the risk of seizure. The full lift of the needle, and the very good lubricity of heavy fuel oil, completely eliminates this risk during normal operation of the engine.

Wet spots have often been found under each slide fuel valve during scavenge port inspections. In a conventional valve the sac volume is emptied after each injection; in the slide valve the fuel stays in the fuel nozzle until the next injection. When turning the engine and after some hours of engine standstill, however, the nozzle will empty and fuel will drip down on the relatively cold piston crown, revealed as wet spots up to 300 mm diameter. After some hours the drops will more or less evaporate, depending on the actual position of the piston in relation to the scavenge ports, leaving only some ash on the piston top. This is not normally an indication of malfunction, advises MAN B&W.

Cylinder cut-out system

A traditional problem with low load operation is engine fouling due to irregular fuel injection and atomization leading to incomplete combustion. The irregular injection may be caused by jiggling of the governor and/or play in the connections in the fuel pump rack control system. The effect in either case is that the fuel pumps, when operating so close to the minimum injection amount, may sometimes have just enough index to inject fuel; at other times, just not enough index to do so.

By introducing a system whereby approximately half of the cylinders are cut out at low speed injection into the remaining working cylinders is considerably improved, yielding more stable combustion and hence stable running and minimizing particle emissions. To avoid excessive amounts of cylinder lube oil collecting in the cylinders that are temporarily deactivated, the cutting out is effected in turns between two groups of cylinders in order to burn surplus lube oil and maintain the same thermal load on all cylinders.

Turns between the groups are made on a time basis, and group separation is determined in order to halve the number of active cylinders and achieve the smoothest possible firing order. To secure a safe start, the cut-out system is disabled during the starting period and until the engine has been stabilised. Stable operation down to 13 rev/min has been logged using the system.

Oil mist detection

As well as applying an oil mist detector, MAN B&W efforts to prevent crankcase explosions have concentrated on designing an engine with ample margins to prevent overheating of components. It was realized, however, that when rare explosions occur they originate from other causes, such as factors related to production, installation or incorrect maintenance of components or lubricating systems. It is therefore recommended that:

- The oil mist detector is connected to the slowdown function.
- The oil mist detector has remote monitoring from the engine control room.
- The relief valves are made with an approved flame arrester function.

Reliance was previously placed on specifying safety equipment that had been type approved by the classification societies; type approval,

however, is not sufficient as it only comprises checking the opening
pressure of the relief valve.

In the event of a crankcase explosion the pressure wave will send a
large amount of oil mist out of the crankcase and into the engineroom,
where it will be moved around by the ventilation. A significant part of
this could be sucked up into the turbocharger air inlet. If the oil mist
meets a hot spot it can be ignited; it is therefore important to keep the
insulation around the exhaust pipes in good condition. The highest
risk of igniting the mist, of course, is if the flame arrester on the relief
valve does not function properly. The relief valve specification has
been updated, and only valves meeting this specification were accepted
on engines ordered after 1 May 1999.

OPERATING ADVICE

Service experience with MC engines of all bore sizes in diverse
propulsion installations underwrites MAN B&W Diesel's assertions
that:

- There is still no substitute for careful monitoring or planned
  maintenance by qualified personnel.
- Maintaining the turbocharging efficiency is essential: a drop of
  even a few points can have adverse effects. Turbochargers and
  air coolers must therefore be well maintained.
- Fuel valve condition and functioning are key factors in overall
  performance. The spindle guide quality is the chief element,
  followed closely by overhaul competence, valve nozzle condition
  and the correct installation of the valve.
- The differential pressure Pmax – Pcomp should not be allowed
  to exceed 25–40 bar in service since this might lead to performance
  problems, especially cylinder condition problems.
- Operating with excessive exhaust gas back pressure may lead to
  difficulties as the heat load increases. This is particularly relevant
  in boiler types with closely spaced finned tubes which use very
  low feedwater temperature and/or low exhaust gas temperatures
  at the boiler outlet, running an increased risk of soot deposits.
- Fuel quality-related difficulties are still encountered today but
  crews have become more conscious of the importance of good
  fuel treatment. Ignition quality is of minor importance for low
  speed engines.

ME ENGINES (ELECTRONICALLY CONTROLLED)

In an MC engine the camshaft mechanically controls fuel injection and exhaust valve operation, a linkage that delivers very limited timing flexibility. MAN B&W ME-series engines, introduced to the market in 2001, dispense with the camshaft and exploit hydraulic-mechanical systems supported by electronic hardware and software for activating fuel injection and the exhaust valves. ME engines are available in bore sizes from 500 mm to 1080 mm offering the same outputs as their MC engine counterparts (Figure 10.35).

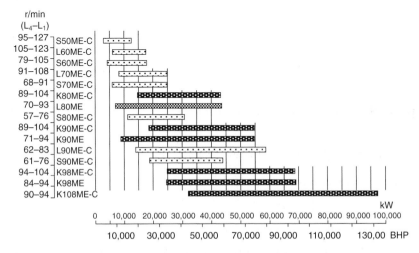

Figure 10.35 The electronically-controlled ME engine programme

Electronically-controlled fuel injection and exhaust valve actuation (Figure 10.36) allows individual and continuous adjustment of the timing for each cylinder, securing these key benefits:

Reduced fuel consumption

- Fuel injection characteristics can be optimized at many different load conditions, while a conventional engine is optimized for the guarantee load, typically at 90–100 per cent maximum continuous rating.
- Constant Pmax in the upper load range can be achieved by a combination of fuel injection timing and variation of the compression ratio (the latter by varying the closing of the exhaust valve). As a result, the maximum pressure can be kept constant over a wider load range without overloading the engine, leading

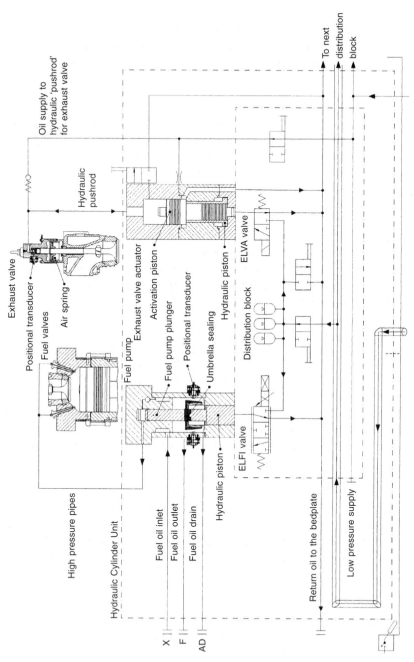

Figure 10.36 Mechanical-hydraulic system for activating the fuel pump and exhaust valve of an ME engine cylinder

to significant reductions in specific fuel consumption at part load.

- On-line monitoring of the cylinder process ensures that the load distribution among the cylinders and the individual cylinder's firing pressure can be maintained at 'as new' standard over the lifetime of the engine.

Operational safety and flexibility

- The engine's crash stop and reverse running performance is improved because the timing of the exhaust valves and fuel injection can be optimized for these manoeuvres as well.
- 'Engine braking' may be obtained, reducing the ship's stopping distance.
- Swifter acceleration of the engine, since the scavenge air pressure can be increased faster than normal by opening the exhaust valve earlier during acceleration.
- Dead slow running is improved significantly: the minimum rev/min is substantially lower than for a conventional engine, dead slow running is much more regular, and combustion is improved thanks to the electronic control of fuel injection.
- Electronic monitoring of the engine (based on MAN B&W's CoCoS-EDS system: *see MAN B&W Medium Speed Engines chapter*) identifies running conditions that could lead to performance problems. Damage due to poor ignition-quality fuel can be prevented by injection control (pre-injection).
- The engine control system incorporates MAN B&W's on-line Overload Protection System (OPS) feature, which ensures the engine complies with the load diagram and is not overloaded (often the case in shallow waters and with 'heavy propeller' operation).
- Maintenance costs will be lower (and maintenance easier) as a result of the protection against general overloading as well as overloading of individual cylinders; and also because of the 'as new' running conditions for the engine, further enhanced by the ability of the diagnosis system to give early warning of faults and thus enable proper countermeasures to be taken in good time.

Exhaust gas emissions flexibility

- The engine can be changed over to various 'low emission' modes, its NOx exhaust emissions reduced below the IMO limits if dictated by local regulations.

- By appropriate selection of operating modes ships may sail with lower exhaust gas emissions in special areas, where this may be required (or be more economical due to variable harbour fee schemes) without having negative effects on the specific fuel consumption outside those areas.

The following components of the conventional MC engine are eliminated in the ME engine: chain drive for camshaft; camshaft with fuel cams, exhaust cams and indicator cams; fuel pump actuating gear, including roller guides and reversing mechanism; conventional fuel injection pumps; exhaust valve actuating gear and roller guides; engine-driven starting air distributor; electronic governor with actuator; regulating shaft; mechanical engine-driven cylinder lubricators; and engine side control console.

These elements are replaced on the ME engine by an electro-hydraulic platform comprising: a hydraulic power supply (HPS); a hydraulic cylinder unit (HCU) with electronic fuel injection (ELFI) and electronic exhaust valve actuation (ELVA); an electronic Alpha cylinder lubricator (*see above and Fuels and Lubes chapter*); an electronically-controlled starting valve; a local control panel; a control system with governor; and a condition monitoring system.

ME engine systems

Valuable experience was gained by MAN B&W Diesel from its 4T50MX research engine at Copenhagen, operated from 1993 to 1997 with a first-generation Intelligent Engine (IE) system. Second-generation IE systems fitted to the engine in 1997 aimed for simplified design, production and installation of the key electronically-controlled fuel injection and exhaust valve actuation systems. Subsequent R&D focused on transforming the electronic elements into a modular system, whereby some of the individual modules could also be applied to conventional engines. This called for the development of a new computer unit and large software packages, both of which had to comply with the demands of classification societies for marine applications.

The second-generation IE system is based on an engine-driven high pressure servo oil system which provides the power for the hydraulically-operated fuel injection and exhaust valve actuation units on each cylinder. Before the engine is started the hydraulic power system (or servo oil system) is pressurized by a small electrically-driven high pressure pump. Fine-filtered main system lube oil is used as the actuating medium supplied by engine-driven multi-piston pumps at around 200 bar (Figure 10.37).

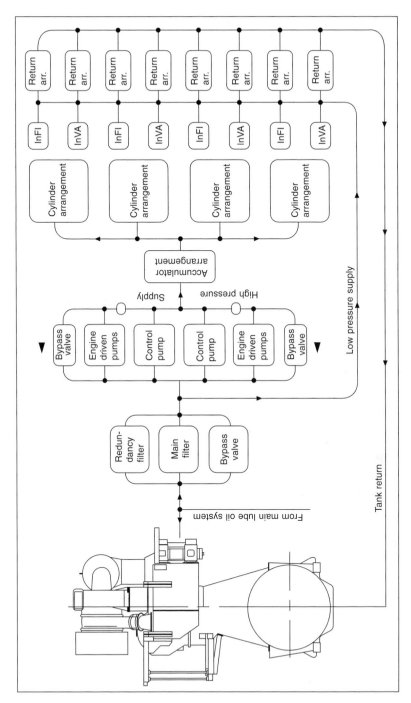

Figure 10.37 Servo oil system flow diagram for ME engine

Fuel injection system

A common rail servo oil system applies this cool, clean and pressurized lube oil to power the fuel injection pump of each cylinder. Each cylinder unit is provided with a servo oil accumulator to ensure sufficiently swift delivery of oil in accordance with the requirements of the injection system, and to avoid heavy pressure oscillations in the associated servo oil pipe system. The movement of the pump plunger is controlled by a fast-acting proportional control valve (a so-called NC valve) which, in turn, is controlled by an electric linear motor that receives its control input from a cylinder control unit. The fuel injection pump features well proven technology and the fuel valves are of a standard design.

Second- and third-generation fuel injection pumps are much simpler than the first-generation design and their components are smaller and easier to manufacture (Figure 10.38). A major feature of the third-generation pump is its ability to operate on heavy fuel oil; the pump plunger is equipped with a modified umbrella design to prevent heavy fuel from entering the lube oil system. The driving piston and injection plunger are simple and kept in contact by the fuel pressure acting on the plunger and the hydraulic oil pressure acting on the driving piston. The beginning and end of the plunger stroke are both controlled solely by the fast-acting NC valve, which is computer controlled.

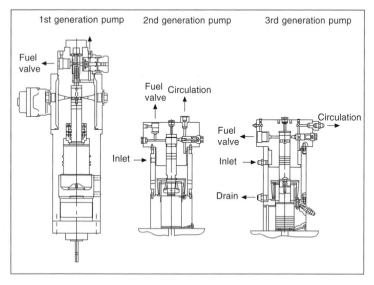

Figure 10.38 Development of fuel injection pump for ME engine

Optimum combustion (and thus optimum thermal efficiency) calls for an optimized fuel injection pattern, which in a conventional engine is generated by the fuel injection cam shape. Large two-stroke engines are designed for a specified maximum firing pressure and the fuel injection timing is controlled so as to reach that pressure with the given fuel injection system (cams, pumps, injection nozzles).

For modern engines, the optimum injection duration is around 18–20 degrees crank angle at full load, and the maximum firing pressure is reached in the second half of that period. To secure the best thermal efficiency, fuel injected after the maximum firing pressure is reached must be injected (and burned) as quickly as possible in order to obtain the highest expansion ratio for that part of the heat released. From this it can be deduced that the optimum 'rate shaping' of the fuel injection is one showing an increasing injection rate towards the end of injection, thus supplying the remaining fuel as quickly as possible. The camshaft of the conventional engine is designed accordingly, as is the fuel injection system of the ME engine. In contrast to the camshaft-based injection system, however, the ME system can be optimized at a large number of load conditions.

MAN B&W Diesel claims the fuel injection system for the ME engines can execute any sensible injection pattern needed to operate the engine. It can perform as a single-injection system as well as a pre-injection system with a high degree of freedom to modulate the injection in terms of injection rate, timing, duration, pressure or single/double injection. In practice, a number of injection patterns are stored in the computer and selected by the control system for operating the engine with optimum injection characteristics from dead slow to overload, as well as during astern running and crash stop. Changeover from one to another of the stored injection characteristics may be effected from one injection cycle to the next.

Exhaust valve actuation system
The exhaust valve is driven by the same servo oil system as that for the fuel injection system, using cool pressurized lube oil as the working medium. The necessary functionality of the exhaust valve is less complex than fuel injection, however, calling only for control of the timing of its opening and closing. This is arranged by a simple fast-acting on/off control valve.

Well proven technology from the established MC engine series is retained. The actuator for the exhaust valve system is of a simple, two-stage design. The first-stage actuator piston is equipped with a collar for damping in both directions of movement. The second-stage actuator piston has no damper of its own and is in direct contact with a gear oil piston transforming the hydraulic system oil pressure into oil pressure in the oil push rod. The gear oil piston includes a damper collar that becomes active at the end of the opening sequence, when the exhaust valve movement will be stopped by the standard air spring.

Control system

Redundant computers connected in a network provide the control functions of the camshaft (timing and rate shaping). The Engine Control System—an integrated element of the intelligent engine—comprises two Engine Control Units (ECU), a Cylinder Control Unit (CCU) for each cylinder, a Local Control Terminal and an interface for an external Application Control System. The ECU and the CCU were both developed as dedicated controllers, optimized for the specific needs of the intelligent engine, Figure 10.39.

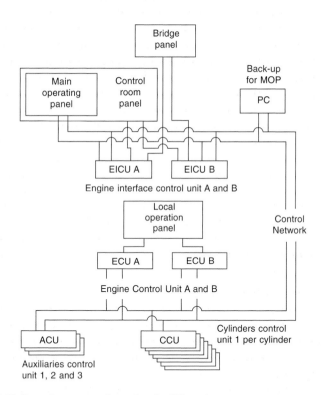

Figure 10.39 Control system configuration for ME engine

The ECU controls functions related to the overall condition of the engine. It is connected to the Plant Control System, the Safety System and the Supervision and Alarm System, and is directly connected to sensors and actuators. The ECU's function is to control the action of the following components and systems:

— The engine speed in accordance with a reference value from the application control system (an integrated governor control).
— Engine protection (overload protection as well as faults).
— Optimization of combustion to suit the running condition.
— Start, stop and reversing sequencing of the engine.
— Hydraulic (servo) oil supply (lube oil).
— Auxiliary blowers and turbocharging.

The Cylinder Control Unit is connected to all the functional components to be controlled on each cylinder, its purpose to control the activation of fuel injection, the exhaust valve, the starting valve, and the cylinder lubricator for a specific cylinder. Since each cylinder is equipped with its own controller (the CCU), the worst consequence of a CCU failure is a temporary loss of power from that particular cylinder (similar to, for example, a sticking fuel pump on a conventional engine). The engine controller (ECU) has a second ECU as a hot standby which, in the event of a failure, immediately takes over and continues the operation without any change in performance (except for the decreased tolerance for further faults until repair has been completed).

In the event of a failure in a controller the system will identify the faulty unit, which can be simply replaced with a spare. As soon as the spare is connected it will automatically be configured to the functions it is to replace, and resume operation. As both the ECU and the CCU are implemented in the same type of hardware only a few identical spares are needed. If failures occur in connected equipment—sensors, actuators, wires—the system will locate the area of the failure and, through built-in guidance and test facilities, assist the engine operating staff in the final identification of the failed component.

Cylinder Pressure Measuring System (PMI)
Reliable measurement of the cylinder pressure is essential for ensuring 'as new' engine performance. A conventional mechanical indicator in the hands of skilled engineroom staff can provide reasonable data on this parameter, but the process is quite time-consuming and the cylinder pressure data derived is not available for analysis in a computer. Some valuable information is therefore less likely to be used in a further analysis of the engine condition.

A computerized measuring system with a high quality pressure pick-up connected to the indicator bore—PMI Off-Line—was developed by MAN B&W Diesel for application to its MC engines. For the Intelligent

Engine, however, on-line measurements of the cylinder pressure are necessary, or at least highly desirable. In this case, the indicator cock cannot be used since the indicator bore would clog up after a few days of normal operation.

The strain-pin type of pressure sensor was applied instead. Here, the pressure sensing element is a rod located in a bottom hole in the cylinder cover, in close contact with the bottom of the hole and close to the combustion chamber surface of the cylinder cover. The sensor thus measures the deformation of the cover caused by the cylinder pressure without being in contact with the aggressive combustion products. The position of the sensor also makes it easier to prevent electrical noise from interfering with the cylinder pressure signal.

The pressure transducer of the above-mentioned off-line system is used for taking simultaneous measurements for calibrating the on-line system. A calibration curve is determined for each cylinder by feeding the two signals into the computer in the calibration mode. The fact that the same high quality pressure transducer is used to calibrate all cylinders means that the cylinder-to-cylinder balance is not at all influenced by differences between the individual pressure sensors.

Both on-line and off-line systems provide the user with valuable assistance in keeping the engine performance at 'as new' standard, extending the time-between-overhauls and reducing the workload of the crew. The systems automatically identify the cylinder being measured without any interaction from the person carrying out the measurement (because the system contains data for the engine's firing order). Furthermore, compensation for the crankshaft twisting is automatic, exploiting proprietary data for the engine design. If there is no such compensation, the mean indicated pressure will be measured wrongly and when the figure is applied to adjust the fuel pumps the cylinders will not have the same true uniform load after the adjustment, although it may seem so. (Crankshaft twisting may lead to errors in mean indicated pressure of some 5 per cent, if not compensated for.)

The computer executes the tedious task of evaluating the 'indicator card' data which are now in computer files; and the cylinder pressure data can be transferred directly to MAN B&W Diesel's CoCoS-EDS engine diagnostic system for inclusion in the general engine performance monitoring. The result presented to engineroom staff is far more comprehensive, comprising a list of necessary adjustments. These recommendations take into account that the condition of the non-adjusted cylinders changes when the adjustments are carried out;

it is not necessary therefore to check the cylinder pressure after the adjustment.

Starting air system

The starting air valves on the ME engine are opened pneumatically by electronically-controlled on/off solenoid valves which replace the mechanically-activated starting air distributor of the MC engine. Greater freedom and more precise control are yielded, while the 'slow turning' function is maintained.

ME ENGINES IN SERVICE

Seeking to confirm the efficiency and reliability of the ME systems in regular commercial service, the Hitachi-MAN B&W 6L60MC main engine of the chemical/product carrier *Bow Cecil*—delivered in October 1998—was prepared for the systems during its production. The mechanical/hydraulic system components were installed on the upper platform of the engine in parallel with the conventional camshaft. The camshaft system was used during the sea trial and the early operating period of the tanker until the electronic components were installed in September 2000. The engine was subsequently commissioned as the first ME engine in service in November 2000.

System flexibility is accessed by different running modes, selected either automatically to suit specific operating conditions or manually by the operator to meet specific goals, such as 'low fuel consumption' or 'limited exhaust gas emissions'. The mode selection screen of the Human Machine Interface in the control room allows the operator to switch from Fuel Economy and Emission Control modes, as well as to switch between governor control modes such as Constant Speed and Constant Torque.

The first engine built from the start as an ME design—a 7S50ME model—was completed by MAN B&W Alpha Diesel in Denmark in early 2003, by which time some 18 ME engines (S50, S60 and S70 versions) were on order or in service. A number of 12 K 98ME engines have since been ordered.

DUAL-FUEL GAS ENGINES

All MAN B&W Diesel two-stroke engines with a bore size of 500 mm or larger can be delivered as dual-fuel engines with high pressure gas injectors for burning volatile organic compounds (VOCs) on shuttle tankers, LNG carriers and crude oil tankers. The performance data for these MC-GI engines are the same as for the diesel versions. (*See Chapter 2 for more details.*)

See Chapter 9 for more details of MAN B&W Diesel's Intelligent Engine development; see Chapter 15 for pioneering work by MAN on electronically controlled fuel injection for low speed engines; and see Introduction for background on historical engines.

11 Mitsubishi low speed engines

The dominance enjoyed for many years by MAN B&W and Sulzer in the low speed two-stroke crosshead engine sector is challenged by Mitsubishi, the only other contender in the international arena. The Japanese group has widened its UEC portfolio in recent years to offer new and refined models across the full propulsion power spectrum. Mitsubishi has traditionally served its home market but a stronger direct-export thrust has earned installations in ships built for European and other overseas owners in Japan and Europe.

Most Mitsubishi engines, unlike MAN B&W and Sulzer engines, are built by the group itself rather than overseas licensees. The complete UEC programme can be handled by its Kobe factory while engines up to 600 mm bore can be built by its Yokohama works and the Japanese licensees Akasaka and Kobe Diesel.

An experimental UEC model built in 1952 yielded a power output 50 per cent higher than its predecessor, the MS engine, which had been developed from the early 1930s. The first commercial UEC engine, a nine-cylinder UEC 75/150 model developing 8830 kW, was introduced in 1955. The UEC engine now shares a common design philosophy with MAN B&W's MC and Sulzer's RTA series, exploiting constant pressure turbocharging and uniflow scavenging through a central hydraulically opened/pneumatically closed exhaust valve in the cylinder cover. (The UEC designation refers to uniflow-scavenged, exhaust gas turbocharged and crosshead type engine.)

The original UEC engines—progressing from A to E type—featured impulse turbocharging, three mechanically-actuated exhaust valves per cylinder and a single centre-injection fuel valve. This arrangement fostered high performance with a smooth exhaust gas outlet and good combustion but exhaust valve lifetime was sometimes short.

A two-stage turbocharging system (unique in production low speed engine practice) was introduced for the UEC-E type in 1977 to boost specific output and achieve a mean effective pressure of 15 kg/cm^2. Low pressure and high pressure turbochargers were linked in series (Figure 11.1). To match this increase in maximum cylinder pressures, the engine frame, bearings and combustion chamber components

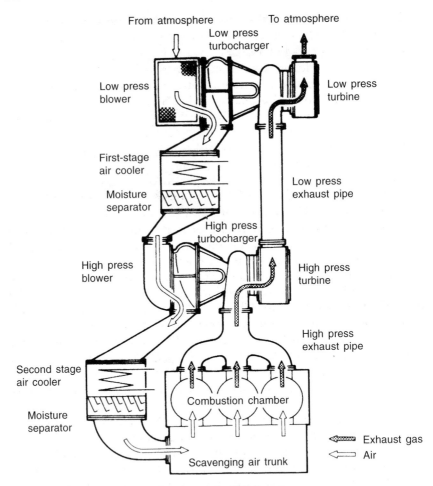

Figure 11.1 Two-stage turbocharging system for UEC-E engines

were made more rigid than those of the UEC-D type predecessor while still retaining the monobloc cast iron construction (three elements held by tie bolts). A new piston crown of specially reinforced molybdenum cast steel was specified with a combination of radial and circular ribs behind the frame plate to withstand the rise in maximum pressure (increased from 90 to 110 kg/cm^2).

The UEC-E engine was comparatively short-lived in the programme, however, as the steep rise in bunker prices in the late 1970s dictated a stronger focus on fuel economy rather than sheer output. Mitsubishi therefore introduced the long stroke UEC-H type model in 1979, abandoning impulse turbocharging in favour of constant pressure turbocharging based on the group's own MET-S turbocharger (see Chapter 7). The H-type also broke with previous UEC engine practice in using one large diameter exhaust valve in the centre of the cylinder cover and twin side-mounted fuel injection valves (Figure 11.2). A significant reduction in fuel consumption was yielded. The H-type is detailed at the end of this chapter. The HA-type, which followed in 1982, adopted the higher efficiency non-water-cooled MET-SB turbocharger and an improved combustion system, further enhancing fuel economy. Output was also raised by 11–14 per cent.

Advances in fuel economy were pursued with the development of longer stroke/lower speed L-type engines from 1983 with successive refinements embodied in the LA-type (1985), the LS-type (1986) and the LSII-type (1987). Stroke–bore ratios were increased to achieve lower direct-coupled propeller speeds and hence higher propulsive efficiency. Efforts to improve specific fuel consumption were reflected in rises in the maximum cylinder pressure.

For most of the 1980s Mitsubishi was content to concentrate on the small-to-medium size ship propulsion sector with just four bores (370 mm, 450 mm, 520 mm and 600 mm). Four- to eight-cylinder models covered outputs up to around 15 000 kW. The UEC-LA models, with a stroke–bore ratio of 3.17:1, were supplemented in 1985 by longer stroke (LS) options for the 520 mm and 600 mm models (3.55 and 3.66:1 ratios, respectively).

A return to the large bore arena, long surrendered to MAN B&W and Sulzer, was signalled in 1987 by the launch of the UEC 75LSII series with a maximum rating of 2940 kW/cylinder. The 750 mm bore/2800 mm stroke design retained the main features of the LS engines but exploited an even longer stroke–bore ratio (3.73:1) and lower running speeds (63–84 rev/min). Mitsubishi stepped up its challenge in the VLCC and large bulk carrier propulsion markets in 1990 with the introduction of an 850 mm bore/3150 mm stroke version. The UEC 85LSII series, available in five- to 12-cylinder models, almost doubled the upper power limit of the portfolio to 46 320 kW at 54–76 rev/min. The design offers a maximum rating of 3860 kW/cylinder.

The propulsion market created by a new generation of large Panamax and post-Panamax containerships requiring service speeds of up to 25 knots was addressed by Mitsubishi in 1992 with its UEC 85LSC series, a shorter stroke (2360 mm) faster running derivative of the UEC 85LSII. The design is the highest rated engine in the group's programme, with an output of 3900 kW/cylinder at 102 rev/min. The series covers a power band up to 46 800 kW from five- to 12-cylinder models.

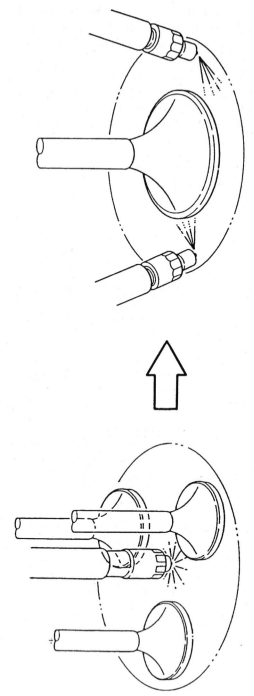

Figure 11.2 The UEC A to E series engines featured three exhaust valves and a centre-injection fuel valve (left). Later series (H to LS) are equipped with one exhaust valve and two side-injection fuel valves (right)

A determination to secure small-ship propulsion business led in 1991 to the introduction of a 330 mm bore/1050 mm stroke LSII design (Figure 11.3). The series succeeded the 370 mm bore LA models which pioneered the modern 'mini-bore' two-stroke engine from the late 1970s for small general cargo tonnage, tankers, colliers, feeder containerships and even large fishing vessels. The longer stroke and more advanced UEC 33LSII engine now develops 566 kW/cylinder at 215 rev/min and covers an output band from 1230 kW to 4530 kW at 162–215 rev/min with four- to eight-cylinder models.

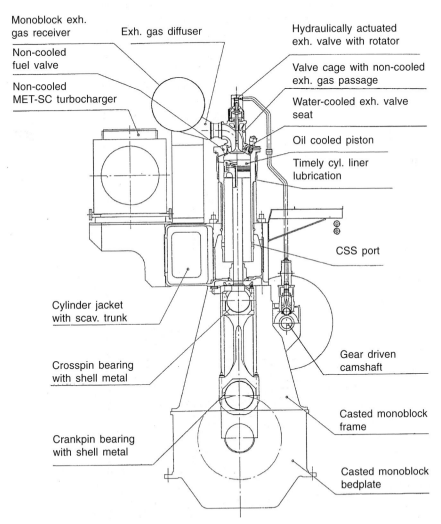

Figure 11.3 Construction of UEC 33LSII engine

Mitsubishi enhanced its medium power programme in 1992 by upgrading the 500 mm and 600 mm bore models to LSII status. Later power upratings of 3–5 per cent for these UEC50LSII/60LSII models, without major design changes, were introduced to meet the demands of larger ships and higher service speeds. The specific outputs respectively rose to 1445 kW and 2045 kW per cylinder. With a stroke/bore ratio of 3.9:1, the 50LSII series embraces four-to-eight-cylinder versions covering an output range up to 11 560 kW at 127 rev/min.

The small bore programme was strengthened in 1995 with 370 mm and 430 mm UEC LSII designs to attract business from handy-sized bulk carriers, feeder containerships, small tankers and gas carriers. The engines plugged a gap between the established 330 mm and 500 mm bore LSII series. The UEC 37LSII engine offers an output of 772 kW/cylinder at 186 rev/min from five-to-eight-cylinder versions, while the UEC 43LSII engine delivers 1050 kW/cylinder at 160 rev/min from four-to-eight-cylinder models. The first UEC 37LSII engine, a six-cylinder model, completed testing in early 2000, and the debut order for a 43LSII engine, a seven-cylinder model, was logged in 2002.

UEC37LSII engine data

Bore	370 mm
Stroke	1290 mm
Stroke/bore ratio	3.49
Output	772 kW/cyl
Cylinders	5–8
Power range	2095–6180 kW
Speed	186 rev/min
Mean piston speed	8 m/s
Mean effective pressure	18 bar
Max. cylinder pressure	147 bar
Specific fuel consumption	175 g/kWh

Development of the LSII design from 1998 resulted in the LSE series, headed by a 520 mm bore model (UEC52LSE) targeting feeder containerships and Panamax bulk carriers. The series was extended by a 680 mm bore model (UEC68LSE) for propelling tonnage including Aframax and Suezmax tankers, Cape-size bulk carriers and containerships. The LSE models share a more compact configuration and a 25 per cent reduction in overall number of components compared with the LSII engines. Benefits were also gained in production costs, reliability and maintainability. The mean effective pressure rating and mean piston speed were respectively raised by five per cent and six per cent to 19 bar and around 8.5 m/s.

UEC-LSE engine data

	52LSE	60LSE	68LSE
Bore, mm	520	600	680
Stroke, mm	2000	2400	2690
Stroke/bore ratio	3.85	4	3.96
Output, kW/cyl	1705	2255	2940
Cylinders	4–12	5–8	5–8
Power range, kW	3700–20 460	7650–18 040	10 050–23 520
Speed, rev/min	127	105	95
Mps, m/s	8.47	–	8.52
Mep, bar	19	–	19
Sfc, g/kWh	160–167	160–166	159–165

An extension of the same 'total economy' LSE concept later resulted in a 600 mm bore version. The current UEC programme embraces models with bore sizes of 330 mm, 370 mm, 430 mm, 450 mm, 500 mm, 520 mm, 600 mm, 750 mm and 850 mm in LA, LS, LSII and LSE versions, covering an output band from 1120 to 46 800 kW at 215 to 54 rev/min (Figures 11.4 and 11.5). Typical performance curves and layout diagrams are shown in Figures 11.6 and 11.7.

Development progress since 1955 is underlined by the LS-type whose specific output is three times that of the original UEC engine and whose weight/power and length/power ratios are reduced by one-third. The advances in key design and performance parameters over the years are illustrated graphically in Figures 11.8 and 11.9.

New 500 mm bore design

A joint venture with the Wartsila Corporation to design and develop a new low speed engine was announced at end-2002. The Japanese group has been a prolific licensed builder of Sulzer engines since 1925, a marque now under the Wärtsilä umbrella. The 500 mm bore/2050 mm stroke design will be available in two versions: the conventional 'mechanical' Sulzer RTA50C and the electronically-controlled Sulzer RT-flex50C. (See Sulzer low speed engines chapter for details of RT-flex technology.)

Each type will offer a maximum continuous rating of 1620 kW/cylinder at 124 rev/min and will be available in five-to-eight-cylinder models covering a power range from 5650 kW to 12 960 kW at 99 rev/min to 124 rev/min. Output and speed ratings will target Handymax and Panamax bulk carriers, large product tankers, feeder containerships

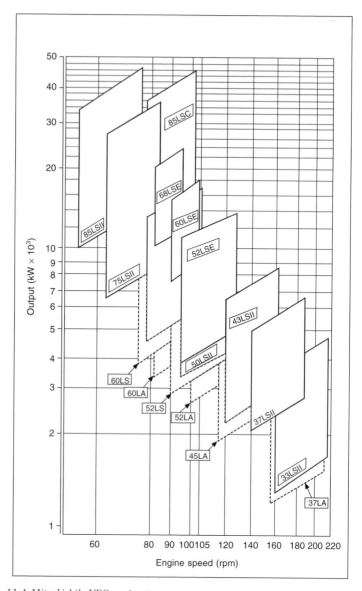

Figure 11.4 Mitsubishi's UEC engine programme

and medium-sized reefer vessels. Both versions will incorporate Sulzer TriboPack measures for enhancing piston-running behaviour, underwriting low cylinder wear rates for three years between overhauls, and minimizing cylinder lube oil consumption.

High efficiency, compactness and environmental friendliness were

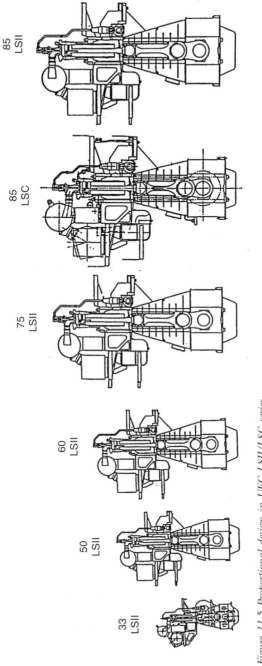

Figure 11.5 Proportional design in UEC LSII/LSC series

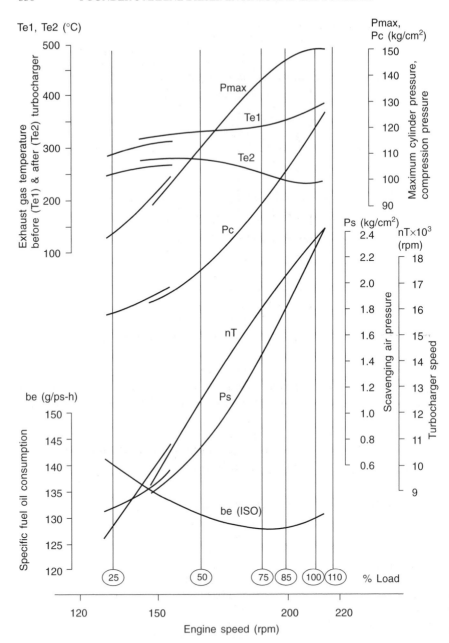

Figure 11.6 Performance curves of 8UEC 33LSII engine (4316 kW at 210 rev/min)

Calculation of fuel consumption at maximum
continuous output of a derated engine

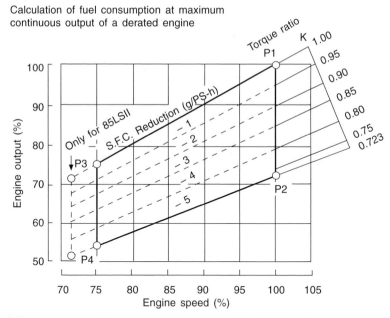

Figure 11.7 Improvement of fuel economy by derating a UEC 85LSII engine

set as design goals for the project to be pursued by a joint working group of engineers from both companies. The new engine was assigned for production in Japan by Mitsubishi and its licensees and by other Japanese licensees of Wärtsilä, and in Korea and China by Wartsila and Mitsubishi licensees. Separate branding—Sulzer or Mitsubishi—will be applied to the engines, depending on the builder. The first examples are due for completion by end-2004.

Mitsubishi-Sulzer RTA50C/RT-flex50C

Bore	500 mm
Stroke	2050 mm
Stroke/bore ratio	4:1
Output, mcr	1620 kW/cyl
Mep (R1)	19.5 bar
Pmax	155 bar
Mean piston speed (R1)	8.5 m/s
Cylinders	5–8
Speed range (R1-R3)	124–99 rev/min
Power range	5650–12 960 kW
Sfc, full load (R1-R2)	165–171 g/kWh

Type	Cylinder bore (cms.)	Turbocharging system	Turbocharging	Fuel system	Valve arrangement
A 1955	85 75 65 52 45 39 33	Impulse	Water cooled	Accumulator	3 exhaust valves 1 FOV
B 1963	85 52 45	Impulse	Water cooled	Bosch	
C 1965	85 65 52 45 39 33	Impulse	Non-cooled MET-OO	Bosch	
D 1970	85 65 52 45	Impulse	MET-OOO	Bosch	Water-cooled seat
E 1975	60 52	Two-stage impulse	+ MET-OOO	Bosch	Water-cooled seat
H 1979	60 52 45 37	Constant pressure	MET-S	Bosch	1 exhaust valve 2 FOVs Water-cooled seat
HA 1982	60 52 45 37	Constant pressure	MET-SB	Bosch	Water-cooled seat
L 1983	60 52 45 37	Constant pressure	MET-SB	(COP)* Bosch	Water-cooled seat
LA 1985	60 52 45 37	Constant pressure	MET-SC	(COP)* Bosch	Water-cooled seat
LS 1986	60 52	Constant pressure	MET-SC	(COP)* Bosch	Water-cooled seat
LSII 1987	85 75	Constant pressure	MET-SC	(COP)* (CIT)† Bosch	Water-cooled seat

* Controllable opening pressure system   † Controllable injection timing

Figure 11.8 Design history of UE engine, 1955–1987

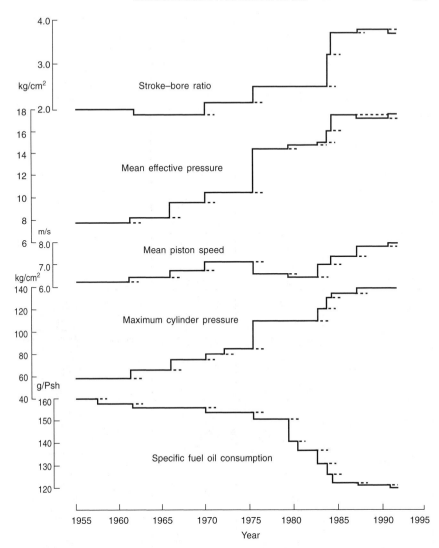

Figure 11.9 Changes in UEC engine parameters and performance since its introduction

UEC DESIGN DETAILS

Structural and design details are exemplified by the UEC 75LSII and 85LSII engines (Figure 11.10). The bedplate and column are of simple welded monoblock construction, and the partition for each cylinder consists of a single welded plate. Side forces are supported by horizontal steel ribs arranged behind the guide bar. The pistons are of the well

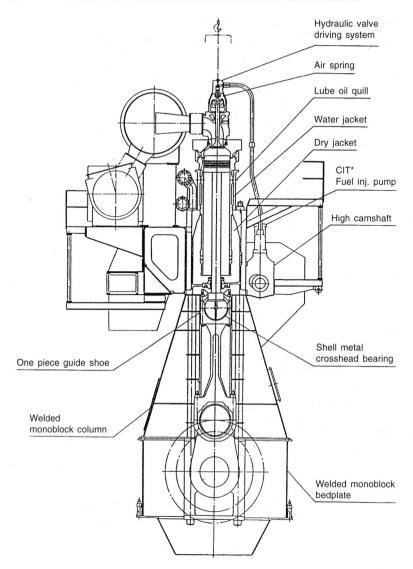

*CIT: Controllable injection timing

Figure 11.10 Construction of UEC 75LSII engine

proven central support type and are fitted with four rings (Figure 11.11). Hard chromium plating applied to the running surfaces of the rings improves wear resistance under high pressure and corrosion resistance in a low temperature corrosive environment. Chromium plating the ring grooves has the same purpose, as well as improving

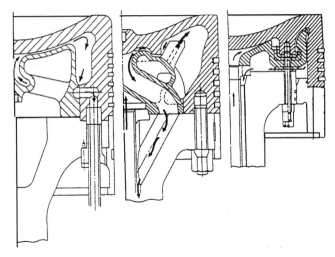

Figure 11.11 Improvements in piston design from the original UE engine component (left) to the HA model (centre) and the current L and LS types (right)

the durability of the rings by ensuring normal movement in the grooves.

The lower part of the cylinder liners feature controlled swirl scavenging (CSS) ports which promote improved scavenging efficiency and hence favourably influence fuel consumption and thermal load. The angle of the lower scavenging port is directed towards the centre of the cylinder to discharge combustion gas on the piston top; and the angle of the upper port is directed sideways to discharge gas around the inner surface of the liner. The CSS configuration (Figure 11.12). reportedly raises scavenging efficiency by 5–7 per cent over the conventional port arrangement. Two-row bore liner cooling is incorporated and cylinder lubrication is timed so as to reduce oil consumption. Corrosive liner wear, caused by low temperatures, is inhibited by controlled positioning of the cooling bores.

The single exhaust valve of each cylinder has a hydraulic rotation system which rotates the valve to improve running control conditions. The rotation wings are fitted on the actuator piston on the exhaust valve side, an arrangement which is said to offer less obstruction to exhaust gas passage than with the traditional rotator wing mounted on the valve spindle. A water-cooled valve seat is specified.

The Bosch fuel pumps have control racks to adjust the quantity of fuel supplied. The controllable injection timing (CIT) pumps also ensure that engine performance is improved at the service load (Figure 11.13). Two side-injection non-water-cooled fuel injection valves (Figure 11.14) are arranged on each cylinder cover; a steam passage is provided for warming the valve body when the engine is stopped.

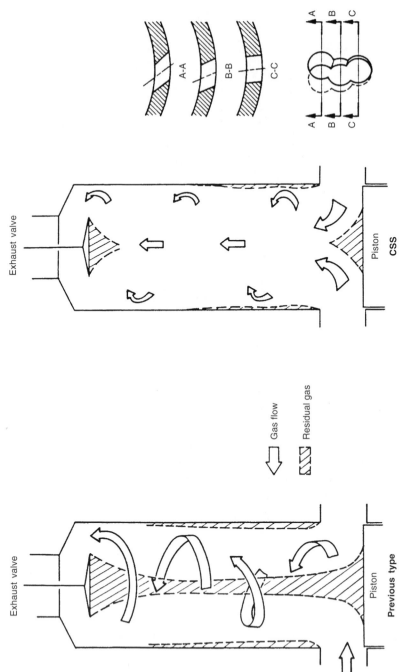

Figure 11.12 Comparison of gas flow in cylinder with original UE engine scavenging and controlled swirl scavenging (CSS)

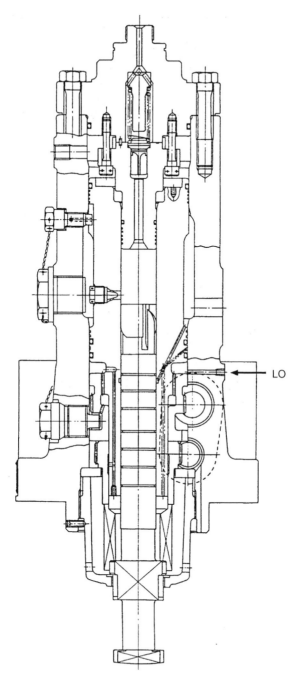

Figure 11.13 Controllable injection timing (CIT) fuel oil pump

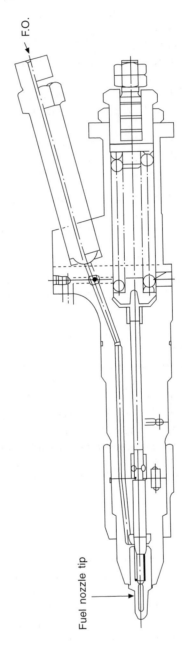

F.O.

Fuel nozzle tip

Figure 11.14 Lightweight (20 kg) non-cooled fuel injection valve for UEC-LSII engine

An optional AFR (automatic flexible rating) control system can be specified to reduce fuel consumption during part-load operation by co-ordinating the engine output with the operational conditions of the ship. The system (Figure 11.15) controls the fuel delivery timing of the injection pump by using the CIT system to produce maximum cylinder pressure, and also controls the turbine nozzle area of the variable geometry (VG) turbocharger to secure an optimum scavenging air pressure. The CIT fuel injection pumps, offered as standard with LSII engines, are operated by a governor when the AFR control system is not installed.

A mean time-between-overhaul of four years was targeted for the LSII engines, Mitsubishi citing contributions from the following key elements:

- Piston rings and cylinder liners, yielding respective wear rates of 0.2 mm/1000 hours and 0.02 mm/1000 hours or less.
- Reliable hydraulically rotated exhaust valves with stellite-coated stems and seats.
- Easily maintained fuel injection valves.
- High performance piston gland packing (stuffing box drainage of less than 10 litres/day/cylinder, and a low wear rate scraper ring).

Mitsubishi's R&D priorities mirror those of its two rivals in the low speed engine arena, with emphases on further upratings of designs, enhancing reliability and durability, and reducing noxious emissions. Its resources include the group's main research centre at Nagasaki. Mitsubishi has investigated all aspects of exhaust gas de-NOxing technology, including the direct injection of water into the combustion chamber. Its stratified injection system-based on a fuel valve designed to inject fuel—water—fuel sequentially—has achieved NOx reductions of around 50 per cent without significantly increasing fuel consumption and is offered as an option for UEC-LSE engines (See Chapter 3).

UEC-H TYPE ENGINE

Although superseded by the LA, LS and LSII designs, and no longer in production, the UEC-H type engine remains in service and warrants more detailed description. The UEC-H type engine (Figure 11.16) is of cast iron construction with the three separate blocks of bedplate, entablature and cylinder block secured to one another by long tiebolts. Unlike other designers of two-stroke engines, Mitsubishi retained the solid cast iron construction (Figure 11.17) to achieve greater rigidity

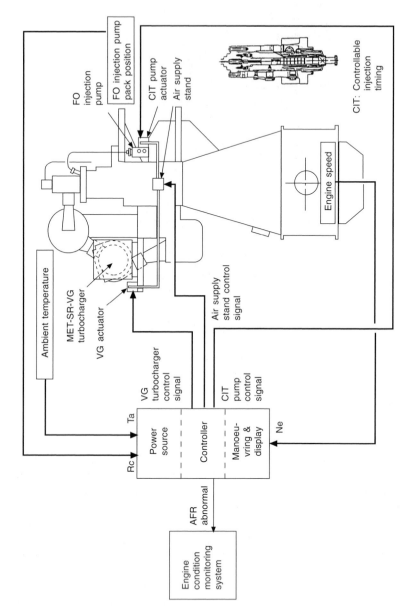

Figure 11.15 A UEC engine served by an automatic flexible rating (AFR) control system

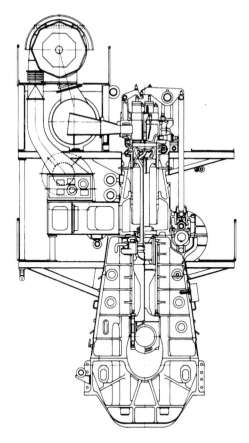

Figure 11.16 Cross-section of the UEC52/125H engine

with less deformation in service and lower vibration and noise levels. The shell type main bearings are supported in housings machined in the bedplate. The running gear is of straightforward conventional design using shell type or whitemetal lined large end bearings and a crosshead of traditional table type (Figure 11.18) with four guide slippers and twin whitemetal-lined top end bearings and the piston rod bolted direct to the crosshead pin. The connecting rod itself is machined from forged steel.

Compared with the two-stage turbocharged E type engine, the constant pressure turbocharged UEC-H engine has a delayed exhaust valve opening and consequently more effective work done during each power stroke.

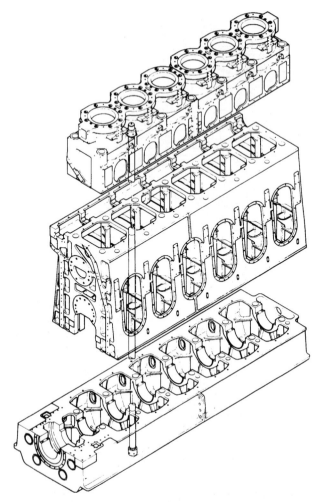

Figure 11.17 Exploded view of UEC-H engine showing cast iron construction

Exhaust and scavenging

Referring to the cross-sectional drawing (Figure 11.16), the exhaust gases leave the engine by way of a diffuser type pipe to enter the exhaust gas manifold. Located above this manifold on a pedestal is the Super MET non-water-cooled turbocharger, or two chargers in the case of a large number of cylinders. The air from this turbocharger is delivered to the intercooler and then the scavenge air trunk by way of non-return valves and then into the cylinder via ports uncovered by the piston. During low load operation the turbocharger air delivery is passed through two auxiliary blowers before entering the cylinders.

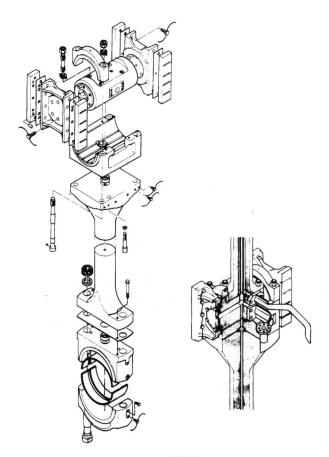

Figure 11.18 Connecting rod and crosshead of UEC-H engine

The diffuser type exhaust pipe is of a three-branch construction where it connects to the engine cylinder. The gas flows from the three cylinder exhaust valves are merged into one to enter the exhaust manifold through the diffuser section. The exhaust manifold is actually made up of many short sections, each corresponding to one engine cylinder, joined together by flexible joints to accommodate longitudinal thermal deformation.

The scavenge air trunk, however, is divided into two compartments. Charge air from the intercooler enters the outer chamber and passes through the multi-cell non-return valves to the scavenge air entablature surrounding the cylinder jackets.

The two motor-driven auxiliary blowers consume approximately 0.6 per cent of the rated engine output and as a matter of standard practice are automatically brought into operation when the engine operates at 50 per cent load or less. Start up of the blowers is by a sensed drop in scavenge air pressure. An emergency feature for engines equipped with only one turbocharger is the facility to operate both auxiliary blowers in series to achieve the required scavenge air pressure for continuous engine operation. This emergency hook-up is a simple process of adjusting some partition plates in the scavenge box.

Figure 11.19 Mitsubishi UEC 85LSII engine in 6-cylinder form on test for VLCC propulsion

12 Sulzer low speed engines

Active in both four-stroke and two-stroke design sectors, Sulzer's links with the diesel engine date back to 1879 when Rudolf Diesel, as a young engineer, followed up his studies by working as an unpaid workshop trainee at Sulzer Brothers in Winterthur, Switzerland. The first Sulzer-built diesel engine was started in June 1898. In 1905 the company built the first directly reversible two-stroke marine diesel engine and, five years later, introduced a valveless two-stroke engine with an after-charging system and spray-cooled pistons. Airless fuel injection was applied to production engines in 1932. and turbocharging from 1954.

Low speed crosshead engine designs from Sulzer after 1956 were of the single-acting two-stroke turbocharged valveless type employing loop scavenging and manifested progressively in the RD, RND, RND-M, RLA and RLB series. Details of the RL-type, many of which are still in service, appear at the end of this chapter. A break with that tradition came at the end of 1981 with the launch of the uniflow-scavenged, constant pressure turbocharged RTA series with a single poppet-type exhaust valve (Figures 12.1 and 12.2).

The original RTA series embraced six models with bore sizes of 380 mm, 480 mm, 580 mm, 680 mm, 760 mm and 840 mm, and a stroke–bore ratio of 2.86 (compared with the 2.1 ratio of the RL series). These RTA38, RTA48, RTA58, RTA68, RTA76 and RTA84 models—collectively termed the RTA-8 series—were supplemented in 1984 by the longer stroke RTA-2 series comprising 520 mm and 620 mm bore models and the RTA84M model. The RTA-2 series was extended again in 1986 by a 720 mm bore model. An uprated 840 mm bore design, the RTA84C, was introduced in 1988 to offer higher outputs in the appropriate speed range for propelling large cellular containerships (Figure 12.3).

A higher stroke–bore ratio (3.47) for the RTA-2 series secured lower engine rotational speeds and hence higher propulsion efficiencies. The ratings of the RTA-2 engines were increased in 1987 in line with the power level already offered by the RTA72 model. Various design improvements, together with higher ratings, were introduced for the

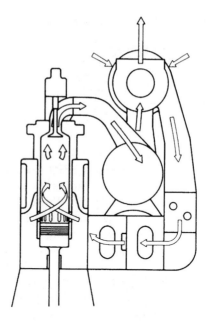

Figure 12.1 Uniflow scavenging system

RTA-2 series in 1988 at the same time that the RTA84C engine was launched. A further upgrading of the RTA-2 series—to RTA-2U status—came in 1992, with a 9 per cent rise in specific power output.

The RTA range was subsequently extended by the introduction in 1991 of the RTA84T 'Tanker' engine design, which, with a stroke-bore ratio of 3.75 and a speed range down to 54 rev/min, was tailored for the propulsion of VLCCs and large bulk carriers. The first production engine entered service in 1994. Even longer strokes (4.17 s/b ratio) were adopted in 1995/96 for RTA-T versions of the 480 mm, 580 mm and 680 mm bore models whose key parameters addressed the power and speed demands of bulk carriers up to Cape-size and tankers up to Suezmax size. These RTA48T, RTA58T and RTA68T designs have more compact dimensions than earlier models, giving ship designers more freedom to create short enginerooms, while reduced component sizes and weights facilitate easier inspection and overhaul. Shipyard-friendly features were also introduced to smooth installation.

In 1994 Sulzer anticipated demand for even higher single-engine outputs from very large and fast containerships by announcing the RTA96C engine, a short stroke (2.6 s/b ratio) 960 mm bore design.

RTA engines of all bore sizes have benefited from continual design refinements to increase power ratings and/or enhance durability and

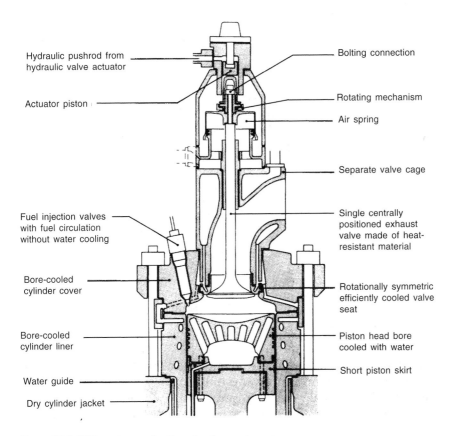

Hydraulic pushrod from
hydraulic valve actuator

Actuator piston

Fuel injection valves
with fuel circulation
without water cooling

Bore-cooled
cylinder cover

Bore-cooled
cylinder liner

Water guide

Dry cylinder jacket

Bolting connection

Rotating mechanism

Air spring

Separate valve cage

Single centrally
positioned exhaust
valve made of heat-
resistant material

Rotationally symmetric
efficiently cooled valve
seat

Piston head bore
cooled with water

Short piston skirt

Figure 12.2 RTA engine combustion chamber components

reliability, based on service experience. In early 1999 the RTA
programme was streamlined by phasing out some older types (such as
the 380 mm bore model) which had fallen out of demand, a measure
also dictated by the need to ensure that all engines could comply with
the IMO Annex VI NOx emissions regulations. In mid-1999 the RTA60C
model was introduced as the first Sulzer engine designed from the
bedplate up to embody RT-flex electronically-controlled fuel injection
and exhaust valve actuation systems. Starting with the 580 mm and
600 mm bore designs, RT-flex series versions became available as options
throughout the RTA programme (see section below).

Sulzer came under the umbrella of the Wärtsilä Corporation in
1997, the Finland-based four-stroke specialist committing to continued
development of low speed two-stroke engines. The current RTA
programme, summarized in Figure 12.4, embraces nine bore sizes

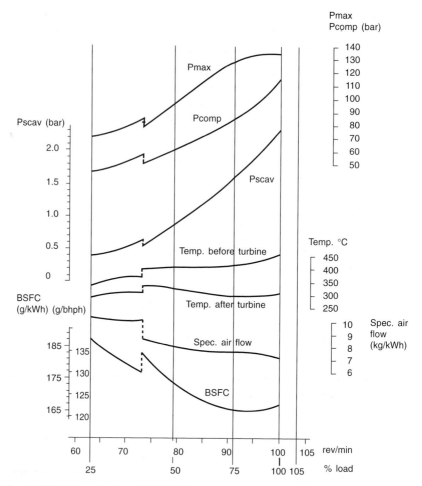

Figure 12.3 Test results from 9RTA84C engine equipped with an exhaust gas power recovery turbine, according to a propeller characteristic. The engine develops 34 380 kW mcr at 100 rev/min

from 480 mm to 960 mm with various stroke–bore ratios, speed and rating options covering an output band from 5100 kW to 80 080 kW with five-to-14-cylinder in-line models.

Engine selection

Selecting a suitable main engine model to meet optimally the power demands of a given project dictates attention to the anticipated load range and the influence that the operating conditions are likely to

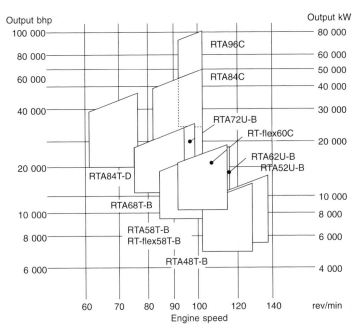

Figure 12.4 Sulzer RTA engine programme; larger bore RT-flex models are also available

have throughout the entire life of the ship. Every RTA engine has a layout field within which the power/speed ratio (= rating) can be selected. It is limited by envelopes defining the area where 100 per cent firing pressure (that is, nominal maximum pressure) is available for the selection of the contract maximum continuous rating (CMCR). Contrary to the 'layout field', the 'load range' is the admissible area of operation once the CMCR has been determined. Various parameters have to be considered in order to define the required CMCR: for example, propulsive power, propeller efficiency, operational flexibility, power and speed margins, possibility of a main engine-driven generator, and the ship's trading patterns.

The layout field, Figure 12.5, is the area of power and engine speed within which the CMCR of an engine can be positioned individually to give the desired combination of propulsive power and rotational speed. Engines within this layout field will be tuned for maximum firing pressure and best fuel efficiency. The engine speed is given on the horizontal axis and the engine power on the vertical axis of the layout field; both are expressed as a percentage of the respective engine's nominal R1 parameters. Percentage values are used so that the same diagram can be applied to all engine models. The scales are

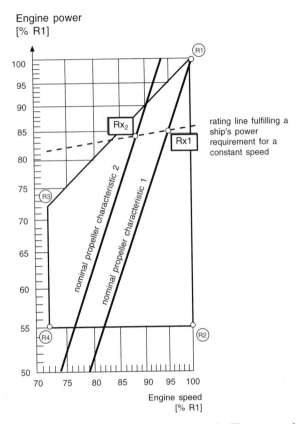

Figure 12.5 Layout field applicable to all Sulzer RTA models. The contracted maximum continuous rating (Rx) may be freely positioned within the layout field

logarithmic so that exponential curves, such as propeller characteristics (cubic power) and mean effective pressure (mep) curves, are straight lines. The layout field serves to determine the specific fuel oil consumption, exhaust gas flow and temperature, fuel injection parameters, turbocharger and scavenge air cooler specifications for a given engine.

The rating points R1, R2, R3 and R4 are the corner points of the engine layout field. R1 represents the nominal maximum continuous rating (MCR). This is the maximum power/speed combination available for a particular engine. A 10 per cent overload of this figure is permissible for one hour during sea trials in the presence of authorized representatives of the enginebuilder. The point R2 defines 100 per cent speed and 55 per cent power. The point R3 is at 72 per cent power and speed. The line R2–R4 is a line of 55 per cent power between 72 and 100 per cent speed. Points such as Rx are power/ speed ratios for the selection of CMCRs required for individual applications. Rating points Rx can be selected within the entire layout field.

Sulzer has smoothed the path for selecting the optimum model for a given propulsion project with its PC computer-based EnSel engine selection program which presents a list of all the models that fulfil the power and speed requirements, along with their main data. The program is offered to ship designers, yards, consultants, owners and licensees. The input data calls for the user to specify the power and speed required, and whether or not an Efficiency-Booster power turbine system is wanted. The output data then lists the engines with the appropriate layout fields and details their MCR power, speed and specific fuel consumption, main dimensions and weight, and other relevant information.

RTA DESIGN FEATURES

The RTA design benefited from principles proven in earlier generations of Sulzer R-type engines. The key elements are:

- A sturdy engine structure designed for low stresses and small deflections comprises a bedplate, columns and cylinder block pretensioned by vertical tie rods. The single-wall bedplate has an integrated thrust block and incorporates standardized large surface main bearing shells. The robust A-shaped columns are assembled with stiffening plates or are of monobloc design. The single cast iron cylinder jackets are bolted together to form a rigid cylinder block (multi-cylinder jacket units for smaller bore engines).
- Lamellar cast iron, bore-cooled cylinder liners with back-pressure timed, load-dependent cylinder lubrication.
- Solid, forged bore-cooled cylinder covers with one large central exhaust valve arranged in a bolted-on valve cage; the valve is made from a heat- and corrosion-resistant material and its seat ring is bore-cooled.
- Semi-built crankshaft divided into two parts for larger bore engines with a large number of cylinders.
- Running gear comprising connecting rod, crosshead pin with very large surface crosshead bearing shells (with high pressure lubrication) and double-guided slippers, piston rod and bore-cooled piston crown using oil cooling. All have short piston skirts.

All combustion chamber components are bore cooled, a traditional feature of Sulzer engines fostering optimum surface temperatures and preventing high temperature corrosion due to high temperatures on one side and sulphuric acid corrosion due to too low temperatures on the other. At the same time, rigidity and mechanical strength are provided by the cooler material behind the cooling bores (Figure 12.6.).

Comfortable working conditions for the exhaust valve are promoted by: hydraulic operation with controlled valve landing speed; air spring; full rotational symmetry of the valve seat, yielding well-balanced thermal and mechanical stresses and deformations of valve and valve seat, as well as uniform seating; extremely low and even temperatures in valve seat areas due to efficient bore cooling; valve rotation by simple vane impeller; valve actuation free from lateral forces, with axial symmetry; and simple guide bushes sealed by pressurized air.

The low exhaust valve seating face temperature reportedly secures an ample safety margin to avoid corrosive attack from vanadium/sodium compounds under all conditions. Efficient valve cooling is given by intimate contact with the bore-cooled seat, together with the appropriate excess air ratio in the cylinder. The specific design features of the valve assembly are also said to deter the build-up of seat deposits, seat distortion, misalignment and other factors which may accelerate seat damage.

- Camshaft gear drive housed in a special double column or integrated into a monobloc column, placed at the driving end or in the centre of the engine for larger bore models with a large number of cylinders.
- Balancer gear can be mounted on larger bore engines, when required, to counter second-order couples for four-, five- and six-cylinder models, and combined first- and second-order couples for four-cylinder models.
- A compact integral axial detuner can be incorporated, if required, in the free end of the engine bedplate.
- The fuel injection pump and exhaust valve actuator are combined in common units for each two cylinders. The camshaft-driven injection pump with double valve-controlled variable injection timing delivers fuel to multiple uncooled injectors. The camshaft-driven actuators impart hydraulic drive to the single central exhaust valve working against an air spring.
- Constant pressure turbocharging is based on high efficiency uncooled turbochargers; auxiliary blowers support uniflow scavenging during low load operation. In-service cleaning of the charge air coolers is possible. A standard optional three-stage charge air cooler unit can be specified for heat recovery.

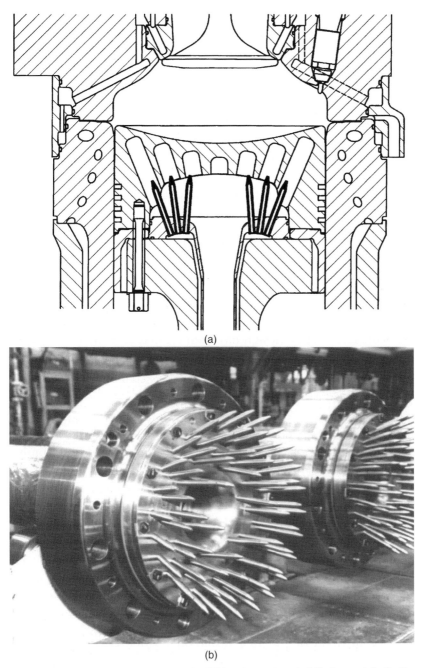

(a)

(b)

Figure 12.6 The combustion chamber of the RTA series engines is fully bore-cooled. Cooling oil spray nozzles on top of the piston rod direct oil into the bores of the piston crown

- A standard pneumatic engine control system is based on a remote manoeuvring stand in the control room and an emergency manoeuvring stand on the engine itself. The optional SBC-7.1 electro-pneumatic bridge control system is matched to the engine and arranged for engine control and manoeuvring from the wheelhouse or bridge wings.
- Standard optional power take-off drives can be arranged either at the side or at the free end for connecting an alternator to the main engine.
- Sulzer's Efficiency-Booster System (see Chapter 7) can be specified for the RTA84C, RTA84T and RTA84M engines, as well as for the RTA72U, RTA62U and RTA52U models, to exploit surplus exhaust gas energy in a power recovery turbine.
- RTA engines can satisfy the speed-dependent IMO limits on NOx emissions from exhaust gases. Tuning is facilitated by the electronic variable fuel injection timing (VIT) system.

RTA DESIGN DEVELOPMENTS

The basic RTA engine design has been refined over the years and the range expanded to address changing market requirements. The improvements have yielded wider power/speed fields, increased power outputs, reduced fuel consumption, lower wear rates and longer times between overhauls. Advances in key performance parameters are illustrated in Figure 12.7. Operating economy has also benefited from the reduced auxiliary power demand of current RTA engines—imposed by electric pumps for cooling water, lubricating oil and fuel supply—which is some 40 per cent less than that of the RL engines.

RTA-2U series

Sulzer's introduction of upgraded RTA-2 engine designs called for some modifications to match the higher power outputs and maximum combustion pressures involved (summarized in Figure 12.8). The shrinkfit of the crankshaft was strengthened to suit the increased torque values; the main bearing was adapted to accommodate the higher loads and ensure optimum oil flow; and the cylinder cover material was changed to take advantage of better fatigue strength at higher loads. Some of the design ideas tested on the 4RTX54 research engine (see Chapter 9) and already applied to the RTA84T engine were also adopted for the RTA-2U series.

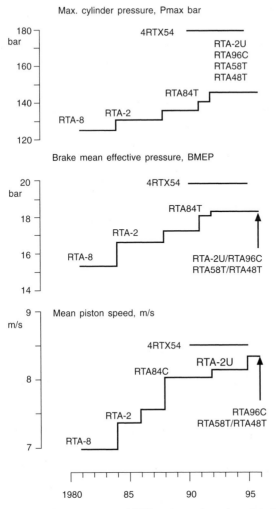

Figure 12.7 Advances in key parameters of RTA series engines since their introduction. (The RTX54 is a Sulzer research engine)

Three fuel injection valves per cylinder were specified for the RTA62U and RTA72U models (though not, because of its smaller bore size, for the RTA52U model which retained two valves per cylinder). The reported benefits of the triple-valve configuration are a more uniform temperature distribution around the principal combustion space components (cylinder cover, liner and piston crown) at the increased maximum combustion pressures, along with even lower temperatures despite the higher loads. Three fuel valves also foster significantly lower exhaust valve and valve seat temperatures.

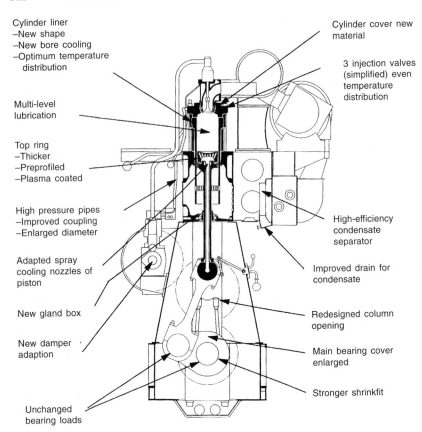

Cylinder liner
–New shape
–New bore cooling
–Optimum temperature
  distribution

Cylinder cover new
material

3 injection valves
(simplified) even
temperature
distribution

Multi-level
lubrication

Top ring
–Thicker
–Preprofiled
–Plasma coated

High pressure pipes
–Improved coupling
–Enlarged diameter

High-efficiency
condensate
separator

Adapted spray
cooling nozzles of
piston

Improved drain for
condensate

New gland box

Redesigned column
opening

New damper
adaption

Main bearing cover
enlarged

Stronger shrinkfit

Unchanged
bearing loads

Figure 12.8 Main improvements introduced in RTA-2U series engines

Other spin-offs from the research engine included a modified cylinder liner bore-cooling geometry whose tangential outlets of the bores aim for optimum distribution of wall temperatures and thermal strains at higher specific loads. The geometry of the oil cooling arrangements of the piston crown was also modified to maintain an optimum temperature distribution. The good piston running behaviour was maintained by retaining established features of the RTA design: multilevel cylinder lubrication; die-casting technology for cylinder liners; and temperature-optimized cylinder liners. Advances in materials technology in terms of wear resistance have permitted engines to run at higher liner surface temperatures. This, in turn, allows a safe margin to be maintained above the increased dew point temperature and thus avoiding corrosive wear.

Some refinements were introduced, however, to match the new

running conditions. Four piston rings are specified for the RTA-2U engines instead of the five previously used. The top ring is now thicker than the other rings and has a pre-profiled running face, as on the RTA84C containership engine. The top ring is also plasma coated. The plasma-coated, pre-profiled thicker rings have demonstrated excellent wear results. The radial wear rates were measured at less than 0.04 mm/1000 hours in trials up to and exceeding 13 000 hours of operation.

An important contribution to low wear rates of liners and pistons results from improving the separation and draining of water borne in the cooled scavenge air before it enters the cylinders, particularly at higher engine loads. The RTA-2U engines were specified with a more efficient condensate water separator in the scavenge air flow after the cooler, along with a more effective drain.

RTA-T series

A number of design modifications to the RTA series were introduced for the ultra-long stroke RTA48T, RTA58T and RTA68T engines to achieve more compact, lower weight models offering reduced production, installation and maintenance costs (Figures 12.9 and 12.10). A time between overhaul for the main components of 15 000 hours was sought. Cutting the manufacturing cost, despite the greater stroke/bore ratio (4.17) than previous RTA series engines, was addressed by a number of measures: reducing the size and weight of components; simplifying the designs of components and sub-assemblies and making them easier to produce; reducing the number of parts; and designing to save assembly time.

An example of the design changes made to reduce the sizes and weights of components is provided by the cylinder cover, along with its exhaust valve, housing and valve actuator. An overall weight saving of 30 per cent was achieved on each new cover, largely due to the smaller dimensions. Reducing the distance between cylinder centres by around 9 per cent allowed components such as the bedplate, monobloc columns and cylinder block to be reduced in size, resulting in weight savings of 13–14 per cent (Figures 12.11, 12.12 and 12.13). The cylinder block is lower in overall height and thus lighter than in equivalent RTA-2U engines. The freedom for ship designers to create short enginerooms is enhanced by a degree of flexibility in the fore and aft location of the turbocharger and scavenge air cooler module.

The hydraulic jack bolts on the main bearings were eliminated and replaced by simple holding-down bolts that fix the bearing cap directly to the bedplate (Figure 12.14). In addition, it was possible to simplify

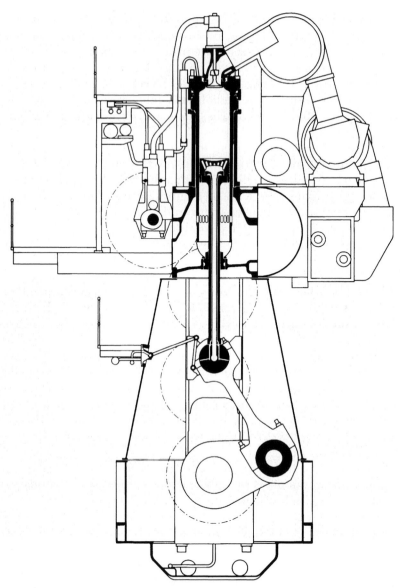

Figure 12.9 Cross-section of RTA58T engine. Note the high camshaft level allowing the use of shorter high pressure fuel injection pipes for better injection control

the column structure in the region where the jack bolts had needed support and thereby omit machining operations. The scavenge air receiver was simplified by using a 'half-pipe' design which is welded to the module incorporating the turbocharger and scavenge air cooler.

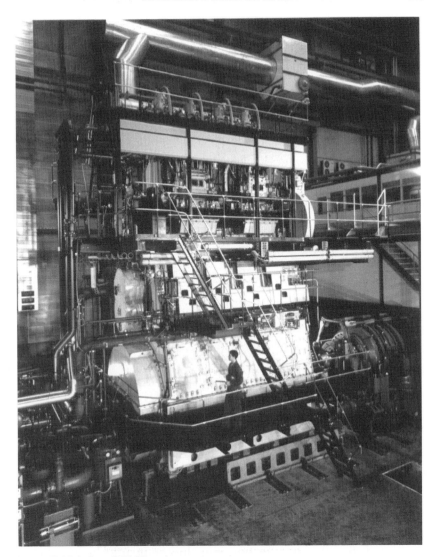

Figure 12.10 Sulzer 4RTA58 engine on test

The result was an easier manufacturing process and significant weight savings compared with equivalent RTA-2U series engines. The single-piece column structure can be manufactured without the need for machining inclined surfaces: all machined surfaces on the columns are either vertical or horizontal by design. Eliminating the jack bolts on the main bearings also allowed the design of the columns to be made more tolerant to welding quality.

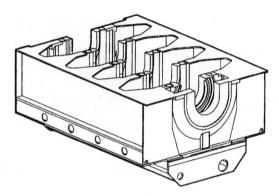

Figure 12.11 Bedplate of RTA58T engine

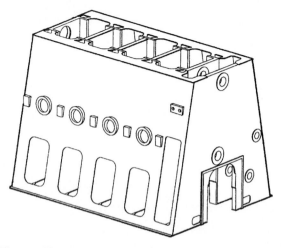

Figure 12.12 The monobloc columns of the RTA48T and RTA58T were designed for simpler welding and machining

The exhaust valve actuator was redesigned for the RTA-T engines. The bush of the actuating piston was eliminated and its function integrated directly into the housing, thus reducing the number of components to be produced and the assembly time. Smaller dimensions and a lighter weight (by 27 per cent) were also achieved. A substantial cost saving can be realized if an electrically driven balancer at the forward end of the engine is specified, allowing the previously necessary coupling between balancer and camshaft to be omitted.

Valve-controlled fuel injection pumps are located at the height of the cylinder blocks. The fuel pump blocks can be directly bolted to the cylinder blocks without using any shims. The intermediate gear wheels of the camshaft drive can be aligned more quickly than before

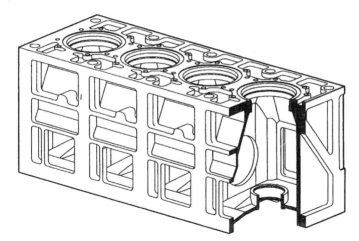

Figure 12.13 The monobloc dry cylinder blocks of the RTA48T and RTA58T were designed for reduced machining, compactness and lower weight

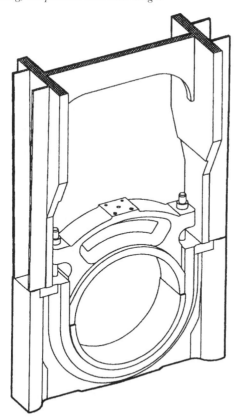

Figure 12.14 The main bearings of the RTA48T and RTA58T use holding-down studs instead of the previous hydraulic jack bolts

by means of a built-in device to move the wheels vertically and horizontally into position.

Engines can be built up from modules, starting with the bedplate and crankshaft, and working up with the columns and cylinder blocks. The modules can be pre-assembled at convenient, separate locations in the workshop. All pipe connections are designed to support this type of modular assembly and achieve the shortest possible assembly time in the works or, if necessary, in the ship itself.

Piston-running technology was based on experience gained and refined from the RTA-2U and RTA84C engines. The cylinder liner is of cast iron with sufficient hard-phase content and a smooth machined running surface. Bore cooling and three fuel injection valves underwrite favourable temperature levels and distributions. The top piston ring is pre-profiled for easy running-in and plasma coated for low wear rates over extended periods. The multi-level cylinder lubrication system fosters optimum distribution of the lubricating oil and adapts to the longer stroke; and condensate water is taken out of the scavenge air efficiently to avoid contamination of the oil film.

RTA60C engine

A new type added to the programme in mid-1999, the RTA60C engine was designed to serve faster cargo tonnage such as medium-sized containerships, car carriers, RoRo ships and reefer vessels (Figure 12.15). A desire for compactness and economical production costs moulded the design, resulting in an engine with a shorter length and lighter weight (5–10 per cent lower) for a given power than others in its class.

A higher rotational speed to suit the intended ship types allowed a reduced piston stroke which, together with the short connecting rod, fostered a significant reduction in engine height: the engine is 8.52 m tall above the shaft centreline and requires a hook height of only 10.4 m for withdrawing pistons (or less by using special tools). The length was minimized by applying measures from the RTA48T and 58T engines, a cylinder distance of 1040 mm being achieved. The six-cylinder RTA60C model has an overall length of 7.62 m, including the flywheel, and weighs 330 tonnes.

The need to keep engine length to a minimum called for particularly good bearing design; thin-walled whitemetal shells are used for crosshead, main and bottom end bearings. The main bearing housing was specifically designed to accommodate and support the thin-walled bearing shell, with four elastic holding-down bolts for each main bearing cap. Two pairs of studs are said to give the most even distribution of

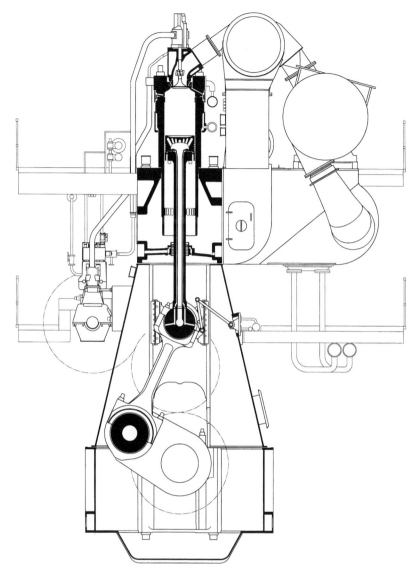

Figure 12.15 Cross-section of Sulzer RTA60C engine, the first RTA model designed from the start to accept RT-flex systems (camshaft version illustrated)

holding-down load, and also allow the tie rods to be located close to the bearing for efficient transfer of firing pressure loads.

Cylinder covers are secured by eight elastic holding-down studs arranged in four pairs, promoting compactness and contributing to a shorter engine length, and also simplifying manufacture. A support ring between the cylinder block and the collar of the cylinder liner carries both liner and cylinder cover; it also passes cooling water to the cooling bores and to the cover. A simple water jacket provides the necessary cooling for the portion of the liner length immediately below the bore-cooled top flange of the liner.

The cylinder jacket is a single-piece iron casting, its height determined by the space required for the scavenge air receiver. Access to the piston underside is possible from the receiver side of the engine to allow maintenance of the piston rod gland and piston ring inspection. On the fuel side, one door per cylinder can be opened for inspection and to support in-engine work from outside. The tilting-pad thrust bearing is integrated in the bedplate, the pads arranged to ensure a safe and uniform load distribution. The thrust bearing girder consists of only two steel cast pieces, omitting welding seams in critical corners; the girder is clearly stiffer than in previous designs.

The piston comprises a forged steel crown with a very short skirt; the four piston rings are all 16 mm thick and of the same geometry. Piston running behaviour benefits from an anti-polishing ring incorporated at the top of the liner, preventing deposits on the piston top land from damaging the liner running surface and its lubrication film. Keeping the top land clean also ensures a good spread of lubricant over the liner surface while using a lower cylinder oil feed rate. (More details of the anti-polishing ring are provided in the TriboPack section below.)

An exhaust system featuring tangential gas inlet and outlet in the manifold allows a smooth flow of gases from the exhaust valve to the turbocharger inlet with an energy-saving swirl along the manifold. The combined turbocharger and scavenge air cooler module is designed to accommodate one or two turbochargers, and allows for different sizes of coolers. The flexibility to locate the turbocharger at the aft (driving end) of the engine is particularly suited to modern all-aft ship designs.

The key operating parameters of the RTA60C (see table) are only slightly higher those of the −B versions of the RTA48T, 58T and 68T models, from whose service experience the new engine benefited.

RTA60C engine data

Bore	600 mm
Stroke	2250 mm
Stroke/bore ratio	3.75:1
Output, R1, mcr	2360 kW/cyl

Speed range, R1-R3	91–114 rev/min
Mean effective pressure at R1	19.5 bar
Mean piston speed at R1	8.55 m/s
Maximum cylinder pressure	155 bar
Number of cylinders	5–8
Power range	8250–18 880 kW
Specific fuel consumption	164–170 g/kWh

The RTA60C engine was designed from the outset to smooth its acceptance of Sulzer RT-flex fully electronically-controlled fuel injection and exhaust valve systems, which eliminate the camshaft and individual fuel pumps (see below). The debut orders for RTA60C engines—in December 2000—specified RT-flex versions.

RTA50C and RT-flex 50C engines

At end-2002 Wärtsilä Corporation announced a joint venture with Mitsubishi to design and develop a new small bore low speed engine. The Japanese group, which designs and builds its own UEC two-stroke engines (Chapter 11), has been a prolific licensed builder of Sulzer engines since 1925.

The new 500 mm bore/2050 mm stroke design was planned for release in two versions: the conventional 'mechanical' Sulzer RTA50C and the electronically-controlled Sulzer RT-flex50C. Each type will offer a maximum continuous rating of 1620 kW/cylinder at 124 rev/min, five-to-eight-cylinder models covering a power range from 5650 kW to 12 960 kW at 99 rev/min to 124 rev/min. Output and speed ratings target Handymax and Panamax bulk carriers, large product tankers, feeder containerships and medium-sized reefer vessels. Both versions will incorporate Sulzer TriboPack measures for enhancing piston-running behaviour, underwriting low cylinder wear rates for three years between overhauls, and minimizing cylinder lube oil consumption (see section below).

High efficiency, compactness and environmental friendliness were set as design goals for the project to be pursued by a joint working group of engineers from both groups. The new engine was assigned for production in Japan by Mitsubishi and its licensees and by other Japanese licensees of Wärtsilä. Separate branding—Sulzer or Mitsubishi—will be applied to the engines, depending on the builder. The first examples were due for completion by end-2004.

RTA50C/RT-flex50C engine data

Bore	500 mm
Stroke	2050 mm
Stroke/bore ratio	4:1
Output, mcr	1620 kW/cyl
Mep at R1	19.5 bar
Pmax	155 bar
Mean piston speed at R1	8.5 m/s
Cylinders	5–8
Speed range, R1-R3	124–99 rev/min
Power range	5650–12 960 kW
Sfc, full load, R1-R2	165–171 g/kWh

RTA84T engine refinements

The piston of the RTA84T engine is similar to that of the earlier RTA84M engine but simplified in design. The skirt is bolted directly to the crown, and the increased rigidity of the piston rod can accept higher loads. The crown is oil cooled by the 'jet/shaker' cooling principle, with the oil spray nozzles matched to secure optimum piston temperatures. The piston rod gland design (Figure 12.16) features bronze rings and a modified scraper/gas sealing package on top compared with previous RTA-series engines. This gland principle confirmed the expected appropriate oil leakage rates in the neutral space. The well-proven RTA exhaust valve design was adopted for the RTA84T, embracing an efficiently bore-cooled valve seat, Nimonic valve, hydraulic actuation and a rotation device.

A deeper cylinder cover, with its lower joint between cover and liner, reduces the portion of the liner exposed to gas pressure and high temperature. The bore cooling principle was adapted to control the temperatures and mechanical and thermal stresses in the components adjacent to the combustion space. The cylinder cover is bolted down with eight studs. Compact uncooled fuel injection valves make it possible to place three nozzles symmetrically in the cover despite the very short cylinder pitch. The nozzle tips are sufficiently long for the cap nut to be shielded by the cylinder cover and hence not exposed to the combustion space, Figure 12.17.

The camshaft was raised to be close to the cylinder covers and is driven via gear wheels. The hydraulic reversing mechanism of the RTA series was retained unchanged. The decision to move the camshaft up towards the engine top was taken after evaluating such factors as engine cost, fuel injection system performance, hydraulic valve actuation and torsional vibration. In particular, with an elongated stroke/bore

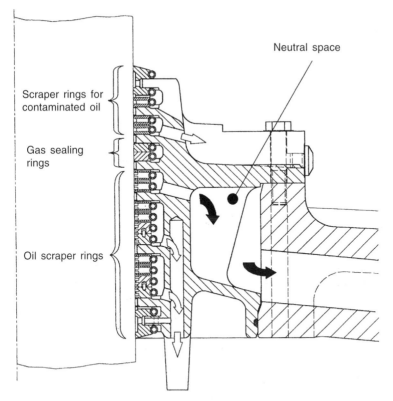

Figure 12.16 Piston rod gland of RTA84T engine; all rings are bronze

ratio of 3.75, the advantage of shorter hydraulic connections for injection and valve actuating pipes was significant in avoiding high pressure losses. The cost of an additional gearwheel (Figure 12.18) is compensated by the slimmer camshaft, the fuel pump arrangement and the shorter high pressure injection pipes. Higher injection pressures, and hence more efficient and cleaner combustion, are thus facilitated.

The camshaft-driven fuel pump is of the valve-controlled design used in the longer stroke RTA-2 engines and located at a high level, bolted directly to the side of the cylinder block (Figure 12.19). The double valve-controlled pump principle, traditional in Sulzer low speed engines, offers high flexibility and stability over time in controlling the injection timing. Timing is controlled by separate suction and spill valves regulated through eccentrics on hydraulically-actuated lay shafts. The pump housing incorporates the compact RTA-series arrangement, with fuel injection pump and exhaust valve actuator modules provided to serve one pair of cylinders.

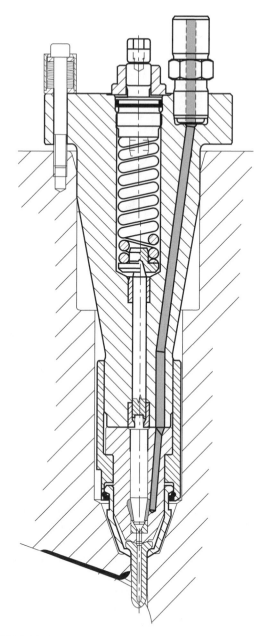

Figure 12.17 RTA84T fuel injection valve. The nozzle is not exposed to the combustion space, thereby avoiding material burning off

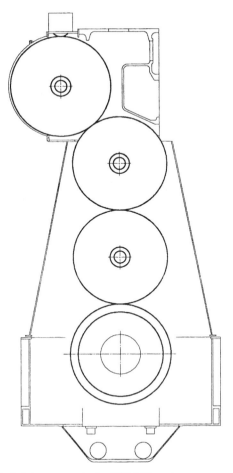

Figure 12.18 Camshaft drive gear train for RTA84T engine

In comparison with a helix type, the valve-controlled fuel injection pump has a plunger with a significantly greater sealing length. The higher volumetric efficiency reduces the torque in the camshaft. Additionally, says Sulzer, injection from a valve-controlled pump is far more stable at very low loads and rotational shaft speeds down to 15 per cent of the rated speed are achieved. Valve control also offers benefits in reduced deterioration of timing over the years owing to less wear and to freedom from cavitation.

A combination of variable exhaust valve closing (VEC) and variable injection timing (VIT) devices provides an improved degree of setting flexibility. In the upper load range, the specific fuel consumption is optimized through the electronically controlled VIT system, maintaining

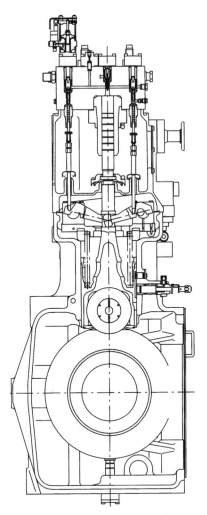

Figure 12.19 RTA84T fuel injection pump with double control valves

the maximum cylinder pressure by injection timing advance. The fuel quality setting (FQS) function is incorporated in the VIT system. Engine efficiency is additionally improved in the lower load range (between 80 and 65 per cent load) via the VEC system (Figure 12.20). The compression ratio is thereby effectively increased by early closing of the exhaust valve.

The working process of the engine can thus be 'shaped' for optimum performance over the whole load range, facilitating its adaptation to a particular ship sailing mode. The resulting benefits promised are

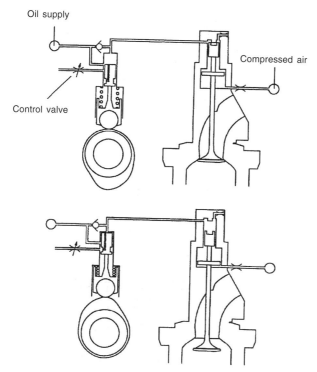

Figure 12.20 RTA84T engine exhaust valve actuating arrangement with variable closing (VEC) system

the lowest possible fuel consumption in the most common load range, and higher exhaust gas temperatures within that load range compared with other solutions, such as variable turbocharger geometry. (Higher exhaust gas temperatures are valued for waste heat recovery.) The VEC, VIT and FQS devices are electronically regulated from the engine control system. The influence of the VIT/VEC combination on the RTA84T engine's performance characteristics are shown in Figure 12.21.

A load-dependent cylinder liner cooling system (Figure 12.22) helps to avoid cold corrosion of the liner surface over the whole load range and hence to reduce wear rates further. The liner temperature is maintained above the prevailing dew point throughout the load range: the liner is efficiently cooled at full load, while over-cooling at part load is avoided. This is achieved by splitting the cooling water supply to the engine into two lines. Only a restricted amount of water is led through the cylinder liner jacket at part load while the main water flow is led directly to the cylinder cover. The water flow is electronically controlled as a function of the engine load.

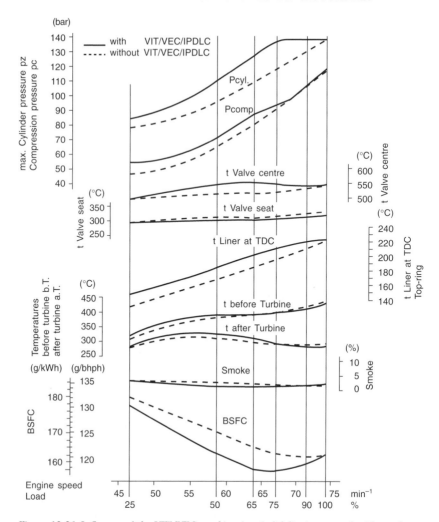

Figure 12.21 Influence of the VIT/VEC combination (solid line) compared with results without VIT/VEC (dashed line) on the performance characteristics of a 7RTA84T engine. The engine has an RI rating of 27 160 kW at 74 rev/min and exploits an exhaust gas power turbine

Cylinder liner lubrication is effected by Sulzer's multi-level accumulator system. This allows the cylinder lubricating oil to be distributed in very small amounts at each engine stroke, thereby creating an optimum oil film distribution. The main improvement for the RTA84T was the introduction of an electronically controlled flexible oil dosage promoting low wear rates and low lubricating oil

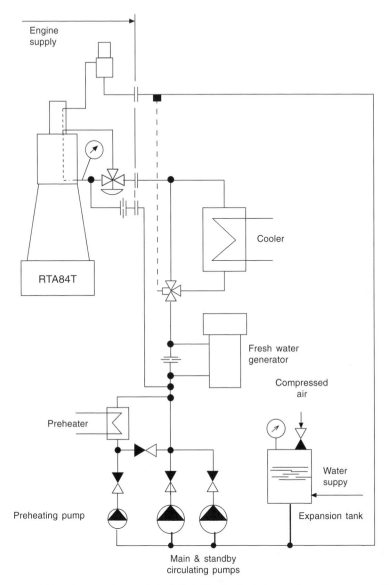

Figure 12.22 Schematic layout of the RTA84T cylinder jacket cooling water system with a load-dependent controlled bypass past the jacket to the cylinder cover

consumption. The oil feed rate is controlled according to engine load and is adjustable as a function of engine condition. A simplified dual-line distributor eliminates the mechanical drive shaft.

Piston-running experience with RTA84T engines in service

demonstrated that engines with very long strokes need higher liner wall temperatures than had previously been required by Sulzer's RTA-2 series (with stroke/bore ratios of around 3.5) to overcome corrosive wear problems when operating on high sulphur fuels. This experience was applied to benefit later RTA84T production models as well as the design of the smaller RTA-T engines. A number of improvements introduced for the latter series have also been applied to the RTA84T, the refinements including design simplifications to make the engine more manufacturing friendly. A significant saving in weight also resulted, the seven-cylinder model being trimmed by some 6.5 per cent to 960 tonnes.

RTA84T-B and -D versions

Following the introduction of the RTA48T and RTA58T models in 1995, the same design concepts were applied to the RTA84T 'Tanker' engine—originally introduced in May 1991—resulting in a Version B in 1996. Easier manufacturing and enhanced service behaviour were realised, with no change in power output. In July 1998 a lower specific fuel consumption (down by 2 g/kWh) was gained by applying 'low port' cylinder liners (scavenging air inlet ports with a reduced height) in combination with higher efficiency turbochargers. There is no penalty in either higher component temperatures or too low exhaust gas temperatures, and the low ports give a longer effective expansion stroke in the cycle (Figure 12.23).

The RTA84T-B engine was uprated at end-1998 from 3800 kW/cylinder at 74 rev/min to 4100 kW/cylinder at 76 rev/min to create the Version D through tuning and turbocharging matching measures. The power available from a seven-cylinder engine was thereby increased from 27 160 kW to 28 700 kW, better addressing the propulsive demands of VLCCs with higher service speeds. Nine-cylinder RTA84T-D models, each developing 36 900 kW, were specified to power a series of 442 500 dwt tankers with a service speed of 16.5 knots.

Structural, running gear, combustion chamber, fuel injection, turbocharging and scavenge air system design details are similar to those of the RTA84C and RTA96C engines described below. The large stroke-to-bore ratio of the RTA84T, however, allows a relatively deeper combustion chamber with more freedom in the layout of the fuel spray pattern. The semi-built crankshaft has to cater for the longest stroke ever applied in a Sulzer engine; to limit the crankshaft weight for production, assembly and transport, the main journals and crankpins are bored. The design of the crank has a good transverse width at the upper part of the web, allowing the latter to be slim longitudinally.

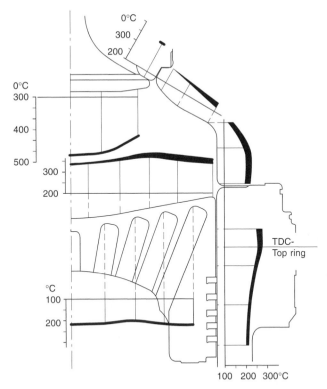

Figure 12.23 Surface temperatures measured on the combustion chamber components of the RTA84T-B engine at full load R1 rating. The thickness of the lines represents the circumferential variation in temperature

The favourable torsional vibration characteristics allow six-cylinder engines to use a viscous damper in many cases instead of a Geislinger damper.

RTA84T-B and -D engine data

	RTA84T-D	*RTA84T-B*
Bore	840 mm	840 mm
Stroke	3150 mm	3150 mm
Stroke/bore ratio	3.75	3.75
Power, mcr	4100 kW/cyl	3880 kW/cyl
Cylinders	5–9	5–9
Speed range	61–76 rev/min	59–74
Mean effective pressure	18.5 bar	18 bar
Mean piston speed	8 m/s	7.77 m/s
Maximum cylinder pressure	144 bar	140 bar
Specific fuel consumption	160–168 g/kWh	160–168 g/kWh

RTA68T-B engine

A further development of the RTA-T programme resulted in the 680 mm bore RTA68T-B engine, the first example of which entered service in November 2000. The design benefited from all the reliability improvements incorporated in the earlier RTA-T engines, with parameters similar to those of the smaller bore RTA48T-B and 58T-B models. TriboPack design measures (see below) contributed to piston-running behaviour, and a new design of piston rod gland featured oil scraper rings with grey cast iron lips: the system oil is fully recirculated and a dry neutral space effected.

RTA68T-B engine data

Bore	680 mm
Stroke	2720 mm
Stroke/bore ratio	4
Power, mcr	2940 kW/cyl
Cylinders	5–8
Speed range	75–94 rev/min
Mean effective pressure	19 bar
Mean piston speed	8.5 m/s
Maximum cylinder pressure	150 bar
Specific fuel consumption	161–169 g/kWh

RTA84C and RTA96C engines

Introduced in September 1988 for propelling the coming generation of larger and faster containerships, the RTA84C was developed from the RTA84, already established in that market. In 1993 its power output was increased by six per cent, the cylinder cover modified and the number of fuel nozzles increased from two to three; these measures contributed to a reduction in the thermal load of the combustion chamber. The current performance of the series, shown in the table, equates to a maximum output of 48 600 kW at 102 rev/min from the 12-cylinder model.

RTA84C engine data

Bore	840 mm
Stroke	2400 mm
Stroke/bore ratio	2.86
Power, mcr	4050 kW/cyl
Cylinders	6–12

Speed range	82–102 rev/min
Mean effective pressure	17.9 bar
Mean piston speed	8.2 m/s
Maximum cylinder pressure	140 bar
Specific fuel consumption	166–171 g/kWh

Increasing sizes of post-Panamax containerships dictated more powerful engines and stimulated the introduction of the RTA96C series in December 1994. The 960 mm bore design was fully based on the RTA84C (Figures 12.24 and 12.25), which it supplemented in the programme. The selection of a stroke (2500 mm) some 100 mm longer than that of the RTA84C engine for the RTA96C design (Figure 12.26) was influenced by demands for the highest reliability. By adopting a longer stroke, the absolute depth of the combustion chamber could be proportionally increased to give more room for securing the best combustion and fuel injection parameters, and better control of temperatures in the combustion chamber components. Additionally, the slightly longer stroke fosters a simplified crankshaft design with enhanced reliability, since the shrunk-in main journals do not cut the journal fillets at the inner sides of the crankwebs.

Figure 12.24 A 12-cylinder Sulzer RTA84C engine on test

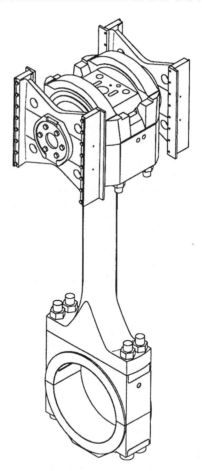

Figure 12.25 Crosshead and connecting rod assembly for RTA84C engine

A time-between-overhaul (TBO) of three years from the RTA96C engine's key components was sought, a goal underwritten by: a cast iron (preferably die cast) cylinder liner with the necessary amount of wear-resistant hard-phase particles and a smoothly machined and honed surface for quick and trouble-free running-in; bore cooling of all the main combustion chamber components; three fuel injection valves symmetrically distributed in the cylinder cover contribute to evenness of temperature distribution; cylinder oil lubrication of the liner surface via two levels of quills to achieve effective and economical distribution; and a top piston ring pre-profiled and plasma coated to secure the lowest wear rate to reach the three-year TBO goal with sufficient margin. Diametral cylinder liner wear rates of around 0.03 mm/1000 hours

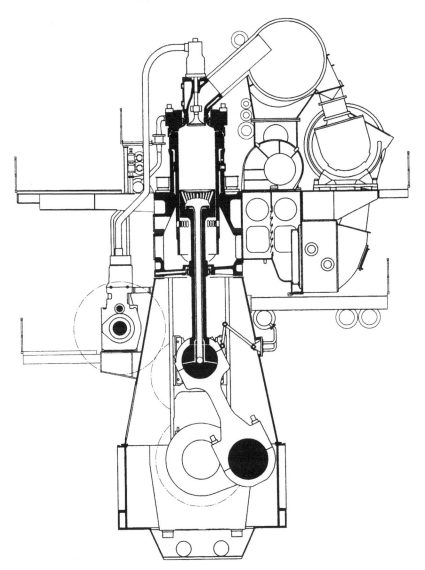

Figure 12.26 Cross-section of RTA96C engine, designed for propelling ultra-large containerships

are reported for RTA96C design, in service. All new engines further benefit from TriboPack measures to enhance piston-running behaviour (see below).

The semi-built crankshaft comprises combined crankpin/web elements forged from a solid ingot, with the journal pins then shrunk into the crankweb. The main bearings have whitemetal shells, and the main bearing caps are held down by a pair of jack bolts in the RTA84C and by a pair of elastic holding-down studs in the RTA96C. The crosshead bearing is designed to the same principles as for all other RTA engines; it also features a full-width lower half bearing. The crosshead bearings have thin-walled shells of whitemetal to yield a high load-bearing capacity. Sulzer low speed engines retain a separate elevated-pressure lube oil supply to the crosshead; this provides hydrostatic lubrication which lifts the crosshead pin off the shell during every revolution to ensure that sufficient oil film thickness is maintained under the gas load: crucial to long term bearing security.

The combustion chamber was recognized as the most important design area because of its key influence on engine reliability and the high power concentration. Component design was based on established practice and benefited from work carried out for the medium bore RTA48T and RTA58T engines. Bore cooling technology, Wärtsilä suggests, provides an escape from the rule that larger components (resulting from a larger bore) when subjected to thermal loading will also have higher thermal strains. With bore cooling, the thermal strains in the cylinder cover, liner and piston crown of the RTA96C can be kept fully within the values of earlier generations of RTA engines, as can the mechanical stresses in those components.

The solid forged steel, bore-cooled cylinder cover is secured by eight elastic studs, and the central exhaust valve of Nimonic 80A material is housed in a bolted-on valve cage. Anti-corrosion cladding is applied to the cylinder covers downstream of the injection nozzles to protect the covers from hot corrosive or erosive attack. The pistons feature a forged steel crown with a short skirt; the crown is cooled by combined jet-shaker oil cooling achieving moderate temperatures on the crown and a fairly even temperature distribution across the crown surface. No coatings are necessary.

A high structural rigidity with low stresses and high stiffness is important for low speed engines. The RTA84C and RTA96C designs exploit a well proven structure with a 'gondola' type bedplate surmounted by very rigid A-shaped double-walled columns and cylinder blocks, all secured by pre-tensioned vertical tie rods. Both bedplate and columns are welded fabrications designed for minimum machining. The cylinder jacket is assembled from individual cast iron cylinder blocks bolted together to form a rigid whole. The fuel pumps are carried on supports on one side of the column and the scavenge air receiver on the other side of the cylinder jacket.

Access to the piston under-side is normally from the fuel pump side but is also possible from the receiver side of the engine, to facilitate maintenance of the piston rod gland and also for inspecting piston rings. The tilting-pad thrust bearing is integrated in the bedplate. The use of gear wheels for the camshaft drive allows the thrust bearing to be very short and stiff, and to be carried in a closed rigid housing.

The three uncooled fuel injection valves in each cylinder cover have nozzle tips sufficiently long for the cap nut to be shielded by the cylinder cover and not exposed to the combustion space. The camshaft-driven fuel injection pumps are of the double-valve controlled type, traditional in Sulzer low speed engines. Injection timing is controlled by separate suction and spill valves regulated through eccentrics on hydraulically-actuated lay shafts. Flexibility in timing is possible through the variable fuel injection timing (VIT) system for improved part-load consumption, while the fuel quality setting (FQS) lever can adjust timing according to the fuel oil quality. The valve-controlled fuel injection pump, in comparison with a helix type, has a plunger with a significantly greater sealing length. The higher volumetric efficiency reduces the torque in the camshaft; additionally, injection from a valve-controlled pump is far more stable at very low loads, and rotational shaft speeds down to 15 per cent of the rated speed are achieved. Valve control also has the benefits of less deterioration of timing over the years owing to reduced wear and freedom from cavitation.

The camshaft is assembled from a number of segments, one for each fuel pump housing. The segments are connected through SKF sleeve couplings, each segment having an integral hydraulic reversing servomotor located within the pump housing. The camshaft drive is a traditional Sulzer arrangement, effected in this case by three gearwheels housed in a double column located at the driving end or in the centre of the engine. The main gearwheel on the crankshaft is in one piece and flange mounted.

Scavenge air is delivered by a constant pressure turbocharging system based on one or more turbochargers, depending on the number of engine cylinders. For starting and during slow running, scavenge air delivery is augmented by electrically-driven auxiliary blowers. The scavenge air receiver incorporates non-return flaps, an air cooler and the auxiliary blowers; the turbochargers are mounted on the receiver, which also carries the fixed foot for the exhaust manifold. Immediately after the cooler, the scavenge air passes through a water separator comprising a row of vanes that divert the air flow and collect the water. Ample drainage is provided to completely remove the condensed water collected at the bottom of the air cooler and separator. Effective separation of condensed water from the stream of scavenge air is thus accomplished, a necessity for satisfactory piston-running behaviour.

RTA engines have simple seating arrangements with a modest number of holding-down bolts and side stoppers (14 side stoppers are needed for a 12-cylinder RTA96C engine). No end stoppers or thrust brackets are needed as thrust transmission is provided by fitted bolts or thrust sleeves applied to a number of the holding-down bolts (Figure 12.27). The holes in the tanktop for the thrust sleeves can be made by drilling or even flame cutting. After alignment of the bedplate, epoxy resin chocking material is poured around the sleeves. The engine is equipped with an integrated axial detuner at the free end of the crankshaft, and a detuner monitoring system developed by Wärtsilä is standard equipment.

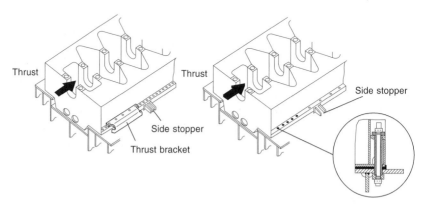

Figure 12.27 Arrangements for transmitting propeller thrust to the RTA84T, RTA84C and RTA96C engine seatings; the inset shows the thrust sleeve for the thrust bolts

A standard all-electric interface is employed for engine management systems—the DENIS (Diesel Engine Interface Specification)—to meet all needs for control, monitoring, safety and alarm warning functions. It matches remote control systems and ship control systems from a number of approved suppliers.

RTA96C uprating and programme expansion

The first RTA96C engine, an 11-cylinder model, started tests in March 1997 and was followed into service by eight-, nine- and 10-cylinder models and numerous 12-cylinder versions. The design was originally offered with an output of 5490 kW/cylinder and up to 12 cylinders but the rating was increased by four per cent in 2000 to 5720 kW/cylinder at 102 rev/min (see table). At the same time, the programme was extended to embrace an in-line 14-cylinder model delivering 80 080 kW (the first time such a low speed engine configuration had featured in a production programme) and suitable for powering single-screw post-Panamax containerships with capacities up to 10 000 TEU and service speeds up to 25 knots. (Figure 12.28).

A 14RTA96C engine would measure 27.31 m long overall to the flywheel flange × 10.93 m high over the shaft centreline (or 13.54 m overall) and weigh 2300 tonnes dry (compared with 2050 tonnes for a 12RTA96C). The individual elements, such as the two crankshaft sections, are close to the maximum lifting capacities of contemporary enginebuilders which impose a limit on engines with higher than 14 cylinders. The camshaft for engines with more than 14 cylinders would need to be split into three parts, and at least two camshaft drives would become necessary. Such engines would therefore have to adopt the Sulzer RT-flex common rail systems for fuel injection and exhaust valve actuation, which eliminate the camshaft (see below).

With a length less than 15 per cent longer than the established 12-cylinder engine, the longitudinal and torsional rigidity of the 14RTA96C model would be adequate for the expected ship structures. The torsional vibration characteristics are also considered acceptable for the envisaged firing order. The crankshaft material and the shrink fit of the journals in the webs, however, would be redefined to suit the increased torque transmitted to the propeller shaft.

RTA96C engine uprating data

	Old rating	New rating
Bore	960 mm	960 mm
Stroke	2500 mm	2500 mm
Stroke/bore ratio	2.6	2.6
Power, mcr	5490 kW/cyl	5720 kW/cyl
Speed	100 rev/min	102 rev/min
Cylinders	6–12	6–12, 14
Mean effective pressure	18.2 bar	18.6 bar
Mean piston speed	8.3 m/s	8.5 m/s
Maximum cylinder pressure	142 bar	145 bar
Maximum output, 12-cyl	65 880 kW	68 640 kW
Maximum output, 14-cyl	–	80 080 kW
Sfc, full load	163–171 g/kWh	163–171 g/kWh

RTA96C service experience

Some cases of main bearing damage were restricted to a specific series of containerships, and involved the local breaking-out of whitemetal in the lower shells and some local fretting. The fretting problem was

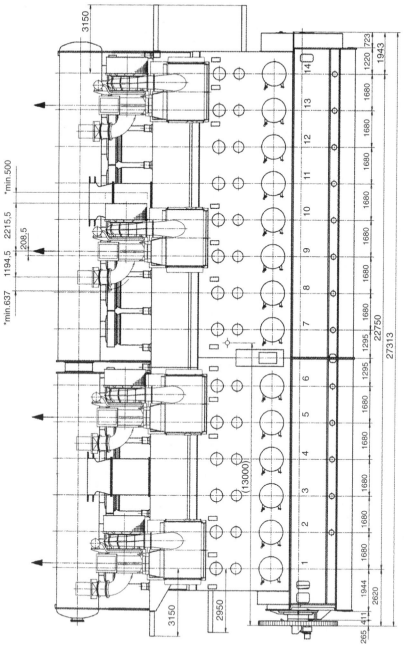

Figure 12.28 Side elevation of a 14-cylinder RTA96C engine

solved by applying a so-called key-slot solution with the aim of increasing the radial forces pressing the bearing shell into the girder, and by some other small design modifications. With the exception of one particular main bearing, no connection could be found with the bearing load and shaft orbit that could explain the breaking-out problem. As a countermeasure, the lower shell of the main bearing next to the thrust bearing is now machined together with its respective bearing cap to ensure better bearing geometry. The more precise alignment of the bearing cap and shell, along with a key between the cap and shell to secure tight fitting of the shell in the bearing girder when tightening down, gradually yielded results.

Some reports of cracks in A-frame columns were first received in mid-1999 after some 5000 to 12 000 running hours. A project team was established to investigate the matter and all other engines in service were inspected. Although the cracks occurred in seven of the eight engines, they were in only some of the double-walled A-frame columns manufactured to the original design, and were of various degrees of severity. The cracks were initiated at the inboard ends of two pairs of horizontal ribs on the columns and propagated vertically and in the direction of the crosshead guide rails. The root of the problem was a combination of lack of welding quality and somewhat higher than anticipated local stress level.

All cracked A-frame columns were repaired by welding. At the same time, the inboard ends of all horizontal ribs in the engines concerned were cut back to limit the stress concentration and thereby eliminate the possibility of cracking. Subsequent measurements showed that the modified shape of the ribs reduces the stress concentrations in the columns down to around half, and thus is not as sensitive to imperfect welding. The same shape of horizontal rib is now applied on all RTA96C engines, whether repaired or built as new; the shape was adapted from the modified rib design introduced in December 1997 but slightly revised for ease of manufacture.

Some leaks were experienced in the cylinder pressure relief valve located in the cylinder cover. These were caused by corrosion which arose because the shortened valve was too cold through being too far from the combustion space. An improved valve design replaced those in existing engines, while later engines were fitted with valves of a traditional design which operate at higher temperatures and hence avoid the original problem.

A number of scavenge air receivers suffered cracks in the form of rupturing of the upper and lower longitudinal welded seams; these were found to have been caused by resonant vibration in the dividing wall. As a remedy, additional stiffening was applied to the dividing walls of the affected receivers (and adopted as standard for new engines) to change the resonant frequency and thereby reduce vibration amplitudes.

A stiffer turbocharger support was developed for new engines after vibration had caused problems in the supports of one of the turbochargers of the first 10RTA96C engine. Excessive transverse movement of the exhaust manifolds on the first such models was also recorded. This was remedied by stiffening the manifold support in the transverse direction, thus halving the manifold movements. Excessive movements of the manifolds also led to the breakage of cooling water discharge pipes; the adoption of flexible pipes of a proprietary make as standard solved the problem.

A number of broken cylinder cover studs have been noted, mostly broken at the first thread at the top but some fractured at the bottom. The lower fractures were initiated by corrosion because of improper sealing; the upper fractures resulted from materials and thread qualities beyond the specification. The design was changed to a wasted stud and a special nut to reduce stress levels and ensure a good safety margin.

TriboPack for extended TBO

The time-between-overhaul (TBO) of low speed marine diesel engines is largely determined by the piston running behaviour and its effect on the wear of piston rings and cylinder liners. Addressing this, Sulzer introduced a package of design measures in 1999, which are now standard on all new RTA engines and retrofittable to existing engines. The TriboPack technology enables the TBO of cylinder components, including piston ring renewal, to be extended to at least three years and also allows a further reduction in the cylinder lubricating oil feed rate. The design measures (Figure 12.29) are:

— Multi-level cylinder lubrication.
— Fully and deep-honed cylinder liner with sufficient hard phase.
— Careful turning of the liner running surface and deep honing of the liner over the full length of the running surface.
— Mid-stroke liner insulation and, where necessary, insulating tubes in the cooling bores in the upper part of the liner.
— Pre-profiled piston rings in all piston grooves.
— Chromium-ceramic coating on the top piston ring.
— RC (Running-in Coating) piston rings in all lower piston grooves.
— Anti-polishing ring at the top of the cylinder liner.
— Increased thickness of chromium layer in the piston ring grooves.

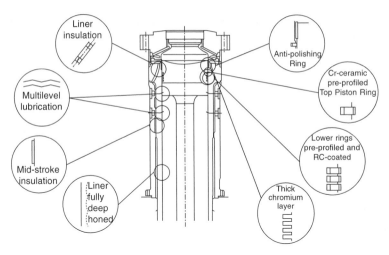

Figure 12.29 The Sulzer TriboPack design measures for improving piston-running behaviour

A key element of TriboPack is the cylinder liner manufactured in cast iron, which needs a controlled hard-phase content and the best grain structure in the running surface for both good strength and running behaviour. Careful machining followed by full deep honing to remove all damaged hard phase from the liner surface reportedly delivers an ideal running surface for the piston rings, together with an optimum surface microstructure. Deep honing of the full liner running surface is a prerequisite for maximizing the benefits of TriboPack, says Wärtsilä: its experience has shown that plateau honing of a wave-cut liner is not adequate because, once the plateau is worn down, the rings run on liner metal whose hard phase structure was damaged during machining. This damaged hard phase must be removed by deep honing.

Pistons have four rings, all of the same thickness. The *chrome-ceramic top ring*, proven in Wärtsilä four-stroke engine practice, has a cast iron base material. The running face is profiled and coated with a layer of chromium as a matrix into which a ceramic material is trapped. High operational safety and low liner and ring wear have been demonstrated, with a much better resistance to scuffing than any other ring material, Wärtsilä asserts. Good performance is conditional, however, on using the chrome-ceramic rings in conjunction with a deep-honed liner. The other piston rings have a running-in and anti-scuffing coating which fosters a safe and swift running-in of the engine when the liners are deep honed.

The *anti-polishing ring* (APR) prevents the build-up of deposits on the top land of the piston which can damage the lube oil film on the liner and cause bore polishing. Deposit build-up can be heavy in some engines, especially those running on very low sulphur content fuel oil (less than one per cent sulphur) combined with an excessive cylinder lube oil feed rate. If such deposits are allowed to accumulate, they inevitably touch the liner running surface over a large part of the piston stroke. The lube oil film can then be wiped off, allowing metal-to-metal contact between the piston rings and liner; in the worst case there can be scuffing.

Applied as standard for some years on Wärtsilä four-stroke engines, the thin alloy steel APR is located in a recess at the top of the liner and has an internal diameter less than the cylinder bore to reduce the clearance to the piston top land. It does not need to be specifically fixed, as the thermal expansion of the hot ring keeps it tightly in place. The steel material was selected to ensure and maintain a high safety margin against thermal yielding. Excessive deposits are scraped off the piston top land at every stroke while they are still soft, thus preventing hard contact between the deposit and the liner wall surface. The oil film on the liner wall remains undisturbed and can fulfil its function. The APR also stops the upward transportation of new lube oil by the layer of deposits to the top of the liner where it is burned instead of being used for lubrication. The ring is thus effective in allowing the lube oil feed rate to be kept down to recommended values.

Load-dependent cylinder lubrication is provided by Sulzer's *multi-level accumulator system*, the lubricating pumps driven by frequency-controlled electric motors. On the cylinder liner, oil distributors bring oil to the different oil accumulators. For ease of access, the quills are positioned in dry spaces instead of in way of cooling water spaces.

It is also important that the liner wall temperature is adapted to keep the liner surface above the dew point temperature over the whole of the piston stroke to avoid cold corrosion and maintain good piston-running conditions. The upper part of the liner is bore cooled with cooling water passing through tangential drillings in the liner collar. The mid-stroke region of the liner is cooled by a water jacket, and only the lower part is uncooled. There is often a tendency for liner temperatures to be too low, thus leading to corrosive wear from the sulphuric acid formed during combustion. Wärtsilä applies two insulating techniques to secure better temperature distributions. For some years, PTFE insulating tubes have been fitted in the cooling bores of the liner. As part of TriboPack, the liner is now also insulated in the mid-stroke region by a Teflon band on the water side. The insulating tubes are adapted according to the engine rating to ensure that the temperature of the liner running surface is kept above the dew point temperature of water over the full length of the stroke and over a wide load range.

Mid-stroke insulation and, where necessary, insulating tubes are therefore used to optimize liner temperatures over the piston stroke. An insulation bandage in the form of Teflon bands with an outer stainless steel shell is arranged around the outside of the liner to raise liner wall temperatures in the mid-stroke region. Mid-stroke insulation is known to be particularly useful for sustained engine operation at low power outputs, while the TriboPack gives an additional safety margin in abnormal operating conditions (for example, against excessive carbon deposits built up on the piston crown).

While trying to avoid corrosive wear by optimizing liner wall temperatures, it is necessary to keep as much water as possible out of the cylinders. Highly efficient vane-type water separators fitted after the scavenge air cooler and effective water drain arrangements are thus vital for good piston running behaviour. Load-dependent cylinder lubrication is provided by the Sulzer multi-level accumulator system, which ensures the timely quantity of lube oil for good piston running. The lube oil feed rate is controlled according to the engine load and can also be adjusted according to the engine condition.

Piston rod glands

Time-between-overhauls of crosshead engines are partly defined by the piston rod glands, in the sense that their removal for exchange of elements is often connected to a withdrawal of the piston and piston rod assembly. The gland elements and piston rods therefore need to have a long life expectation (TBO of three years or more). At the same time, they have to assure sealing of the crankcase from the piston underside, limit contamination of crankcase system oil by combustion residues, and keep the oil consumption at a reasonable level for maintaining oil quality.

Recent improvements have introduced additional gas-tight top scraper rings, stronger springs for the other scraper rings, enlarged drain channels for the scraped-off oil, and the exclusive application of bronze scraper rings on fully hardened rods. The drain quantities from the neutral space were reduced by a factor of three. Additionally, the scraped-off oil is reusable without any treatment and therefore can be directly fed back internally in the gland box to the crankcase. System oil consumption figures were significantly reduced. The design of the gland box housing was modified, allowing it to be dismantled either upwards during piston overhaul or downwards without pulling the piston.

Complete retrofit packages available for all RTA engines in service comprise the newly-designed upper scraper, new middle sealing and new lower scraper groups, and some modifications on the gland box housing. The upper scraper group consists of a two-piece housing with newly-designed oil scraper rings made of bronze, newly-designed gas-tight sealing rings and modified tension springs. The oil scraper rings consist of four segments conforming better to the piston rod. Two new seal rings in three parts and adapted tension springs were introduced for the middle seal group. For the lower scraper group, all rings are of bronze, since Teflon has an inferior performance when there is an increased amount of hard particles in the oil residues coming from the piston underside. Here, the new scraper rings comprise three slotted segments for adaptability to the piston rod; they are provided with grooves at the top to promote draining of the scraped oil.

The actual surface condition and shape of the piston rod is of paramount importance, Wärtsilä advises. Ideally, the new glands should be used with hardened piston rods. Existing rods can be retained, however, providing their surface condition and geometry are acceptable, before introducing any new stuffing box elements. If the rod is worn down, roughened or otherwise surface damaged it can be ground to standard diameters of 2 mm or 4 mm undersize and then surface hardened.

Exhaust valve behaviour

The exhaust valve is subjected to hot gases and the temperature resistance of its seat and body is therefore crucial. Nimonic valves combined with proper seat cooling have yielded excellent service behaviour and long life times. When the RTA96C engine was introduced, and its shallow combustion space created difficult conditions for combustion chamber components, some exhaust valves were additionally coated with Inconel alloy. After limited running times, however, there was some cracking of the coating originating from the centre hole, with loosened material, making removal by grinding necessary. Non-coated valves, on the other hand, showed excellent performance, remaining free of cracks after over 14 000 running hours, without any loss of material. Today's standard therefore is the non-coated valve.

Piston crowns

Burning off of material on piston crowns is very dangerous as hole formation leads to direct contact of the combustion flame with the piston cooling oil system and dire consequences. The use of the combined shaker and jet cooling system in RTA engines assures piston crown temperatures below 400 °C and thus eliminates such burning.

Condition monitoring

In seeking longer times-between-overhauls it is desirable to keep the engine 'closed' for as long as possible; this calls for additional systems to monitor the condition inside the cylinder without having to dismantle the engine for access. A growing family of products has been developed by Sulzer over many years to support the engine operator, these MAPEX tools complementing and expanding the functions of standard remote control systems. They include features for monitoring, trend analysis and planning, as well as for management support for spare parts ordering and stock control and maintenance, and are applicable to new engines and retrofit to existing engines.

A basic tool, MAPEX-CR (Combustion Reliability), continuously monitors the pressure in the cylinder. It can therefore detect and eliminate various malfunctions that may occur during engine operation while allowing the engine to run at optimum conditions. Previously, cylinder pressure could only be measured periodically by either a traditional indicator or a peak pressure indicator, or more recently by temporarily attached electronic sensors. Newly developed pressure sensors are sufficiently robust to be installed permanently for continuous on-line measurement.

MAPEX-CR monitors several key parameters (peak pressure, pressure gradient, compression pressure, pressure ratio and mean indicated pressure) along with their deviations. Evaluations of these parameters enables defects—such as permanently overloaded running gear, unbalanced cylinder load share, injection problems or broken piston rings—to be detected at an early stage. The system also generates an automatic alarm signal when permissible limits are exceeded.

MAPEX-PR (Piston Running) is a useful tool for checking piston-running behaviour without opening the engine. It continuously monitors the temperatures of the cylinder liners and the cooling water, and analyses them according to several alarm criteria which also take into account the engine load. Alarm signals are generated in the case of abnormal events. The past history of temperatures can also be reconstructed to enable relevant conclusions to be drawn. A unique ability is claimed to detect high friction on the cylinder liner at an early stage, this ensuring the reduction or even prevention of major consequential damage through piston scuffing. Other conditions indicated by the system include increased scavenge air temperature, excessive fluctuations in cooling water temperature, too high or too low average temperature of the whole engine, and excessive deviation of individual cylinder temperatures. Irregularities in the engine can be detected before serious damage occurs.

Condition-based maintenance

A new tool for managing condition-based maintenance of Sulzer two-stroke engines combines the expert knowledge and wide experience of Wärtsilä Corporation engineers. The CBM Management tool is offered in three different solutions, depending on the extent that customers want to outsource this activity.

A wide range of products for diesel engine monitoring, trend analysis, diagnosis and management already support condition-based maintenance. CBM Management takes these a step further, with the aim of optimizing the balance between long TBOs, low spare part costs, less time off-hire and high reliability from the engine. Wärtsilä appreciated that, although customers use the same engine types in their fleet, they experience quite different maintenance costs and varying engine reliability. This arises from various factors that influence the condition of the engine, such as the quality of fuel, lube oil and spare parts, and the quality of routine inspections and overhauls.

Analysis of cases of extensive damage to engine components had shown that long before the damage occurred it would have been possible to foresee that something was going wrong. With the appropriate expert knowledge and a systematic analysis of available information, it is often possible to predict problems and to prevent expensive maintenance costs. Wartsila's solution was to collect the expert knowledge of engineers from its technical service organisation worldwide and store it in a common database.

Some data are specific to an individual engine: shop trial results, sea trial results, electronically stored information from general service reports made by Wärtsilä, and performance data collected onboard. All these were combined with the database of expert knowledge, enabling the resulting expert system to automatically generate advice on the condition of the specific engine. The CBM Management tool comprises a Data Collector (hardware and software) and an Expert System (hardware and software).

The *Data Collector* is a hand-held computer whose main purpose is to enable the ship's engineers to collect data in a standardized way. It is automatically loaded with engine performance data from the control system through a standard interface but manually fed with maintenance and inspection data. For systematic and efficient manual input, the Data Collector contains a structured inspection guide for the engine, supported by interactive templates for measurement records and performance sheets. This guide exists for the following twelve function groups: piston performance; piston rod gland condition; scavenge air flow; combustion performance; fuel injection pumps; camshaft; engine structure; gears and wheels; bearings; crankshaft; engine control system; and vibration dampers.

The Data Collector takes into account different conditions of important components, such as piston rings, cylinder liners and bearings. A picture gallery helps the ship's engineers to understand the terminology used. The Data Collector reduces inspection time and partly replaces the necessary paperwork through the automatically created templates.

The *Expert System* is the heart of the tool and runs on a PC using data from the Data Collector. There are three steps from collected data to real expert advice: calculate trends; assess engine condition by comparing the data with a weighted set of reference cases based on the expert knowledge of Wärtsilä engineers; and assess the condition of the engine components and parts, and offer advice to the ship's engineers on achieving optimized TBOs. The system supports the operator with graphs, pictures, trends and text recommendations. Communication is possible between ships, the operator's headquarters and Wärtsilä's CBM centre in all directions.

Three different solutions are offered to suit the individual requirements of owners and operators:

- CBM Virtual Expert: the operator performs data collection and analysis, and Wärtsilä's expert knowledge is given as far as possible without further involvement. The onboard engineers carry out data collection, the expert system being used on the ship as well as at the operator's HQ. Such a solution typically suits the needs of owners and operators who have their own technical experts.
- CBM Report Agreement: the operator collects the data and Wärtsilä undertakes the analysis. The owner or operator uses only the Data Collector, while the data is analysed at Wärtsilä's Two-Stroke CBM centre in Switzerland, which sends the customer a regular report. This type of agreement typically suits owners and operators without their own technical office.
- CBM Inspection Agreement: Wärtsilä carries out both data

collection and analysis. Wärtsilä service engineers visit the ship in port, collect the data and send it to the CBM centre, from where the customer receives a report. Such a solution suits owners and operators without a technical office and who want to outsource inspections.

CBM solutions were expected to be generally available by mid-2003. A rapid growth in expert knowledge was anticipated from service engineers using the data collecting tool as a standard instrument.

RT-FLEX ELECTRONIC ENGINES

The Sulzer RT-flex system, which will be progressively offered as an option for all models in the RTA programme, resulted from a project originated in the 1980s to develop an electronically-controlled low speed engine without the constraints imposed by mechanical drive of the fuel injection pumps and exhaust valve actuation pumps (see the introductory chapter on Low Speed Engines.) Traditional jerk-type fuel injection systems combine pressure generation, timing and metering in the injection pump with only limited flexibility to influence the variables. In contrast, Sulzer's common rail system separates the functions and gives far more flexibility for optimizing the combustion process with injection and valve timing.

RT-flex engines are essentially the same as their standard RTA equivalents but dispense with the camshaft and its gear drive, jerk-type fuel injection pumps, exhaust valve actuator pumps and reversing servomotors. Instead, they are equipped with common rail systems for fuel injection and exhaust valve actuation, and full electronic control of these functions (Figure 12.30).

The following benefits for operators are cited for the RT-flex system:

- Smokeless operation at all running speeds.
- Lower steady running speeds (in the range of 10–12 per cent nominal engine speed) obtained smokelessly through sequential shut-off of injectors while continuing to run on all cylinders. Very steady running at 12 rev/min has been demonstrated.
- Reduced running costs through lower part-load fuel consumption and longer times-between-overhauls.
- Reduced maintenance requirements, with simpler setting of the engine; the 'as new' running settings are automatically maintained.
- Reduced maintenance costs through precise volumetric fuel injection control, leading to extendable TBOs. The common rail fuel system and its volumetric control yields excellent balance

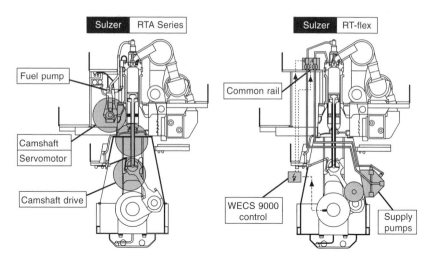

Figure 12.30 The fuel pumps, valve actuator pumps, camshafts and drive train of the standard Sulzer RTA engine are replaced by a compact set of supply pumps and common rail fuel system on the RT-flex engine

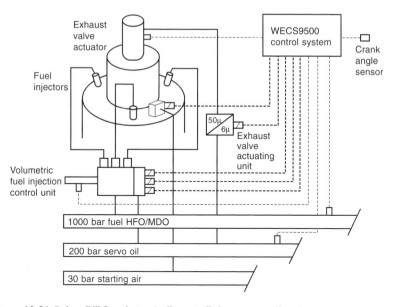

Figure 12.31 Sulzer RT-flex electronically-controlled common rail systems

in engine power developed between cylinders and between cycles, with precise injection timing and equalized thermal loads.
• Reliability underwritten by long term testing of common rail

system hardware and the use of fuel supply pumps based on proven Sulzer four-stroke engine fuel injection pumps.

- Higher availability resulting from integrated monitoring functions and from built-in redundancy: full power can be developed with one fuel pump and one servo oil pump out of action. High pressure fuel and servo oil delivery pipes, and electronic systems, are also duplicated.
- A reduced overall engine weight: approximately two tons per cylinder lower in the case of a 580 mm bore RT-flex engine compared with its conventional RTA counterpart.

The common rail for fuel injection is a manifold running the length of the engine at just below the cylinder cover level; the rail and other related pipework are arranged on the top engine platform with ready accessibility from above (Figure 12.32). The common rail is fed with heated fuel oil at a high pressure (nominally 1000 bar) ready for injection into the engine cylinders. The fuel supply unit embraces a number of high pressure pumps mechanically driven from the crankshaft and running on multi-lobe cams, which increase their supply capacity and hence reduce the number needed. A four-pump set is sufficient for a six-cylinder RT-flex 58T-B engine. The pump design, based on fuel injection pumps used in Sulzer four-stroke engines, has suction control to regulate the fuel delivery volume according to engine requirements (Figure 12.33).

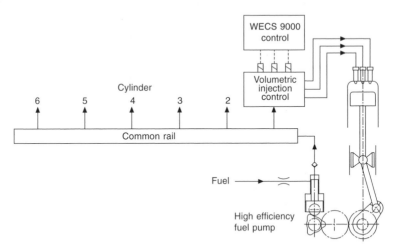

Figure 12.32 Schematic layout of Sulzer RT-flex common rail fuel system

Heated fuel oil is delivered from the common rail through a separate injection control unit for each engine cylinder to the standard fuel

Figure 12.33 Three of the six fuel oil pumps for a Sulzer 7RT-flex60C engine

injection valves which are hydraulically operated in the usual way by the high pressure fuel. The control units, exploiting quick-acting Sulzer solenoid rail valves, regulate the timing of fuel injection, control the volume of fuel injected, and set the shape of the injection pattern. The three fuel injection valves in each cylinder cover are separately controlled so that, although they normally act in unison, they can also be programmed to operate separately as necessary. The key features of the common rail system are defined as:

- Precise volumetric control of fuel injection, with integrated flow-out security.
- Variable injection rate shaping and free selection of injection pressure.
- Stable pressure levels in common rail and supply pipes.
- Possibility for independent control and shutting-off of individual fuel injection valves.
- Ideally suited for heavy fuel (up to 730 cSt at 50°C) through clear separation of the fuel oil from the hydraulic pilot valves.
- Proven standard fuel injection valves.
- Proven high efficiency common rail fuel pumps.

The fuel injection pressure can be freely selected up to more than 1000 bar over the whole load range. In combination with different injection patterns, this provides the opportunity to optimize the engine in several ways: for example, for low emission levels or improved fuel efficiency at non-optimum loads. (See *Environmental Performance* section below.)

The injection system can be adapted to different patterns, such as: pre-injection, with a small part of the fuel charge injected before the main charge; triple injection, with the fuel charge injected in three separate short sprays in succession; and sequential injection, with individual actuation of the fuel injection nozzles so that injection timing is different for each of the three nozzles in a cylinder. Different shapes of cylinder pressure profile during the engine cycle can thus be created which, with free selection of the rail pressure, allows the optimum pattern to be selected in each case for the loads and performance optimization target of the engine.

Selective shut-off of single injectors is valuable for low manoeuvring speeds or 'slow steaming' as this facility fosters better injection and atomization of the small quantities of fuel needed. In such modes, the common rail system is controlled to use the three injection valves in sequence. Regulated by an electronic governor, the RT-flex engine demonstrated very steady running at a lowest speed of 12 rev/min.

Exhaust valves are operated in much the same way as in conventional Sulzer RTA engines by a hydraulic 'pushrod', but with the actuating energy coming from a servo oil rail at 200 bar pressure. The servo oil is supplied by hydraulic pumps mechanically driven from the same gear train as the fuel supply pumps. An electronically-controlled actuator unit for each engine cylinder—operated by hydraulic pressure from the servo oil rail—gives full flexibility for valve opening and closing timing. Two redundant sensors inform the WECS-9500 control system (see below) of the current position of the exhaust valve. Lube oil from the engine is used as servo oil to keep the system simple and compatible. Before entering the servo oil circuit the oil is directed through an additional six-micron filter with an automatic self-cleaning device to ensure reliability and a long lifetime of the actuator units and solenoid valves.

All functions of the RT-flex system are controlled and monitored through the integrated Wärtsilä WECS-9500 electronic control system. This modular system has separate microprocessor control units for each cylinder, with overall control and supervision by duplicated microprocessor control units which provide the usual interface for the electronic governor and remote control and alarm systems.

The full load efficiency of RT-flex engines is the same as their conventional RTA engine equivalents but improvements in part-load fuel economy are gained. This results from the freedom allowed in selecting the optimum fuel injection pressure and timing, and exhaust valve timing, at all engine loads or speeds, while maintaining efficient combustion at all times, even during dead slow running. A similar freedom in exhaust valve timing allows the RT-flex system to keep the combustion air excess high by earlier closing as the load/speed is reduced. Such a facility is not only beneficial for fuel consumption; it also limits component temperatures, which normally increase at low load. Lower turbocharger efficiencies at part load normally result in low excess combustion air with fixed valve timing.

Another contribution of the RT-flex system to fuel economy cited by Wärtsilä is the capability to easily adapt the injection timing to various fuel properties influencing poor combustion behaviour. Variable injection timing (VIT) over load has been a traditional feature of Sulzer low speed engines for many years, using a mechanical arrangement primarily to keep the cylinder pressure high for the upper load range. This is much easier to arrange in an electronically-controlled engine.

Environmental performance. A very wide flexibility in optimizing fuel injection and exhaust valve processes enables RT-flex engines to comfortably meet IMO limits on NOx emissions. The most visible benefit cited is smokeless operation at all ship speeds, underwritten by superior combustion. The common rail system allows the fuel injection pressure to be maintained at an optimum level irrespective of engine speed. In addition, at very low speeds, individual fuel injectors are selectively shut off and the exhaust valve timing adapted to help keep smoke emissions below the visible limit. In contrast, engines with traditional jerk-type injection pumps have increasing smoke emissions as engine speed is reduced because the fuel injection pressure and volume decrease with speed and power, and they have no means of cutting off individual injection valves and changing exhaust valve timing. (Figure 12.34).

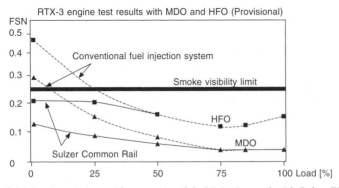

Figure 12.34 Smoke emissions with conventional fuel injection and with Sulzer RT-flex common rail technology for engines burning heavy fuel and marine diesel oil

As all settings and adjustments within the combustion and scavenging processes are made electronically, future adaptations are possible simply through changes in software, which could be easily retrofitted to existing RT-flex engines. A possibility is to offer different modes for different emissions regimes. In one mode, the engine would be optimized for minimum fuel consumption while complying with the global NOx limit; then, to satisfy local emissions regulations, the engine could be switched to an alternative mode for even lower NOx emissions while the fuel consumption is allowed to rise.

RT-flex engines in service

The RT-flex system was first applied to a full-size research engine in June 1998 at Wärtsilä's facilities in Switzerland and represented the third generation of electronically-controlled Sulzer diesel engines. The first RT-flex engine, a six-cylinder RT-flex58T-B model, was tested in January 2001 and later installed as the propulsion plant of a 47 950 dwt bulk carrier, which was handed over in September that year. The engine's slow-running capability was demonstrated by steady operation at speeds down to 12 rev/min. Subsequent orders called for seven-cylinder RT-flex60C models, the first Sulzer low speed engine designed from the bedplate up with electronically-controlled common rail systems (Figure 12.35). The RT-flex option was extended to all bore sizes in

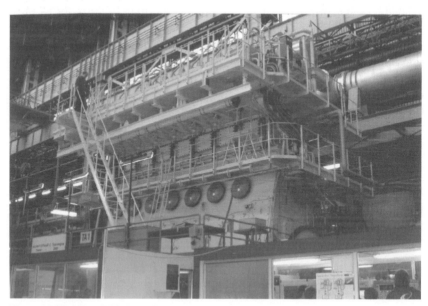

Figure 12.35 Sulzer 7RT-flex60C engine on test at Wärtsilä's Trieste factory

the RTA series, the first contract for a 12-cylinder RT-flex96C engine developing 68 640 kW being booked for containership propulsion in early 2003.

RL TYPE ENGINES

The RTA engine's immediate predecessor in the Sulzer low speed programme was the loop-scavenged RL series, of which examples remain in service and merit attention here.

The RLA56, a small bore two-stroke engine introduced in 1977, incorporated the basic design concept of the successful RND and RNDM series but extended the power range at the lower end. This engine, of comparatively long stroke design, was the first model in the RLA series. It retained many of the design features of the then most recent economical loop-scavenged RLB type (Figures 12. 36 and 12.37), both engines using many features of the earlier RND-M series.

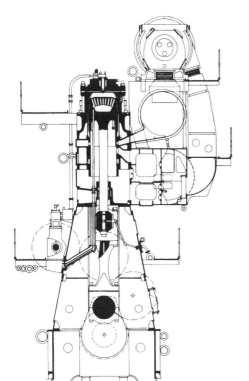

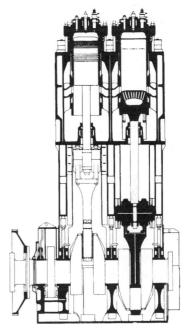

Figure 12.37 Longitudinal section of RLB90 engine

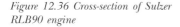

Figure 12.36 Cross-section of Sulzer RLB90 engine

A number of RND-M engine features were retained for the RLA and RLB type, namely:

- Constant pressure turbocharging and loop scavenging.
- A bore-cooled cylinder liner and one-piece bore-cooled cylinder cover and water-cooled piston crowns.
- Double guided crossheads.
- New cylinder liner lubrication system with accumulators for the upper liner part and thin-walled aluminium–tin crosshead shell bearings.

Major new design features peculiar to the RL types were:

- A new bedplate design with an integrated thrust block; a new box type column design.
- A semi-built or monobloc type crankshaft without a separate thrust shaft.
- Location of the camshaft gear drive at one end for engines of four to eight cylinders.
- Multiple cylinder jackets.
- A bore-cooled piston crown.
- Modified crosshead.
- A new design of air receiver.

Bedplate with thrust block

For the RL type engine bedplate a new concept of great simplicity was applied. Both the crossgirders and the longitudinal structure are of single-wall fabricated design giving very good accessibility for the welded joints. The central bearing saddles are made of cast steel and only one row of mounting bolts on each side is used to secure the bedplate to the ship's structure.

The completely new design feature was the method of integrating the thrust block into the bedplate (Figure 12.38) allowing an extremely compact design and saving engine length. The first crankshaft bearing on the driving end is a combined radial-axial bearing. The bedplate is a one-piece structure for all engines from four to eight cylinders but, if required, it can be bolted together from two halves.

Thick-walled whitemetal lined bearing shells are used to support the crankshaft and guarantee an optimum safety for the running of the crankshaft.

Crankshaft

In addition to the traditional semi-built type crankshaft, a monoblock continuous grain flow forged type can be used. From four to eight

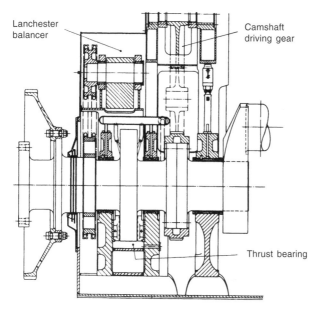

Lanchester balancer

Camshaft driving gear

Thrust bearing

Figure 12.38 Integrated thrust bearing and camshaft drive for RL engine

cylinders, the crankshaft is a one-piece component with an integrated thrust collar section. Only one journal and pin diameter was used for all engines up to eight cylinders. For RLA engines larger than the RLA56, the crankshaft was semi-built, and the thrust shaft separately bolted to the crankshaft.

Engine frame

For the small RLA56 engine a new method of frame construction was used. Cast iron central pieces, on which the crosshead guides are bolted, are sandwiched between two one-piece fabricated side-frame girders. This replaced the traditional construction of 'A' frame bolted to the bedplate with side plates attached with access doors forming the enclosure. Larger RL type engines use 'A' shaped columns of fabricated double wall design, assembled with longitudinal stiffening plates to constitute a rigid structure between the bedplate and cylinder blocks (Figures 12.39 and 12.40).

Cylinder jackets

The cylinder jackets are made of fine lamellar cast iron produced as single block units bolted together in the longitudinal plane to form a

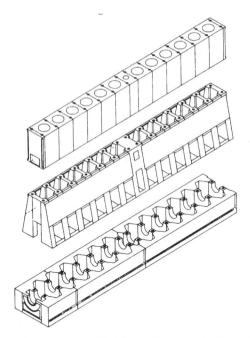

Figure 12.39 Structural arrangement of bedplate, columns and cylinder jackets for 12-cylinder Sulzer RL engine

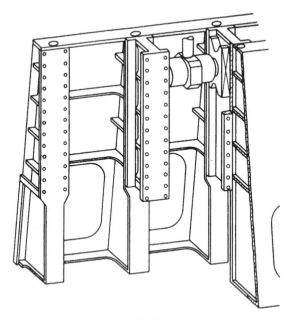

Figure 12.40 Arrangement of columns for RLA56 engine

single rigid unit, and held on top of the frame section by long tie bolts secured in the bedplate. For the small RLA56 engine, multi-cylinder blocks consisting of two or three cylinders were standard and provide great rigidity of the structure. The arrangement of the cooling water passages in the jackets ensures forced water circulation and optimal water distribution around the exhaust canal.

Because of the steadily rising charge air pressures, the design of the air receiver was modified to allow the use of automatic welding techniques. Instead of a rib-stiffened plain side plate, a semi-circular pressure containment was fitted with an integral air inlet casing. The auxiliary blower is mounted on the front end of the receiver, thus eliminating inclined ducts.

Combustion chamber components

The combustion chamber of RL type engines is principally of the same shape as that of the RND-M. The cylinder cover is basically a one-piece steel block with cooling bores, while an identical arrangement of bore cooling is applied to the upper collar of the cylinder liner which is made from lamellar cast iron, a material of good heat conductivity and wear resistance. Figure 12.41 shows the arrangement with cover, cylinder liner and piston.

A new type of piston crown was introduced for the RL series engines. This combustion chamber component similarly uses a bore cooling arrangement (Figures 12.42 and 12.43) as previously only applied to the liner and cover. Water cooling was retained but the piston bore cooling uses a somewhat different mechanism to the force flow system of the liner and cover.

The cooling space of the piston crown is approximately half filled with cooling water and, as a result of piston acceleration and deceleration, a 'cocktail shaker' effect is produced to provide excellent heat removal. This effect is capable of ensuring efficient heat transfer under all prevailing load conditions in order to keep the vital temperatures on the piston crown and around the piston rings within suitable limits.

As a result of this new design the crown temperatures were lowered compared with the previous construction. Forged steel blocks were specified for piston crown manufacture but cast steel versions were also used.

Cylinder lubrication

Lubrication of the cylinder is through six quills mounted in the upper area of the liner just above the cylinder jacket, as shown in Figure

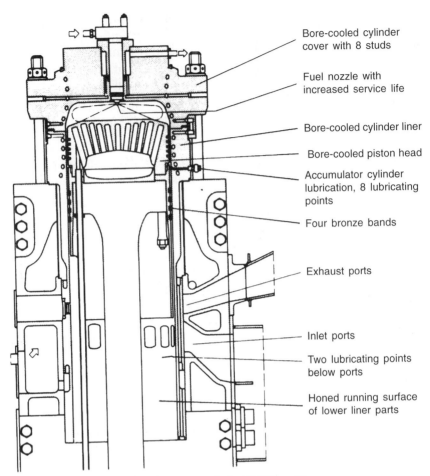

Bore-cooled cylinder
cover with 8 studs

Fuel nozzle with
increased service life

Bore-cooled cylinder liner

Bore-cooled piston head

Accumulator cylinder
lubrication, 8 lubricating
points

Four bronze bands

Exhaust ports

Inlet ports

Two lubricating points
below ports

Honed running surface
of lower liner parts

Figure 12.41 Combustion chamber components for RLA90 engine

12.44. The oil distribution grooves have a very small angle of inclination to avoid blow-by over the piston rings and small but regular quantities of cylinder oil supplied by an accumulator system. Two further lubrication points are provided below the scavenge ports on the exhaust side with the necessary oil pumps positioned on the front side of the engine above the camshaft drive.

Crosshead design

The crosshead is similar to that used for RND engines and has double guided slippers. The pin size was increased for safety and thin-walled half shells of the aluminium–tin type were used. The pin itself is of

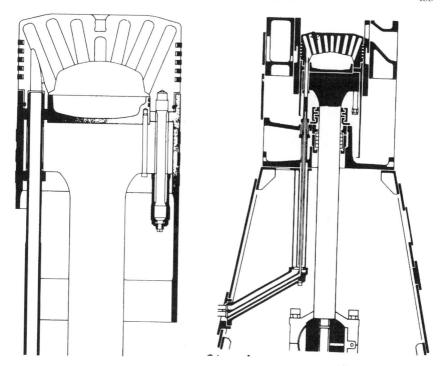

Figure 12.42 Detail view of RL engine piston with bore cooling

Figure 12.43 Arrangement of telescopic pipes for RL engine piston cooling

forged homogeneous steel of symmetrical design, and can therefore be turned around in case of damage. The piston rod is connected to the pin by a single hydraulically tightened nut, while the slippers, made of cast steel and lined with whitemetal, are bolted to the ends of the pin. The cast iron double guide faces are fitted to the engine columns.

The connecting rods of traditional marine type have a forged normalized steel bottom end bearing, lined with white metal, held in place by four hydraulically tensioned bolts. Compression shims are provided between the bottom end bearing and the palm of the connecting rod.

Piston assembly

The piston (Figure 12.42) consists of a water-cooled cast steel crown, a cast iron skirt with copper bandages and a forged steel piston rod, with the piston rings fitted in chromium plated grooves.

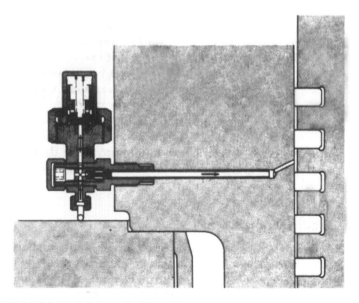

Figure 12.44 Cylinder lubricator for RL engine

The water-cooled piston has proved very reliable when running on heavy fuels and the use of water cooling has resulted in practically negligible system oil consumption. (With oil cooling, oil is consumed usually as a result of thermal ageing on hot piston walls. Oil leaks from oil-cooled pistons may also occur on other engine types.) The two-part gland seals for the piston rod and telescopic piston cooling pipes can be inspected while the engine is running and can be dismantled without removing the piston. The double-gland diaphragm around the piston rod completely separates the crankcase from the piston undersides, preventing contamination of the crankcase oil by combustion residues or possible cooling water leakages. In addition, the fresh water piston cooling water system is served by an automatic water drain-off when the circulating pumps are stopped. This avoids leakages when the ship is in port.

Turbocharging arrangement

Sulzer RL engines all employ the constant pressure turbocharging system and with the high efficiency Brown Boveri series 4 turbochargers include provision for automatic cleaning. The layout of the turbocharging arrangement is shown in Figure 12.45.

The use of the piston undersides to provide a scavenge air impulse

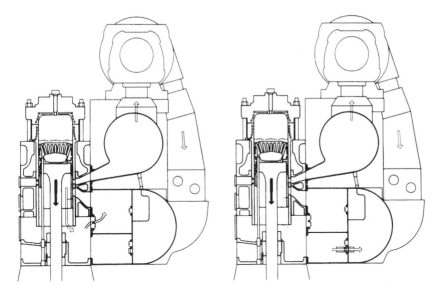

Figure 12.45 Constant pressure turbocharging and operation of under-piston supercharging (RL engine)

eliminates the need for large electrically driven auxiliary blowers. At higher loads a simple flap valve opens to cut out the piston underside pumping effect with a consequent improvement in fuel consumption, whilst a small auxiliary fan incorporated in the scavenging system improves the smoke values at the lowest loads. The piston underside scavenge pump facility allows the engine to start and reverse even with total failure of all turbochargers. The auxiliary fan and engine will even operate at up to 60 per cent load, thus giving a 'take home' facility.

The turbochargers are mounted on top of the large exhaust gas receiver with the scavenge air receiver which forms part of the engine structure beneath. The charge air is passed down through seawater cooled intercoolers which are mounted accessibly alongside the scavenge air receiver. Flap-type valves which operate according to the scavenge air pressure direct the air inside the three-compartment air receiver to the piston underside (at low scavenge air pressure) or direct to the scavenge ports when the higher air pressure from the turbochargers keeps the underside delivery flap valves shut.

For RL type engines equipped with only one turbocharger a separate turbocharger/air intercooler module is available as a standard option. This unit can be located adjacent to the engine at either the forward or aft end, thus considerably reducing the mounting height and width of the main engine. The turbocharger module consists of a base frame onto which the turbocharger, charge air cooler, air ducts and cooling water pipes are solidly mounted with flexible connections to the engine.

Upgrading RLB engines

Based on field tests carried out between 1991 and 1995, a retrofit package of modifications was introduced for Sulzer RLB90 engines in 1993 and for RLB66 engines in 1995. The main element is the change to loop-cooling in the cylinder liners, an improved cooling technique that reduces the thermal and mechanical stresses in the top collar of the liner (Figure 12.46). By optimizing the running surface temperature, the system also reduces corrosive wear and extends the time-between-overhauls up to two years.

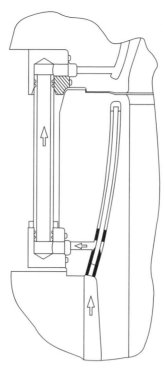

Figure 12.46 Loop cooling of the cylinder liner benefits Sulzer RLB90 and RLB66 engines

The retrofit package comprises the loop-cooled cylinder liners as well as:

- A new fully gas-tight, vermicular top piston ring of increased thickness.
- A modified top piston ring groove with increased height and increased thickness of chromium layer.
- Multi-level lubrication.
- A new water guide jacket to suit the new liner.
- A modified condensed water drain from the scavenge air receiver.

As RLB engines exploit loop scavenging with both scavenge and exhaust ports in the lower part of the liner, the fully gas-tight top ring significantly reduces the lateral forces at the piston skirt in addition to reducing the blow-by of hot combustion gases—both of which increase the piston skirt life. The package can be applied to the engine either as a full package or step-by-step (intermediate package) depending on service experience with the liners and whether the general piston-running behaviour meets today's expectations. It is also possible to replace individual original liners with loop-cooled units whenever they need to be renewed; there is no problem in running loop-cooled and original design liners together in the same engine.

See Chapter 9 for details of Sulzer research engines, and see Introduction for Sulzer's early low speed engine designs.

13 Burmeister & Wain low speed engines

A pioneering Danish designer and builder of low speed two-stroke crosshead engines, Burmeister and Wain (B&W) focused in the post-War era on high pressure turbocharged single-acting uniflow-scavenged models with one centrally located exhaust valve. The uniflow system fosters effective scavenging of the exhaust gases from the cylinder in an even progressive upwards movement with low flow resistance through to the exhaust manifold via a large diameter poppet-type exhaust valve (Figure 13.1).

Until 1978 all B&W engines operated on the impulse turbocharging system (Figure 13.2), but the search for enhanced economy in the wake of contemporary fuel crises dictated a change to the constant pressure system, yielding a saving of around 5 per cent in specific fuel consumption (Figure 13.3). In the years before its takeover by MAN of Germany in 1980, B&W introduced a number of engine series—successively the K-GF, L-GF, L-GFCA and L-GB programmes—before launching the current MC design portfolio now developed and marketed under the MAN B&W Diesel banner (see Chapter 10).

All these designs were improved variants of the K-GF series which itself introduced many notable design innovations, such as box frame construction, intensively cooled cylinder components and hydraulically actuated exhaust valves.

A major change to the structure of the company followed its purchase by MAN, resulting in Copenhagen becoming the centre for two-stroke engine development of all designs bearing the MAN B&W trade mark.

K-GF TYPE ENGINES

B&W two-stroke engines based on the K-GF series featured many new design features. Notable was the box-type construction of the crankcase and the use of a hydraulically actuated exhaust valve instead of the conventional mechanical rocker arm system used in the earlier K-EF types. A sectional sketch of the K-GF is shown in Figure 13.4. As this

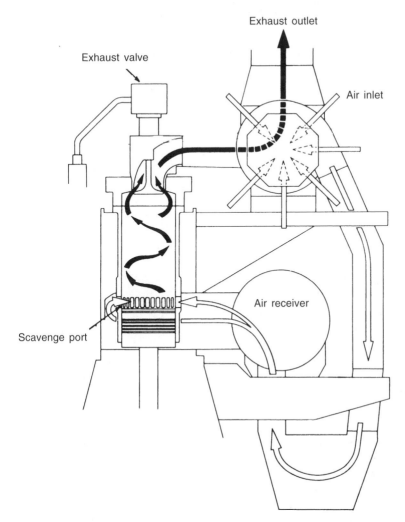

Figure 13.1 B&W uniflow scavenging system

model was rapidly superseded by the L-GF and L-GB models, only brief details of the K-GF construction are given here.

Bedplate and frame

The standard K-GF engine bedplate is of the high and fabricated design with cast steel main bearing housings welded to the longitudinal side frames. The standard thrust bearing is of the built-in short type and separated from the crankcase by a partition wall (Figure 13.5).

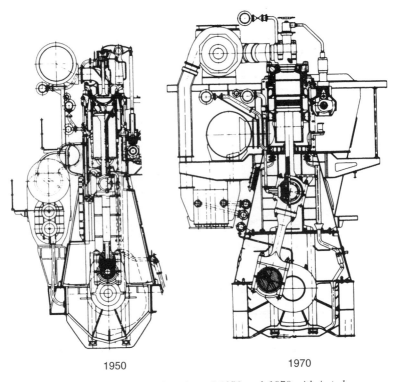

1950   1970

Figure 13.2 B&W uniflow scavenged engines of 1950 and 1970 with impulse turbocharging

The frame or entablature section consists of three units: a frame box of a height corresponding to the length of the crosshead guides, bolted together in the longitudinal direction in the chain drive section only, and two longitudinal girders, also bolted together in this section. This method of construction gives rigid units designed for easy handling and mounting, both on the testbed and when erecting the engine in the ship. The rigid design with few joints ensures oil tightness, and large hinged doors provide easy access to the crankcase.

The crosshead guides, consisting of heavy I sections, are attached at top and bottom in the frame boxes, with the lower attachment flexible in the longitudinal direction.

Cylinder jackets

Cast iron one-piece cylinder sections connected in the vertical plane by fitted bolts are held to the frame section by long tie-bolts which are secured in the main bearing saddles.

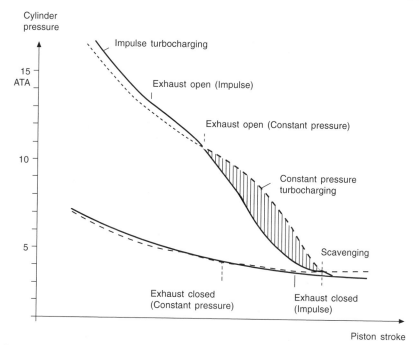

Figure 13.3 Working diagrams for impulse and constant pressure turbocharging

The cylinder liner is provided with a water-cooled flange low down to secure low operating temperatures in the most heavily loaded area. The liners are made of alloyed cast iron with ports for scavenge air and bores for the cylinder lubrication system.

Crankshaft

The crankshaft is semi-built for all cylinder numbers with cast steel throws for six- to ten-cylinder engines. Balancing of the engine is undertaken by varying the crankpin bore holes, thus entirely eliminating bolted-on counterweights.

Lubrication of the crank bearing is from the crosshead through the connecting rod, thus eliminating stress raising bores in the crankshaft.

Crosshead

The crosshead (Figure 13.6) is short and rigid and the bearings are so constructed that the bearing pressure between journal and bearing is distributed evenly over the entire length of the bearing. The bearing

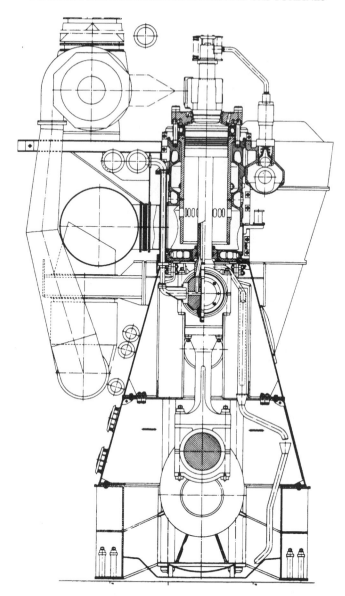

Figure 13.4 Cross-section of K90GF engine

pressure is smaller than previously and the peripheral speed is higher, which improves the working conditions of the bearings.

Interchangeable bearing shells of steel with a 1 mm whitemetal lining are fitted to the crosshead bearings. The shells are identical

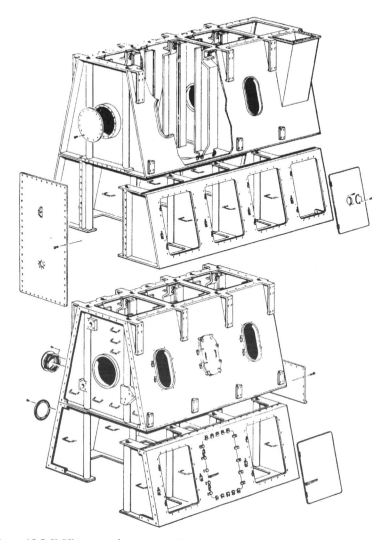

Figure 13.5 K-GF structural arrangement

halves, precision bored to finish size and can be reversed in position during emergencies when a damaged or worn lower half can be temporarily used as an upper shell.

Piston

The piston is oil cooled and consists of the crown, cooling insert and the skirt. The cooling insert is fastened to the upper end of the piston

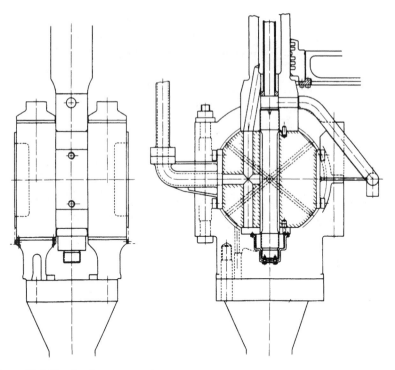

Figure 13.6 Crosshead arrangement

rod and transfers the combustion forces from the crown to the rod. The crown and skirt are held together by screws while a heavy Belleville spring is used to press the crown and the cooling insert against the piston rod. The crown has chromium plated grooves for five piston rings while the rod has a longitudinal bore in which is mounted the cooling oil outlet pipe. The oil inlet is through a telescopic pipe fastened to the crosshead, and the oil passes through a bore in the foot of the piston rod to the cooling insert (Figure 13.7).

Cylinder cover

The cylinder cover is in two parts, a solid steel plate with radial cooling water bores, to which a forged steel ring with bevelled cooling water bores is bolted. The insert has a central bore for the exhaust valve cage and mountings for fuel valves, a safety valve and an indicator valve.

The cover is held against the cylinder liner top collar by studs screwed into the cylinder block and the nuts are tightened by a special

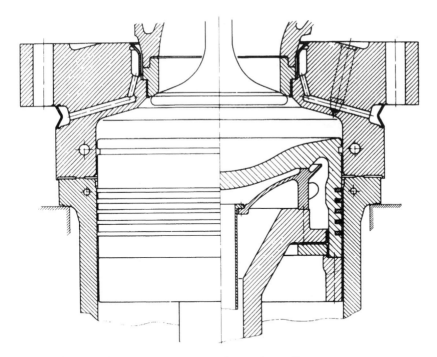

Figure 13.7 Combustion chamber and piston showing bore-cooling arrangement

hydraulic tool to allow simultaneous tightening and correct tension of the studs.

Exhaust valve

The valve is similar to earlier types but with hydraulic actuation, a feature originally unique to the K-GF series (Figure 13.8). The valve consists of a cast iron cage and a spindle, with steel seats and the valve mushroom faced with Stellite for hard wearing. Studs secure the valve housing to the cylinder cover.

The hydraulic actuation system is based on a piston pump driven by a cam on the camshaft, and the pressure oil is led to a working cylinder placed on top of the exhaust valve housing. The oil pressure is used to open the valve and closing is accomplished by a ring of helical springs.

Camshaft

The camshaft is divided into sections, one for each cylinder, enclosed and suspended in roller guide housings with replaceable ready-bored

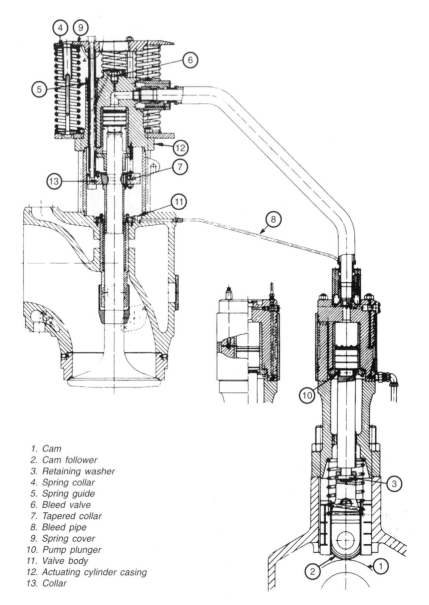

1. Cam
2. Cam follower
3. Retaining washer
4. Spring collar
5. Spring guide
6. Bleed valve
7. Tapered collar
8. Bleed pipe
9. Spring cover
10. Pump plunger
11. Valve body
12. Actuating cylinder casing
13. Collar

Figure 13.8 Components of hydraulically actuated exhaust valves

bearing shells. The cams and couplings are fitted to the shaft by the
SKF oil-injection pressure method and the complete camshaft unit is
driven by chain from the crankshaft. For reversing of the engine, the
chain wheel floats in relation to the camshaft which is driven through

a self-locking crank gear; reversing crankpins are turned by built-on hydraulic motors.

Turbocharging

Until 1978 all B&W engines were equipped with turbochargers operating on the impulse system, as was the case for the K-GF when first introduced. All B&W engines (including the K-GF) were subsequently specified

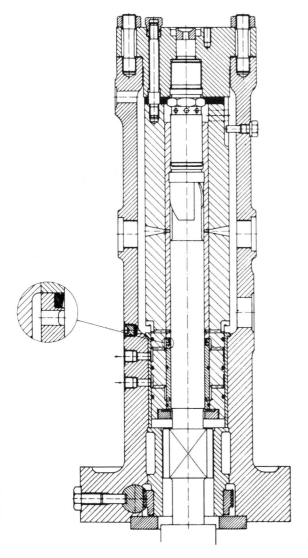

Figure 13.9 Fuel pump

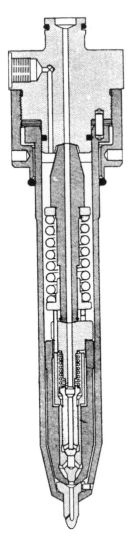

Figure 13.10 Fuel injector

with a constant pressure charging system in which the exhaust pulses from the individual cylinders are smoothed out in a large volume gas receiver before entering the turbine at a constant pressure. This system gives a fuel consumption which is some 5 per cent lower. For part load operation and starting electrically driven auxiliary blowers are necessary. The arrangement of the constant pressure turbocharged K-GF engine is shown in Figure 13.4.

L-GF AND L-GB ENGINES

The oil crisis in 1973 and the resulting massive increase in fuel prices stimulated enginebuilders to develop newer engines with reduced specific fuel consumption. B&W's answer was the L-GF series combining constant pressure turbocharging with an increase in piston stroke: an increase in stroke of about 22 per cent results in a lowering of the shaft speed by around 18 per cent, leading to greater propulsive efficiency when using larger diameter propellers and around 7 per cent increase in thermal efficiency.

The design of the L-GF series engine was heavily based on that of the K-GF and the necessary changes were mainly those relating to the increase in piston stroke and modifications to components to embody thermodynamic improvements (Figure 13.11). A major component design change was the cylinder liner, made longer for the increased stroke, and featuring cooling bore drillings forming generatrices on a hyperboloid to ensure efficient cooling of the high liner collar without cross borings with high stress concentration factors.

The cylinder covers are of the solid plate type used for the K-GF, while the pistons also are of the original oil cooled type but with

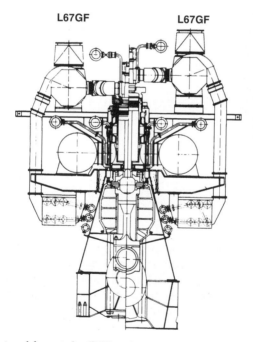

Figure 13.11 Short and long stroke 67GF engines

somewhat improved cooling caused by the stronger 'cocktail shaker' effect resulting from the greater quantity of oil in the elongated piston rod. The design of the crankshaft, crosshead, bearings, and exhaust valve actuating gear are, in the main, similar to the components introduced with the K-GF series.

Figure 13.12 12L90GFCA engine of 34 800 kW at 97 rev/min

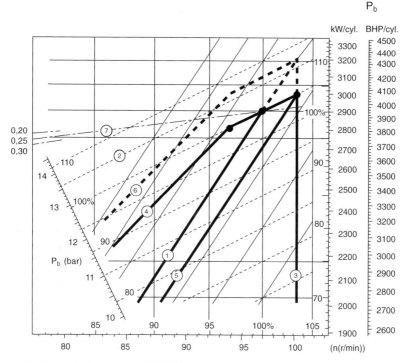

Figure 13.13 Load diagram for L90GFCA engine

Constant pressure turbocharging

The increased stroke and thus reduced rev/min of L-GF engines compared with K-GF engines was not aimed at developing more power but an improvement in ship propulsive efficiency. The uniflow scavenging system has the advantage of a good separation between air and gas during the scavenging process, and the rotating flow of air along the cylinder contributes to the high scavenging efficiency and clean air charge.

As an engine's mean indicated pressure increases the amount of exhaust gas energy supplied during the scavenging period, relative to the impulse energy during the blow-down period, constant pressure turbocharging is advantageous; also for uniflow scavenged engines with unsymmetrical exhaust valve timing. Theoretical calculations for improved fuel economy showed a possible gain of 5–7 per cent in specific fuel consumption by using constant pressure turbocharging.

The most obvious change with constant pressure turbocharging is that the exhaust pipes from each valve body led to a common large exhaust gas receiver instead of to the turbochargers as in the impulse system. When the cylinders exhaust into a large receiver the outflow of gas is quicker because the large gas impulse at the commencement of the exhaust period is levelled out in the gas receiver and the outflow of gas will not be retarded, as in the case of impulse turbocharging where a pressure peak is built up in the narrow exhaust pipe before the turbocharger. The opening of the exhaust valve can be delayed about 15°, thereby lengthening the expansion stroke and improving the efficiency and reducing the fuel consumption. The energy before the turbine is less than for impulse turbocharging but as the pressure and temperature before the turbine is nearly constant the turbine can be adapted to run at peak efficiency and the blower can supply sufficient air above 50 per cent of engine load. The scavenging air pressure is increased and the compression pressure somewhat decreased compared with impulse turbocharging.

A small auxiliary blower is necessary for satisfactory combustion conditions at loads up to about 50 per cent. Two blowers, of half capacity, are used for safety, and even with one of these out of action the other with half the overall capacity is satisfactory for starting and load increase. The engine will still run at down to 25 per cent load but with a smoky exhaust.

The change from K-GF to L-GF resulted in a 2 per cent improvement in the specific fuel consumption and the lower speed accounts for a 5 per cent improvement in propeller efficiency. Constant pressure turbocharging adds a further 5 per cent to the improvement, resulting in a reduction by 12 per cent in specific fuel consumption between the L-GF and K-GF types. The next engine model of the constant pressure turbocharged type was the L-GFCA, shortly succeeded by the L-GB type.

L-GB TYPE ENGINES

A further improvement in specific fuel consumption was yielded by the L-GB and L-GBE series engines. By using the optimum combination of longer stroke, higher output and higher maximum pressure, and the newest high efficiency turbochargers, much lower fuel consumption rates were achieved. The L-GB series had an mep of 15 bar at the same speed as the L-GFCA to give an increase in power of 15 per cent and an increase in the firing pressure from 89–105 bar. Accordingly, the important Pmax/mep ratio is almost the same but the specific fuel consumption is some 4 g/kWh lower than the L-GFCA series. A further economy rating is obtained with engines of the L-GBE type by holding the Pmax at reduced engine output (so-called 'derating', as outlined

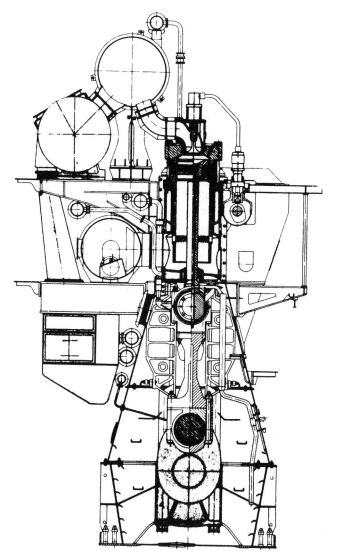

Figure 13.14 L-GB/GBE engine cross-section

in Chapter 5: the practice of offering propulsion engines in both normal and derated versions was adopted by many engine manufacturers, even for four-stroke medium speed engines).

Bedplate and main bearing

The bedplate consists of high, welded longitudinal girders and welded crossgirders with cast steel bearing supports. For the four- and five-cylinder engines the chain drive is placed between the aftermost cylinder and the built-in thrust bearing. For the six- to twelve-cylinder engines the chain drive is placed at the assembling between the fore and aft part. For production reasons, the bedplate can be made in convenient sections. The aft part contains the thrust bearing. The bedplate is made for long, elastic holding-down bolts tightened by hydraulic tool.

The oil pan is made of steel plate and welded to the bedplate parts. The oil pan collects the return oil from the forced lubricating and cooling oil system. For about every third cylinder it is provided with a drain with grid.

The main bearings consist of steel shells lined with whitemetal. The bottom shell can, by means of hydraulic tools for lifting the crankshaft and a hook-spanner, be turned out and in. The shells are fixed with a keep and the long elastic studs tightened by hydraulic tool.

Thrust bearing

The thrust bearing is of the B&W–Michell type. Primarily, it consists of a steel forged thrust shaft, a bearing support, and segments of cast iron with whitemetal. The thrust shaft is connected to the crankshaft and the intermediate shaft with fitted bolts.

The thrust shaft has a collar for transfer of the 'thrust' through the segments to the bedplate. The thrust bearing is closed against the crankcase, and it is provided with a relief valve.

Lubrication of the thrust bearing derives from the system oil of the engine. At the bottom of the bearing there is an oil sump with outlet to the oil pan.

Frame section

The frame section for the four- and five-cylinder engine consists of one part with the chain drive located aft. The chain drive is closed by the end-frame aft. For six- to twelve-cylinder engines the frame section consists of a fore and an aft part assembled at the chain drive. Each part consists of an upper and a lower frame box, mutually assembled with bolts.

The frame boxes are welded. The upper frame box is on the back of the engine provided with an inspection cover for each cylinder. The lower frame box is on the front of the engine provided with a large hinged door for each cylinder.

The guides are bolted onto the upper frame box and offer possibility for adjustment. The upper frame box is provided on the back side with a relief valve for each cylinder and on the front side with a hinged door per cylinder. A slotted pipe for cooling oil outlet from the piston is suspended in the upper frame box.

The frame section is attached to the bedplate with bolts. The stay bolts consist of two parts assembled with a nut. To prevent transversal oscillations the assembly nut is supported. The stay bolts are tightened hydraulically.

Cylinder frame, cylinder liner and stuffing box

The cylinder frame unit is of cast iron. Together with the cylinder liner (Figure 13.15) it forms the scavenging air space and the cooling

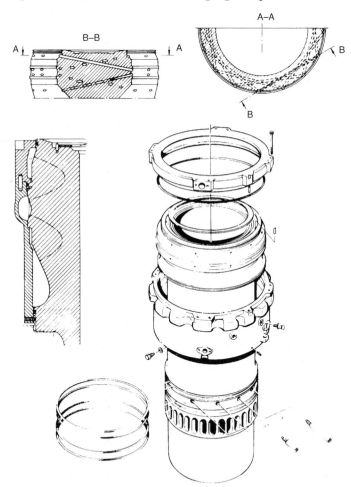

Figure 13.15 Components of L-GB cylinder liner

water space. At the chain drive there is an intermediate piece. The stay bolt pipes and the double bottom in the scavenging air space are water cooled. On the front the cylinder frame units are provided with a cleaning cover and inspection cover for scavenging ports. The cylinder frame units are mutually assembled with bolts.

Housings for roller guides, lubricators and gallery brackets are suspended on the cylinder frame unit. Further, the outside part of a telescopic pipe is fixed for supply of piston cooling oil and lubricating oil. At the bottom of the cylinder frame unit there is a piston rod stuffing box. The stuffing box is provided with sealing rings for scavenging air and oil scraper rings preventing oil from coming up into the scavenging air space.

The cylinder liner is made of alloyed cast iron and is suspended in the frame section with a low located flange. The uppermost part of the liner has drillings for cooling water and is surrounded by a cast iron cooling jacket. The cylinder liner has scavenging ports and drillings for cylinder lubrication.

Cylinder cover

The cylinder cover is made in one piece of forged steel and has drillings for cooling water. It has a central bore for the exhaust valve and bores for fuel valves, safety valve, starting valve and indicator valve (Figure 13.16).

The cylinder cover is attached to the cylinder frame with studs tightened by a hydraulic ring covering all studs.

Exhaust valve and valve gear

The exhaust valve consists of a valve housing and a valve spindle. The valve housing is of cast iron and arranged for water cooling. The housing is provided with a bottom piece of steel with Stellite welded onto the seat. The spindle is made of heat resistant steel with Stellite welded onto the seat. The housing is provided with a spindle guide. The exhaust valve housing is connected to the cylinder cover with studs and nuts tightened by hydraulic jacks. The exhaust valve is opened hydraulically and closed by a set of helical springs. The hydraulic system consists of a piston pump mounted on the roller guide housing, a high pressure pipe, and a working cylinder on the exhaust valve. The piston pump is activated by a cam on the camshaft.

Cover mounted valves

In the cylinder cover there are three fuel valves, one starting valve, one safety valve, and one indicator valve.

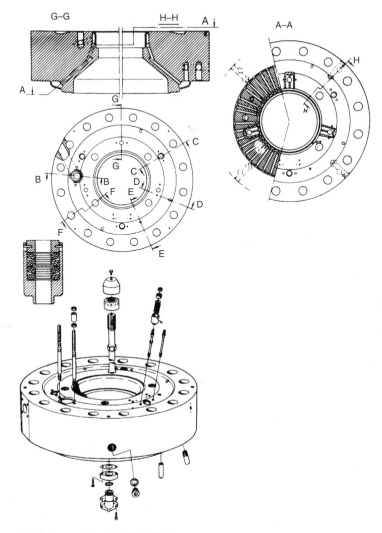

G–G   H–H   A
A–A

Figure 13.16 Components of L-GB cylinder cover

The fuel valve opening is controlled by the fuel oil pressure and it is closed by a spring. An automatic vent slide allows circulation of fuel oil through the valve and high pressure pipes and prevents the compression chamber from being filled up with fuel oil in the event of a sticking spindle in a stopped engine. Oil from venting and other drains is led away in a closed system.

The starting valve is opened by control air from the starting air distributor and closed by a spring. The safety valve is spring loaded. The indicator valve is placed near the indicator gear.

Crankshaft

The crankshaft for four- and five-cylinder engines is made in one part, and for six- to twelve-cylinder engines it is made in two parts assembled at the chain drive with fitted bolts. The crankshaft is semi-built with forged steel throws.

The crankshaft has in the aft end a flange for assembling with the thrust shaft. The crankshafts are balanced exclusively by borings in the crankpins, though in some cases supplemented by balance weight in the turning wheel.

Connecting rod

The connecting rod is of forged steel. It has a Tee-shaped base on which the crank bearing is attached with hydraulic tightened bolts and nuts with Penn-securing. The L90GBE engine has shims placed between the base and the crank bearing, as this engine type needs a smaller compression chamber because of a higher compression ratio. The top is square shaped on which the crosshead bearings are attached with hydraulic tightened studs and nuts with Penn-securing. The bearing parts are mutually assembled with bolts and nuts tightened by hydraulic jacks.

The lubrication of the crank bearing takes place through a central drilling in the connecting rod.

The crank bearing is of steel cast in two parts and lined with whitemetal. The bearing clearance is adjusted with shims. The crosshead bearings are of cast steel in two parts and provided with bearing shells.

Piston—piston rod—crosshead

The piston consists of piston crown, piston skirt and cooling insert for oil cooling (Figure 13.17). The piston crown is made of heat-resisting steel and is provided with five ring grooves which are hard-chrome plated on both lands. The piston skirt is of cast iron. The piston rings are right- and left-angle cut and of the same height.

The piston rod is of forged steel. It is fixed to the crosshead with a hydraulic tightened stud. The piston rod has a central bore which, in connection with a cooling oil pipe and the cooling insert, forms inlet and outlet for cooling oil.

The crosshead is of forged steel and is provided with steel cast guide shoes with whitemetal on the running surfaces. A bracket for oil inlet from the telescopic pipe and a bracket for oil outlet to slit pipe are mounted on the crosshead.

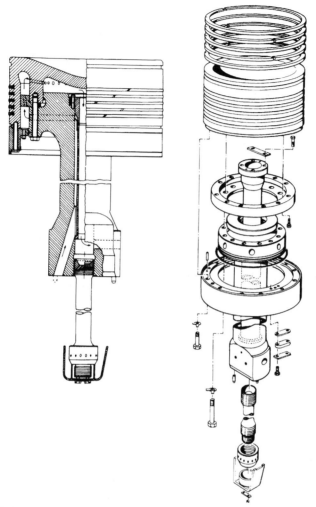

Figure 13.17 Components of L-GB piston

Fuel pump and fuel oil high pressure pipes

The fuel pump consists of a pump housing of nodular cast iron and a centrally placed pump cylinder of steel with sleeve and plunger of nitrated steel. The plunger has an oblique injection edge which will automatically give an optimum fuel injection timing. There is one pump for each cylinder. In order to prevent fuel oil from being mixed into the separate lubricating system on the camshaft the pump is provided with a sealing device.

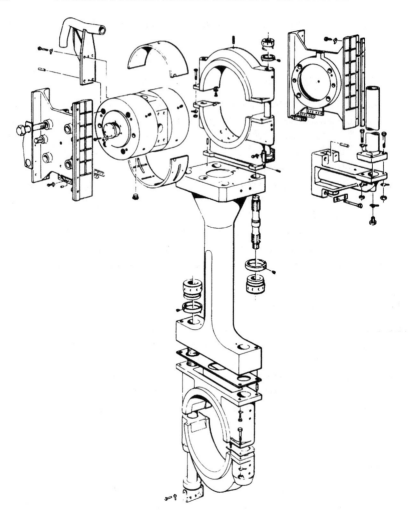

Figure 13.18 Components of L-GB connecting rod and crosshead

The pump gear is activated by the fuel cam, and the injected volume is controlled by turning the plunger by a toothed bar connected to the regulation mechanism. Adjustment of the pump lead is made with shims between top cover and pump housing.

The fuel pump is provided with a pneumatic lifting device: this can, during normal operation and during turning, lift the roller guide roller free of the cam.

The fuel oil high pressure pipes have protecting hoses. The fuel oil system is provided with a device which, through the pneumatic lifting tool, disconnects the pump in case of leakage from the high pressure pipes.

Camshaft and cams

The camshaft is divided into sections for each cylinder. The individual sections consist of a shaft piece with one exhaust cam, one fuel cam, one indicator cam, and coupling parts. The exhaust and fuel cams are of steel with a hardened roller race, and are shrunk on the shaft. They can be adjusted and dismounted hydraulically.

The indicator cams, which are of cast iron, are bolted onto the shaft. The coupling parts are shrunk on the shaft and can be adjusted and dismounted hydraulically.

The camshaft is located in the housing for the roller guide. The camshaft bearings consist of two mutually interchangeable bearing shells, which are mounted in hydraulically tightened casings.

Chain drive and reversing

The camshaft is driven from the crankshaft by two $4^1/_2$-inch chains. The chain drive is provided with a chain tightener and guidebars support the long chain strands. The camshaft is provided with a hydraulically actuated reversing gear turning the camshaft to the position corresponding to the direction of rotation of the crankshaft.

Starting air distributor, governor and cylinder lubricators are driven by separate chain from the intermediate wheel.

Moment compensator

Four-, five- and six-cylinder engines are prepared for moment compensators, which can be fixed to the fore and aft ends of the frame section and are driven by the camshaft through flexible couplings. The moment compensator will reduce the second order external moments to a level between a quarter of the original figure and zero.

Governor

The engine rev/min is controlled by a hydraulic governor. For amplification of the governor's signal to the fuel pump there is a hydraulic amplifier. The hydraulic pressure for the amplifier is delivered by the camshaft lubricating oil system.

Cylinder lubricators

The cylinder lubricators are mounted on the cylinder frame, one per cylinder, and interconnected with shaft pieces. The lubricators have built-in adjustment of the oil quantity. They are of the 'Sight Feed Lubricator' type and each lubricating point has a glass. The oil is led to the lubricator through a pipe system from an elevated tank. A heating element rated at 75 watt is built into the lubricator.

Manoeuvring system (without bridge control)

The engine is provided with a pneumatic manoeuvring and fuel oil regulating system which transmits orders from the separate manoeuvring console to the engine.

By means of the regulating system it is possible to start, stop, reverse and control the engine. The speed control handle in the manoeuvring console activates a control valve which gives a pneumatic speed-setting signal to the governor dependent on the desired number of revolutions. The start and stop functions are controlled pneumatically. At a shut-down function the fuel pumps are moved to zero position independent of the speed control handle.

Reversing of the engine is controlled pneumatically through the engine telegraph and is effected via the telegraph handle.

Reversing takes place by moving the telegraph handle from 'Ahead' to 'Astern' and by moving the speed control handle from 'stop' to start' position. Control air then moves the starting air distributor and, through the pressurizer, the reversing mechanism to the 'Astern' position.

Turning gear and turning wheel

The turning wheel has cylindrical teeth and is fitted to the thrust shaft; it is driven by a pinion on the terminal shaft of the turning gear which is mounted on the bedplate. The turning gear is driven by an electric motor with built-in gear and brake. Further, the gear is provided with a blocking device that prevents the main engine from starting when the turning gear is engaged. Engagement and disengagement of the turning gear is executed by axial transfer of the pinion.

Gallery brackets

The engine is provided with gallery brackets placed at such a height that the best possible overhaul and inspection conditions are obtained. The main pipes of the engine are suspended in the gallery brackets.

A crane beam is placed on the brackets below centre gallery manoeuvring side.

Scavenging air system

The air intake to the turbocharger takes place directly from the engine-room through the intake silencer of the turbocharger. From the turbocharger the air is led via charging air pipe, air cooler and scavenging air pipe to the scavenging ports of the cylinder liner. The charging air pipe between turbocharger and air cooler is provided with a compensator and insulated.

Exhaust turbocharger

The engine is as standard arranged with MAN or BBC turbochargers. The turbochargers are provided with a connection for Disatac electronic tachometers, and prepared for signal equipment, to indicate excessive vibration of the turbochargers. For water cleaning of the turbine blades and the nozzle ring during operation, the engine is provided with connecting branches on the exhaust receiver in front of the protection grid.

Exhaust gas system

From the exhaust valves the gas is led to the exhaust gas receiver where the fluctuating pressure will be equalized and the gas led further on to the turbochargers with a constant pressure. After the turbochargers, the gas is led through an outlet pipe and out in the exhaust pipe system.

The exhaust gas receiver is made in one piece for every cylinder and connected to compensators. Between the receiver and the exhaust valves and between the receiver and the turbocharger there are also inserted compensators.

For quick assembling and dismantling of the joints between the exhaust gas receiver and the exhaust valves, a clamping band is fitted. The exhaust gas receiver and exhaust pipe are provided with insulation covered by a galvanized steel plate.

Between the exhaust gas receiver and each turbocharger there is a protection grid.

Auxiliary blower

The engine is provided with two electrically driven blowers which are mounted in each end of the scavenging air receiver as standard. The suction sides of the blowers are connected to the pipes from the air coolers, and the non-return valves on the top of the outlet pipes from the air coolers are closed as long as the auxiliary blowers can give a supplement to the scavenging air pressure.

The auxiliary blowers will start operating before the engine is started and will ensure complete scavenging of the cylinders in the starting phase, which gives the best conditions for a safe start.

During operation of the engine the auxiliary blowers will start automatically every time the engine load is reduced to about 30–40 per cent, and they will continue operating until the load is again increased to over approximately 40–50 per cent.

In cases when one of the auxiliary blowers is out of service, the other auxiliary blower will automatically function correctly in the system, without any manual readjustment of the valves being necessary. This is obtained by automatically working non-return valves in the suction pipe of the blowers.

Starting air system

The starting air system contains a main starting valve (two ball valves with actuators), a non-return valve, a starting air distributor and starting valves. The main starting valve is combined with the manoeuvring system which controls start and 'slow turning' of the engine. The 'slow turning' function is actuated manually from the manoeuvring stand.

The starting air distributor regulates the control air to the starting valves so that these supply the engine with starting air in the firing order.

The starting air distributor has one set of starting cams for 'Ahead' and 'Astern' respectively, and one control valve for each cylinder.

14 Doxford low speed engines

The last British-designed low speed two-stroke engine—the distinctive Doxford opposed-piston design—was withdrawn from production in 1980 but a few of the J-type remain in service and merit description. In its final years the company also designed and produced the unusual three-cylinder 58JS3C model, which developed 4050 kW at 220 rev/min and was specified to power several small containerships. The 58JS3C design (Figure 14.16) was based on the J-type but with refinements addressing the higher rotational speed and relatively short piston stroke.

Figure 14.1 A 76J4 engine on the testbed

465

DOXFORD J-TYPE

Doxford J-type engines were built in long and short stroke versions in bore sizes of 580, 670 and 760 mm, and with three to nine cylinders delivering up to around 20 000 kW (Table 14.1). The engine is a single-acting two-stroke opposed-piston type with each cylinder having two pistons which move in opposite directions from a central combustion chamber.

The pistons in each cylinder are connected to a three-throw section of the crankshaft, the lower piston being coupled to the centre throw by a single connecting rod, crosshead and piston rod. Each pair of side cranks is connected to the upper piston by two connecting rods, crossheads and side rods. As the pistons move towards each other air is compressed in the cylinder and, shortly before reaching the point of minimum volume between the pistons (inner dead centre), fuel at

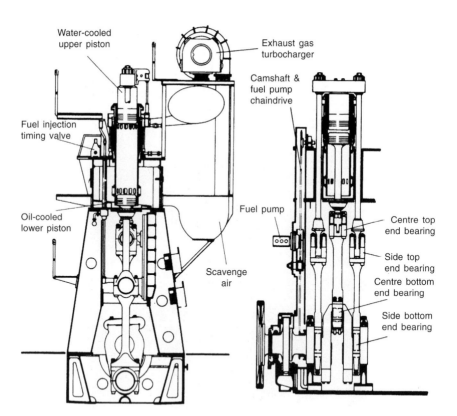

Figure 14.2 Doxford J-type configuration

Table 14.1

Engine type	58J4	67J4	67J5	67J6	76J3	76J4	76J5	76J6	76J7	76J8	76J9
Number of cylinders	4	4	5	6	3	4	5	6	7	8	9
Cylinder bore (mm)	580	670	670	670	760	760	760	760	760	760	760
Stroke upper piston (mm)	480	500	500	500	520	520	520	520	520	520	520
Stroke lower piston (mm)	1 370	1 640	1 640	1 640	1 660	1 660	1 660	1 660	1 660	1 660	1 660
Combined piston stroke (mm)	1 850	2 140	2 140	2 140	2 180	2 180	2 180	2 180	2 180	2 180	2 180
Number of turbochargers	1	1	2	2	1	1	2	2	3	3	4
Weight, (excluding oil and water) tonnes	175	270	320	370	270	335	400	470	540	600	680
Ratings											
Max. continuous rating (MCR) bhp metric	7 500	9 000	11 000	13 500	9 000	12 000	15 000	18 000	21 000	24 000	27 000
Engine speed, rev/min	160	127	127	127	123	123	123	123	123	123	123
Brake mean effective pressure (bmep) bars	10.57	10.34	10.34	10.34	10.81	10.81	10.81	10.81	10.81	10.81	10.81
Recommended max service power (90% MCR)	6 750	8 100	9 900	12 200	8 100	10 800	13 500	16 200	18 900	21 600	24 300
Engine speed r.p.m.	155	123	123	123	119	119	119	119	119	119	119
Brake mean effective pressure (bmep) bars	9.82	9.61	9.61	9.61	10.06	10.06	10.06	10.06	10.06	10.06	10.06

Consumption

Fuel consumption 150 to 153 g/bhp—h with H.V. fuel of 10 000 K cal/Kg

Cylinder lubricating oil 0.45 g/bhp—h

high pressure is injected through the injector nozzles into the combustion space.

During the first stages of combustion the pressure in the cylinder continues to rise until a maximum value is reached soon after the pistons begin to move apart. After combustion is completed the hot gases continue to expand, thereby forcing the pistons apart until the exhaust ports in the upper liner are uncovered by the upper piston. As the exhaust ports open, the hot gases in the cylinder, now at a reduced pressure, are discharged to the turbine of the turbochargers, so causing the pressure in the cylinder to drop to a level just below that of the scavenge air. At this point the air inlet ports in the lower liner are uncovered by the lower piston, so allowing air under pressure, which is delivered to the scavenge space by the turbocharger, to flow through the cylinder and expel the remaining burnt gases (Figure 14.3(a)).

During the inward compression stroke of the pistons the air inlet ports are closed just before the exhaust ports. The air in the cylinder is then compressed and the cycle repeated. The opening and closing of the exhaust and air inlet ports are shown diagrammatically in the typical timing diagram (Figure 14.3(b)). This diagram is constructed with reference to the inner dead centre position, the unsymmetrical timing of the ports being due to the fact that the side cranks are at an angle of more than 180° from the centre crank. The angle of lead of the side cranks varies with the size of engine.

The upper piston of the engine requires its own running gear and crank throws. The advantages obtained with this concept are considerable. For equal mean indicated pressure, mean piston speed and cylinder bore, the Doxford opposed-piston engine will develop 30–40 per cent higher power per cylinder than a single-piston engine. The first order inertia force from the lower piston is balanced against the corresponding force from the upper pistons and a well-balanced engine is obtained. All the forces from the upper piston are transferred through the running gear.

Long tension bolts, as used in single-piston engines to transfer the forces from the cylinder covers to the bedplate, are not required in the opposed-piston engine. The engine structure is therefore simple and relatively free from stresses. No valves are required in the scavenge-exhaust system and the scavenge efficiency of the cylinders is high. The flow areas through the exhaust ports are substantial, and fouling, which inevitably takes place in service, has little effect. The large port areas also make the engine well suited for turbocharging.

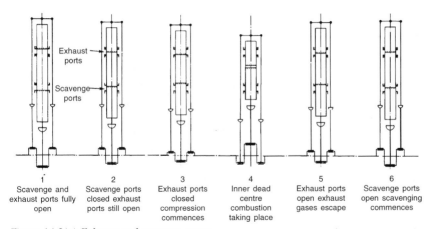

1	2	3	4	5	6
Scavenge and exhaust ports fully open	Scavenge ports closed exhaust ports still open	Exhaust ports closed compression commences	Inner dead centre combustion taking place	Exhaust ports open exhaust gases escape	Scavenge ports open scavenging commences

Figure 14.3(a) Exhaust and scavenge events

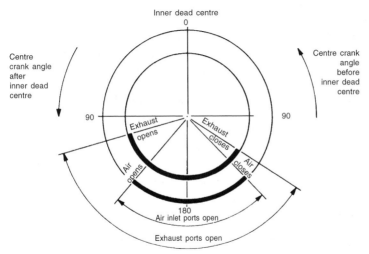

Figure 14.3(b) Engine type 76J4C port timing diagram

Engine construction

The bedplate is built up of two longitudinal box girders extending over the full length of the engine. The transverse girders, which incorporate the main bearing housings, are welded to these longitudinal members. A semi-circular whitemetal lined steel shell forms the lower half of each main bearing and is held in place by a keep secured by studs to the bedplate.

The entablature is also a welded steel box construction, arranged to carry the cylinders which are bolted to the upper face. It is bolted to the tops of the columns and to the crosshead guides. The entablature

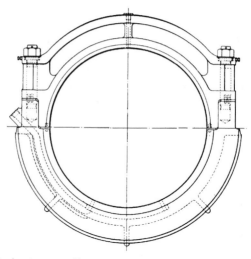

Figure 14.4 Main bearing assembly

forms the air receiver from which the cylinders are supplied with air and its volume is supplemented by the air receiver on the back of the engine, the top face of which forms the back platform. The intercoolers are mounted on this receiver and the air deliveries from the turbochargers are connected to these coolers. Alternatively, in the case of engines with end-mounted turbochargers, the air from the turbocharger is supplied directly to the entablature after passing through the end-mounted intercoolers.

Crankshaft

Each cylinder section of the crankshaft is made up of three throws, the two side cranks being connected to the upper piston and the centre crank to the lower piston. The side crankwebs are circular and form the main bearing journals. Each centre crank is made of an integral steel casting or forging for semi-built shafts. For fully built shafts these units are made of two slabs shrunk onto the centre pin. The centre cranks are shrunk onto the side pins.

   Lubricating oil is fed to the main bearings and through holes in the crankshaft to the side bottom end bearings. At the after end, the thrust shaft is bolted to the crankshaft. The thrust bearing of the tilting pad type is housed in the thrust block, which is bolted to the end of the bedplate. A turning wheel, which engages with the pinion of the electrically driven turning gear, is bolted to the after coupling of the thrust shaft.

Cylinders

The cylinder liner (Figure 14.5) is a one-piece casting and incorporates the scavenge and exhaust ports in the lower and upper sections respectively. A special wear resistant cast iron is used as cylinder liner material. An exhaust belt round the exhaust ports and a water cooling jacket are clamped to the top and bottom faces respectively of the combustion section of the liner by long studs. The cylinders are secured

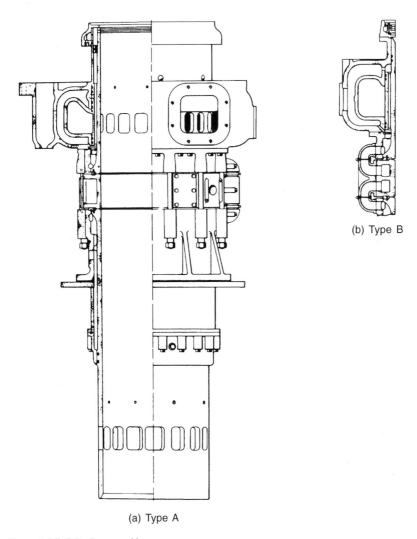

(b) Type B

(a) Type A

Figure 14.5 Cylinder assembly

to the entablature by means of a flange on the jacket passing over
studs and secured with nuts.

The cylinders are water cooled. The cooling water enters the jacket
(Figure 14.6) and circulates round the top of the lower part of the
liner. It then passes through holes drilled at an angle through the
liner round the combustion chamber. Above the combustion chamber
some of the water is taken directly into the water jacket of the exhaust
belt and the remainder passed through the exhaust bars first and

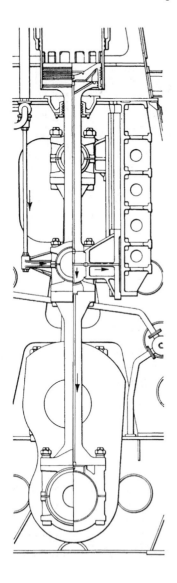

Figure 14.6 Lubricating and cooling oil circuits

then into the exhaust belt water jacket. Finally, the water is returned to the cooling water tank through sight flow hoppers.

Lubricator injectors are provided for the supply of lubricating oil to both the upper and lower cylinder liners (Figure 14.7). The lubricating points are equally distributed around the liners and supplied with oil from timed distributor-type lubricators.

The central combustion chamber section of the cylinder liner supports the fuel injectors, the air starting valve, the relief valve and the cylinder indicator connection.

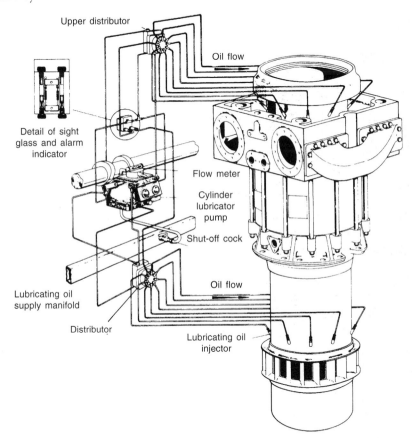

Figure 14.7 Cylinder lubrication system

The exhaust belts convey the exhaust gas from the cylinders to cast iron exhaust pipes which are connected to the turbochargers. Flexible expansion pieces are fitted between the exhaust belts and the exhaust pipes as well as between the exhaust pipes and the turbochargers. A grid is fitted at the inlet to prevent broken piston rings from damaging the turbocharger. A scraper ring carrier at the upper end of each liner prevents the passage of the exhaust gases past the upper piston skirt and scrapes the lubricating oil downwards.

Pistons

The upper and lower piston heads are identical. They have dish-shaped crowns to give a spherical form to the combustion space and are designed to be free to expand without causing undue stresses. The piston heads are attached to the rods by studs in the underside of the piston crowns so that the gas loads are transmitted directly to the piston rods. The under faces of the piston heads are machined to form cooling spaces between the piston heads and the upper faces of the piston rods, the cooling medium being supplied and returned through drilled holes in the rods.

A cast iron ring is fitted around each piston head to form a bearing surface. Four compression rings are fitted into grooves above this bearing ring and there is one ring below it to act both as a compression ring and as a lubrication oil spreader ring. The ring grooves are chromium plated to minimize wear of the surfaces in contact with the rings.

The lower piston rods have spare palms formed on their lower ends for bolting to the crossheads. The upper ends are cylindrical and form the faces to which the piston heads are bolted. Oil for cooling the lower piston is transmitted through the centre crosshead and up the piston rod to the piston head and is returned in a similar way (Figure 14.8). The centre crosshead bracket also carries the telescopic pipes for the piston cooling oil and lubricating oil to the centre connecting rod bearings.

Glands attached to the underside of the entablature, through which the lower piston rods pass, form a seal between the crankcase and the scavenge air space. The glands contain a number of segmental rings held to the body of the piston rod by garter springs. These rings are so arranged that the lower ones scrape oil from the rod back into the crankcase, while the upper ones provide an air seal and also prevent the passage of any products of combustion from the cylinders into the crankcase.

The upper piston rods are bolted to the upper piston heads and to the transverse beams which carry the loads from the pistons to the side rods. A cast iron skirt is provided around each upper piston rod to shield the exhaust ports and prevent the exhaust gases passing back into the open end of the cylinder.

Water is used as the cooling medium for the upper pistons, this

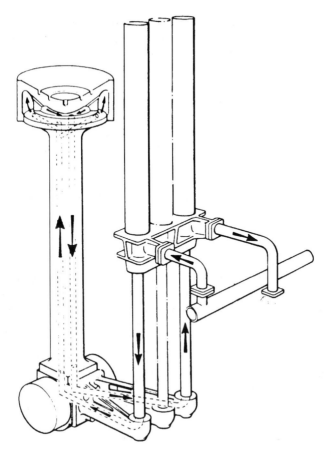

Figure 14.8 Lower piston cooling system

being conveyed to and from the piston heads through holes in the upper piston rods. Brackets attached to the transverse beams carry telescopic pipes for the cooling water.

Connecting rods

The centre connecting rod has a palm end at the lower end to which the bottom end bearing keeps are bolted, whereas the upper end of the rod has an integral continuous lower half keep to which the upper half bearing keeps are bolted. The side connecting rod is formed with palm ends at both ends of the rod to which the top and bottom end bearings are bolted. The centre connecting rod top end bearings consist of continuous whitemetal lined shells for the lower (loaded) halves, whereas two bearing keeps over the ends of the crosshead pins form the upper halves. The side connecting rod bottom end bearings are whitemetal lined and are supplied with lubricating oil from the main bearings through holes passing up the connecting rods to the side top end bearings. The centre connecting rods also have cast steel whitemetal lined bottom end bearings which are supplied with oil through holes in the rods from the top end bearings. The side top end bearings and centre top end bearings are fitted with whitemetal lined thin-shell bearings.

The centre crosshead pins are made of nitriding steel and hardened by this process. At the top the pins are bolted to the palm end of the piston rods with the crosshead brackets sandwiched in between. Also, two long studs pass through the crosshead pins horizontally and secure the pins to the brackets at the back (Figure 14.9). The telescopic pipes for lubricating and cooling oil are supported by the crosshead brackets at the front of the piston rods and the guide shoes are bolted to these brackets at the back. Each side crosshead (Figure 14.10) is made up of a steel casting into which is shrunk the crosshead pin, the side rod being screwed into the top of the casting.

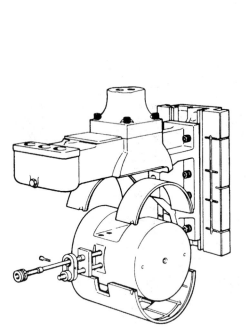

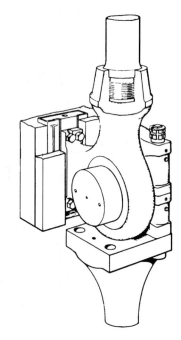

Figure 14.9 Centre crosshead assembly   *Figure 14.10 Side crosshead assembly*

Camshaft

The camshaft is mounted on the top of the entablature and is driven through a roller chain from the crankshaft. It operates timing valves for controlling the fuel injection to each cylinder, cylinder lubricators and starting air distributors for controlling the starting air supply to the cylinders; it also drives the governor through a step-up gear. A drive to the fuel pump is also taken from the camshaft driving chain.

Fuel injection

The fuel injection system operates on the common rail principle in which timing valves, operated by cams on the camshaft, control the injection of fuel from a high pressure manifold through spring-loaded injectors to the cylinders (Figure 14.14). Fuel is delivered to the high pressure main by the multi-plunger pump fitted at the after end of the engine (Figure 14.11), the pressure being maintained at the desired value by means of a pneumatically operated spill valve.

Two fuel injectors are fitted to each cylinder and these open when the timing valves are operated by the cams on the camshaft. The duration of opening of the timing valves and hence the period of

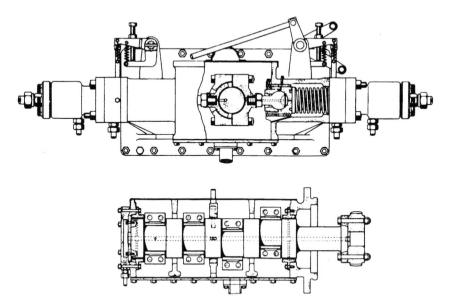

Figure 14.11 High pressure fuel pump

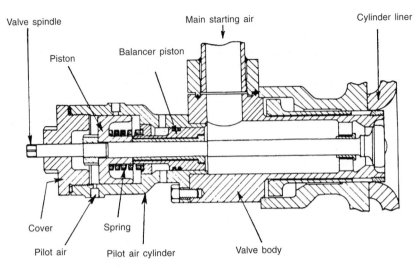

Figure 14.12 Air starting valve

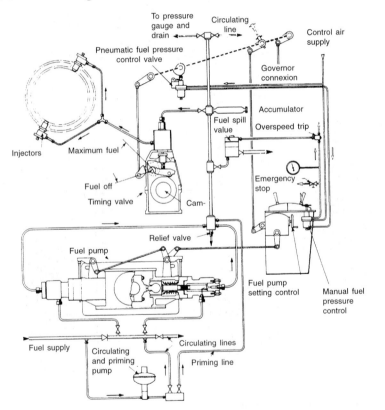

Figure 14.13 Air starting system

injection of fuel into the cylinders is controlled by the governor. Either a centrifugal governor with a hydraulically operated output or an electronic governor with a pneumatic actuator can be fitted. In the former case the governor input is set pneumatically from the control station; in the latter case it is set electrically. In this way the speed of the engine can be adjusted as required. Means of adjusting the setting of the timing valves by direct mechanical linkage are also provided. The timing of injection is determined by the positions of the timing valve cams. Only one cam is required for each cylinder for both ahead and astern running.

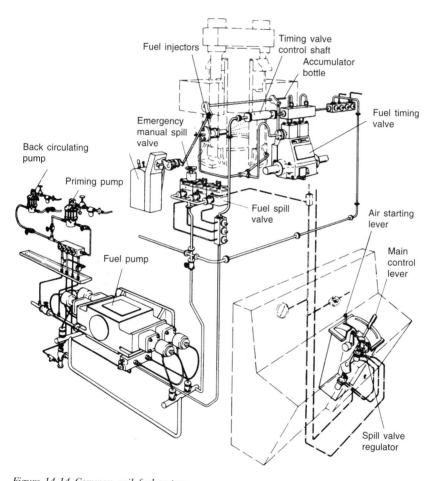

Figure 14.14 Common rail fuel system

Turbocharging

The last J-type engines were turbocharged on the constant pressure system following a changeover from the original impulse charging system. Three- and four-cylinder engines have only one turbocharger, mounted at the forward or aft end, while two or three chargers have been fitted to seven-, eight- and nine-cylinder engines. Between each turbocharger and the engine entablature is a finned-tubed seawater-cooled aftercooler. An electric auxiliary blower is provided for slow running or emergency duties.

Starting of the engine is by compressed air, which is admitted to the cylinders through pneumatically operated valves (Figures 14.12 and

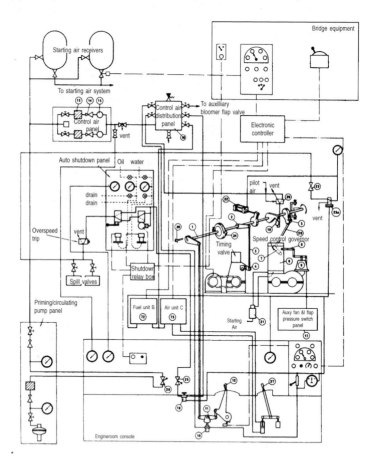

Figure 14.15 Diagrammatic arrangement of speed controls and other pneumatic controls, including remote control

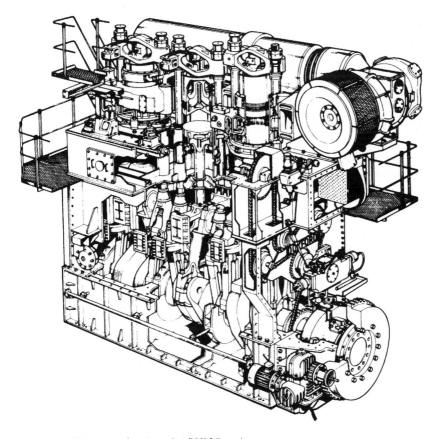

Figure 14.16 Cutaway drawing of a 58JS3C engine

14.13). The valves are controlled by a rotary air distributor driven
from the camshaft to govern the timing and duration of opening for
starting the engine ahead or astern. Levers for starting and speed
control are grouped in a control box near the engine; when bridge
controls are fitted pneumatic valves to control engine movements are
actuated by moving the bridge telegraph (Figure 14.15).   C.T.W.

15 MAN low speed engines

Low speed marine diesel engines built by Maschinenfabriek Augsburg-Nurnberg (MAN) were of single-acting two-stroke crosshead design exploiting loop scavenging. The programme was phased out after the German company's acquisition of Danish-based rival Burmeister & Wain in 1980 and replaced by the MAN B&W MC series (Chapter 10). MAN's KSZ design (Figure 15.1) was introduced in the mid-1960s and followed progressively by the KSZ-A, KSZ-B and KSZ-C types, each embodying refinements to promote greater reliability and fuel economy. The KSZ-C and KSZ-CL types represented the last examples of MAN loop-scavenged engine design technology.

MAN engines had some notable features to promote simplicity of design, ease of maintenance and low specific fuel consumption, not least the pioneering application—albeit only on the testbed—of electronic fuel injection to a marine diesel engine.

MAN two-stroke engines employ constant pressure turbocharging (Figure 15.2) and are scavenged according to the loop scavenging system. The cylinder exhaust ports are located above the scavenging ports on the same side of the liner, occupying approximately one-half of its circumference. The scavenging air is admitted through the scavenge ports, passing across the piston crown and ascending along the opposite wall to the cylinder cover, where its flow is reversed. The air then descends along the wall in which the ports are located, expelling the exhaust gases into the exhaust manifold. The piston closes the scavenging ports and then, on its further upward travel, also closes the exhaust ports compressing the charge of pure air in the cylinder.

When the KSZ series was introduced it followed the traditional two-stroke engine construction technique of cylinder blocks mounted on 'A' frames which sat on a cast bedplate, the three structural items held by a series of long tie-rods. However, this method of erection was abandoned when MAN introduced the KSZ-B type with the so-called box-type construction which was retained for the later KSZ-C and KEZ-B types. (This latter type was identical to the KSZ models but featured electronic fuel injection—described below—instead of the traditional mechanical fuel pump arrangement; however, it never

482

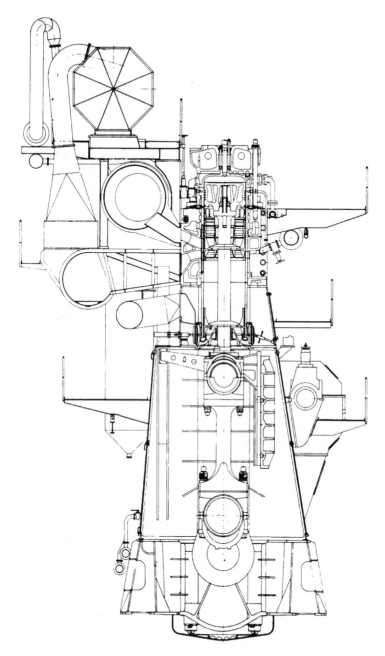

Figure 15.1 Cross-section of KSZ 90/160 engine with older constant pressure parallel injector supercharging system

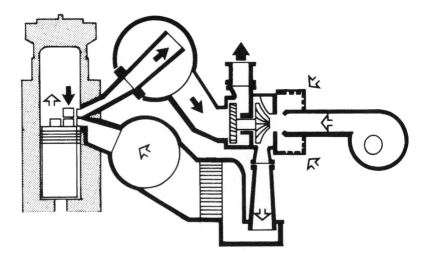

Figure 15.2 Constant pressure turbocharging of KSZ-B and KSZ C/CL engines

Figure 15.3 K8SZ 90/160 B/BL engine on the testbed

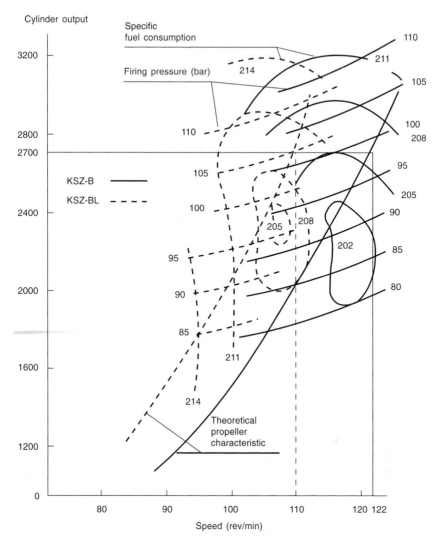

Figure 15.4 Performance curves for the K8SZ 90/160 engine

entered production.) In general design details the KSZ-B and KSZ-C
engines are basically similar but for the longer piston strokes of the
latter series and other refinements to achieve a considerable reduction
in the specific fuel consumption.

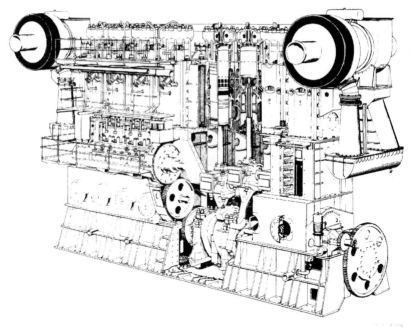

Figure 15.5 Cutaway drawing of K8SZ 90/160B/BL engine

DESIGN PARTICULARS

MAN's KSZ-B/BL and KSZ-C/CL ranges are both of identical design, but with the later model giving a 10 per cent reduction in the speed at the same horsepower rating, this being achieved by raising the cylinder mean effective pressure. Minor design changes are only where the increase in piston stroke made this necessary.

Engine structure

The significant constructional characteristic of the KSZ-B and -C type engines is the box-shaped longitudinal girders comprising a deep-section single-walled bedplate, a frame which is either in one piece or split (depending on the number of cylinders) on cast, high cylinder jackets. In the largest (900 mm bore) engine the frame box is split horizontally (Figure 15.6) while smaller engines of 700 and 520 mm bore have the frame as a single-piece steel fabrication. For engines of seven to 12 cylinders the frames of engines of the two largest bore sizes are vertically split, while smaller engines with the camshaft gear drive located at the coupling end have no vertical split in the frame structure. The cylinder jackets are cast in either two or three units.

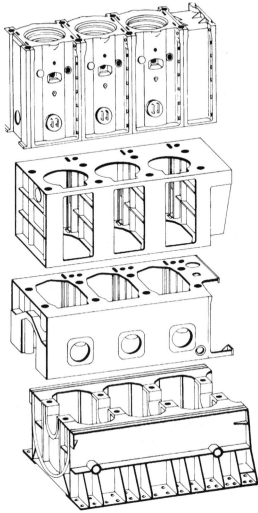

Figure 15.6 Construction method

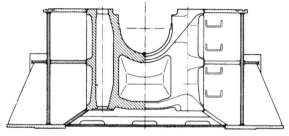

Figure 15.7 Bedplate design

The box frame design allows a high degree of stiffness and contributes to a gradual reduction in the deformations induced by the ship's double bottom and transmitted to the bedplate. These stiffness characteristics reduce the risk of uncontrolled transverse loading on the crankshaft main bearings, and cylinder liners at their clamping points are only slightly deformed by the outside structure. This contributes to keeping liner wear rates as low as possible.

Running gear

The crankshafts of the larger engines are of built-up type with forged steel crank throws pressed on to forged steel journals. Crankshafts for the smallest (520 mm bore) engine have one-piece forged shafts up to the highest number of cylinders. The main bearings are of the so-called block bearing type comprising three-metal shells which can be replaced without subsequent fitting. The crankpin bearing is also of three-metal type with a comparatively thick whitemetal layer.

The crosshead (Figure 15.8) features hydrostatic lubrication, a principle which ensures an adequate oil film between the crosshead pin and the single whitemetal lined bearing shell. The high oil pressure required for this purpose is generated by double plunger high pressure lubricators which are directly driven by linkages from the connecting rod to force oil under pressure into the lower crosshead bearing shell.

Camshaft drive

The box-shape construction of the frame forms a deformation-resistant casing to accommodate the gear train of the camshaft drive. All engines have a gear train with only one intermediate wheel. The wear pattern and backlash on the intermediate wheel can be adjusted by draw-in and thrust bolts which act upon the frame at the bearing points of the intermediate wheel axle. The crankshaft gear wheel is clamped directly to the shaft as a two-piece component.

Combustion chamber components

In order to control thermal stresses, thin-wall intensively cooled components are used for the combustion chamber parts (Figure 15.9). The piston crown is of high-grade heat-resistant steel and incorporates the MAN jet-cooled honeycomb system. An internal insert in the piston crown serves only to direct the cooling water flow and, in particular, to generate water jets which enter at the edge of each honeycomb element, thereby intensifying the conventional 'cocktail shaker' effect.

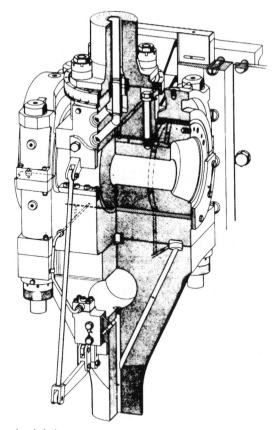

Figure 15.8 Crosshead design

The structural arrangement of the KSZ-C piston has the steel crown, insert, and piston rod and the cast iron skirt all held together by a single row of waisted studs. The fresh water for piston cooling (Figure 15.12) is fed into the piston rod with return through the co-axial bores to telescopic pipes.

The cylinder cover comprises a stiff grey cast iron cap fitted with a thin-walled forged heat-resistant steel bottom. The pre-tension between the cap and bottom, which is necessary to prevent micromotion, is achieved by an inner row of studs. An outer row of larger studs secures the cover to the cylinder block.

The cylinder liners are short single-piece units with intensive cooling of the upper part achieved by bore cooling of the liner flange (Figure 15.14). The vertical lands between the exhaust ports are also water cooled; the combined techniques ensure that sufficient heat is removed from the piston rings, the upper portion of the cylinder liner and the

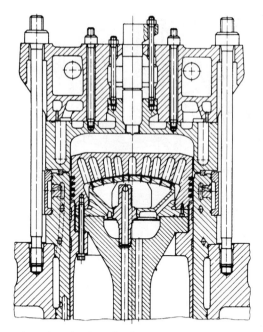

Figure 15.9 Cross-section of combustion chamber

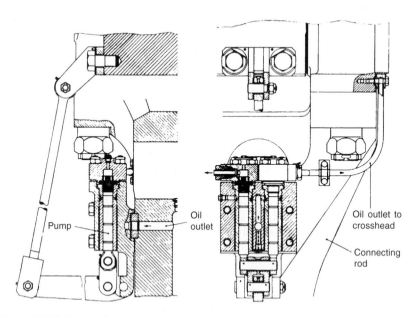

Figure 15.10 Crosshead oil pump

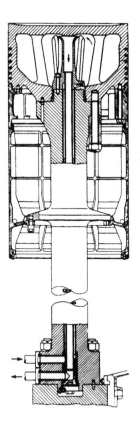

Figure 15.11 Piston and piston rod

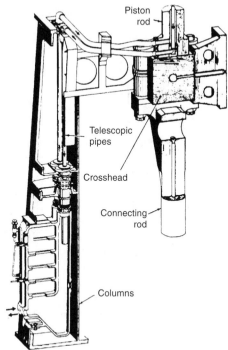

Figure 15.12 Piston cooling arrangement

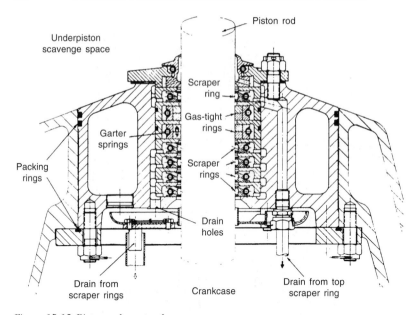

Figure 15.13 Piston rod scraper box

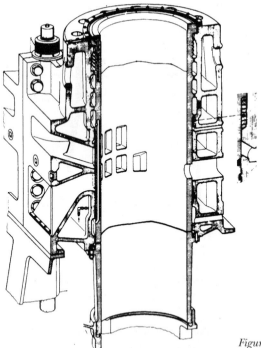

Figure 15.14 Cylinder liner and jacket

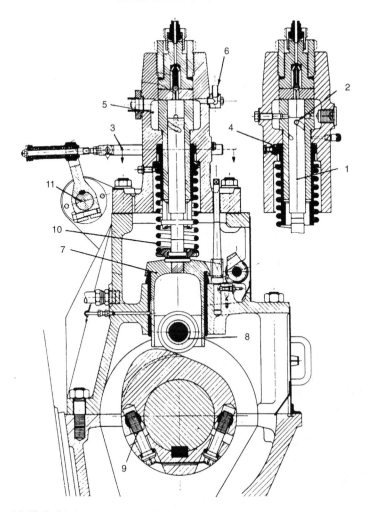

Figure 15.15 Fuel injector pump

port lands to obtain a stable lubricating oil film, to reduce the wear rate and not to interfere with ring performance in way of the porting.

Turbocharging

An important feature of MAN engines is the use of the space beneath some of the pistons to supply charge air to assist the turbocharger at low loads and particularly on starting. This feature, known as 'injector drive turbocharging', was introduced on the KSZ-type engines but was replaced by a system of electrically driven blowers for the KSZ-A and subsequent engine models. These blowers cut in and out as a function of scavenge air pressure and usually remain in operation until the engine reaches 50 per cent load. The blowers are placed ahead of the turbocharger and need to be in operation before the engine will start.

Turbocharging is based on the constant pressure system and embodies low flow resistance air and exhaust piping with suitable diffusers between the ports and manifolds. For engine outputs up to 8000 kW only one turbocharger is used. All engines above this power require two chargers, one at each end of the engine at the cylinder head level, though turbochargers have been mounted on special baseframes together with the auxiliary blower adjacent to the engine.

ELECTRONIC FUEL INJECTION

A system developed for MAN engines—though never applied to production models—was electronically controlled fuel injection (Figure 15.16). The 'heart' of the electronic injection system was a microprocessor to which the actual engine speed and crankshaft position were transmitted as output signals. When the desired speed is compared with the actual speed, the processor automatically retrieves the values of injection timing and pressure for any load under normal operating conditions. In the event of deviations from normal operating conditions additional signals are transmitted to the microprocessor, either manually or by appropriate temperature and pressure sensors. The reported result was an engine operating at optimum injection pressure and timing at all times. Adjustments to the system during operation were possible. Corrections could be made for: adapting the injection system to varying conditions, such as ballast voyages and bad weather; matching the engine to different fuels, ignition lag and cetane numbers; readjusting for operation in the tropics or in winter; controlling the injection characteristics of individual cylinders, such as when running in; and reducing the minimum slow running speed (for example, during canal transits).

Injection system components

The most noticeable feature of a KEZ type engine with electronic fuel injection is the absence of a camshaft and individual fuel pumps. The fuel is delivered to a high pressure accumulator by normal fuel injection plunger pumps driven by gears and cams from the crankshaft. These pumps are located above the thrust bearing and, depending on the size and cylinder number of the engine, two or more are used. To

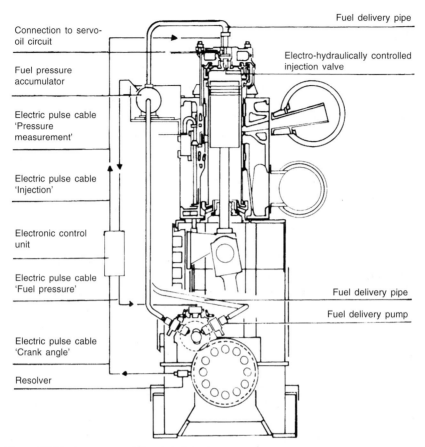

Figure 15.16 Components of electronic injection system

safeguard against failure of a pump or a pressure pipe, two independent fuel delivery pipes are placed between the high pressure pumps and the accumulator.

The high pressure accumulator generates a buffer volume to prevent the pressure in the injection valve from dropping below an acceptable limit when the fuel is withdrawn during injection into the cylinder. For the electronically controlled injection system metering of the fuel volume for one cylinder is not effected by limiting the delivery volume per stroke of the injection pump plunger, as in conventional systems, but by limiting the duration of injection. This is initiated by an electronic controller: a low voltage signal which controls the opening and closing action of an electro-hydraulic servo valve. The electro-hydraulic valve is actuated by oil under pressure from two pumps and an air cushion-type accumulator.

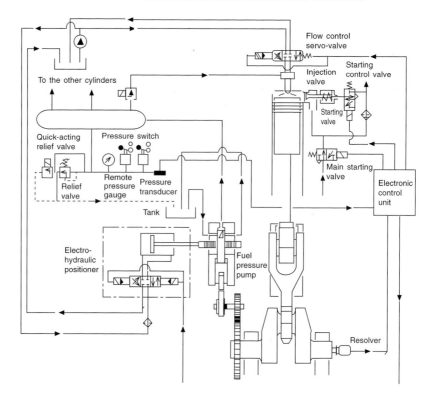

Figure 15.17 Schematic arrangement of electronic injection systems

Sequential control ensures that, according to the engine firing sequence, the correct engine cylinder is selected in the proper crank angle range and that the electronically controlled injection valve is actuated at the exact point of injection.

For engine starting and reversing the correct injection timing is linked with the electronically controlled starting air valves. Since electronically controlled starting valves have no time delay between the switching pulse and opening of the air valve, and since there are no flow losses due to the elimination of the control air pipes, the starting air requirement is reduced in comparison with the former conventional air start system.

The constant pressure maintained during fuel injection with the electronically controlled fuel injection system, and the fact that the injection pressure can easily be matched to each load condition, foster lower fuel consumption rates over the entire load range. It also means satisfactory operation at very slow speeds, which contributes favourably to the manoeuvrability of a vessel; a K3EZ 50/105 C/CL engine on the testbed was operated at a dead slow speed of 30 rev/min, which had never been achieved before. The full load speed is 183 rev/min.

C.T.W

Electronic fuel injection has since been applied to production low speed engines by MAN B&W Diesel (Chapter 10) and Sulzer (Chapter 12).

16 Medium speed engines—introduction

New designs and upgraded versions of established models have maintained the dominance of medium speed four-stroke diesel engines in the propulsion of smaller ships as well as larger specialist tonnage such as cruise vessels, car/passenger ferries and ro-ro freight carriers. The larger bore designs can also target the mainstream cargo ship propulsion market formed by bulk carriers, containerships and tankers, competing against low speed two-stroke machinery. The growth of the fast ferry sector has benefited those medium speed enginebuilders (notably Caterpillar and Ruston) who can offer designs with sufficiently high power/weight and volume ratios, an ability to function reliably at full load for sustained periods, and attractive through-life operating costs. Medium speed engines further enjoy supremacy in the deepsea genset drive sector, challenged only in lower power installations by high speed four-stroke engines.

Significant strides have been made in improving the reliability and durability of medium speed engines in the past decade, both at the design stage and through the in-service support of advanced monitoring and diagnostic systems. Former weak points in earlier generations of medium speed engines have been eradicated in new models which have benefited from finite element method calculations in designing heavily loaded components. Designers now argue the merits of new generations of longer stroke medium speed engines with higher specific outputs allowing a smaller number of cylinders to satisfy a given power demand and foster compactness, reliability, reduced maintenance and easier servicing. Progress in fuel and lubricating oil economy is also cited, along with enhanced pier-to-pier heavy fuel burning capability and better performance flexibility throughout the load range.

Completely bore-cooled cylinder units and combustion spaces formed by liner, head and piston combine good strength and stiffness with good temperature control which are important factors in burning low quality fuel oils. Low noise and vibration levels achieved by modern medium speed engines can be reduced further by resilient mounting systems, a technology which has advanced considerably in recent years.

IMO limits on nitrogen oxides emissions in the exhaust gas can generally be met comfortably by medium speed engines using primary measures to influence the combustion process (in some cases, it is claimed, without compromising specific fuel consumption). Wärtsilä's low NOx combustion technology, for example, embraces high fuel injection pressures (up to 2000 bar) to reduce the duration of injection; a high compression ratio (16:1); a maximum cylinder pressure of up to 210 bar; and a stroke/bore ratio greater than 1.2:1. Concern over smoke emissions, particularly by cruise ship operators in sensitive environmental areas, has called for special measures from engine designers targeting that market, notably electronically-controlled common rail fuel injection and fuel-water emulsification.

Ease of inspection and overhaul—an important consideration in an era of low manning levels and faster turnarounds in port—was addressed in the latest designs by a reduced overall number of components (in some cases, 40 per cent fewer than in the preceding engine generation) achieved by integrated and modular assemblies using multi-functional components. Simplified (often plug-in or clamped) connecting and quick-acting sealing arrangements also smooth maintenance procedures. Channels for lubricating oil, cooling water, fuel and air may be incorporated in the engine block or other component castings, leaving minimal external piping in evidence. Compact and more accessible installations are achieved by integrating ancillary support equipment (such as pumps, filters, coolers and thermostats) on the engine. Lower production costs are also sought from design refinements and the wider exploitation of flexible manufacturing systems to produce components.

The cylinder unit concept is a feature of the latest four-stroke designs, allowing the head, piston, liner and connecting rod to be removed together as a complete assembly for repair, overhaul or replacement by a renovated unit onboard or ashore. This modular approach was adopted by MAN B&W for its L16/24, L21/31and L27/38 designs, by MTU for its Series 8000 engine, by Rolls-Royce for its Bergen C-series, and by Hyundai for its H21/32 design: all detailed in subsequent chapters.

Compactness and reduced weight remain the key attractions of the medium speed engine, offering ship designers the opportunity to increase the cargo capacity and lower the cost of a given newbuilding project, and the ability to achieve via reduction gearing the most efficient propeller speed. Medium speed enginebuilders can offer solutions ranging from single-engine plants for small cargo vessels to multi-engine/twin-screw installations for the most powerful passenger ships, based on mechanical (geared) or electrical transmissions (see Chapter 6). Multi-engine configurations promote plant availability and operational flexibility, allowing the number of prime movers engaged at any time to match the service schedule. The convenient direct drive of alternators and other engineroom auxiliary plant (hydraulic power packs, for example) is also facilitated via power take-off gearing.

Among the design innovations in recent years must be noted Sulzer's use of hydraulic actuation of the gas exchange valves for its ZA50S engine, the first time that this concept (standard on low speed two-stroke engines for many years) had been applied to medium speed four-stroke engines (Figure 16.5). In conjunction with pneumatically

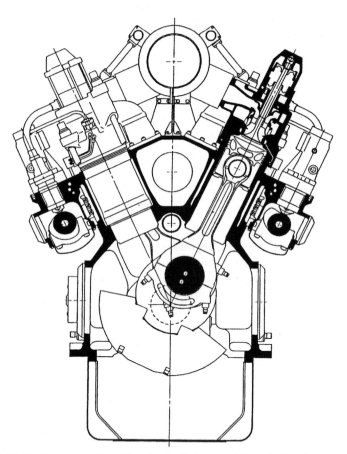

Figure 16.1 The Japanese ADD 30V engine is distinguished by a single-valve gas exchange system comprising a main valve located at the centre of the cylinder and a control valve placed co-axially against the main valve. The control valve switches the air intake and exhaust channels in the cylinder cover. Both valves are driven hydraulically. Side-mounted fuel injectors are arranged around the cylinder periphery

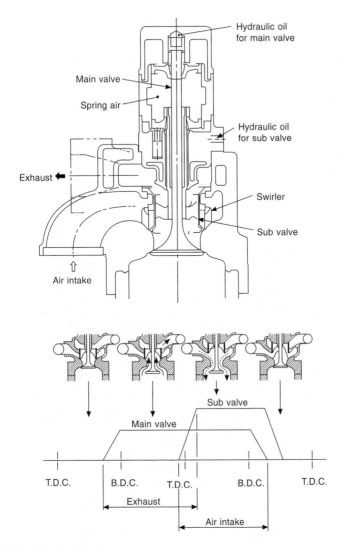

Figure 16.1   (Continued)

controlled load-dependent timing to secure variable inlet closing, hydraulic actuation on the ZA50S engine allows flexibility in valve timing, fostering lower exhaust gas emissions and improved fuel economy.

Variable inlet closing, combined with optimized turbocharging, contributes to a very flat fuel consumption characteristic across the load range of the engine as well as a considerable reduction in smoke levels in part-load operation. The ZA50S engine, like the smaller bore

ZA40S design it was derived from, features Sulzer's rotating piston which was also exploited in GMT's upgraded 550 mm bore medium speed engine.

Wärtsilä's 46 engine exploited a number of innovations in medium speed technology, originally including a twin fuel injection system (featuring pilot and main injection valves), thick-pad bearings (large bearings with thick oil films) and pressure-lubricated piston skirts. The twin injection system was later superseded as advances in fuel injection technology allowed a single-valve system to be applied.

The operating flexibility of MAN B&W's L32/40 design benefits from the provision of separate camshafts, arranged on either side of the engine, for the fuel injection and valve actuating gear. One camshaft is dedicated to drive the fuel injection pumps and to operate the starting air pilot valves; the other serves the inlet and exhaust valves. Such an arrangement allows fuel injection and air charge renewal to be controlled independently, and thus engine operation to be more conveniently optimized for either high fuel economy or low exhaust emissions mode. Injection timing can be adjusted by turning the camshaft relative to the camshaft driving gear (an optional facility (Figure 16.2)). The valve-actuating camshaft can be provided with different cams for full-load and part-load operations, allowing valve timing to be tailored to the conditions. A valve camshaft-shifting facility is optional, the standard engine version featuring just one cam contour (Figure 16.3).

A carbon cutting ring is now a common feature of medium speed engines specified to eliminate the phenomenon of cylinder bore polishing caused by carbon deposits and hence significantly reduce liner wear. It also fosters a cleaner piston ring area, low and very stable lubricating oil consumption, and reduced blow-by.

Also termed an anti-polishing or fire ring, a carbon cutting ring comprises a sleeve insert which sits between the top piston ring turning point and the top of the cylinder liner. It has a slightly smaller diameter than the bore of the liner, this reduction being accommodated by a reduced diameter for the top land of the piston. The main effect of the ring is to prevent the build-up of carbon around the edges of the piston crown which causes liner polishing and wear, with an associated rise in lubricating oil consumption. A secondary function is a sudden compressive effect on the ring belt as the piston and carbon cutting ring momentarily interface. Lubricating oil is consequently forced away from the combustion area, again helping to reduce consumption: so effectively, in fact, that Bergen Diesel found it necessary to redesign the ring pack to allow a desirable amount of oil consumption. Lubricating oil consumption, the Norwegian engine designer reports,

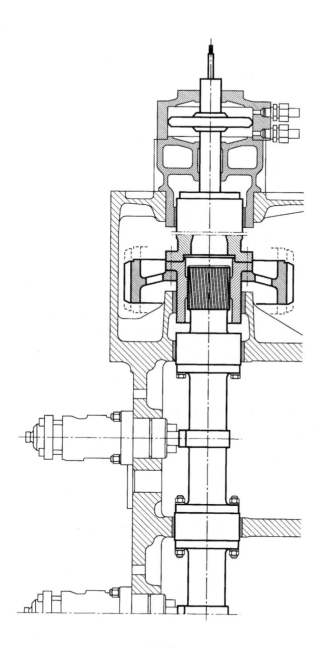

Figure 16.2 Optimization of fuel injection timing on MAN B&W's L32/40 engine is facilitated by turning the dedicated camshaft relative to its driving gear

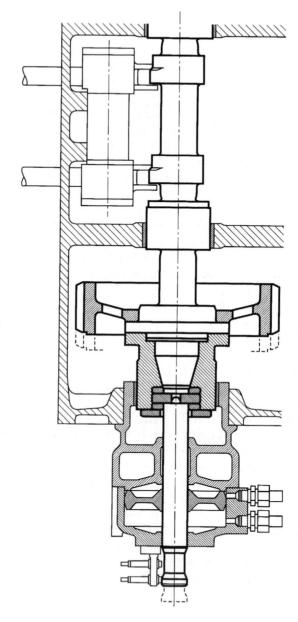

Figure 16.3 Variable valve timing on MAN B&W's L32/40 engine is secured by different cams for part- and full-load operations

is cut by more than half and insoluble deposits in the oil reduced dramatically, significantly extending oil filter life. Carbon cutting rings can be retrofitted to deliver their benefits to engines in service. Removal prior to piston withdrawal is simply effected with a special tool.

Designers now also favour a 'hot box' arrangement for the fuel injection system to secure cleaner engine lines and improve the working environment in the machinery room thanks to reduced temperatures; additionally, any fuel leakage from the injection system components is retained within the box.

The major medium speed enginebuilders have long offered 500 mm-bore-plus designs in their portfolios. MAN B&W still fields the L58/64 series and SEMT-Pielstick its 570 mm bore PC4.2 and PC40L series, but MaK's 580 mm bore M601 and the Sulzer ZA50S engines have been phased out, as was Stork-Wärtsilä's TM620 engine in the mid-1990s.

In the 1970s MAN and Sulzer jointly developed a V-cylinder 650 mm bore/stroke design (developing 1325 kW/cylinder at 400 rev/min) that did not proceed beyond prototype testing.

Wärtsilä's 64 series, launched in 1996, took the medium speed engine into a higher power and efficiency territory, the 640 mm bore/900 mm stroke design now offering an output of 2010 kW/cylinder at 333 rev/min. A V12-cylinder model delivers 23 280 kW at 400 rev/min. The range can therefore meet the propulsive power demands of virtually all merchant ship tonnage types with either single- or multi-engine installations. The key introductory parameters were: 10 m/s mean piston speed; 25 bar mean effective pressure; and 190 bar maximum cylinder pressure. The Finnish designers claimed the 64 series to be the first medium speed engine to exceed the 50 per cent thermal efficiency barrier, and suggested that overall plant efficiencies of 57–58 per cent are possible from a combi-cycle exploiting waste heat to generate steam for a turbo-alternator.

At the other end of the medium speed engine power spectrum, the early 1990s saw the introduction of a number of 200 mm bore long stroke designs from leading builders, such as Daihatsu, MaK and Wärtsilä Diesel, contesting a sector already targeted by Sulzer's S20 model. These heavy fuel-burning engines (typically with a 1.5:1 stroke/bore ratio) were evolved for small-ship propulsion and genset drive duties, the development goals addressing overall operating economy, reliability, component durability, simplicity of maintenance and reduced production costs. Low and short overall configurations gave more freedom to naval architects in planning machinery room layouts and eased installation procedures (Figure 16.4).

The *circa*-320 mm bore sector is fiercely contested by designers

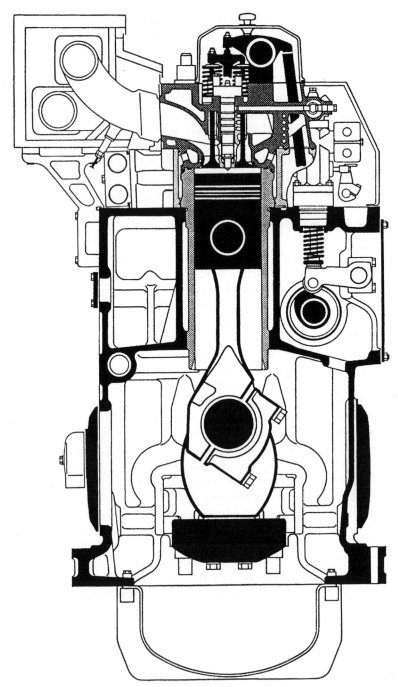

Figure 16.4 MaK's M20 engine, representing a new breed of 200 mm bore designs

serving a high volume market created by propulsion and genset drive demands. A number of new designs—including Caterpillar/MaK's M32 and MAN B&W Diesel's L32/40 —emerged to challenge upgraded established models such as the Wärtsilä 32.

A Japanese challenger in a medium speed arena traditionally dominated by European designer/licensors arrived in the mid-1990s after several years' R&D by the Tokyo-based Advanced Diesel Engine Development Company. The joint venture embraced Hitachi Zosen, Kawasaki Heavy Industries and Mitsui Engineering and Shipbuilding. The 300 mm bore/480 mm stroke ADD30V design, in a V50-degree configuration, developed up to 735 kW/cylinder at 750 rev/min, significantly more powerful than contemporary medium speed engines of equivalent bore size. A mean effective pressure of around 25 bar and a mean piston speed of 11.5 m/s were exploited in tests with a six-cylinder prototype, although an mep approaching 35 bar and a mean piston speed of 12 m/s are reportedly possible. In addition to a high specific output, the developers sought a design which was also over 30 per cent lighter in weight, 10–15 per cent more fuel economical and with a better part-load performance than established engines. Underwriting these advances in mep and mean piston speed ratings are an anti-wear ceramic coating for the sliding surfaces of the cylinder liners and piston rings, applied by a plasma coating method. A porous ceramic heat shield was also developed for the combustion chamber to reduce heat transfer to the base metal of the piston crown.

A key feature is the single-valve air intake and exhaust gas exchange (Figure 16.1), contrasting with the four-valve (two inlet and two exhaust) heads of other medium speed engines. The greatly enlarged overall dynamic valve area and the reduction in pressure losses during the gas exchange period promote a higher thermal efficiency. The system is based on a heat-resistant alloy main poppet valve located over the centre of the cylinder and a control valve placed co-axially against this main valve. The control valve switches the air intake and exhaust channels in the cylinder cover. Both main and control valves are driven hydraulically.

Fuel injection is executed from the side through multiple injectors arranged around the cylinder periphery instead of a conventional top-mounted central injection system. Combustion characteristics were optimized by raising the fuel injection pressure to around 2000 bar, thus enhancing the fuel-air mixture formation and fostering low NOx emissions without sacrificing fuel economy. A computer-based mechatronics system automatically controls the timing of fuel injection and valve opening/closing to match the operating conditions. The first 6ADD30V production marine engines, built by Mitsui, were specified

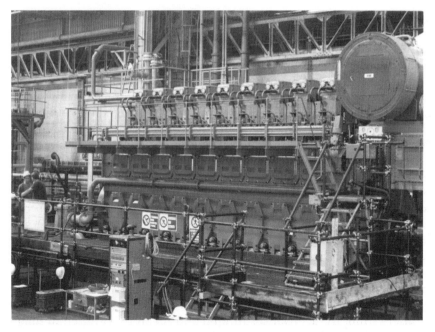

Figure 16.5 The first example of the Sulzer ZA50S engine, a nine-cylinder model, on test. The design (no longer produced) was distinguished by hydraulic actuation of the gas exchange valves

as the prime movers for the diesel-electric propulsion plant of a large Japanese survey vessel.

Offshore industry market opportunities—and the potential of mainstream shipping interest—have encouraged a number of medium speed enginebuilders to develop dual-fuel and gas-diesel designs offering true multi-fuel capabilities with high efficiency and reliability, and low carbon dioxide emissions. The engines can run on gas (with a small percentage of liquid pilot fuel) or entirely on liquid fuel (marine diesel oil, heavy fuel or even crude oil). Switching from one fuel to another is possible without interrupting power generation. (See Chapter 2.)

The high cost of R&D to maintain a competitive programme and continuing investment in production resources and global support services have stimulated a number of joint ventures and takeovers in the four-stroke engine sector in recent years. Most notable have been Wärtsilä's acquisition of the former New Sulzer Diesel and Caterpillar's takeover of MaK. Earlier, Wärtsilä had acquired another leading medium speed enginebuilder, the Netherlands-based Stork-Werkspoor Diesel. This trend towards an industry comprising a small number of major multi-national players contesting the world market has continued with the absorption of the British companies Mirrlees Blackstone, Paxman and Ruston (formerly part of Alstom Engines) into the MAN B&W Diesel group. Rolls-Royce inherited the Bergen Diesel interests in Norway through the takeover of Vickers-Ulstein, adding to its Allen programme.

Considerable potential remains for further developing the power ratings of medium speed engines, whose cylinder technology has benefited in recent years from an anti-polishing ring at the top of the liner, water distribution rings, chrome-ceramic piston rings, pressurized skirt lubrication and nodular cast iron/low friction skirt designs.

The pressure-lubricated skirt elevated the scuffing limit originally obtaining by more than 50 bar, reduced piston slap force by 75 per cent and doubled the lifetime of piston rings and grooves. Furthermore, it facilitated a reduction in lube oil consumption and, along with the simultaneous introduction of the nodular cast iron skirt, practically eliminated the risk of piston seizure. The anti-polishing ring dramatically improved cylinder liner lifetime beyond 100 000 hours, and lube oil consumption became controllable and stable over time, most engines today running at rates between 0.1 and 0.5 g/kWh. A further reduction in piston ring and groove wear was also achieved, and the time-between-overhauls extended to 18 000–20 000 hours. The ring itself is a wear part but is turnable in four positions in a four-stroke engine, fostering a lengthy lifetime for the component.

Such elements underwrite a capability to support a maximum cylinder pressure of 250 bar, of which 210–230 bar is already exploited in some engines today. Leading designers such as Wärtsilä suggest it may be possible to work up towards 300 bar with the same basic technology for the cylinder unit, although some areas need to be developed: bearing technology, for example, where there is potential in both geometry and materials. A steel piston skirt may become the most cost effective, and cooling of the piston top will probably change from direct oil cooling to indirect. The higher maximum cylinder pressure can be exploited for increasing the maximum effective pressure or improving the thermal efficiency of the process. Continuing to mould the development of the medium speed engine will be: NOx emissions; CO_2 emissions; fuel flexibility; mean time-between-failures; and reduced maintenance.

DESIGNING MODERN MEDIUM SPEED ENGINES

Investment in the development of a new medium speed engine may be committed for a number of reasons. The enginebuilder may need to extend its portfolio with larger or smaller designs to complete the available range, to exploit recently-developed equipment and systems, or to enhance the reliability, availability and economy of the programme. As an alternative to a new design, it may be possible to upgrade an existing engine having the potential for improved power ratings, lower operational costs, reduced emissions and weight. A recent example is the upgrading by MAN B&W Diesel of its 48/60 engine, whose new B-version offers a 14 per cent higher specific output with considerably reduced fuel consumption and exhaust emission rates, a lower weight, and the same length and height but a narrower overall width than its predecessor (Figure 16.6).

Figure 16.6 MAN B&W Diesel's 48/60 engine, shown here in V14-cylinder form, benefited from a redesign that raised specific output by 14 per cent and reduced fuel consumption and emissions

Having set the output range of the new engine, a decision must be made on whether it should be built as an in-line cylinder version only or as a V-type engine as well. In the case of the largest and heaviest medium speed engines, the efforts and costs involved in adding a V-type to the existing portfolio might not be justified if there is a limited potential market for engines above 20 000 kW. Other restraining factors could be the capacity of the foundry for casting very large crankcases or difficulties in transporting engines weighing 400 tonnes or more over land.

Once the power per cylinder of the proposed new engine is known, the first indication of its bore size is determined by the piston load, which relates the specific output to the circular piston surface. Piston loads have increased with the development of better materials: the original 48/60 engine has a piston load of 58 kW/cm^2, while the 48/60B engine achieves 66.4 kW/cm^2. The bore and stroke of an engine are interrelated by the stroke/bore ratio, which in turn is based on the designer's experiences with earlier engines. Some 25 years ago it was not uncommon to design medium speed engines with very similar bore and stroke dimensions (so-called 'square' engines); more recently, the trend has been towards longer stroke designs, which offer clear advantages in optimizing the combustion space geometry to achieve lower NOx and soot emission rates. A longer stroke can reduce NOx emissions with almost no fuel economy penalty and without changing the maximum combustion pressure. The compression ratio can also be increased more easily and, together with a higher firing pressure, fuel consumption rates will decrease. Finally, long stroke engines yield an improved combustion quality, a better charge renewal process inside the cylinder and higher mechanical efficiency.

A good compromise between an optimum stroke/bore ratio and the costs involved, however, is an important factor: it is a general rule that the longer the engine stroke, the higher the costs for an engine per kW of output. The trend towards longer stroke medium speed engine designs is indicated by the table showing the current MAN B&W Diesel family; the first four models were launched between 1984 and 1995, and the remaining smaller models after 1996.

Model	Stroke/bore ratio
58/64	1.1
48/60	1.25
40/54	1.35
32/40	1.25
27/38	1.41
21/31	1.47
16/24	1.5

The above is a simplified outline of the initial design process and the reality is much more complex. The final configuration results from considering a combination of choices, which may have different effects on fuel consumption rates and emissions. Nevertheless, the usual trade-off between fuel economy and NOx emissions can be eliminated. The use of high efficiency turbochargers is also essential.

A decision on engine speed is the next step. The mean piston speed in metres per second can be calculated from the bore and speed of an engine using the following formula: mean piston speed (m/s) = bore (m) × speed (rev/min)/30. The upper limit of the mean piston speed is primarily given by the size and mass of the piston and the high forces acting on the connecting rod and crankshaft during engine running. A mean speed between 9.5 m/s and slightly above 10 m/s is quite common for modern large bore medium speed engines. Any substantial increase above 10 m/s will reduce operational safety and hence reliability. Since medium speed engines may be specified to drive propellers and/or alternators, the selection of engine speed has to satisfy the interrelation between the frequency of an alternator (50 Hz or 60 Hz) and the number of pole pairs.

In designing the individual engine components, two main criteria are addressed: reducing manufacturing costs and reducing the number of overall engine parts to ease maintenance.

A valuable summary of the design process is also provided by Rolls-Royce, based on its development of the completely new Allen 5000 series medium speed engine in conjunction with Ricardo Consulting Engineers of the UK. The initial functional specification defined a target power of 500 kW/cylinder and a wide application profile embracing industrial and standby power generation, pumping, marine propulsion and auxiliary power. It was considered essential to design-in sufficient flexibility to accommodate all key parameters necessary to ensure superior performance in these sectors well into the 21st century. The following basic concepts were adopted by the design team:

- A power range of 3000 kW to 10 000 kW in steps of 1000 kW.
- An output of 525 kWb per cylinder + 10 per cent for one hour in 12 hours required for power generation (equivalent to 500 kWe energy at both 720/750 rev/min synchronous speeds).
- An output of 500 kWb per cylinder maximum continuous rating + 10 per cent for one hour in 12 hours required for propulsion and pumping over a speed range of 375–750 rev/min + 15 per cent overspeed.
- Fuel injection equipment to be of advanced technology, with the flexibility to satisfy future emissions legislation and specified fuel consumption requirements.
- Fuel flexibility covering heavy fuel oil and distillates.

- High standards of reliability, durability and maintainability.
- Condition monitoring capability essential.
- Low number of parts.

The choice of brake mean effective pressure (bmep) was crucial to the competitiveness of the design. It was essential to choose a bmep that would allow the engine to be sold at a competitive price in £ per kW output. Before commencing design work, therefore, a survey of competitive engines was carried out. The survey showed that bmep levels increased on average by 0.25/0.33 bar per year, and predicted that in 1998 the highest rated engines would have a rating of 25/25.3 bar. On this basis, coupled with engine economics, a rating of 26 bar bmep was selected for the new design. Cycle simulation studies showed that this was achievable with a boost pressure ratio of 4.5:1, provided that the combustion and injection systems were properly matched. Any significant increase in bmep would dictate the use of boost pressures of 5:1 and, while turbochargers were available to deliver this level of performance, the special features incorporated in them to achieve it commanded a premium price and increased the cost of the engine.

To secure the specified power per cylinder of 525 kW at a speed of 720/750 rev/min (60 Hz and 50 Hz synchronous speeds) a cylinder bore between 310 mm and 330 mm was required. It was decided at the onset of the design process to limit the maximum piston speed to 10.5 m/s, which equates to a stroke of 420 mm. The bore/stroke ratio therefore lay between 1.2 and 1.4, typical for an engine in the anticipated market sector.

Combustion system design is governed by the customer's need for compliance with exhaust emissions legislation and excellent fuel economy. This dictates that the engine should have a high air-fuel ratio, a shallow open bowl combustion chamber and a very high compression ratio. The fuel injection system must be capable of generating at least 1600 bar line pressure; an injection system with built-in capability to vary the injection timing was a key feature of its specification.

In order to achieve a high trapped air-fuel ratio, the valve area should be high and the ports designed for maximum efficiency with a small amount of swirl (around 0.2 to 0.4 swirl ratio). The bowl shape should be wide and open with no valve recesses, and the top ring should be placed reasonably high so that the crevice volume is minimized to secure a high air utilization factor. A maximum cylinder pressure target of 210 bar was chosen with the aim of achieving excellent fuel economy; this in turn allows a high compression ratio to be used (in the range 15.5 to 16.5:1). An injector nozzle with eight to 10 holes was specified. The combustion bowl design was optimized to ensure excellent mixing of the fuel sprays to keep the air-fuel ratio as homogeneous as possible and minimize the peak cycle temperatures and hence NOx formation.

The choice of bore size was influenced by considering the following: power requirement (525 kW/cylinder); bmep (30 bar maximum); mean piston speed (10.5 m/s); overall length/width/height; specific weight; and specific cost.

To fulfil the design concept, the power requirement for the engine was defined as 500 kWe/cylinder continuous rating with a 10 per cent overload capacity. The maximum continuous bmep is 26.6 bar, a limit set by the available boost pressure level from single-stage turbocharging. The limit of 10.25 m/s mean piston speed was selected to allow reliable operation at the maximum continuous speed without significant risk of piston, ring and liner scuffing and wear problems. By selecting high values for rated bmep and mean piston speed, the size of the cylinder components was minimized; this kept the overall engine size small, with consequent benefits in production costs.

Having set the piston speed and bmep at rated speed, it is possible to derive the piston area and hence the cylinder bore: 320 mm. Having limited the mean piston speed to a maximum of 10.25 m/s at the maximum rated synchronous speed of 750 rev/min, the maximum allowable stroke was determined as 410 mm. Supporting these decisions, simulation work using the WAVE program was carried out to examine bore sizes from 315 mm to 330 mm and stroke sizes from 380 mm to 420 mm. The results confirmed the selected values.

Factors influencing the choice of cylinder configuration (in-line or V-form) are: weight; cost; installation limitations; stress limitations (crank); vibration levels; production tooling; and market acceptance. From manufacturing, weight, cost and installation considerations, a V-form family appeared attractive. V6 and V8-cylinder designs, however, have inherently poor vibration characteristics and attempts to reduce this to acceptable levels would add cost and complexity to the engines. The following configurations for the new engine were therefore initially selected: in-line six and eight, and V10, 12, 14 and 16-cylinders (V18 and V20 models were subsequently added to the production programme).

The range of V-angles for competitive engines was from 45 degrees to 60 degrees, with a compromise angle of 50 degrees favoured by several manufacturers. Angles less than 45 degrees were considered impractical due to close proximity of the lower end of the left and right bank cylinder liners. For even firing intervals and low torque fluctuations to give acceptable cyclic speed variations, optimum V-angles vary from 72 degrees for a V10 to 45 degrees for a V16 engine. An investigation of the effects of V-angle on cyclic torque fluctuations concluded that a common angle of 52 degrees was suitable for V12, 14 and 16 engines, while the V10 required an angle of 72 degrees to optimize torque fluctuation and balance.

Simulating the performance of the Allen 5000 series engine was conducted using a WAVE engine performance simulation model based on the in-line six-cylinder design. For the concept design stage, the basic engine design parameters, such as fuel injection characteristics and valve events, were chosen and then fixed. These were based on previous experience and data from engines of a similar type with the aim of achieving good performance and emissions characteristics. The dimensions of the manifolding and air chest were typical for this size of engine.

For the baseline simulation an inlet cam period of 248 degrees and an exhaust period of 278 degrees were chosen, with an injection period of 30 degrees and start of injection at 13 degrees before top dead centre. The port flow data was typical of a current port design with a good flow performance. The trapped air-fuel ratio was set at 31:1 and a compression ratio of 14.5:1 applied.

The initial simulation model used a bore of 320 mm and a stroke of 390 mm. The results of the simulation showed that at the 10 per cent overload condition the predicted Pmax was 227 bar and the boost pressure 4.7 bar, which were considered to be excessive. Increasing the stroke to 400 mm reduced the Pmax to 202 bar and the boost pressure to 4.2 bar. Based on this initial bore and stroke investigation, larger bore and stroke sizes were examined. This was considered necessary since there was a requirement for the engine to operate at 720 rev/min at the same rating of 525 kW/cylinder and also at the 10 per cent overload condition, both of which result in higher cylinder pressures and boost pressure requirements. Combinations of bore sizes of 320 mm, 325 mm and 330 mm and strokes of 410 mm, 415 mm and 420 mm were assessed for boost pressure, Pmax and fuel consumption predictions.

Based on these simulations, it was decided to retain the bore size at 320 mm and increase the stroke to 410 mm; this would enable the engine to run at full rating at 720 rev/min at the standard injection timing without exceeding the Pmax limit of 210 bar. Having fixed the bore and stroke dimensions, variations in compression ratio and start of combustion were investigated for their effect on fuel consumption and maximum cylinder pressure, see Figure 16.7. Finally, the effect of valve size on performance was assessed before a 96 mm diameter inlet

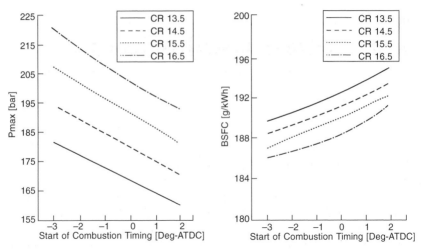

Figure 16.7 Development of Allen 5000 series engine: effect of variations in compression ratio and start of combustion on maximum cylinder pressure (left) and fuel consumption (right) at a power rating of 525 kW/cylinder

valve and an 89 mm diameter exhaust valve were selected as the optimum choice.

Structural and running gear details of the resulting Allen 5000 series engine, which covers an output band from 3150 kW to 10 500 kW with in-line six to V20-cylinder models, are given in the chapter on Allen (Rolls-Royce) designs.

Acknowledgements: MAN B&W Diesel, Rolls-Royce (Allen) and Wärtsilä.

17 Allen (Rolls-Royce)

The British designer Allen, a Rolls-Royce company, fields the S12/VS12 and the S37/VS37 medium speed engine series which have evolved over many years. The S12 is a 241 mm bore/305 mm stroke design produced in four-, six-, eight- and nine-cylinder in-line versions covering a power band from 507 kW to 1584 kW at 720–1000 rev/min. The 325 mm bore/370 mm stroke S37/VS37 design (Figure 17.1) is available in six, eight and nine in-line and V12- and V16-cylinder forms to deliver outputs up to 5255 kW. A completely new 320 mm bore design, the Allen 5000 series, was introduced in 1998 and the S12, VS12 and S37/VS37 designs were respectively re-branded as the 2000 series, 3000 series and 4000 series.

The VS12 engine (now the 3000 series) was first introduced in the 1950s but benefited from a major redesign in 1990 to enhance compactness and raise the power-to-weight and power-to-space ratios. V12- and 16-cylinder versions of the 240 mm bore/300 mm stroke design cover a power band from 1584 kW to 4160 kW at 900–1000 rev/min for propulsion and auxiliary power duties. Its rigid spheroidal graphite cast iron crankcase, designed to carry the underslung forged steel crankshaft and main bearing caps, has integral mounting feet. The assembly is enclosed by a fabricated steel sump and a cast iron gearcase. The single camshaft is carried in bearings centrally located in the 60-degree vee between the cylinder banks and is driven by a gear train from the crankshaft.

The one-piece crankshaft is arranged with a solid or flexible coupling through the flywheel which carries a gear ring for air motor starting. The steel shell main bearings have a lead-bronze lining with overlay plating. Thrust rings either side of the flywheel-end main bearing locate the crankshaft. The main bearing studs pass deep into the crankcase and are hydraulically tensioned; bolts also pass through the crankcase into the sides of the main bearing caps to maximize rigidity. The camshaft gear drive at the free end of the crankshaft and the auxiliary drive hub are fitted on a taper using the oil injection method.

Individual wet-type cylinder liners feature a thick top collar, drilled to provide bore cooling of the upper section. Three Viton O-rings

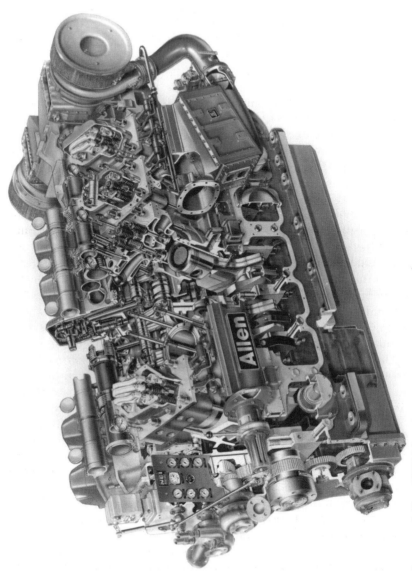

Figure 17.1 Cutaway of Allen 12VS37G engine

make a watertight joint between the lower liner land and the crankcase. A steel ring, sealed by a Viton O-ring, is fitted around the top collar of the liner to direct water from drillings to the six cylinder head transfer holes, each of which is also sealed by O-rings. A thick section mild steel ring makes the gas joint between the liner and cylinder head. Pistons are of composite construction, with a heat-resistant steel crown and forged aluminium body. Three pressure rings are fitted in the crown (the top two featuring hardened flanks) and an oil ring in the body. The pressure rings and spring-loaded conformable oil scraper ring all have chrome-plated running surfaces. The pistons are intensively cooled by lubricating oil supplied from fixed jets in the crankcase, this oil also being used to lubricate the small end bearings.

The connecting rods are arranged in pairs, fitting side-by-side on the crankpins. Each rod has a detachable cap secured by four set-screws and located by dowels; the joint is inclined at 45 degrees to enable the rod to be withdrawn through the cylinder bore. Steel shell bearings lined with lead bronze and overlay plating are fitted at the big end; a steel-backed lead bronze-lined bush is fitted at the small end. Individual four-valve cylinder heads are secured to the crankcase by four studs and hydraulically tensioned nuts. The two inlet and two exhaust valves seat on renewable inserts in the head, those for exhaust service being intensively cooled and sealed with Viton O-rings. The inlet valves are manufactured from EN 52 and the exhaust valves from Nimonic. Each valve has two valve springs, a rotator fitted below them to ensure even wear distribution. The thick flame plate of the cylinder head is cooled by drilled water passages close to the combustion face. A combined fuel pump/injector is fitted centrally in the cylinder head, the fuel inlet and return connections being via drilled passages in the cylinder head sealed by Viton O-rings.

A cast iron rocker housing supports the fulcrum for the fuel and valve levers, encloses the valve gear and also forms the cylinder head water outlet connection and a part of the water outlet main. An aluminium cover is fitted over the rocker housing. Lubricating oil is fed to the fuel pump and valve levers through hollow pushrods and is drained to the sump via an external pipe. A drilled passage is provided in the cylinder head to allow an adaptor and indicator cock to be fitted for performance checks.

The camshaft is built up from two-cylinder sections, each machined from a solid forging and complete with two inlet valves, two exhaust valves and two fuel pump cams. The sections are rigidly bolted together and registered by dowels to ensure precise alignment. The finger-type cam followers are pivoted from a central fulcrum shaft. Lubricating oil flows from each camshaft bearing into the fulcrum through a

Figure 17.2 VS12 unit fuel pump/injector

drilled hole in the follower lever and then to the hollow pushrod. An oil spray from the lever lubricates the roller and pin.

A cast aluminium housing encloses the camshaft and cam followers, and also carries brackets for the two control shafts. The cover on the housing incorporates a heat shield.

A unit fuel pump injector with removable nozzle is arranged in each cylinder head and operated from the camshaft by the fuel cam via cam follower, pushrod and lever in a similar manner to the inlet and exhaust valves. It is sealed into the head by four O-rings. An engine-driven fuel pressurizing pump supplies fuel via a filter to the fuel supply main and is controlled conventionally using an electronic governor and electric actuator. A relief valve in the fuel return main regulates the system pressure, with excess fuel returning to the pump suction. Engine lubrication is effected by a wet sump system. A gear-type lubricating oil pump, mounted on the gearcase and driven from the crankshaft gearwheel, draws oil from the sump through a filter

Figure 17.3 VS12 cylinder head

strainer. A relief valve on the pump delivery safeguards the system against excessive pressure. Oil under pressure is cooled and filtered before being fed into oil mains bolted to the sides of the crankcase. The mains supply oil to the piston cooling jets and to crankcase drillings feeding the crankshaft and camshaft bearings and the intermediate gear.

External pipes supply oil to the turbochargers, water pumps and intermediate gears mounted in the gearcase. The oil filters are of the replaceable cartridge type. A motor-driven priming pump is incorporated in the system to ensure that all parts are supplied with oil before the engine is started.

Cooling water from the primary system circulates through the engine jackets, turbochargers and heat exchanger or radiator. A secondary water system supplies the charge air coolers and lubricating oil cooler. Primary and secondary water pumps are gear driven from an auxiliary gear train connected to the camshaft by a quill shaft.

Each cylinder bank is served by its own turbocharger, charge air cooler and inlet and exhaust manifolds. The turbocharger compressor discharges via the air shut-off butterfly valve into the air inlet manifold

Figure 17.4 Allen S12 engine in-line cylinder form

through the charge air cooler. A single exhaust manifold leads to the turbocharger inlet and a restricting throat is incorporated in the exhaust branch at each cylinder head. Bellows units and V-clamps connect the individual exhaust branches to form the exhaust manifold.

Engine starting is effected by a compressed air-powered starter motor which engages automatically with the gear ring on the flywheel. An electronic governor is fitted, the standard system including a crankshaft speed sensing pick-up installed near the flywheel gear ring, a control box mounted off-engine, and an actuator unit mounted on the engine gearcase connected through a collapsible link to the fuel injection pump control shaft. A stop/run handle is connected to the fuel control shafts to ensure that manual shutdown is always possible. The levers operating the control racks of the unit injectors are spring loaded so that an engine can always be shut down even if a control rack should stick.

Overspeed protection on the engine comprises two air shut-off butterfly valves fitted between each turbocharger compressor discharge and air manifold. The valves close on overspeed and cut off the combustion air supply to the engine. They are held in the open (running) position by an air-operated actuator equipped with a spring return arranged to close when the air supply is lost. The air supply is fed via a solenoid valve energized to pass air to the butterfly actuators. The solenoid valve is fed from a 24V dc supply controlled by an Amot speed monitor which takes its speed signal from the magnetic pick-up mounted over the engine flywheel gear ring. Other contacts in the speed monitor are connected to the electronic governor so that the actuator also moves to zero fuel on overspeed. The overspeed trip is set to operate when the engine speed reaches 15 per cent above normal.

Explosion relief valves are fitted to the crankcase to relieve any build-up of excess pressure inside the engine; they are also designed to prevent the admission of air. The valves open if crankcase pressure rises, and the oil mist mixture directed to atmosphere via a flame trap and deflector cowl. As soon as the crankcase pressure returns to normal a coil spring closes the valve.

Allen 5000 series

A completely new medium speed engine was launched in 1998, the 320 mm bore/410 mm stroke Allen 5000 series having an output of 525 kW/cylinder to cover a power range from 3150 kW to 10 500 kW at 720/750 rev/min from in-line six and eight and V10, 12, 14, 16, 18 and 20-cylinder models. The V12 and V16 versions were released in 1998 (Figure 17.5), the other configurations planned to follow

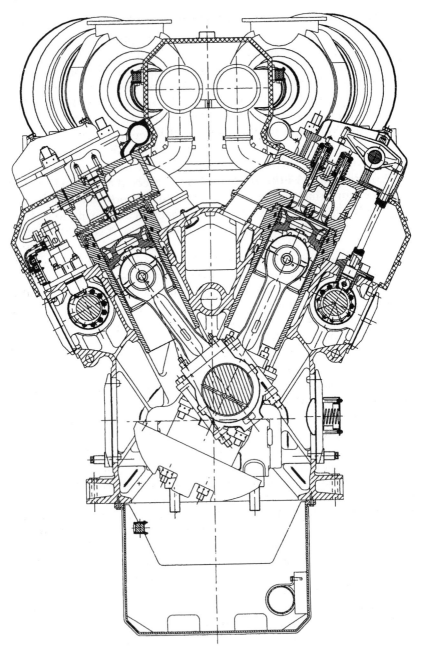

Figure 17.5 Cross-section of Allen 5000 series engine

progressively. The initial applications focused on land power generation but marine propulsion and auxiliary power opportunities were anticipated later.

With no carry over of components from existing Allen engines, the brief was to produce a robust design that was economical to manufacture, incorporated multi-functional components, had the minimum number of external pipes, used hydraulic tensioning for critical fasteners, and was easy to service in the field. Excellent fuel economy, smoke-free combustion and low emissions of other pollutants were also targeted during the design process.

The crankshaft design is typical for a medium speed engine: a one-piece steel slab forging, hardened and tempered; the material has an ultimate tensile strength of 900 Mpa; and the front end and flywheel end flanges are forged integrally with the shaft. In order to minimize the cylinder centre distance a crankpin diameter of 310 mm and a main journal diameter of 330 mm were specified, together with recessed fillet radii for both crankpins and journals. A torsional vibration analysis of the six- to V16-cylinder versions was carried out, indicating that a crankshaft torsional vibration damper would be required, the choice of a tuned or viscous damper depending on the cylinder configuration. As a result of this analysis a cylinder bank angle of 52 degrees was chosen for the subsequent calculations.

Both classical and finite element stress analyses for the crankshaft were executed. Safety factors of 1.7 or greater under combined bending and torsional loads were predicted for engines used in fixed speed generating applications. The external imbalance of the in-line six, V12 and V16 models is zero for all primary and secondary forces and couples. The in-line nine, V10 and V14 models all have non-zero primary couples. The V10-14 models show inherent vertical and horizontal couples increasing as the bank angle is narrowed. By applying a suitable counter-weighting arrangement these couples can be partially balanced to leave a minimum possible imbalance. Further reduction is possible through the use of balancer shafts or reductions in reciprocating mass. In order to achieve the optimum balance for the V10 and V14 engines, extra balance weight must be applied to the crankshaft to counter both the primary rotating and mean primary reciprocating couples. As a result of the detailed analysis, the 52 degree bank angle was confirmed as the best choice for the V-cylinder engines.

Bi-metallic aluminium tin bearings were selected for both main and large end bearings. The crankpin diameter of 310 mm made it impossible to design a two-piece connecting rod arrangement that could be withdrawn up the cylinder bore (320 mm). A three-piece rod design of the marine-type was therefore selected; this enables the piston and rod to be removed through the liner while leaving the big end assembly bolted undisturbed on the crankpin. An advantage of this rod type is that the overhaul height requirement is less than for a two-piece design.

All the joint faces of the connecting rod are machined at 90 degrees to the longitudinal axis of the rod. No serrations are used and the three parts of the assembly are located to each other with fitted dowels arranged to prevent incorrect assembly. The shank of the rod is bolted to the big end block with four 24 mm studs, and the two halves of the big end block are secured with two 39 mm studs; both sets of fasteners are hydraulically tensioned. The majority of the surfaces on the rod remain in the 'as forged' condition. A full finite element analysis of the connecting rod showed that, apart from an area around the small end of the rod, the safety factors were generous. More material was added to the small end, this change increasing the safety factor at this point to an acceptable value.

The cylinder head is a rigid design of sufficient depth to evenly distribute the clamping load from the four hydraulically-tensioned head studs over the gas sealing face between the head and the cylinder liner; it is cast in compacted graphite cast iron. The MAGMASOFT casting simulation program was used to optimize the casting process. The inlet and exhaust ports are located on the same side of the head, with the inlet ports designed to impart a low level of swirl.

Both inlet and exhaust valve seats are intensively cooled, as is the rest of the flame face of the head. The cooling strategy of both head and liner was optimized using the VECTIS CFD simulation program, and a complete model of the coolant passages was made. The flow regime was evaluated throughout the model, with particular attention given to ensure there was a positive water flow velocity of between 1.5 m/s and 3 m/s around all the valve seats and drillings in the flame face of the head. As a result of this optimized cooling, the predicted temperatures at the full load condition are evenly distributed around the combustion face. These temperature values, together with stresses and safety factors, were calculated from a full 3D finite element model of the head. After some minor refinements to the design to increase the stiffness of the head stud bosses, all the temperatures, stresses and safety factors were within the capabilities of the compact graphite material.

At the conceptual phase of the design it was decided that the cylinder liner should be strategically cooled, and several different layouts were analysed using finite element techniques to calculate the temperature profile in the upper region of the liner. Layouts 1 and 2 (Figure 17.6) were rejected on the grounds that the predicted temperature at the

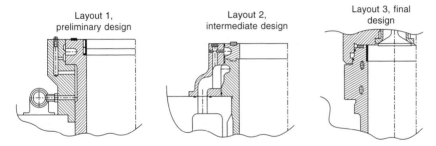

Figure 17.6 Cylinder liner cooling layouts investigated for the Allen 5000 series

top ring reversal was too high or that the temperature profile was unsatisfactory. Layout 3, which uses a conventional bore cooling arrangement, was selected on the basis that the predicted liner surface temperature at the top ring reversal point was 169°C while 100 mm down from that point it was 138°C.

The liner is made from centrifugally cast iron. An anti-polishing ring is fitted at its top to eliminate bore polishing and ensure both a low and stable lube oil consumption over extended engine operation. The coolant flow around the liner was optimized using the VECTIS CFD simulation code. The coolant passages of both liner and cylinder head were incorporated into a single model.

A two-piece piston design, with steel crown and nodular cast iron skirt, was considered the best choice for securing a robust long life when operating at high cylinder pressures (210 bar design limit). The crown is cooled by oil supplied via drillings in the connecting rod and piston pin. Three compression rings and one oil control ring are fitted and, in conjunction with the anti-polishing ring in the liner, are expected to ensure low and stable lube oil consumption over a long period, especially in heavy fuel operation.

The crankcase is a one-piece nodular iron casting comprising the cylinder blocks and crankcase. The crankshaft is underslung and the main bearing caps are retained by hydraulically-tensioned vertical studs and horizontal side bolts. A gear case is incorporated at both ends of the crankcase, that at the flywheel end having an extra main bearing to support the mass of the flywheel. The air manifold is formed in the space in the centre of the vee. At the bottom of the crankcase are substantial mounting rails which impart additional stiffness in that area. Many of the oil and water transfer passages are either cast in or drilled to minimize the number of pipes used.

A full 3D finite element analysis of the V12-cylinder crankcase was carried out to confirm the design goals of high rigidity in both bending and torsion, together with modest stress levels, to ensure trouble-free operation during the life of the engine. Before casting the first examples of the crankcase, the MAGMASOFT casting simulation program was used to optimize the design of the runners and risers, and to confirm that the crankcase design was such that good quality castings would be produced without the need for corrections.

A conventional push rod operated valve gear was selected. The camshaft is mounted high in the cylinder block to minimize the length of the push rods and keep the valve gear stiffness high, and also to keep the length of the high pressure fuel injection pipe as short as possible. The camshaft is created from individual cylinder sections, which are then bolted to cylindrical sections (Figure 17.7). Such a design allows single sections of the camshaft to be removed sideways from the engine; it also enables solid bearing housings to be used, improving the stiffness of the cylinder block in that area. The proportions of the camshaft segments are deliberately generous to ensure moderate fuel cam operating stresses and minimum bending and torsional deflections when operating at the full design pressure of the fuel injection system (2000 bar). The camshaft is driven by a train of spur gears mounted at the flywheel end of the engine.

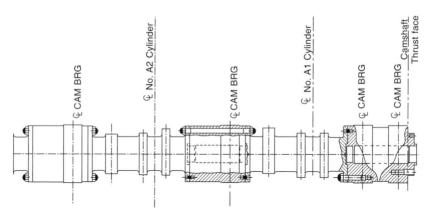

Figure 17.7 Camshaft assembly for Allen 5000 series engine

In specifying the fuel injection system, it was believed that fuel tolerance and injection flexibility to manage the conflicting requirements of low fuel consumption and low emissions could only be achieved by using digital electronics. R&D by Bosch (Hallein) removed the need for mechanical systems to alter the entire camshaft or parts that act on the fuel cams by exploiting proven digital electronic technology in a configuration co-developed by Rolls-Royce and Bosch. High efficiency and low emissions at all loads are promised, each cylinder calibrated for optimum performance. Individual pump and injector units for each cylinder are served by a side entry double-skinned high pressure fuel pipe. The pumping system is based on a conventional arrangement of camshaft and roller-operated plunger providing injection pressure capability over 2000 bar, with injection controlled electronically by a solenoid servo system via a spill valve.

By applying Ricardo engine cycle simulation software (WAVE) during both conceptual and definitive design phases, the geometries and thermodynamics of the engine performance were continually developed to allow full optimization of the power cycle objectives. ABB Turbo Systems' high efficiency TPL axial turbochargers are used. The charge air system comprises an integrated casing onto which the turbocharger can be mounted, and allows a high performance intercooler to be fitted within the casting. Two-stage intercooling is applied to provide charge air heating at lower loads and give improved heat recovery when this is required.

See Medium Speed Engines Introduction chapter for the evolution of the Allen 5000 series design parameters.

18 Alpha Diesel (MAN B&W)

Since their introduction in the early 1970s MAN B&W Alpha Diesel's kindred 23- and 28-type medium speed engines have been refined and uprated a number of times to address changing requirements from small-ship propulsion and genset drive markets. Upgrading programmes by the Frederikshavn, Denmark-based member of the MAN B&W group have tapped extensive service experience with both designs which respectively feature 225/300 mm and 280/320 mm bore/stroke dimensions. The current 28/32A engine covers a power range from 1470 kW to 3920 kW at 775 rev/min with 6L- to V16-cylinder models, and is optimized for service intervals of 24 to 30 months based on an annual running time of 6000–7000 hours. Outputs from 800 kW to 1920 kW at 825 or 900 rev/min are offered by 6L, 8L and V12-cylinder variants of the 23/30A series.

Early experience with small medium speed engines burning heavy fuel oil, over 20 years ago, found that exhaust valves could function for around 2000 hours (although often less) and had to be scrapped after 3000 hours' service. In a few engines, whose fuel was less aggressive and the load moderate, the interval could be extended to 6000 hours. Generally, however, the valve lifetimes were considered too short in relation to the cost savings made from using heavy fuel and owners opted for marine diesel oil instead.

Subsequent 28/32A engine developments raised the maintenance interval to at least 15 000 hours, representing a time-between-overhaul of two years. Operation on heavy fuel is thus cost-effective for almost all ships, according to MAN B&W Alpha Diesel, the exceptions being tonnage whose load profiles dictate frequent starting and stopping and low annual running hours.

Advances in performance have been sought over the years along with improvements in durability and reliability. The specific power rating has risen by almost 40 per cent while the brake mean effective pressure has remained at around 19 bar and the specific fuel consumption at 190 g/kWh.

The general design features of the current 28/32A engine date from 1990. The engine's overall appearance is much the same as it was when launched in 1974, with closed-in fuel piping and pumps. But the cross-section (Figure 18.1) shows that many of the key elements—such as the frame, cylinder liner, piston, cylinder cover and valves—have been improved.

Frame

The frame has been reinforced and is now cast in nodular cast iron. The support and cooling of the cylinder liner have been improved, and the staybolts for the cylinder head have a deeper and enhanced grip in the frame to carry the load from higher combustion pressures.

Combustion

Contributions to reduced fuel consumption and lower exhaust gas emissions are made by: accurate start of fuel injection timing; precision-made injection nozzles; fuel pumps with precise fuel metering; modified combustion chambers yielding a compression ratio of 13.9; precisely defined fuel spray geometry; and increased injection pressure. Also influential is the MAN B&W turbocharger introduced in the 1980s and matched to the 28/32A engine.

Fuel valves and pumps

Good interaction between the engine and its fuel injection system determines the engine's torque and power output as well as its emissions and noise generation, the designer asserts. Single-plunger injection operates with fuels of diverse viscosities and with peak pressures up to around 1100 bar. The plunger and barrel assembly is provided with a leakage oil return and an extra oil block for leaked fuel. A ring-shaped groove is machined into the pump barrel. Lube oil is forced into the fuel pump housing at the pump foot, preventing the spring guide from sticking in asphaltenes. The very small leakage quantity involved is drawn off through a separate outlet and led into a collecting tank.

Combustion chamber

Development of the 28/32A engine during 1988/90 yielded an improved combustion chamber in which the fuel spray mixes better with the air and avoids spraying at the valves and piston top. In combination with an adjustment of the pressures, the combustion

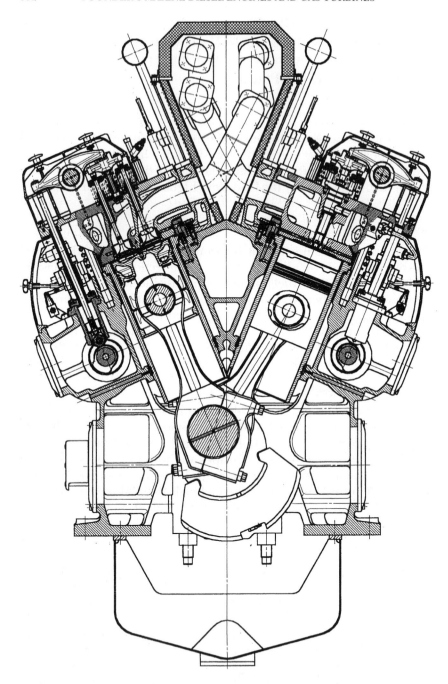

Figure 18.1 Cross-section of MAN B&W Alpha V28/32A engine

process fostered reduced fuel consumption and lower exhaust gas temperatures, along with minimized exhaust smoke.

Cylinder head, valves and ducts

Nodular cast iron is specified for the cylinder head to secure high strength and stiffness. The head features a bore-cooled flame deck and is fitted with four rotating valves and cooled valve seats. Air and gas ducts are dimensioned to ensure minimum resistance to an optimized air flow and to achieve exhaust valve temperatures of below 400°C. Overhaul intervals exceeding 18 000 hours on heavy fuel oil have been confirmed in service. The performance of valves and seats has been closely monitored. Well-cooled seats, a rotating mechanism and the construction material mean that it is not the valves that dictate the time for maintenance. Regrinding can be carried out twice to give a valve lifetime up to 40 000 hours (but sometimes shorter if the exhaust gases have attacked the valve stem with sulphuric acid).

Cylinder liner

This key component is centrifugal cast in fine-grained perlitic cast iron. The improved combustion chamber made it possible to introduce a step in the bore diameter at the top of the cylinder with an uncooled flame ring. The inside diameter of the ring is somewhat smaller than the liner bore. The piston crown is therefore made correspondingly smaller, allowing carbon coke on the piston top to be kept well clear of the running surface of the cylinder liner. The flame ring—introduced on all new engines from 1992—has proved beneficial (Figure 18.2). It avoids liner polishing and micro-seizure, and helps to maintain the piston rings in a good condition; a sound seal against the gas and combustion forces results in a low and stable lube oil consumption.

Piston

There have been two suppliers of pistons for the 28/32A engine. The designs are different but have both generally performed well. One type is a monobloc piston of nodular cast iron with a splash-cooled piston top and a thin-walled skirt to keep the weight down. The other type is a composite piston with a steel top and a nodular cast iron skirt. Each has an intensively cooled top to keep the rings in good condition and foster a low wear rate. The ring grooves in the pistons are hardened and ground. The top piston ring was a focus of development, the aim being to find the best compromise between low

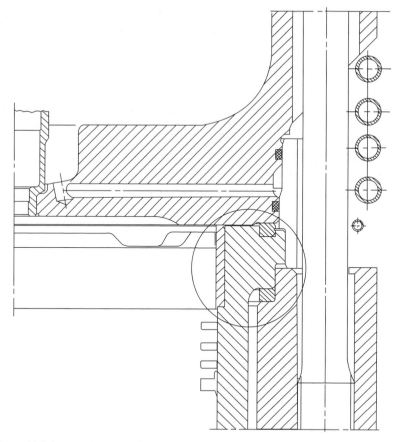

Figure 18.2 Section through 28/32A cylinder, highlighting the flame ring

lube oil consumption and good lubrication. All the rings have a chrome layer on the running surface.

Camshafts

The cam shape has been optimized and the camshaft material and hardening procedure improved. The rollers have been 'flexible shaped' to avoid edge loading.

Bearings

The application of 'Rillenlager' (Mibi's multi-metal layer bearing) solved the problem of damaged bearings due to catfines in the fuel and increased bearing lifetime to over 30 000 hours. This means they can be re-installed at 12 000–15 000 hours service intervals and run for a similar second period.

Turbocharger

The MAN B&W turbocharger, introduced in the 1980s, was matched to the 28/32A engine and in conjunction with the compression ratio and valve timing fostered optimized conditions for the valves. Inboard floating bearings, with a reported lifetime of over 24 000 hours, are lubricated by the engine lubrication system.

19 Caterpillar

The largest model in Caterpillar's own-design portfolio—and the US-based group's only medium speed engine outside the MaK programme—is the 3600 series. Since its introduction in the early 1980s the 280 mm bore/300 mm stroke design has earned numerous references in commercial and naval propulsion and auxiliary sectors as well as in power station, locomotive and industrial applications. Higher rated, lighter weight versions were developed in the 1990s and successfully applied to the propulsion of large fast ferries.

The 3600 series programme embraces six- and eight-cylinder in-line and 50-degree V-configuration 12, 16 and 18-cylinder models. An output band from 1490 kW to 4020 kW at rated speeds of 750–1000 rev/min was initially covered but the upper limit from the V16-cylinder model was raised to 4320 kW and then to 5420 kW through improvements in combustion efficiency. Another 15 per cent rise in specific output over the standard models in the mid-1990s addressed fast ferry propulsion demands, taking the maximum continuous ratings of the V12 and V16 models to 4250 kW and 5650 kW, respectively. Power-related modifications included a new high pressure ratio turbocharger, a higher capacity aftercooler, new pistons and cylinder liners, and a new bearing design. Further refinements for high performance duty, progressed in conjunction with Bazan Motores of Spain, resulted in the 1998 launch of the V18-cylinder 3618 engine offering up to 7200 kW at 1050 rev/min. The same year brought a rating rise for the 3616 engine, now released with an output of 6000 kW.

3600 series

A standard specification 3600 series engine (Figure 19.1) features the following key components:

Cylinder block: a one-piece casting of heavily ribbed, weldable grey iron alloy. The intake plenum runs the full engine length, providing even air distribution to the cylinders.

Crankshaft: a press forging, induction hardened and regrindable. A

536

counterweight for each cylinder is welded to the crankshaft and then ultrasonically examined to assure weld integrity. The crankshaft end flanges are identical to enable full power to be taken from either end of the engine.

Bearings: the main, connecting rod and camshaft bearings are of steel-backed aluminium with a lead–tin overlay copper-bonded to the aluminium.

Connecting rods: these are forged, heat treated and shot peened before machining. The special four-bolt design allows for an extra-large bearing which reduces bearing load, extends bearing life and improves crankshaft strength and stiffness.

Cylinder liners: high alloy iron castings, induction hardened, plateau honed, and water jacketed over their full depth. The flange cooling provided for heavy fuel-burning versions reduces the top piston ring groove temperature.

Pistons: two-piece components comprising a steel crown and forged aluminium skirt for strength and durability with light weight. Two rings are arranged in hardened grooves in the crown, and another two rings in the skirt. The top compression ring is barrel faced and plasma coated for greater hardness. The two middle rings are taper faced and chrome plated, while the lower oil control ring is double rail chrome faced with a spring expander. Cooling oil is jet sprayed into passageways within the piston.

Valves: these seat on replaceable induction-hardened inserts. Positive rotators are provided to increase valve life by maintaining a uniform temperature and wear pattern across the valve face and seat. The exhaust valves specified for heavy fuel engines received special attention to extend their lifetime. Vanadium-induced corrosion is significantly minimized, Caterpillar reports, by using a Nimonic 80A material in the valves and reducing the exhaust valve temperature to 400°C. A low temperature is maintained by exploiting valve overlap, water cooling the insert seats, and applying a ceramic coating to the valves.

High efficiency turbochargers (one set is mounted on the in-line cylinder models and a pair on the vee-models): equipped with radial flow compressors and axial flow turbines. The axial flow turbine combined with a Caterpillar-supplied washing device is said to be well suited to heavy fuel operation. The turbochargers are water cooled and their bearings lubricated with engine oil. The turbocharger technology combined with a large cylinder displacement and efficient after-cooling yield a high air/fuel ratio. The results are more complete burning for maximum efficiency and better cooling of the combustion chamber and valves.

Cooling system: a single external cooling system supplies water to the two water pumps. The system sends cool water to the after-cooler and oil cooler, and hot water to the jacket water system to combat fuel sulphur-induced corrosion. The option of separate circuit cooling is offered for heat recovery.

Lubricating system: this features a pre-lube pump and a priority valve which regulates oil pressure at the oil manifold rather than at the pump. Bearings are thus assured continuous lubrication if the filters plug. Duplex filters with replaceable elements allow servicing without shutdown.

Fuel injection: Caterpillar's unit injector system has reportedly proven itself in distillate, marine diesel oil and heavy fuel burning operation. Efficient combustion is promoted by a high injection pressure (1400 bar) and precise timing. The mounting of the unit injector pumps directly in the cylinder head, however, allows the external fuel lines to carry only 5 bar pressure, reducing the danger of line breakage. Injector tip cooling—reducing carbon formation on injectors—is available optionally for heavy fuel burning engines. Individual control racks for each cylinder foster precise injection timing and minimize fuel wastage.

Exhaust and air intake systems: these contribute to fuel efficiency and response. The air induction system is routed from the turbocharger to the intake plenum with only one right-angle bend to the after-cooler housing. The exhaust manifold system is dry shielded to secure a low surface temperature.

The pumps, filters and coolers are engine mounted. An accessory module—which can support an expansion tank, heat exchanger, instrument panel, heater and other support elements—simplifies installation. It can be front engine mounted or remotely installed up to 8 m from the engine.

Ease of maintenance was addressed by arranging the gear train components for removal or replacement without taking off the front or rear engine cover. The crankshaft damper is mounted outside the engine housing for optimum cooling and accessibility; bolted covers and replaceable nylon bearings permit rebuilding of the damper on site. Four inspection covers for each cylinder allow convenient inspection and service access to the connecting rod and main bearings, the segmented camshaft (one per cylinder), the pushrods and fuel linkage, the injector and valve train.

The 3600 engine family is released for operation on fuels with a viscosity and contaminants up to CIMAC class K55 (700 cSt at 50°C) at speeds of 800 rev/min and below, and CIMAC class G35 (380 cSt) above 800 rev/min. The following features distinguish the heavy fuel engine from its distillate fuel-burning counterpart: water-cooled valve inserts; ceramic-coated exhaust valves; fuel injector nozzle tips optimized

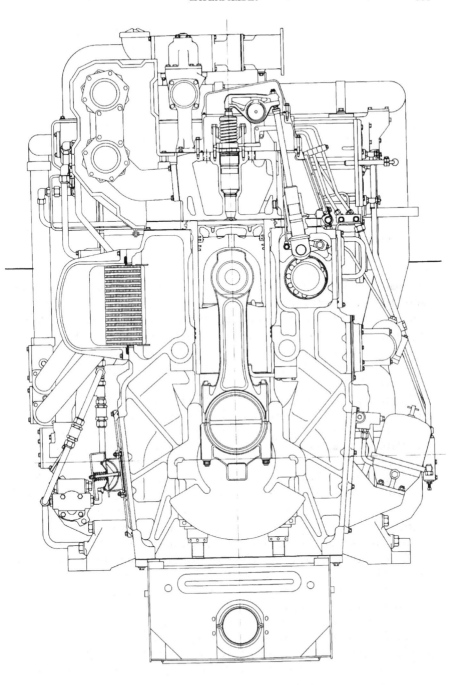

Figure 19.1 Caterpillar 3600 series medium speed engine in in-line cylinder configuration

for high viscosity fuel; cooled fuel injector nozzle tips; higher mass flow rate turbocharger components; remotely mounted heated fuel filter; flange-cooled cylinder liners; and a turbocharger water wash system.

Caterpillar 3600 series engine data

Bore	280 mm
Stroke	300 mm
Displacement	18.5 litres/cyl
Cylinders	6/8L, 12/16/18 V
Compression ratio	13:1
Rated speed	750–1000 rev/min
Mean piston speed	7.5–10 m/sec
Mep, mcr	22–23.7 bar
Mep, csr	20–21.5 bar
Output range	1490–7200 kW

Figure 19.2 High performance version of Caterpillar 3600 series engine

3618 (BRAVO) ENGINE

A V18-cylinder derivative of the 3600 series was jointly developed by Caterpillar with the Spanish enginebuilder Bazan Motores to target high speed ferry propulsion opportunities (Figures 19.2 and 19.3). The 3618 engine was initially released in 1998 with a fast commercial vessel rating of 7200 kW at 1050 rev/min; higher sprint ratings were available for military applications. The designers sought improved engine reliability and maintenance-friendliness in addition to a 17 per cent rise in specific output (to 400 kW/cylinder). An improved power-to-weight ratio was also yielded despite increased component robustness, primarily by using nodular cast iron for the block, adding strength without increasing wall thickness and weight.

All critical operating temperatures—coolant, lube oil and exhaust gases—are lower than previous 3600 series engines, despite the higher power output. An 80°C reduction in the inlet temperature to the turbocharger was secured, promising improved engine reliability and durability. A new seawater pump provided 33 per cent more flow and 60 per cent more suction lift. A significant contribution to higher performance came from an ABB TPL65E turbocharger, with a compressor bypass system closely matching the amount of intake air

Figure 19.3 One of four 7200 kW Caterpillar 3618 engines installed in a fast catamaran ferry

to the operating requirements and eliminating mid-speed instability. The new aftercooler was twice as large as before.

Other new elements include: a simplified and improved rear engine gear train (the idler gear removed to form a direct V-drive to the camshafts); new stainless steel fabricated exhaust manifold sections with permanent lagging to reduce maintenance as well as bulk; high strength Nimonic valves for both air intake and exhaust gas flows; a deep bowl steel piston crown connected to a new nodular iron skirt for added strength; interior piston cooling and wrist pin lubrication doubled with two oil jets per cylinder; and a new crankshaft counter-weighted to optimize rotary mass balance. Oblique split connecting rods ease servicing: rods are pulled with the piston, allowing cylinder overhauls to be executed without draining the engine.

Some of these features were applied to an uprated 3616 engine, released in 1998 with an output of 6000 kW at 1020 rev/min (equivalent to 375 kW/cylinder) for fast ferry propulsion. A six per cent power rise over the standard V16 model's 5650 kW was achieved without increasing peak cylinder pressure. Smoother acceleration, higher efficiency, lower emissions and more flexible installation requirements were also gained. Optional fore and aft turbocharger mounting was offered, along with off-engine lube oil coolers and a compressor bypass air induction system. The bypass system promotes more efficient engine operation at full power without causing surging at part load, such as during acceleration.

See Chapter 30 for Caterpillar high speed engines.

20 Deutz

A wide range of high and medium speed engine designs has been produced over the years by Deutz, which absorbed another German enginebuilder, MWM, and was latterly known as Deutz MWM. The medium speed programme is currently focused on the 240 mm bore 628 series. The 250 mm bore 632 series, co-developed with US-based General Electric Transportation Systems and introduced from 1994, is now available only as a gas engine covering outputs from 3000 kW to 4000 kW. Higher specific ratings were yielded by larger engines in the earlier programme: the 330 mm bore 645 series (described below) and the older 370 mm bore 640 series.

628 SERIES

The 240 mm bore/280 mm stroke 628 design is available in six, eight and nine-cylinder in-line and V12 and 16 configurations covering a power range from 995 kW to 3600 kW at 750–1000 rev/min.

A one-piece crankcase of nodular cast iron with suspended main bearings transfers gas and mass forces directly to the crankshaft bearings. Two horizontal and two vertical bolts for the bearing cap form the integration between the main bearing and the crankcase. Cast-on mounting straps enable flexible engine mounting, and a high degree of crankcase stiffness is achieved through deep main bearing caps tightened cross-wise.

The drop-forged steel crankshaft is dimensioned for mechanical security. Inductively hardened journals and an oil supply to the main and connecting rod bearings independent of piston cooling help to minimize bearing wear.

A composite piston incorporates a forged steel upper part and forged aluminium lower part. A side-mounted injection spray nozzle arranged in the crankcase continuously supplies the piston with cooling oil, intensively cooling its head, fire land and ring section by the shaker effect.

A nodular cast iron cylinder head, hydraulically bolted to the crankcase at four points, incorporates two inlet and two exhaust valves. The charge air pipe is split between the cylinder heads and connected by sliding pieces. Ease of maintenance was sought by the designers: it takes no more time to disassemble the entire cylinder head than it takes to remove the valve cages. The gas exchange valves are therefore arranged directly in the head, with the advantage of providing better cooling of the seat rings.

Engines for marine applications are equipped with single-circuit mixed cooling, with a coolant pump mounted and return cooling system for engine, lube oil and charge air cooling. A raw water pump can be attached.

TBD 645 SERIES

Introduced in 1993, the Deutz 645 engine is a 330 mm bore/450 mm stroke design produced in six-, eight- and nine-cylinder in-line versions to serve propulsion plant and genset drive requirements from 2550 kW to 4140 kW at 600/650 rev/min (Figures 20.1 and 20.2).

The monobloc crankcase is made of ductile cast iron, the side walls extending far down below the crankshaft centreline. It is provided with suspended crankshaft bearings, large service access openings and continuous mounting rails. The main bearing caps are bolted in the lower part to either side of the crankcase to ensure high crankcase stiffness. The crankshaft is of heat-treated forged steel and machined on all surfaces. The main and big end bearings are designed as grooved bearings. The die-forged diagonally split connecting rods feature a toothed joint and a stepped small end. The shank cross-section is of double-T configuration, with cooling oil supplied to the piston through a bore.

A forged steel crown and a forged aluminium body form the composite piston. Three compression rings are carried in hardened grooves in the crown, and one oil control ring in the body section. All the piston rings have a chromium-plated working surface.

The thick-walled centrifugally cast cylinder liner is salt bath-nitrided to minimize wear. A separate water guide ring with four passages to the cylinder head ensures intensive cooling of the upper liner section.

The ductile cast iron cylinder head is of double-bottom configuration and arranged to accommodate two inlet and two exhaust valves, each fitted with Rotocaps. The exhaust valves are installed in easily removable cages and can be exchanged without draining the cooling water. The water-cooled seat insert is armoured with Stellite F; the valve cones are

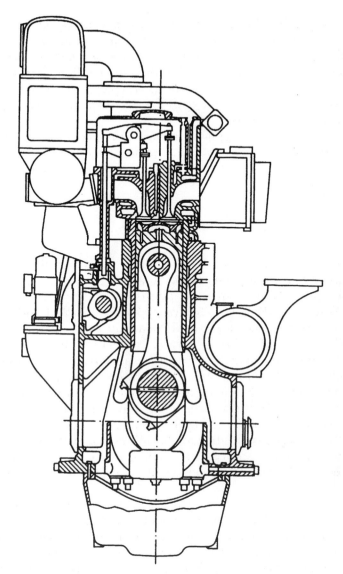

Figure 20.1 Cross-section of Deutz 645 engine. Note how the main bearing caps are bolted in the lower part to both sides of the crankcase, imparting high crankcase stiffness

armoured with Stellite 12 or, optionally, with Colmonoy 56 for heavy fuel operation.

Valve timing is actuated by a one-piece camshaft with hydraulically pressed-on individual cams, roller tappets, hollow pushrods and forked

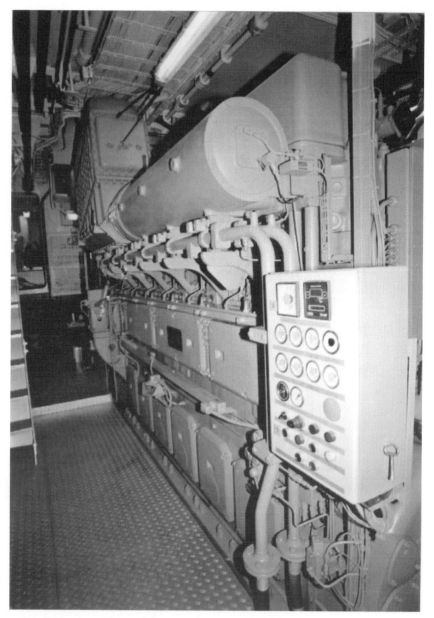

Figure 20.2 Deutz 645F engine installed in a Dutch trawler

rocker arms. The camshaft can be removed from the side to simplify servicing procedures.

A L'Orange fuel injection system comprises individual fuel injection pumps, very short injection lines ducted through the lower part of the cylinder head, and injectors which are cooled by the engine lube oil system. The injection pump element and roller tappet are separated by a lip seal. The commencement of delivery is adjusted according to the load applied.

Single-stage pulse turbocharging based on ABB turbochargers is associated with a two-stage charge air cooler allowing preheating of the air under low load conditions.

21 MaK (Caterpillar Motoren)

The German designer MaK, formerly a member of the Krupp group, came under the umbrella of Caterpillar in 1997 as MaK Motoren, adding a valuable medium speed portfolio to the US-based high speed engine specialist's programme. The company was renamed Caterpillar Motoren in the year 2000. At the time of the takeover, the MaK range embraced five designs with bore sizes from 200 mm to 580 mm offering outputs up to 10 000 kW: the 'new generation' M20, M25 and M32 and the longer-established M552C and M601C models. The latter pair of designs have since been phased out and a fourth new generation series, the M43, added in 1998. V-cylinder versions of this 430 mm bore design were launched in 2002, extending the power range of the MaK portfolio to 16 200 kW.

The new generation of long stroke medium speed engines was headed in 1992 by the 200 mm bore/300 mm stroke M20 design for small ship propulsion and genset drive duties. A power range from 1020 to 1710 kW at 900/1000 rev/min is covered by six-, eight- and nine-cylinder models. The M20 was joined in 1994 by the 320 mm bore/480 mm stroke M32 (detailed below to typify the new generation engines). The power gap between the M20 and M32 models was bridged in 1996 by a 255 mm bore/400 mm stroke M25 derivative (Figure 21.1) which covers an output band from 1800 to 2700 kW with six-, eight- and nine-cylinder models running at 720/750 rev/min. All three engines shared common development goals: robustness, high reliability, low overall operating costs, simple maintenance and extended intervals between overhauls.

Operational reliability was sought by avoiding corrosion, reducing mechanical wear and tear, increasing component strengths, and reducing the number of interfaces. Low operating costs were addressed by low fuel and lube oil consumptions, and low grade heavy fuel-burning capability. Modest maintenance costs were sought from a reduced number of components, a long service life from components, simple regulation and adjustment procedures, and plug-in connections.

Reduced installation work was achieved by providing simple and easily accessible interfaces for the shipbuilder. All connection points

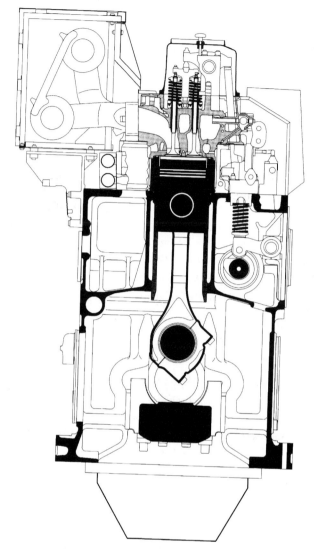

Figure 21.1 The M25 design typifies the new generation of engines from MaK (contrast with Figure 21.8)

for fuel, lube oil and cooling water systems are arranged for convenient access at the free end of the engine; and the fuel and lube oil filters are already mounted on the engine. The turbocharger mounting arrangement is variable, either at the flywheel end or free end of the engine.

High functional integration of components achieved engines with

40 per cent fewer components than their predecessors. A reduced amount of piping work in particular (and hence fewer connection points) yielded benefits in assembly, installation and maintenance. The remaining connections, where possible, are simple plug-in arrangements. Air, water, lube oil and fuel are guided through bores and plug-in connections inside the individual components, leaving minimal external piping in evidence. The charge air duct, for example, is integrated in the crankcase. The air flows through channels in the crankcase and water guide ring into the cylinder head to the inlet valves. Thus, when maintenance work on the head is necessary, no charge air pipe has to be removed and any danger of leakage or working loose is eliminated. The camshaft is also integrated into the crankcase. Another example is the integration of a slide valve gear into the fuel injection pump, eliminating starting air distributor and control air pipes.

Summarizing the common design features, MaK highlights:

- Underslung crankshaft.
- Stiff engine block with integrated charge air and lubricating oil ducts.
- Dry cylinder block; cooling water only where necessary.
- Long stroke (stroke/bore ratio of 1.5 in the case of the M32 in-line cylinder engine).
- Up to 40 per cent fewer components than earlier designs.

A high stroke/bore ratio fosters good fuel injection and combustion in a large combustion space; additionally, the high compression ratio underwrites low fuel consumption and emission figures.

M32 ENGINE

The M32 engine (Figure 21.2) superseded the well-established M453C series, a 320 mm bore/420 mm stroke design which had been offered as an in-line engine with six, eight and nine cylinders or as V12 and V16 versions with an output per cylinder of 370 kW. A substantial increase in power was yielded by the 320 mm bore M32 design whose modular construction also achieved an engine with 40 per cent fewer parts than its predecessor.

The in-line six-, eight- and nine-cylinder M32 models have a 480 mm stroke and originally yielded an output per cylinder of 440 kW at 600 rev/min, covering an output band from 2400 kW to 3960 kW at the economic continuous rating (ECR). The rating was raised to 480 kW/cylinder in 1998 to meet market requirements. V-configuration

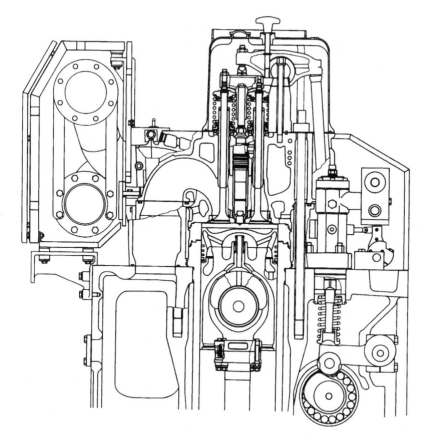

Figure 21.2 Upper part of M32 engine

models (12 and 16 cylinders) introduced in 1997 retained the 420 mm stroke of the M453C design and offered up to 480 kW per cylinder at 750 rev/min. The ECR outputs of the V-versions range from 4800 kW to 7700 kW.

The designer highlights the following key features of the in-line cylinder M32 models which mainly target marine applications:

• *Optimum thermodynamic conditions:* a stroke/bore ratio of 1.5 promotes a favourable air/fuel mixture and combustion in a spacious combustion chamber. Simultaneously, a high compression ratio (14.5:1) fosters low fuel consumption—around 180 g/kWh at full load—and low noxious exhaust emission values.

• *Integral construction:* advanced machining centres allow components to be fashioned to serve several functions. The engine frame, for example, embodies cast-on boxes for the camshaft drive on the flywheel end, and the vibration damper and secondary pinion gear for auxiliaries on the free end. The frame also conveys the charge air to the cylinders via a cast-in duct. The camshaft runs directly in the frame.

• *Stiff backbone:* the integral engine frame permits the firing and mass forces to flow through the frame, matching the flux of the lines of force and with low deformation. Girder-like cross-sections run through the engine housing in a longitudinal direction yielding, together with stiff walls, a system with high bending and torsional rigidity. Rigid as well as resilient foundation work is easily effected, and structure- and air-borne sound emissions are kept at a low level.

The 'supporting' and 'cooling' functions are clearly separated in the engine frame which is immune from any corrosion damage because no cooling water flows within it. The cylinder liner, however, is cooled where it is most important: in the upper region outside the engine frame.

• *Robust running gear:* the train from piston to crankshaft is designed for a high load-carrying capacity to secure operational reliability and provide reserves for future development in line with market demands. The integral construction ensures a safe bedding for the crankshaft, camshaft, camshaft drive and cam followers, fostering long service life values (Figure 21.3).

• *Modular system:* the integrated sub-assemblies, such as the cylinder head and engine frame, are complemented by support system modules designed with a reduced number of components for easy pre-assembly and flexibility of application. An example is provided by the modular turbocharging package whose charge air cooler is housed drawer-like in a console screwed onto the frame (Figure 21.4).

• *Functional groupings:* arranged to facilitate operation and maintenance procedures. The camshaft side of the engine is easily accessible because there are no obstructions from charge air ducts and other mountings. The exhaust gas pipes are arranged on the opposite side. An optimized flow in the cylinder head ensures low flow coefficients in the tandem conduits for the inlet air and the exhaust gas. Simplified maintenance was addressed by specifying plug-in type connections throughout, with special tools only required where conventional tools could not guarantee an adequate degree of mounting security.

A six-element hydraulic tool is used to release simultaneously the six nuts securing the cylinder head studs (Figure 21.5). A single bolt tightens the clamped joint between the cylinder head and the exhaust

Figure 21.3 M32 crankshaft installation

Figure 21.4 The M32 engine's charge air cooler is housed drawer-like in a console screwed onto the frame

Figure 21.5 A six-element hydraulic tool is used to secure or release the M32 cylinder head nuts

gas pipe. The fuel injection pump and nozzle can be disconnected by releasing just one screwed connection.

The connecting rod (Figure 21.6) is split just below the lower edge of the piston, promoting a low removal height for the running gear. The bearings do not have to be opened when a piston is drawn.

• *Tough materials:* essential castings, such as the engine frame and cylinder head, derive stiffness and security against fracture from high tensile nodular cast iron structures. The material inhibits elastic movement of the frame when subjected to firing forces. The circularity of the liners and the bedding of the plain bearings thus remain virtually unaffected and the axes of the gearwheels are maintained in parallel.

The material stiffness and stable design of the cylinder head secure a firm seating for the valves (Figure 21.7). The high Young's modulus of the nodular cast iron, the thick combustion chamber bottom and the adjacent cooling bores combine to foster operational reliability from the chamber and valves. The piston skirt is also of nodular cast iron, while the steel crown is served by numerous cooling bores.

Figure 21.6 The connecting rod of the M32 engine is split just below the lower edge of the piston

Figure 21.7 M32 cylinder head

• *Low thermal load:* the measured exhaust valve temperature of 390°C means that the Colmonoy armouring of the valves is highly resistant to inter-metallic corrosion attacks. A temperature of below 300°C measured on the outside of the piston crown permits only soft deposits.

Fuel injection can be adjusted for commencement, duration and pressure. Specially shaped control edges on the injection pump plunger secure correct commencement of injection as a function of output and speed. The M32 engine can also be provided with eccentric control of the cam follower between the camshaft and the injection pump. This arrangement facilitates simple adjustment of the firing pressure to suit the particular fuel grade or the climatic conditions and achieve optimum economy and operation on heavy fuel. Control of injection pressure is effected by an MaK-patented system which is said to foster part-load running with particularly low emissions and to enhance the ignition of heavy fuels.

Conventional injection pump systems deliver fuel in pulses which can cause vibrations in the associated piping and damage to upstream systems. The M32 engine features a system based on larger volumes with intermediate throttles for damping the delivery process without restraining it. The remaining vibrations in the fuel admission and return pipes are reportedly so low that elastic transition elements can be fitted between the pumps.

Adjustable valve timing is also required for special operating modes or to influence emission behaviour. The pushrods for the inlet and exhaust valves are therefore arranged on cam followers which can be adjusted by an optional eccentric. This arrangement reportedly allows valve overlap and the turbocharging group characteristic to be adapted to secure good part-load performance.

Like its predecessors, the M32 design is served by a pulse charging system allowing permanent operation on the propeller curve. As an option for particularly difficult applications, the in-line engine can be specified with control of the turbocharger turbine surface by MaK's Variable Multi Pulse (VMP) system. Faster acceleration, lower temperatures and higher part-load torques are possible. The designer has also developed systems for air bypass, exhaust gas blow-off and air blow-off.

MaK M32 ENGINE

Bore	320 mm
Stroke, in-line models	480 mm
Stroke, vee-models	420 mm
Cylinders	6, 8, 9L/12, 16V
Speed, in-line models	600 rev/min
Piston speed	9.6 m/sec
Speed, vee-models	750 rev/min
Piston speed	10.5 m/sec
Output/cyl	480 kW
Mep	22.8 bar
Output range	2400–7700 kW

M32C ENGINE

A C-version of the M32 engine was introduced in 2000 with design refinements promising cuts in operating costs, easier maintenance and installation, and performance and fuel economy boosted by a higher efficiency turbocharger. Among the numerous modifications are simple plug-in connections for the cooling water system, improved access to each cylinder unit in the camshaft region, single-pipe exhaust ducting and a newly-designed exhaust pipe casing with a smaller installation volume, and a nodular cast iron crankcase with integrated ducts.

The entire exhaust system was simplified in such a way that only one exhaust pipe (instead of two) is now required for the six-cylinder

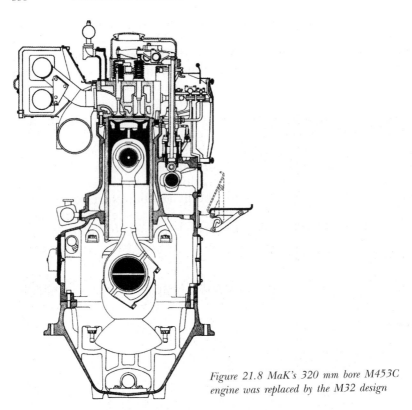

Figure 21.8 MaK's 320 mm bore M453C engine was replaced by the M32 design

engine; the number of pipes for the eight- and nine-cylinder models was reduced from three or four to one. Improvements were also made to the cooling water pipework at the supercharge air cooler. The cooler was modified by providing cast elbows as interfaces and simplifying accessibility to the cooler for cleaning. The application of stainless steel components on the 'cold side' also fosters an extended lifetime for these elements. The V12 and V16-cylinder versions of the M32C were introduced with two power ratings: 480 kW/cylinder at 720 rev/min and 500 kW/cylinder at 750 rev/min.

M43 ENGINE

A fourth series of new generation long-stroke engines, the M43, was introduced by MaK in 1998, effectively replacing the 450 mm bore M552C model and exploiting the same design philosophy as that of the M20, 25 and 32 series. With a specific rating of 900 kW/cylinder at 500/514 rev/min, the initial in-line six, seven, eight and nine-cylinder models covered an output range up to 8100 kW. The power band was extended in 2002 with the launch of V12, 16 and 18-cylinder (VM 43) models taking the upper output of the series to 16 200 kW (Figure 21.9).

Figure 21.9 A V12-cylinder version of the 430 mm bore M43 design (The VM43)

Eighty per cent component commonality is shared by the in-line and vee-cylinder designs, and simplicity of assembly was sought by combining as many functions as possible into single structural elements, thus reducing the overall number of components in the engine. Common components and component groups are: the complete cylinder head; valve drive; cylinder liner with flame ring; exhaust gas system (without compensator); camshaft bearings; fuel injection pump; high pressure connection; injection nozzles; cooling water distributor housing and line (without insertion tube); connecting rod and bearings; and vibration damper.

Cylinder head: a proven double-ended nodular cast iron design ensuring a high level of shape retention, the four-valve head is held on the cylinder liner by six hydraulically tightened studs (all the bolts are tightened simultaneously). Simple dismantling and assembly is fostered by insertion tube connections for all the media. The two exhaust valves are arranged to rotate, and the valve seating rings and ducts are aerodynamically optimized.

Piston: the two-part component comprises a steel crown and steel skirt. The crown is equipped with two compression rings and an oil stripper ring; the first ring is chromium plated on the flanks. The oil-cooled shaker space of the piston ensures effective cooling at a high mean effective pressure and contributes to a long piston life.

Connecting rod: the split-shaft design allows low piston installation/withdrawal heights. The rod bolts and shaft bolts are tightened hydraulically. (The size and weight of the hydraulic tools used for tightening bolts on the engine are minimized by operating with 2500 bar pressure technology.)

Special features of VM 43 engine:

Crankcase: a nodular cast iron design whether specified for resilient or rigid mounting. Stiffness in bending and in torsion is achieved by strong base strips and by integration of the boost air and camshaft casings; the transverse bolting of the basic cover also contributes. The multi-functionality of the crankcase is increased by its integration of the gearing space and damper space, which avoids numerous bolted connections and contributes to noise and vibration damping. Integrating the oil pipes in the crankcase also contributes here and facilitates maintenance work. Good accessibility to the main bearings and connecting rod bearings is provided via large space openings. The crankcase is naturally free from cooling water and hence protected against corrosion. The first main bearing is of tandem design, reducing the engine length.

Crankshaft: manufactured from heat treated alloy steel, with unhardened bearing trunions and concave fillets. A flange is provided for the vibration damper at the end opposite to the coupling; a gearwheel for the gear drive is flanged on the coupling end.

Exhaust gas installation: designed as a single-pipe system with gas outlets and pipe cross-sections optimized for efficiency. The elements are the same as those used in the in-line cylinder engines. The different distance between cylinders is made possible by compensators of varying lengths.

Turbochargers: configured as a sub-assembly and equipped with a block aftercooler, the turbocharger group can be arranged at either end of the engine, facilitating electric power generation or propulsion drives. A 10 000 kW power take-off drive can be specified optionally at the front end of the engine. The turbocharger arrangement contributes to a minimized centre distance in twin-engine installations, a low vibration level, reduced engine height, a maintenance-friendly block cooler and optimized exhaust gas flow. ABB Turbo Systems TPL73 turbochargers serve the V12 engine, and TPL77 models the V16 and V18 models.

Contributing to even lower emissions, the optional Flex Cam Technology system was applied to an MaK engine for the first time, promising NOx emissions of less than 8 g/kWh and soot emissions below the visibility limit in any operational mode. This is accomplished by adjusting fuel injection and inlet valve timing relative to the given load.

M43 engine design data

Bore	430 mm
Stroke	610 mm
Output	900 kW/cyl
Speed	500/514 rev/min
Cylinders	6,7,8,9L/V12,16,18
Power range	5400–16 200 kW
Mean piston speed	10.2/10.5 m/s
Mean effective pressure	24.4/23.7 bar
Maximum combustion pressure	190 bar
Specific fuel consumption*	175 g/kWh
Specific lube oil consumption	0.6 g/kWh

* 100 per cent mcr

Expert system support

MaK offers intelligent expert systems to lower operational costs, minimize downtime and increase availability. DICARE is a Windows-based expert system providing real-time engine diagnosis. It describes and diagnoses the actual engine condition and provides references for condition-based maintenance. Remote data transfer to a shipowner's shore base is facilitated by a communication module; additionally, MaK offers a link to its service department for evaluating and interpreting the data.

DIMOS is a computer-based maintenance and service system covering MaK engines. It delivers service interval supervision and immediate access to maintenance and component information via an electronic operations handbook. Bus-technology underwrites safe digital signal transmission and represents a clearly defined connecting point for all ship management systems.

EARLIER MaK MODELS

Numerous examples of earlier MaK medium speed designs are in service in propulsion and auxiliary power installations based on 240 mm bore M282C/M332C, 320 mm bore M453C, 450 mm bore M552C and 580 mm bore M601C engines. The 'C' designation models were introduced in the 1980s with refinements enhancing the capability to operate on heavy fuel (Figure 21.8).

The robust engines feature housings of high duty cast iron which, in the case of the 320 mm bore and larger models, take the form of multi-sectional blocks connected by steel tension rods. The high pre-stressing of the rods, which keep the cast components under compression stress, prevents any tensile stress in the cast material, even under ignition loads. The 240 mm bore series features an underslung crankshaft.

The cylinder liners of all the engines are salt-bath-nitrated for high wear resistance and anti-corrosion properties; crankshafts are forged in one piece from wear-resistant, quenched and tempered alloy steel. Single fuel injection pumps are used throughout the series, the injection system and inlet/exhaust valves being controlled by a one-part camshaft.

22 MAN B&W Diesel

A long tradition in medium speed engine design was exploited by MAN B&W Diesel to create a new generation of 400 mm, 480 mm and 580 mm bore models sharing common principles. The 40/54, 48/60 and 58/64 series (Figure 22.2), progressively launched during the 1980s, were joined in 1992 by the 32/40 design which inherited similar features but modified to suit its smaller bore. The four engines form the core of the Augsburg, Germany-based group's programme offering outputs from 2880 kW to 21 600 kW (see Table 22.1), which now also embraces a more powerful B-version of the 48/60 engine and a V40/50 design.

Figure 22.1 An eight-cylinder version of MAN B&W Diesel's largest medium speed engine design, the L58/64

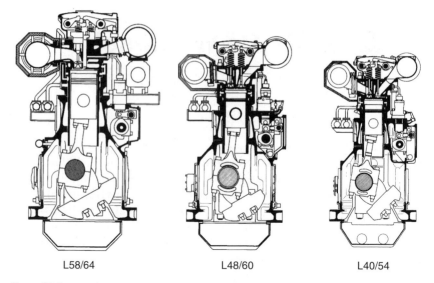

L58/64   L48/60   L40/54

Figure 22.2

Table 22.1 MAN B&W 'new generation' medium speed engines

Parameter	Model			
	32/40	*40/54*	*48/60B*	*58/64*
Bore (mm)	320	400	480	580
Stroke (mm)	400	540	600	640
Cylinders	5–9L/12–18V	6–9L	6–9L/12–18V	6–9L
Output (kW/cyl)	480	720	1200	1390
Speed (rev/min)	720/750	550	514	428

L58/64 ENGINE

The 'new generation' programme was headed in 1984 by the 580 mm bore/640 mm stroke L58/64 series whose designers sought a high overall operating economy, reliability, ease of maintenance, component durability and unrestricted heavy fuel compatibility. The rugged (Figures 22.1 and 22.3) result can be contrasted with MAN B&W Diesel's previous large bore engine, the 520 mm bore 52/55B (Figure 22.4).

Among the main features of the L58/64 and its 400 mm and 480 mm bore derivatives is a stiff monobloc frame casting with continuous tie-rods from the underslung main bearings to the top edge of the engine frame, and tie-rods from the cylinder head to the frame diaphragm plate (Figures 22.5 and 22.6). This arrangement was associated with: an optimum flow of forces from the cylinder head

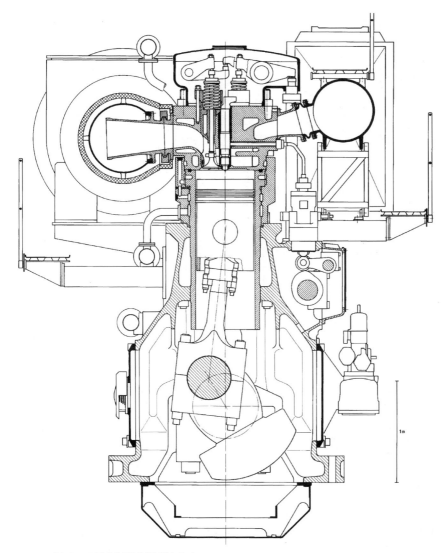

Figure 22.3 MAN B&W L58/64 design

down to the crankshaft; minimal deformation of the cylinder liner;
more reliable piston performance; all parts of the frame under pre-
stress; cross-braced crankshaft bearings; an amply dimensioned outboard
bearing for absorbing high radial forces at the power take-off end of
the crankshaft; no cooling water passage in the casing (eliminating
the potential for corrosion); sturdy landings to prevent plant

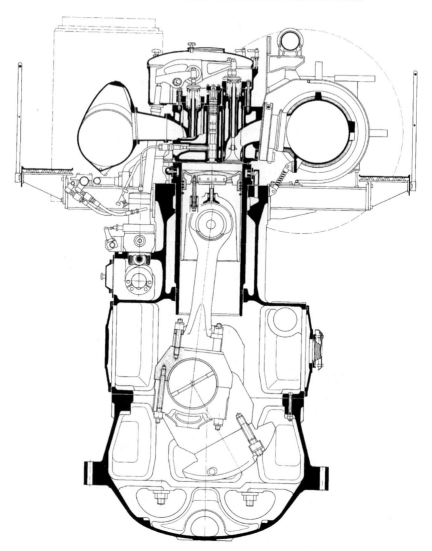

Figure 22.4 The L52/55B design preceded the 'new generation' range headed by the L58/ 64 engine

deformation; and camshaft drive gear and vibration damper integrated in the engine casing

Minimal deformation of the liner, reliable piston performance, no deformation of the neighbouring cylinder and reduced deformation over the full length of the engine are further underwritten by individual cylinder jackets.

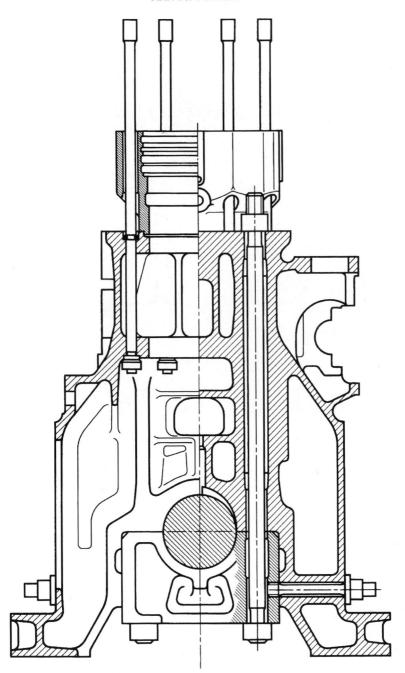

Figure 22.5 L58/64 engine frame

Figure 22.6 Monobloc frame of L58/64 engine

Cavitation of the cylinder liner (Figure 22.7) is prevented by thick walls giving a high resistance to deformation, while virtually uniform temperatures over the entire liner surface result from intensive cooling of just the upper part of the liner (cooling is not considered necessary elsewhere) thanks to a special arrangement of the cooling water spaces. Good lubricating conditions result and low temperature corrosion is inhibited.

A thin flame plate for the cylinder head promotes good heat conductivity while high stiffness is imparted by a strong deck plate absorbing the vertically acting gas forces (Figures 22.8, 22.9 and 22.10). A long valve life and clean, gas-tight seats are promoted by: water-cooled exhaust valve cages which are withdrawable without removing the rocker arms; armoured exhaust valves with gas-driven propeller rotators; and caged inlet valves rotated by a Rotocap device. Pollution of operating media is avoided in the cylinder head by separate spaces

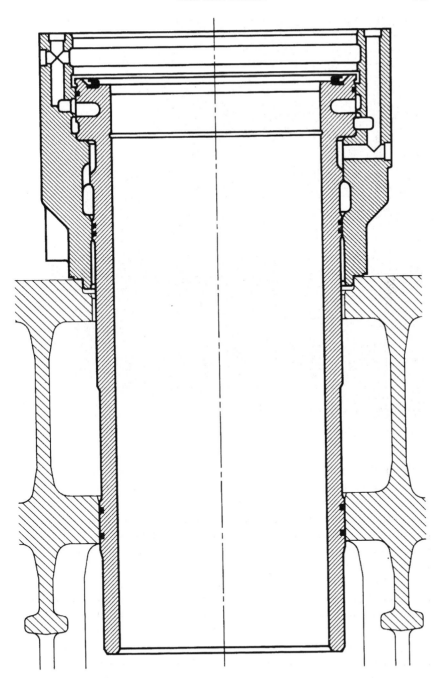

Figure 22.7 Cylinder liner of original L58/64 engine. A flame ring was specified in later models (see Figure 22.20)

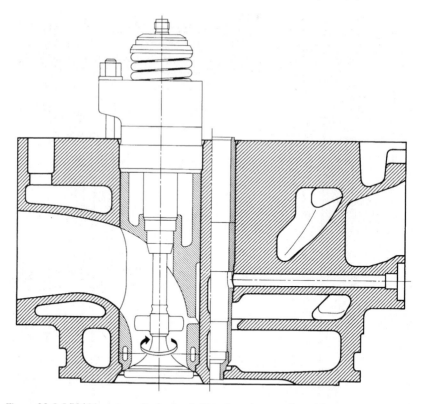

Figure 22.8 L58/64 engine cylinder head. Note the exhaust valve with rotator

for lubricating oil, fuel and cooling water. A simple connection attaches the exhaust manifold to the head.

A composite piston (Figure 22.11) is formed from a forged aluminium skirt and a forged thin-walled crown of high grade steel (the smaller bore engines in the programme feature a nodular cast iron skirt). Effective cooling of the crown is sought from the cocktail shaker method (splash effect) which nevertheless maintains temperatures in way of the ring grooves high enough to prevent wet corrosion. Cooling oil flow arrangements are optimized from inside to outside. A piston pin without cross-bore enhances component reliability. A reduction of mechanical loads on the piston rings was addressed by a narrow piston clearance, keeping out abrasive particles and protecting the lubricating film. A long piston groove life was sought from induction hardening, with the depth of hardening sufficient to allow re-machining several times.

All three compression rings are mounted in the crown, their durability

Figure 22.9 Cylinder jacket, head and cover of L58/64 engine

enhanced by plasma coating of the first ring and chrome plating of
the second and third rings. A consistently low lubricating oil
consumption is reported from the optimized ring set configuration.

The L58/64 engine was reportedly the first to feature a connecting
rod with a joint in the upper part of its shaft (Figure 22.12). A compact
and very stiff flange is permitted, and motion of the joint under mass
forces (observed in traditional connecting rods) is not encountered.
A merit is the very low overhead space required for piston removal,
with no need to open the crankpin bearing when pulling the piston
(Figure 22.13).

A fuel-optimized injection system with economy plunger is installed
(Figures 22.14, 22.15 and 22.16). A modified control edge on the
pump plunger keeps the firing pressure virtually constant between 85
and 100 per cent of maximum continuous rating, fostering favourable
consumption rates throughout the part-load range. A pressure-

Figure 22.10 Cylinder head of L48/60 engine

equalizing valve prevents the formation of partial vacuums (and thus cavitation) as well as pressure fluctuations in the system (and thus fuel dribbling). A high injection pressure yields efficient atomization of the fuel, while an advantageous firing pressure/mean effective pressure ratio underwrites low fuel consumption. A rocking lever between pump plunger and cam—horizontally adjustable by eccentric shaft—allows simple optimization of injection timing. The injection nozzles can be preheated for operation on heavy fuel oil.

High efficiency and high air flow rates at full and part loads are delivered by a constant pressure turbocharging system based on uncooled MAN B&W turbochargers. There is no need to preheat the air for part-load operation.

L40/54 and L/V48/60 engines

The L58/64 engine was followed on the market in 1987 by the L40/54 derivative, MAN B&W Diesel's experience with 400 mm bore designs at the time extending back over 20 years. The 40/45 design was introduced in the 1970s as an updated shorter stroke version of the RV40/54 series which made its debut in the mid-1960s as a 'first generation' modern medium speed engine. These older long stroke

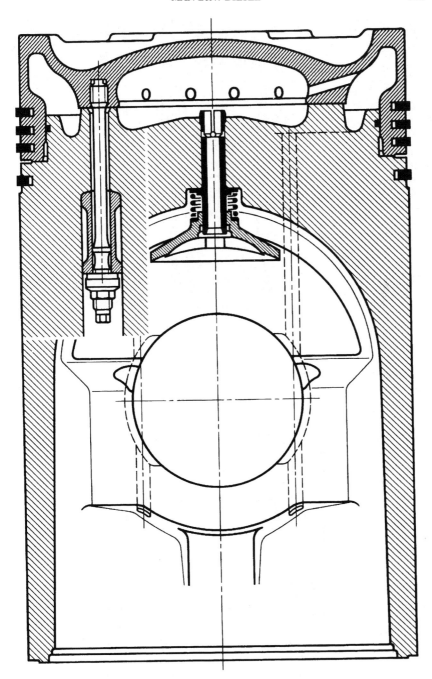

Figure 22.11 L58/64 engine composite piston

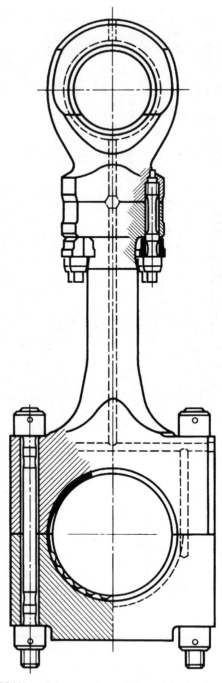

Figure 22.12 The L58/64 engine connecting rod has a joint in the upper shaft

Figure 22.13 A low piston dismantling height is achieved by the joint in the upper part of the connecting rod shaft (L58/64, L48/60 and L40/54 engines)

(540 mm) models were still produced by some licensees after phase-out from the official programme.

A third member of the family, the L48/60 series, became available in in-line cylinder form from end-1989. The prototype V-cylinder version, jointly developed with MAN B&W Diesel's associate company SEMT-Pielstick of France, was started up at end-1990 and production engines were offered from early 1992.

32/40 ENGINE

The 'new generation' family was extended in September 1992 by a smaller member to contest the competitive 320 mm bore propulsion

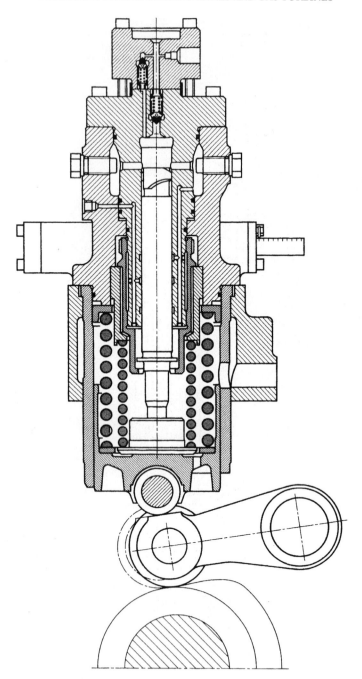

Figure 22.14 L58/64 engine fuel injection pump

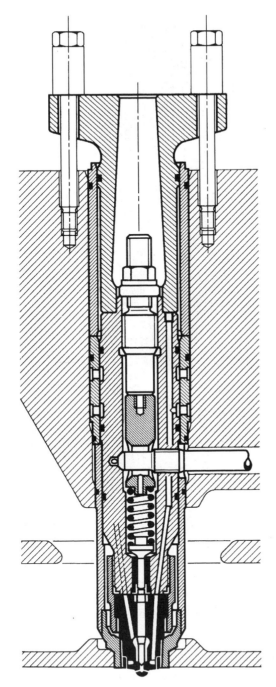

Figure 22.15 L58/64 engine fuel injection valve

Figure 22.16 Main components of L58/64 engine fuel injection valve

and auxiliary power arenas. The 32/40 series took the MAN B&W medium speed engine programme down to cover an output band from 2200 kW to 7920 kW with five, six, seven, eight and nine in-line and V12, 14, 16 and 18-cylinder models. The five-cylinder model was later removed.

The 320 mm bore/400 mm stroke design (Figure 22.17) originally yielded an output of 440 kW/cylinder at 720/750 rev/min on mean effective pressures and mean piston speeds at these rev/min ratings of respectively 22.8/21.9 bar and 9.6/10 m/s. A specific fuel

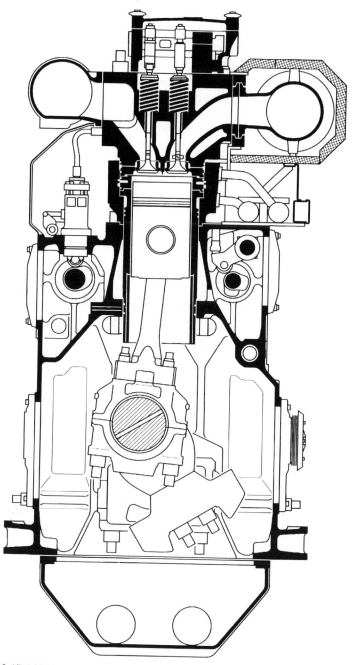

Figure 22.17 L32/40 engine. Note the separate camshafts for fuel injection and valve actuation, respectively, on the left and right sides of this cross-section

consumption of just under 180 g/kWh was quoted at 85 per cent maximum continuous rating. The 32/40 engine is currently released with a rating of 480 kW/cylinder, the series now spanning a power band from 2880 kW to 8640 kW. The mean effective pressure is 23.9 bar (750 rev/min) and 24.9 bar (720 rev/min), the mean piston speed around 10 m/s and the compression ratio 14.5.

Proven elements of the larger bore brothers were down-scaled for the 32/40 engine: a stiff one-piece frame with tie-rods; cylinder liners in the jackets; load-dependent cylinder lubrication; water-cooled liner fire rings with stepped pistons; and variable injection timing for optimum adaptation of the engine to various fuel grades (for both improved fuel consumption and reduced NOx emissions). Fuel economy was targeted by a high stroke/bore ratio, constant pressure turbocharging, the fuel-optimized injection system with 'economy plunger', pressure-equalizing valves, high intensity atomisation and high compression ratio.

An innovatory contribution to operating flexibility was made by the adoption of separate camshafts (arranged on either side of the engine) for the fuel injection and valve actuating gear. One camshaft is dedicated to drive the fuel injection pumps and to operate the starting air pilot valves; the other camshaft, on the opposite exhaust side, serves the inlet and exhaust valves. This arrangement allows fuel injection and air charge renewal to be controlled independently, and thus engine operation to be optimized conveniently for either high economy or low exhaust emissions.

In the larger bore MAN B&W medium speed engines fuel injection timing may be influenced by a laterally displaceable rocking lever arranged between camshaft and pump tappet. The timing of the 32/40 engine, however, is adjusted by turning the camshaft relative to the camshaft driving gear. Such a solution can be applied without restriction since the air charge renewal cams are mounted on a separate camshaft. The relative movement between driving gear and camshaft is via a helically toothed sleeve integrated in the gearwheel, meshing with the corresponding counterpart on the camshaft. By shifting the gear coaxially the camshaft (which is fixed in the coaxial direction) can be turned infinitely relative to the gear by the required and/or desired angle (Figure 22.18).

The camshaft shifting facility is optional; a simpler design is specified for engines whose projected operational profiles do not justify adjustable injection timing. Retrofitting of the facility is possible at a later date, however, if service conditions dictate.

The valve actuating camshaft is provided with different cams for full-load and part-load engine operation, allowing valve timing to be

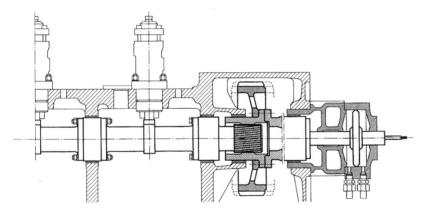

Figure 22.18 An L32/40 engine camshaft arranged for adjustment of fuel injection timing

tailored optimally to the conditions. As with a reversible engine, the camshaft can be shifted from one final position to the other to engage the two different cam configurations. Shifting is performed with the engine running. The load point at which switchover from one cam contour to the other takes place can be pre-set as required via the control system of the shifting mechanism (see Chapter 16/Figure 16.3).

Load-dependent adaptation of valve timing promotes low pollutant combustion in both high- and part-load operating modes. Clean combustion with little emission of soot particles is reportedly secured over the full load range.

The valve camshaft shifting facility—like its fuel injection counterpart—is an optional device: the standard engine version features just one cam contour. The conversion to an adjustable system for the charge renewal side, however, can be effected at a later date. Independent camshafts, along with the ability to fine-tune the timing and fuel pumps, reportedly underwrite almost constant firing pressures over the 85–100 per cent rating range, optimized performance down to partial loads, and the ability temporarily to handle fuels with poor ignition qualities. Efficient operation on low grade fuels with viscosities of up to 700 cSt is also fostered. An optimized cam- and camshaft-shifting facility on the fuel injection side allows IMO requirements on NOx emission levels to be met; and changeover from open sea to coastal and harbour service (from fuel-optimized to emissions-optimized/SOx-reduced operating mode) is relatively simple.

The 32/40 design is based on a very stiff cast monobloc single-part frame which is suitable for direct resilient mounting or semi-resilient mounting of the engine, with individual cylinder jackets as well as through bolts for the main bearings and lengthened cylinder cover bolts: a principle applied first in the 58/64 and subsequently in the 40/54 and 48/60 engines. The concept was modified for the smaller bore engine (Figure 22.19).

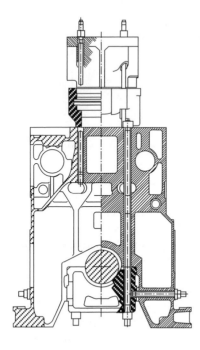

Figure 22.19 L32/40 engine frame

Individual cylinder jackets allow the geometry of the cylinder liner to be maintained under all load conditions, avoiding disturbing influences from neighbouring cylinders. The main bearing through bolts and the cylinder cover bolts extending to the middle frame range place the engine frame under compressive pre-stress, promoting high component reliability despite the high firing pressures.

The mounting arrangement of the cylinder liner in the jacket and engine frame is shown (Figure 22.20). Cooling water is supplied to the cylinder jackets only; none circulates in the engine frame structure, eliminating the risk of corrosion and water penetration of the lubricating oil. Individual cylinder jackets foster low distortion of the thick-walled laser-hardened liners. A uniform temperature pattern over the entire liner surface is reportedly secured through intensive cooling by a special water feed in its upper area. A separate fire ring (anti-polishing or carbon cutting ring) for the liner in association with an appropriately stepped piston inhibits the formation of combustion residue deposits

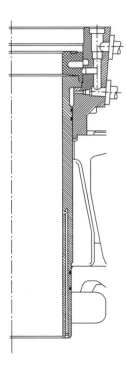

Figure 22.20 L32/40 engine cylinder liner section. Note the flame ring mounted on the top

on the liner; liner surface polishing is thereby prevented and low lubricating oil consumption promoted. The fire ring features jet bore cooling like the previous one-piece liner design. Experience has shown that removing the ring—necessary for piston withdrawal—can be carried out quite easily with the simple tools provided.

A four-bolt cylinder cover was designed relatively deep to ensure an excellent flow of power from the nut contact faces of the cover bolts to the seal between the cylinder head and the liner, as well as a very uniform distribution of pressure at the circumference of this seat face (Figure 22.21). The charge air and exhaust lines are arranged on opposite sides of the cover, the cylinder head depth allowing the flow ducts to be configured to enhance the overall efficiency of the turbocharging system. A bore-cooled flame deck yields low temperatures at the combustion chamber side and a high mechanical strength permitting the high firing pressures to be absorbed with a good safety margin. Valve cages were omitted. Both exhaust and inlet valves are accommodated in cooled seat rings, the exhaust valves featuring a rotator with a gas-driven propeller vane at the valve stem. The cylinder head, rocker arm casing and a section of the charge air manifold form a unit for maintenance purposes which can be dismantled as a whole when a piston has to be withdrawn.

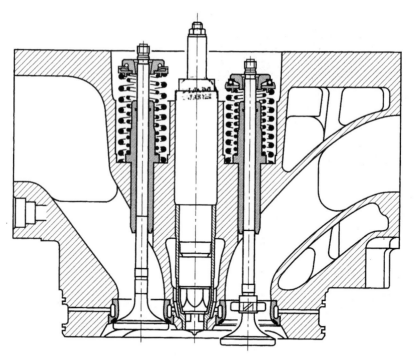

Figure 22.21 L32/40 cylinder head

The composite piston, similar to 40/54 and 48/60 engine practice, consists of a nodular cast iron skirt topped by a forged steel crown whose diameter is reduced in way of the top land to match the fire ring (Figure 22.22). The crown is cooled by the cocktail shaker method (splash effect) in its inner and outer areas, the latter also served by bore cooling. The configuration of the cooling chambers ensures a good shaking effect with the cooling oil supplied via the connecting rod.

Chromium/ceramic coatings for the top piston ring and chromium-plated second and third rings are specified to contribute to ring time-between-overhauls of two years or more. A nodular cast iron skirt allows a smaller clearance with the rigid liner to be adopted compared with aluminium, this creating favourable preconditions for piston ring running behaviour.

Smaller overall dimensions and more restricted space in the crankcase of the 32/40 engine precluded the use of the connecting rod design used in the larger members of the MAN B&W medium speed family which features a joint in the upper part of the shank. The traditional marine head design was therefore adopted, with a joint arranged just

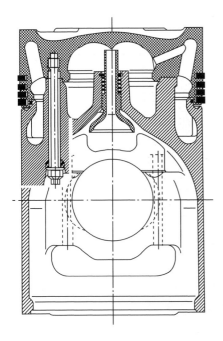

Figure 22.22 L32/40 engine composite piston

above the big end bearing box (Figure 22.23). No drawback is suffered with respect to the required headroom for piston withdrawal since the distance between the four cylinder cover bolts is sufficient for the connecting rod shank to pass without difficulty.

Design and material specifications of other key wearing components target extended lifetimes and maintenance intervals of at least two years for the liners and exhaust valves. A very low wear rate of the valves is promised from the armoured seats and directly water-cooled seat rings. A clean gas-tight seat is maintained by the exhaust valve rotator.

Constant pressure turbocharging is based on a MAN B&W uncooled turbocharger which, along with the charge air and exhaust gas ducting arrangements, promote a positive charge renewal performance and hence fuel economy, the designer asserts. MAN B&W NR26 or NR34 radial-flow turbochargers were typically specified for all cylinder number versions of the engine. Streamlined transitions from the cylinder heads to the exhaust gas line, as well as the double diffusor arranged downstream of the compressor, help to keep pressure losses in the exhaust gas system at a low level; and part of the kinetic energy of the charge air leaving the compressor at high speed is transformed into pressure on the air side. Both properties contribute to a substantial increase in overall efficiency.

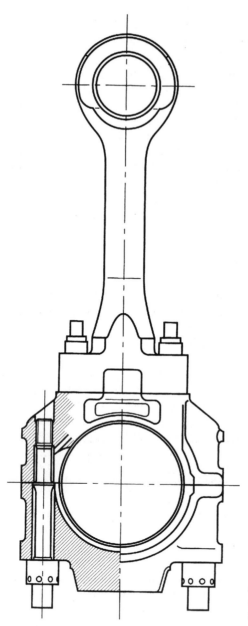

Figure 22.23 The L32/40 engine connecting rod is a traditional marine head design

If an engine's load profile dictates, warm charge air can be withdrawn directly after the compressor and supplied to the last section of the exhaust gas line upstream of the turbine via a bypass. The section is provided with a special belt to ensure perfect mixing of the bypassed air with the exhaust gas; a relief valve can also be fitted in the belt if required.

The charge air cooler can be of the single-stage or two-stage type. The latter is specified when the heat contained in the high temperature stage is to be exploited or if, in the case of frequent part-load operation, the charge air is to be preheated by the engine cooling water in the lower load range to ensure favourable combustion conditions.

Ease of inspection and overhaul—a focus of medium speed engine designers in an era of low manning—is addressed by the four-bolt cylinder cover with quick-acting seals and clamped and plug-in pipe connections. Only two different types of hydraulic tool are required to handle all the main screwed connections on the engine. The marine-type crankpin bearings allow the connecting rod to be withdrawn up the bore for piston and ring inspection without disturbing the bearings. The exhaust gas line features quick-release couplings that permit swift disconnection from the cylinder head.

MAN B&W Diesel originally anticipated a 550 man-hour annual maintenance workload for an eight-cylinder L32/40 engine running for 6000 hours on heavy fuel. The basic regular jobs are: replacing the fuel nozzles every 3000 hours; overhauling the exhaust valves (removing the cylinder cover) every 6000 hours; and overhauling the piston (including the replacement of rings and honing the liner) every 12 000 hours. Service lives and maintenance intervals for the key components of current 32/40 engines are indicated in Table 22.2.

Table 22.2 Service lives and maintenance intervals for 32/40 engine components

Component	Service life (× 1000h)	Maintenance intervals (× 1000h)
Cylinder liner	80–100	15–20
Piston crown	60–80	30–40
1st piston ring	15–20	–
Inlet valve cone	30–40	15–20
Exhaust valve cone	30–40	15–20
Fuel injection nozzle	6–10	3

32/40 engine refinements

Continual development of the 32/40 engine has pursued reductions in smoke emissions throughout the operating range without undermining fuel economy, easier maintenance, and improved flexibility in setting the engine's parameters for specific projects. Many components were modified in meeting these goals, such as the cylinder head, fire ring, support ring, rocker arms, valve cones, pistons, camshaft and gas exchange cams. Improved efficiency, particularly at loads below 50 per cent, was gained from a modified MAN NR-type turbocharger.

Measures to enhance emission values included a new injection nozzle design with a smaller sac volume minimizing the amount of fuel sprayed into the cylinder after the needle is closed. The benefits are reduced dripping and hence reduced smoke formation. A new cylinder head design incorporates an improved combustion chamber geometry promoting unhindered atomization of the fuel jets. The fuel injection pressure was also increased from its base of 1350 bar, and closing of the injection needle improved.

Efforts were also directed at reducing engine cost and weight, and facilitating maintenance. A redesigned rocker arm casing, for example, is lighter and allows faster nozzle replacement. A new steel piston was released in 2001. The lubricating oil and cooling water systems were combined into a single unit—the frame auxiliary box—to create a compact, vibration-free package calling for reduced piping.

48/60B ENGINE

The 48/60 engine was originally introduced to series production in 1989 with a rating of 885 kW/cylinder, which was progressively increased to 1050 kW/cylinder. Further development of the 48/60 design was stimulated by the higher propulsion power demands of cruise ships, ferries and large multipurpose cargo ships, as well as stricter environmental standards. The resulting B-version was launched in 2002 with a 14 per cent rise in specific output to 1200 kW/cylinder at 500/514 rev/min from a lighter and more compact design. The 48/60B programme spans a power band from 7200 kW to 21 600 kW with six, seven, eight and nine in-line and V12, 14, 16 and 18-cylinder models. The V-cylinder engines were released first, with the in-line models planned to follow in 2003 (Figure 22.24).

Retaining the original 480 mm bore/600 mm stroke dimensions, the designers aimed to: reduce fuel consumption; optimise combustion with a view to reducing NOx and soot emissions; boost the specific power output; reduce engine width; lower engine weight; optimize a range of engine components including the big end bearing; and reduce the overall number of engine components. Easier maintenance, higher reliability and lower manufacturing costs were also pursued. The goals were achieved by developing new sub-assemblies and/or improving components.

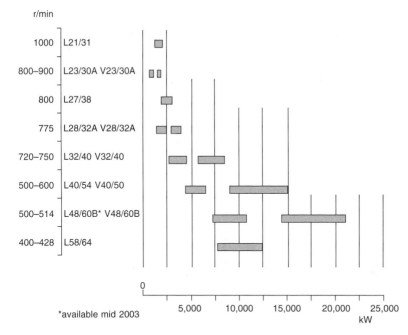

Figure 22.24 MAN B&W medium speed propulsion engine programme

Combustion benefited from: a simplified exhaust gas system with only one manifold and only one turbocharger on the V-cylinder engine; a redesigned combustion space; a new fuel injection system with high injection intensity; modified valve timings; and the new MAN TCA-series turbocharger (see Chapter 7). The compression ratio was increased from 14.4 to 15.3, a rise facilitated by a combustion space optimized by avoiding dead volumes, edges and ribs in the combustion side of the cylinder cover. The fuel injection pressure was increased from 1300 bar to 1600 bar, the higher intensity helping to reduce visible flue gas emissions, particularly in the part load range. Higher NOx emissions were neutralized by a slightly delayed injection timing.

A significant contribution to fuel economy came from the new MAN TCA77 turbocharger through an increased flow rate and an efficiency rise of 8–10 per cent (Figure 22.25). The minimum specific fuel consumption of 173 g/kWh (for the V-engine at 85 per cent maximum continuous rating) is 7 g/kWh less than that of the original 48/60 design, while the emission values on the IMO NOx test cycle are eight per cent less at 12 g/kWh (Figure 22.26).

Accompanying the higher output, fuel economy and environmental friendliness is a lower specific engine weight (reduced by 16 per cent

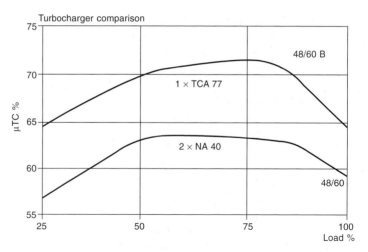

Figure 22.25 Comparison of turbocharger efficiencies: the MAN TCA77 on the 12V48/60B engine and the MAN NA40 on the original 12V48/60 engine

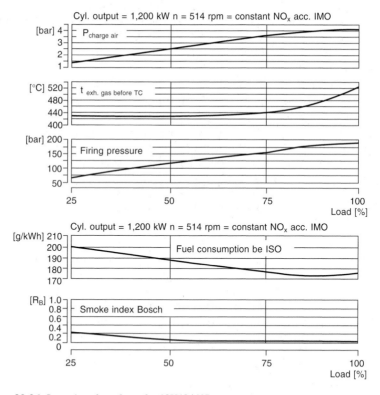

Figure 22.26 Operating data from the 12V48/60B test engine

to 11.9 kg/kW). The V12-cylinder 48/60B model, for example, develops 14 400 kW and weighs 181 tons, while its predecessor yields 12 600 kW from a weight of 193 tons. For the V18-cylinder engines, the reduction in weight is approximately 15 tons and the extra power around 2700 kW. The dimensions were also reduced: the height remains the same but the B-engine is 800 mm (15 per cent) narrower, thanks to a new exhaust system concept. A pair of 48/60B engines can thus be mounted closer together than before, giving an installation width saving of one metre. The turbocharger and two charge air coolers were combined to form a single module on the V-engine.

Internal refinements included an improved piston, with steel crown and nodular cast iron (or steel, if preferred) skirt. The cylinder head incorporates a re-shaped combustion chamber, no valve cages, an innovative rocker arm concept and a modified rocker arm cover. The connecting rod shank features optimized bearing shells, an increased oil clearance, reduced rotating masses and a strengthened bearing cover design.

Among the elements retained from the original design, with proven reliability in heavy fuel operation, were the stepped piston with chrome-ceramic first ring; exhaust valve with propeller (to ensure a clean and gas-tight seat); and MAN B&W's constant pressure charging system with exhaust gas wasting and bypass control.

IS (Invisible Smoke) variants of the 48/60 became available in early 2001 incorporating measures to achieve low Bosch smoke figures throughout the load range. The development was stimulated by requirements imposed on cruise ships in environmentally sensitive regions such as Alaska. The refinements—including a turbocharger optimized for lower loads, charge air preheating and a new cooled part-load fuel nozzle—can be applied to new engines and retrofitted

MAN B&W 48/60B engine data

Bore	480 mm
Stroke	600 mm
Speed	500–514 rev/min
Output	1200 kW/cyl
Cylinders	6,7,8,9L/12,14,16,18V
Power range	7200–21 600 kW
Mean piston speed	10–10.3 m/s
Mean effective pressure	25.8–26.5 bar
Maximum firing pressure	200 bar
Specific fuel consumption*	173 g/kWh
Specific lube oil consumption	0.6–0.8 g/kWh

* at 85 per cent mcr
(See Chapter 16 for details of 48/60B engine development procedure.)

to engines in service (see Chapter3/Exhaust Emissions and Control for details). MAN B&W favoured the fuel-water emulsion route to reduced smoke over direct water injection (Figure 22.27).

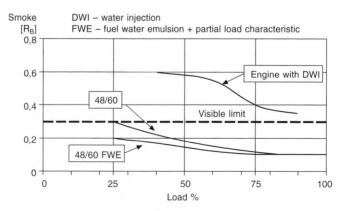

Figure 22.27 Smoke reduction from a 48/60 engine equipped with fuel-water emulsion (FWE) system

V40/50 ENGINE

A contender for ships valuing compact, high powered propulsion installations—notably the new breed of fast displacement-hulled RoPax ferries—was introduced by MAN B&W Diesel in 2001. The V40/50 series was evolved from group member SEMT-Pielstick's established 400 mm bore/500 mm stroke PC2.6B design (see chapter 25) and develops 750 kW/cylinder at 600 rev/min on a mean effective pressure of 23.9 bar. Outputs from 9000 kW to 15 000 kW are covered by V12, 14, 16 and 20-cylinder models, and a specific fuel consumption of 179 g/kWh at 85 per cent maximum continuous rating is claimed.

Improvements to the French designer's PC2 design were blended with MAN B&W Diesel's own proven medium speed engine features to yield a heavy fuel-burning model with a power/weight ratio of 9.3–11.6 kg/kW and a power/volume ratio of 71–75 kW/m^3 (Figure 22.28). Measures were taken to enhance the service life and maintenance intervals of key components, as well as to reduce overhaul times. Beneficial here is the patented modular pulse converter (MPC) pressure charging system which features only one turbocharger. IMO requirements on NOx emissions can be met by internal engine measures alone but much lower values—reportedly below 4.5 g/kWh—can be achieved by fitting a Munters Humid Air Motor (HAM) system based

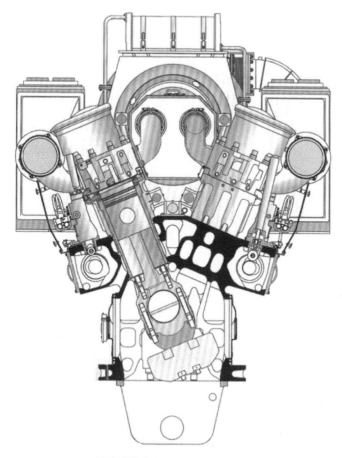

Figure 22.28 Cross-section of V40/50 design

on seawater for humidifying the charge air (see Chapter 3/Exhaust Emissions and Control).

L27/38 AND L21/31 ENGINES

A new generation of smaller medium speed designs for propulsion and genset drives—the L27/38 and L21/31 series—extrapolated the principles of the successful L16/24 auxiliary engine (see MAN B&W Holeby in High Speed Engines chapter), including the twin camshaft configuration pioneered by the L32/40 series. The L27/38 series (Figure 22.29) was introduced in 1997 and the L21/31 series in 2000.

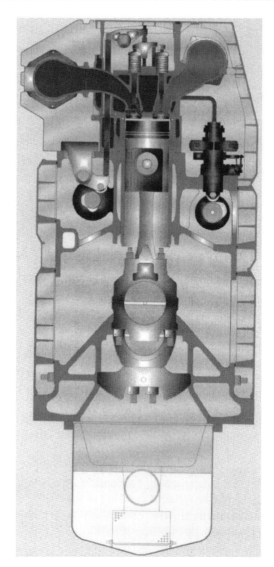

Figure 22.29 Cross-section of L27/38 engine, a larger version of the L21/31 design

The monobloc cast iron engine frame is designed for a maximum combustion pressure of 200 bar and direct resilient mounting. A static preloading of the casting is maintained by through-going main bearing tie rods and deeply positioned cylinder head tie rods, thus absorbing the dynamic loads from gas and mass forces with a high safety margin. All the rods are tightened hydraulically. Well supported main bearings carry the crankshaft with generously dimensioned journals. The combination of a stiff box design and carefully balanced crankshaft promotes smooth and vibration-free running. Large noise-dampening covers on the frame sides offer good access for inspection and overhaul. A 'pipeless engine' philosophy resulted in integrated oil ducts, with no water flowing through the frame; risk of corrosion or cavitation is thereby eliminated.

An innovation inherited from the L16/24 design, the front end box, is arranged at the free end of the engine. It contains connecting ducts for cooling water and lube oil systems as well as pumps (plug-in units), thermostatic valve elements, the lube oil cooler and automatic back-flushing lube oil filter. External pipe connections are arranged on the sides of the front end box to reduce engine length. A small optional power take-off is located on the forward side.

Less advanced but pursuing the same philosophy as the front end box, an aft end box is arranged at the flywheel end. This carries the turbocharger and incorporates a two-stage charge air cooler and integrated ducts for high and low temperature cooling systems. A regulating valve controls the water flow to the second stage of the cooler, adjusting to the operating conditions to secure an optimum charge air temperature.

High performance from the piston and ring pack depends on the cylinder liner geometry over the entire load range, the temperature in the TDC position of the top piston ring, and measures to avoid bore polishing. The liner is of the thick-walled high collar design, which protects the sealing between cylinder head and liner from the influence of engine frame distortions. Only the collar is water cooled, ensuring a stable geometry under variable load conditions. The temperature in the TDC position of the top piston ring is optimized to prevent cold corrosion, which is especially important in heavy fuel oil operation.

Years of experience with the flame ring concept successfully applied to the 28/32A and 23/30A engine series (see chapter 18/Alpha Diesel) was adopted for the L27/38 engine. The flame ring, inserted directly in the top of the cylinder, has a smaller inside diameter than the cylinder bore. The piston has a similar reduction in its top land diameter, allowing the flame ring to scrape away coke deposits and avoid bore polishing of the liner; optimum ring performance and low lube oil consumption over a long period are fostered.

Monobloc pistons in nodular cast iron had been state of the art for a number of years for applications imposing a maximum combustion pressure of around 160 bar. With pressures up to 200 bar, however, a monobloc piston was not suitable for the new engine generation. A composite piston, with bore-cooled steel crown and nodular cast iron body, was therefore specified. The different barrel-shaped profile of the piston rings and the chromium-ceramic layer on the top ring target low wear rates and extended maintenance intervals.

The connecting rod is of the marine head-type, with the joint above the head fitted with hydraulically tightened nuts; the head remains on the crankshaft when the nuts are loosened for removal. A low overhaul height is secured. If engineroom space allows, the complete cylinder unit (cylinder head, water collar, liner, piston and connecting rod) can be withdrawn as a single assembly. A complete unit can thus be sent ashore for overhaul and replaced with a new or overhauled unit. The thin-walled main bearing shells are dimensioned for moderate bearing loads and a thick oil film. The crankshaft is a one-piece forged component provided with counter-weights on all crank webs (attached by hydraulically-tightened nuts) to yield suitable bearing loads and vibration-free running.

A combustion pressure of 200 bar and a mean effective pressure of 23.5 bar influenced the choice of nodular cast iron as the material for the four-stud cylinder head. The rocker arms for the two inlet and two exhaust valves are mounted on a single shaft supported in the casting. The cylinder segment charge air receiver is integrated in the casting, with the exhaust gas outlet arranged on the opposite side. This cross-flow design, together with flow-optimised inlet and outlet ducts, creates a swirl effect benefiting charge air renewal and combustion.

Inlet and exhaust valve spindles are armoured with heat-resistant hard metal on the seats. The exhaust valve spindles feature integrated propellers for rotation by the gas flow in order to equalize and lower the seat temperature, and to prevent deposits forming on the seat and on the water-cooled valve seat ring. The inlet valve spindles have valve rotators to minimize seat wear.

Two independent camshafts separate the functions of fuel injection and charge air renewal, one operating the injection pumps and one actuating the valve gear. Optimum adjustment of the gas exchange without influencing the injection timing is thus secured. Cam followers on the valve camshaft each actuate two valves via pushrods, rocker arms and a yoke.

A dedicated camshaft for fuel injection makes it possible to alter the timing, when required, for a low NOx emission mode while the engine is in operation; the engine can also be easily adapted for running on different fuel oil qualities. The injection pump has an integrated roller guide directly activated by the fuel cam (Figure 22.30). Pipe systems are virtually eliminated thanks to slide connections between the pumps, promoting easy assembly/disassembly and safety against

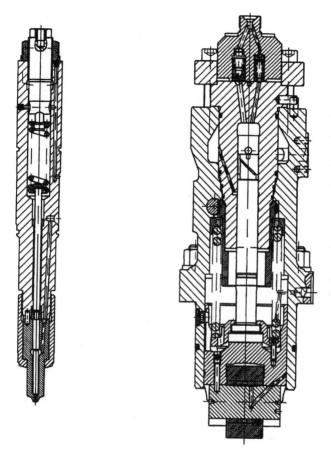

Figure 22.30 Fuel pump and injector of L27/38 engine

vibration and leakage. An uncooled injection valve contributes to simplicity of the injection system, which is designed for a pressure of 1600 bar; effective atomization is thus fostered even at part load operation with small injection volumes, contributing to a low exhaust smoke value. The complete injection equipment is enclosed behind easily removed covers. A fuel feed pump and filter are standard on propulsion engines.

The constant pressure turbocharging system incorporates a bypass connection from the turbocharger air outlet to the gas inlet side in order to increase the charge air pressure at part load. For certain operating conditions requiring high engine torque at reduced engine speed, a waste gate arrangement is incorporated in the exhaust system.

An entirely closed lube oil system eases installation and avoids the risk of dirt entering the lube oil circuit. A helical gear-type lube oil pump is mounted in the front end box and draws oil from the wet sump. The oil flows via a pressure regulator through the lube oil plate cooler and the automatic back-flushing filter; this solution eliminates the exchange of filter cartridges as well as the waste disposal problem. The back-flushing filter is drained to the sump and a purifier connected to maintain the lube oil in a proper condition. An integrated thermostatic valve ensures a constant lube oil temperature to the engine.

A cooling water system based on separate low temperature (LT) and high temperature (HT) systems is standard, both circuits cooled by freshwater (Figure 22.31). In the LT system, water is circulated by the LT pump through the second stage of the charge air cooler, the lube oil coolers for engine and gearbox, the HT cooler, through the

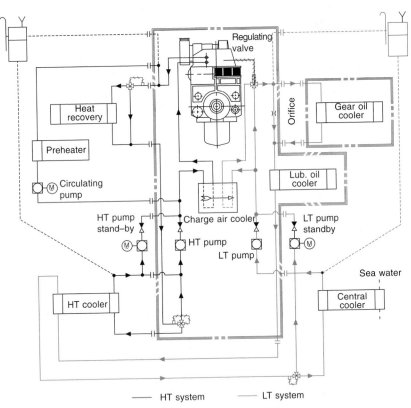

Figure 22.31 Cooling water system of the L27/38 engine, with the high temperature (HT) circuit on the left and low temperature (LT) circuit on the right

central cooler and via a thermostatic valve back to the LT pump. In the HT system, water is circulated by the HT pump through the first stage of the charge air cooler, the cooling water collar, cylinder heads and thermostatic valve, through the HT cooler and back to the HT pump. Almost 100 per cent of the heat removed from the HT system can be exploited for heat recovery. If required by the customer, a mixed LT/HT cooling system can also be arranged.

Engine speed is controlled by a mechanical or electronic governor with hydraulic actuator. All media systems are equipped with temperature and pressure sensors for local and remote reading. The number and type of parameters to be monitored are specified in accordance with (but not limited to) the classification society requirements. Shut-down functions for lube oil pressure, overspeed and emergency stop are provided as standard. The monitoring, control and safety system is arranged for compatibility with MAN B&W Diesel's CoCoS system for engine diagnosis, maintenance and planning, and spare parts handling (see below).

L21/31 and L27/38 propulsion engine data

	L21/31	L27/38
Bore	210 mm	270 mm
Stroke	310 mm	380 mm
Cylinders	6–9L	6–9L
Speed, rev/min	1000	800
Output, kW/cyl	215	340
Power range, kW	1290–1935	2040–3060
Mean effective pressure, bar	24.1	23.5
Mean piston speed, m/s	10.3	10.1
Combustion pressure, bar	200	200

(Five-cylinder versions are also offered for genset applications)

CoCoS system

Computer Controlled Surveillance (CoCoS) is a software family developed by MAN B&W Diesel and SEMT-Pielstick for engine surveillance, logistics and maintenance support. The operational benefits are cited as: optimized engine control and performance; automatic documentation of all events during the engine's life cycle; improved maintenance planning and execution; efficient spare parts stock keeping and ordering; and elimination of unscheduled, time consuming and costly repairs. The full CoCoS package comprises four modules:

- CoCoS-EDS (Engine Diagnostics System), which handles data logging, monitoring, trends and diagnostics. It can be operated with manual data entry or with operating values automatically transferred from the engine's monitoring and alarm system. The comparison of actual running performance with the ideal situation permits the identification of problems and irregular running conditions. Data (presented as graphs, bar charts or alphanumerically) include operating and reference values, pressure curves, load diagrams, characteristic maps and performance curves.
- CoCoS-MPS (Maintenance Planning System), which provides the data needed for managing all aspects of maintenance: scheduling, working instructions, reporting and the allocation of manpower, spare parts and tools. It automatically allocates dates for periodic maintenance based on recommended service intervals, and also suggests adding planned maintenance to repair work if this is time- and cost-saving.
- CoCoS-SPC (Spare Parts Catalogue), for easy and intuitive identification of access to information on individual parts. Information is organized into catalogues. It can be accessed either via a graphic over-view or in a list format; in the former, the user can identify particular components and display information about them by clicking on context-sensitive areas in a product drawing.
- CoCoS-SPO (Spare Parts Ordering) is a combined stock handling and spare parts ordering system which, together with the MPS module, automates spare parts availability/purchasing. It keeps track of parts in stock, making allowance for a safety margin and onshore stocks, and generates proposals for orders (if required, grouping the parts in question by manufacturer). The report generator creates statistics for use in evaluating and optimizing spare parts consumption.

See separate sections for other MAN B&W Diesel group four-stroke engine designs (Alpha, Holeby, Paxman, Ruston and SEMT-Pielstick).

23 Rolls-Royce Bergen

The diverse shipbuilding and marine engineering interests of Norway's Ulstein group for many years embraced the design and production of Bergen medium speed engines, now a key element of the UK Rolls-Royce group's diesel interests. Two Bergen engine families for propulsion and genset applications—the 250 mm bore K-series which dates from the early 1970s, and the 320 mm bore B-series introduced in 1986—have been progressively upgraded and were joined in 2001 by a completely new design, the 250 mm bore C-series, jointly developed with Hyundai Heavy Industries of South Korea.

K-SERIES

First introduced in 1971, the 250 mm bore/300 mm stroke K-engine was designed mainly to be a genset drive but numerous propulsion plant applications have been logged in tugs, offshore supply ships and fishing vessels. Engines are available as in-line three-to-nine-cylinder or V12-, 16- and 18-cylinder models covering an output range from 600 kW to 4010 kW at speeds of 720/750/825/900 rev/min.

Design improvements introduced over the years have sought to maintain market competitiveness, the maximum brake mean effective pressure being raised from 20 to 22 bar to increase the output to 220 kW/cylinder. The later adoption of a 'hot box' arrangement for the fuel injection system followed a trend in larger engine designs. It provides the engine with cleaner lines and improves the working environment in the machinery room: temperatures are reduced and any fuel leakage from the injection system components is retained within the box.

Other improvements have included a revised timing chain, sprockets and camshaft system to extend service life and accommodate a higher injection pressure. The valve stem rotators have been strengthened and the valve guides and guide seals improved to ensure better operational results running at 900 rev/min on heavy fuel.

The provision of a carbon cutting (or anti-polishing) ring at the top of the cylinder unit significantly reduced lube oil consumption and the amount of insoluble deposits in the oil (hence extending filter life). The ring comprises a sleeve insert which sits between the top piston ring turning point and the top of the cylinder liner. It has a slightly smaller diameter than the bore of the liner, this reduction being accommodated by a reduced diameter piston top land. The main effect of the ring is to prevent the build-up of carbon around the edges of the piston crown which causes liner polishing and wear with an associated increase in lube oil consumption.

A secondary function is a sudden compressive effect on the piston ring belt as the piston and carbon cutting ring momentarily interface; this forces lube oil away from the combustion area and again helps to reduce consumption. The efficiency of the rings was such that Ulstein Bergen found it necessary to redesign the ring pack to secure a desirable degree of oil consumption. The carbon cutting ring, refined over a two-year period, is now standard on all new Bergen engines and numerous engines in service have benefited from retrofits.

B-SERIES

The 320 mm bore/360 mm stroke B-series engine was conceived as a compact, heavy fuel-burning main engine and launched in 1986 with special features facilitating operation on lower price, poor quality fuel oils. The design's tolerance of fuel quality has been demonstrated by an oil rig installation running on crude direct from the well after degassing. A nominal rating of 360 kW/cylinder at 750 rev/min was initially quoted but some of the first engines sold were contracted at around 400 kW/cylinder, equivalent to a 10 per cent overload.

An uprating in 1991/92 took the output to 440 kW/cylinder at 750 rev/min which yielded a power band of 1870 kW to 3970 kW from in-line six-, eight- and nine-cylinder BR models. A decision to double the upper limit of the range—to 7940 kW—was announced in 1996 through the introduction of V12-, 16- and 18-cylinder models; only the V12 version is currently in the programme, however, offering an output of 5300 kW. These BV-engines (see below) were planned to benefit from Bergen's exhaust emissions reduction R&D by adopting faster fuel injection, a change in compression ratio and different valve timing to the existing in-line cylinder engines.

The B-engine was reportedly the first of its size to feature a completely bore-cooled cylinder unit and combustion space (initially, a bore-cooled liner and head and, later, bore-cooled pistons). This arrangement yields good strength and stiffness combined with good temperature control, which is important in heavy fuel operation.

Early in the engine's development it was discovered that the temperature profile on the liner was not the optimum. The resulting design change, benefiting subsequent production engines, has reportedly proved effective in the long term for both propulsion and auxiliary power duties, even at medium load. A drawback to the original bore-cooling concept soon became apparent, however: the use of electrically driven cooling water pumps. Electrical black-outs could not be tolerated and the fluctuation in cooling water flow away from the optimum resulted in some cracking of liners. An early change therefore saw a switch to engine-driven pumps and the incorporation of a new strengthened liner design.

In Arctic areas it was also found that piston rings were vulnerable to poor ignition qualities and high temperature gradients, resulting, in some cases, in ring cracking. This problem was solved by optimizing the clearances and the layout of the ring pack.

Other refinements following early installations included an improved bearing arrangement for the timing gears and, in line with other enginebuilders, a new fuel injection system was introduced in 1991 to counter an industry-wide problem of leakages. The complete injection system was enhanced by bigger pumps, better seals, larger nozzles and improved high pressure pipes. The improvements coincided with the 1991/92 decision to uprate the engine's output to 440 kW/cylinder while retaining the same engine speed of 750 rev/min.

Contributing to this power rise—and a brake mean effective pressure increase to 24.4 bar—was the first application of ABB's VTR-4P high efficiency turbocharger with a titanium compressor wheel, and a new type of composite piston with a bore-cooled forged steel crown and nodular cast iron skirt. A carbon cutting ring (described above in the K-series section) was introduced to the B-engine in 1994.

The one-piece design cylinder block of nodular cast iron is arranged for an underslung crankshaft and features large crankcase doors. A receiver and supply ducts for oil and water are cast in. The centrifugally cast cylinder liner is bore cooled in the upper part only where cooling is needed; its running surface is specially treated to enhance wear resistance. The bore-cooled cylinder head is provided with a thick bottom for control of mechanical and thermal loads, and six holding bolts for good load distribution. Heavy fuel service is underwritten by exhaust valves with a Nimonic head and welded-on hard seat facing.

A fully forged, continuous grain flow crankshaft has large diameter journals and pins for low bearing loads, and is fitted with hydraulically tightened counterweight bolts. The connecting rods are forged in

Figure 23.1 Ulstein Bergen's 320 mm bore series was extended in 1997 by a V-cylinder programme. The BVM-12 prototype is shown on test

alloy steel and machined all over, and feature an obliquely split big end with hydraulically tightened bolts. The tri-metal thin-walled bush bearings are formed from steel backed with lead/bronze bearing material and soft overlay. The two-piece piston is fitted with three compression rings and an oil scraper ring, all chromium plated to secure low wear.

A L'Orange fuel injection system was developed for an injection pressure of 1400 bar (endurance tested at 1700 bar) with constant pressure unloading for cavitation-free operation at all loads and speeds. A Bergen-developed cleaning/lubricating system cleans the lower pump parts of heavy fuel residues and lubricates the moving parts, helping to reduce the risk of fuel rack sticking problems.

An impulse turbocharging system based on ABB VTR..4 turbochargers is specified for six- and nine-cylinder engines, and a modified pulse converter system for the eight-cylinder model. Operation at low loads is fostered by elevated charge air temperatures through two-stage charge air cooling, high fuel injection pressures and optimized valve timing.

BV engine

The first BV engines—twin V12-cylinder models, each developing 5294 kW at 750 rev/min—were delivered in 1998 for powering a large anchor handling/tug/supply vessel. The BV design is based on a single-piece engine block cast in GGG500 nodular iron carrying two banks of cylinders in a 55-degree V-configuration, an underslung crankshaft and two camshafts; it also incorporates the charge air receiver between the cylinder banks (Figures 23.2 and 23.3). The camshafts are located outside each bank and housed in open-sided recesses in the block, allowing the complete camshafts to be removed sideways. At the front of the block is an opening for the charge air cooler and another for the auxiliary gear drive; the timing gears are arranged at the rear of the engine. The whole structure is designed for firing pressures in excess of 200 bar. An advantage of the block material is that it can be repaired by welding in the event of accidental damage.

A new cylinder liner design specified for the BV engines features a thicker upper wall section than the in-line cylinder BR models and a revised bore cooling layout (Figure 23.4). The liner, rated for mean effective pressures up to 32 bar and peak pressures in excess of 220 bar, was subsequently standardized for all B-series engines. Small changes were made to the BV cylinder heads, mainly a new head gasket matching the redesigned liner. The two-piece piston is essentially the same as that used in the BR engines, with a nodular iron skirt and bore-cooled

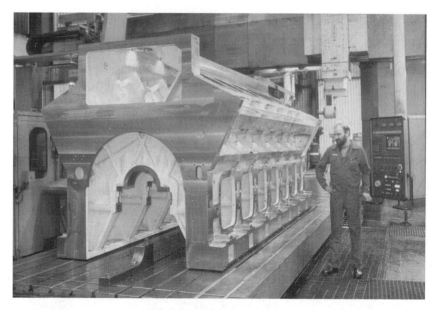

Figure 23.2 The BV engine block is a nodular iron casting

steel crown. The connecting rods were lengthened but used the same bearings as before.

Shorter fuel injection periods dictated strengthened camshafts to meet the increased loads, with a larger diameter both for the shaft and for the cam base circles. The fuel pumps are the same as those for BR engines but the fuel supply system was modified, chiefly by increasing pipe volumes and changing the layout to avoid cavitation and smooth out vibration caused by pressure pulses from the pumps.

Twin turbochargers mounted above the two insert-type charge air coolers operate on the impulse system, the pipework enclosed in an insulated box between the cylinder banks. A choice of electronic governors is offered, operating in conjunction with a standardized hydro-mechanical actuator. All electrical transducers on the BV engine are linked to a common electrical rail, one on each side of the engine, in a neat layout enabling faulty transducers to be quickly changed.

B32:40 ENGINE

The B-series benefited from another rejuvenation in the shape of the longer stroke (400 mm) B32:40 design offering an output of 500 kW/cylinder at 750 rev/min with a mean effective pressure of 24.9 bar.

Figure 23.3 Crankshaft of the BV engine

Production engines became available in the year 2000, supplementing the standard B-series in the programme. The B32:40 was initially offered in six, eight and nine-in-line and V12-cylinder configurations to span a power band from 3000 kW to 6000 kW; outputs up to 9000 kW were planned by extending the V-cylinder programme.

A new crankshaft design was incorporated to achieve the longer stroke and provide increased bearing areas, with new bearing technology

Figure 23.4 BV engine cylinder liner with bore-cooled upper part, also applied to the in line cylinder BR models

addressing the larger bearing loads. Modified cylinder liners sought good control over temperatures at all points, and the piston/ring designs were also changed to minimize lube oil consumption and cylinder wear. The cylinder head design and material specification was refined to allow a small rise in the maximum combustion pressure and to handle the greater combustion air throughput from the upgraded turbocharging system. A new fuel injection system provided both an increase in capacity and a higher injection pressure. These changes contributed to a specific fuel consumption of 183 g/kWh at the rated speed of 750 rev/min and NOx emissions under the IMO limit. Improved reliability and maintainability were promised by a new pump end module and turbocharger support structure.

Cylinder block: one-piece nodular cast iron design of rigid structure with underslung crankshaft, cast-in charge air manifold and large crankcase doors. The main bearing bolts are hydraulically tightened.

Cylinder liner: centrifugal cast, bore-cooled design with a running surface treated to improve wear resistance; a carbon cutting ring is incorporated at the top of the liner.

Cylinder head: bore-cooled flame plate; six hydraulically-tightened securing bolts to ensure even distribution; new cooled exhaust valve seat.

Crankshaft: continuous grain flow forged; large diameter journals and pins; hydraulically-tightened counterweight bolts.

Connecting rod: forged of alloy steel and fully machined; obliquely split and serrated big end; hydraulically-tightened big end bolts.

Bearings: steel backed with Sn/Al bearing material.

Piston: improved oil-cooled two-piece design; two compression rings and one oil scraper ring, all chromium plated.

Fuel injection system: pumps designed for 2000 bar injection pressure; totally enclosed in heat-insulated compartment; constant pressure unloading for cavitation-free operation at all loads and speeds.

Turbocharging system: multi pulse converter system based on ABB TPL series turbochargers; easily removable insulation panels for inspection and maintenance.

C-ENGINE (C25:33L)

A completely new design, the C25:33L, was launched in mid-2001 after a joint development project between Bergen and Hyundai Heavy Industries of South Korea, which markets and builds the engine as the HiMSEN H25/33. Flexibility in terms of power and speed ranges was a key development criterion to allow the 250 mm bore engine to target heavy fuel-burning gensets and smaller propulsion plants. Outputs from 1200 kW to 2700 kW at speeds from 720 rev/min to 1000 rev/min are covered initially by five- to nine-cylinder in-line models, with V-engines planned to double the power threshold. The C25:33L was expected eventually to replace the popular but ageing K-series, particularly in the genset market. The first four production engines (nine-cylinder models) were due for delivery in spring 2002 as the core of a diesel-electric propulsion plant for an offshore service vessel.

In drawing up the design parameters, the stroke (330 mm) was determined by the piston speed which, in turn, was set by the desired time-between-overhauls of 15 000 hours for the top end and 30 000 hours for the bottom end, running on heavy fuel oil. The running speed was determined by the frequencies in genset applications, focusing on 900 rev/min to 1000 rev/min for 60 Hz and 50 Hz requirements. These ratings sought the best compromise between an industry preference for a moderate speed in heavy duty gensets and the potentially lower price per kW from a faster running set. The engines can also be supplied, however, for 720/750 rev/min operation.

The C25:33L engine, Figure 23.5, is based on a compact and stiff nodular cast iron frame, with the charge air receiver, lube oil channel and coolant transfer channel incorporated in the casting to eliminate pipework. A continuous grain flow forged steel crankshaft with steel plate balance weights allowed the cylinder centre distance to be kept the same as the K-series engine in the interests of compactness and rigidity. Full power can be taken off either end of the crankshaft, and an additional main bearing allows single-bearing alternators to be driven.

A key feature is the cylinder unit system, allowing a complete liner, piston, upper connecting rod, water jacket and cylinder head to be drawn as an assembly for servicing or exchange. The components are

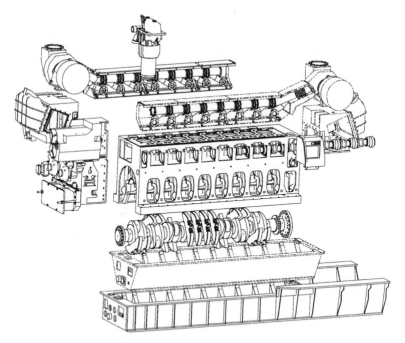

Figure 23.5 Primary modules of Bergen C25:33L engine

clamped together by the cylinder jacket and held down by four cylinder head bolts. A duct transfers air, exhaust and cooling water to and from the head, each cylinder unit being connected to its neighbour by quick-acting couplings to speed assembly and dismantling. The cylinder liner and water jacket combination is of the 'open deck' type, designed for intensive cooling and high strength without stress raisers in the critical top-end zone.

The piston comprises a steel crown with three rings and a nodular cast iron skirt. An extended life and low controlled lube oil consumption are fostered by a chrome-ceramic piston ring coating, an anti-polishing ring at the top of the cylinder liner and special honing of the liner surface. A three-part connecting rod enables its upper part to be detached when drawing pistons without disturbing the big end bearing.

One mechanical fuel pump is provided for each cylinder, with a simple two-step electronic timing feature controlled by the engine management system. (A common rail fuel system was not selected by Bergen because it was not seen as essential for the genset applications envisaged, nor as durable enough for unrestricted heavy fuel operation.)

An impulse turbocharging system is based on an uncooled radial turbocharger which can be mounted at either end of the engine to suit the particular installation. A swift response to load changes, with minimal smoke emissions under rapidly changing load and speed conditions, is fostered by the high efficiency turbocharging, two-stage charge air cooling with electronic temperature control, and a high speed electronic governor.

Ease of maintenance was addressed by providing good access to key components and a small number of hydraulic tools able to handle all the main bolts. Ancillaries are grouped in a front end module accommodating pumps, charge air cooler, lube oil cooler and filters. The components are designed for plugging-in, with a minimum of joints avoiding leakages.

Bergen C25:33 engine data

Bore	250 mm
Stroke	330 mm
Speed	720–1000 rev/min
Output	220–300 kW/cyl
Cylinders	5,6,7,8,9L
Power range	1200–2700 kW
Mean effective pressure	22.6–24.7 bar
Mean piston speed	7.9–11 m/s
Firing pressure	190 bar
Specific fuel consumption	182–186 g/kWh

See Chapter 2 for Bergen gas-diesel engines.

24 Ruston (MAN B&W)

A programme embracing three medium speed designs—the RK215, RK270 and RK280 series—was developed and produced by Ruston Diesels, a pioneering British enginebuilder which was a member of the Alstom Engines group until acquired by MAN B&W Diesel AG of Germany in June 2000.

The RK270 engine entered service in 1982, the 270 mm bore/305 mm stroke design inheriting the pedigree of the 254 mm bore RK series which originated in the 1930s and continued in the programme as an upgraded RKC series also with a 305 mm stroke. The RK270 retained well-proven features of its predecessor but exploited new components to underwrite a higher performance and achieve enhanced lightness and compactness. A fast ferry version of the engine has been progressively refined since entering service in 1990, with aluminium alloy specified in non-stressed areas and fabrications replacing castings where feasible. Attention was also paid to the weight of the ancillary equipment serving the engine. Later design refinements included revised cylinder heads with larger valves and upgraded fuel injection systems.

Early fast ferry propulsion installations were based on 16RK270 engines with individual ratings of 3650 kW at 750 rev/min (Figure 24.1). The output was raised in succeeding ferry generations to 4050 kW at 760 rev/min, 4440 kW at 782 rev/min and 5500 kW at 1000 rev/min. Demand for higher powers stimulated Ruston in early 1995 to introduce a V20- cylinder model to the programme, supplementing the established six and eight in-line and V12 and V16 engines. An output of 6875 kW at 1000 rev/min on a mean effective pressure of 23.6 bar is quoted for the 20RK270 design; some fast ferries feature engines specified with continuous ratings of 7080 kW at 1030 rev/min; and a naval rating of 7550 kW at 1032 rev/min is sanctioned (maximum power required for no more than five per cent of the operating profile). The V-engines are arranged with 45-degree cylinder banks. (Figures 24.2 and 24.3).

Design refinements have included the introduction of a revised configuration of compact pulse converter exhaust system in stainless

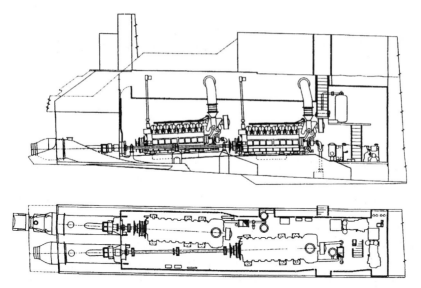

Figure 24.1 Two 16RK270 engines are arranged in each hull of the fast catamaran ferry Hoverspeed Great Britain

steel, with a larger bore and optimized area ratios, fostering a better gas flow that improves fuel efficiency. The system became standard on new engines and has been retrofitted to existing V16- and V20-cylinder installations. Engine tests confirmed the performance simulation prediction that the larger flow area, if applied alone, would in fact reduce turbine energy in the turbocharger and actually increase exhaust temperature without any significant improvement in fuel economy. A turbocharger rematch, however, resulted in an overall reduction in turbine inlet temperature by up to 30°C and a reduction in specific fuel consumption of around three per cent.

New 20RK270 engines also benefited from two TPL 65 turbochargers, the latest generation from ABB Turbo Systems (see Chapter 7). The higher efficiency of these units, compared with the previous VTC 304 type, yielded a further reduction in exhaust temperature and specific fuel consumption at full power. When applying such high efficiency turbochargers to engines operating over a propeller law power curve, Ruston explains, it is desirable to introduce an air bypass between the air manifold and the turbine inlet manifold over the lower part of the speed range. Such an arrangement delivers a further significant reduction in turbine inlet temperature in the region around 30 per cent power at 700 rev/min, and also improves the stability of the turbocharger compressor. The bypass valve is an integral part of the engine, programmed by the Regulateur Europa Viking 22 governor

Figure 24.2 One of four V20-cylinder RK270 engines installed in a fast ferry, each rated at 7080 kW

to open and close at pre-selected engine speeds: typically opening on a rising engine speed at 580 rev/min and closing at 920 rev/min.

Two main modifications were made to ensure compliance with IMO's Marpol Annex VI NOx emission requirements:

- The fuel injection timing was retarded by approximately five degrees, this change alone resulting in a NOx emissions reduction of some 20 per cent when calculated in accordance with the E3 test cycle appropriate to a propeller law-operated engine. Unfortunately, the specific fuel consumption was adversely affected.
- The compression ratio was increased from 12.3 to 13.3 by a simple adjustment that had been incorporated in the engine design; this allowed advantage to be taken of the increase in maximum

Figure 24.3 V20-cylinder version of RK270 engine

cylinder pressure resulting from the fuel injection retardation. The change causes a rise in NOx emission but of less than three per cent, and the resulting emissions are compliant with the IMO requirements; furthermore, most of the fuel consumption 'lost' by the injection retardation is restored. The increased compression ratio also improves starting on poorer quality fuels and curbs particulate and smoke emissions, particularly at part load.

Ruston gained valuable experience with selective catalytic reduction (SCR) installations required to satisfy extremely low NOx emission limits dictated in special regions. The four 20RK270 engines powering the 110 m long fast monohull ferry *Gotland* exhaust via Siemens SINOx systems which reduce emissions to below 2 g/kWh – less than 20 per cent of the IMO standard.

RK270 ENGINE DETAILS

The bedplate is machined from a high grade iron casting. Transverse diaphragms for each main bearing provide rigid support for the crankshaft, and angled joint faces ensure positive locking of the main bearing caps in the bedplate. The crankshaft is machined from a single-piece alloy steel forging with bolted-on balance weights. There is an integral forged coupling flange for the flywheel, and the camshaft drive gear is split to facilitate replacement.

The main bearings are pre-finished, steel-backed shells with bi-metal linings and are retained in the housing by caps drilled to direct oil to the bearings. The caps are located transversely in large registers in the bedplate and held in position by studs and nuts. The bearings are easily removable through the crankcase doors.

The crankcase housing has transverse diaphragms between each cylinder to provide water compartments around the cylinder liners. It incorporates an integral air chest and is machined from a high grade ductile spheroidal graphite iron casting. Explosion relief valves are fitted to the crankcase doors which allow easy access to the connecting rods and main bearings and to the camshaft and drive.

Separate wet-type liners cast in alloy iron are flanged at their upper ends and secured by the cylinder heads. Cutting rings are fitted at the top of the liner to prevent the build-up of carbon on the piston crowns. The lower ends of the liners are located in the crankcase and sealed by synthetic rubber rings. The liners are machined all over and the bore is hone finished to secure good piston/liner compatibility.

A single camshaft for each cylinder bank is installed from the side of the engine and run in generously proportioned bearings. The camshafts are of modular construction with single-cylinder sections joined by bearing journals. Individual sections can be replaced with the camshaft *in situ* or the complete camshaft removed from the side. The final timing is adjustable through a slotted driven gear and hub assembly. The camshafts are driven by a train of hardened and ground steel spur gears from the crankshaft split gear.

A two-piece piston, comprising a steel crown and an aluminium skirt, features a combustion bowl of the Hesselman design. Cooling oil is fed from the connecting rod through the small end bush and drillings in the gudgeon pin and piston body to the cooling gallery. The connecting rods are manufactured from steel forgings. The shell bearings are carried in obliquely split big ends, with the cap located by serrations on the joint face (a feature reducing the bending and shear loads across the joint). The stepped small end has a large diameter gudgeon pin with bronze bushings. A drilling through the shank of the connecting rod delivers oil for small end bearing lubrication and piston cooling.

Individual cylinder heads are machined from iron castings which accommodate two inlet and two exhaust valves surrounding a central fuel injector. Both the air inlet port and the exhaust port are located on the same side of the cylinder head. The inlet port draws air from the integral air chest, and the exhaust port feeds into a manifold system mounted above the air chest. Each pair of valves is operated via short stiff pushrods and conventional rockers. The pushrods are driven from the side entry camshafts via roller cam followers.

The fuel injection system features individual pumps directly operated from the camshaft, and injectors for each cylinder.

Air motor starting is standard, using one or two motors operating via spur gears on the flywheel rim.

A sensitive hydraulic governor is bevel driven from the camshaft. Overspeeding is prevented by a separate safety trip mechanism which returns the control shaft to the 'No-Fuel' position. A digital electronic governor is available as an option.

Auxiliaries are driven from the free end of the engine through a spring drive and spur gears. The standard arrangement embraces one water pump for the jacket water, one water pump for the charge air coolers and lubricating oil cooler circuits, a fuel lift pump and lubricating oil pumps. A secondary water pump may be fitted optionally. An extension shaft may be fitted to allow power to be taken from the free end of the engine.

The lubricating oil system includes single or twin engine-driven pumps, full flow filtration, thermostat and oil cooler. The oil pressure is controlled by a single or double spring-loaded relief valve. The jacket cooling system is thermostatically controlled and includes an engine-driven water pump.

The turbocharger(s) is mounted as standard at the free end of the engine, although a flywheel end location can be arranged if required.

RK215 ENGINE

The 215 mm bore/275 mm stroke RK215 engine (Figure 24.4) was originally produced in in-line six and V8 versions but V12 and V16-cylinder models later extended the upper power limit to 3160 kW at 720–1000 rev/min. RK215 engines have earned references in commercial and naval propulsion and genset applications, Ruston summarizing the benefits of the design to operators as: cost effective; high power-to-weight ratio; reduced space requirements; ease of installation; fuel efficient; and minimized environmental impact.

The general layout of the engine is conventional with wet cast iron liners and an underslung crankshaft. The latter feature was a departure for Ruston which had formerly produced bedplate-type engines (see

Figure 24.4 Ruston RK215 engine in V8-cylinder form

RK270 model above). The change in design philosophy reflected the high cylinder pressures (up to 180 bar), dictating a level of crankcase strength that could only be obtained from an underslung configuration.

The combustion space is sealed by a steel ring, and each cylinder is held down by six bolts. The underslung crankshaft is mounted to the block with rigid bearing caps, each having two studs; in addition, the V-cylinder engine uses cross-bolts for cap location. Hydraulic tensioning is applied to secure the vertical studs. The alloy steel crankshaft is a die forging with bolted-on balance weights, two for each throw, to maximize bearing oil film thicknesses. The bearing shells are of bi-metal construction with an aluminium–tin–silicon running surface which has a very high strength. In spite of the high cylinder pressures, the designer claims, the bearings are not highly loaded: the large end being under 41 MPa.

A two-piece die forging forms the connecting rod, with the large end split at 50 degrees to the vertical to allow removal through the cylinder bore. The split angle was optimized through model tests and finite element analysis to provide minimum bending at the split position under both firing and inertia load conditions, while maintaining an adequate bearing diameter and large end bearing housing rigidity. Each cap is retained by four bolts. The small ends of the connecting rods are stepped to minimize pressures on both small end bearing and piston boss.

The pistons are one-piece nodular iron castings with an integral cooling gallery fed with oil through drillings in the connecting rod and gudgeon pin. A three-ring pack is specified for the piston, all the rings having chrome running faces; the top ring is asymmetrically barrelled and the second ring is taper faced. The grooves for the compression rings have hardened surfaces. The oil control ring is one piece with a spring expander. Six drilled holes in the piston drain excess oil from the oil control ring back inside the piston.

Oil, water and fuel lift pumps are all driven by hardened steel gears located at the non-flywheel end of the crankshaft. The rotor-type oil pump delivers up to 400 litres/minute for lubrication and cooling via an engine-mounted cooling and filtration system which incorporates a change-over valve to enable continuous operation. The standard freshwater pump may be supplemented by one or, in special cases, two additional pumps to supply heat exchangers. The standard cooling system incorporates a freshwater thermostat and water bypass. Also standard is the fuel lift pump and pipework incorporating a pre-filter, fine filter and pressure regulating valve. The pump is sized to give a 3:1 excess feed capacity to provide cooling for the injection pumps.

Unit injectors for each cylinder were specified (reportedly for the first time in British medium speed engine practice), operating at pressures up to 1400 bar. The injectors, purpose-developed for the RK215 by L'Orange in conjunction with Ruston, are operated from the camshaft via a pushrod. A characteristic cited for unit injector engines is a low level of exhaust smoke in all operating conditions.

Fuel is supplied to the injectors through drillings in the cylinder head, dispensing with high pressure fuel lines and removing a potential cause of engineroom fire.

The iron cylinder heads are of two-deck construction with a very thick bottom deck which is cooled by drilled passages. Inlet and exhaust valve pairs are generously proportioned at 78 mm and 72 mm diameter respectively, and also feature wide seats to foster low seating pressures and good heat transfer from the valves to the seat inserts. This is particularly important because the exhaust seat insert incorporates a water-cooled cavity to maximize heat flow from the valves. The cooled exhaust seat is made from steel with a hardened surface, while the inlet valve uses the more common high chromium content iron.

Both inlet and exhaust ports are located on the same face of the cylinder head. The valves, which have hardened seat faces, are operated by rocker levers and bridge pieces. A third rocker lever between the valve rockers operates the unit fuel injector. All rocker levers on each line are actuated by pushrods and lever-type followers from the induction-hardened alloy steel camshaft. All bearings, including those for the follower roller and the spherical pushrod ends, are positively lubricated from the main oil supply at full pressure.

Two Garrett TV94 turbochargers were specified for the six-cylinder in-line model, each mounted on a very short manifold served by three cylinders. Charge air is fed from both units via a manifold to a single, integrally mounted air/water charge cooler. This arrangement is facilitated by the positioning of the inlet and exhaust on the same side of the engine. The resulting small volumes of both inlet and exhaust systems, plus the low turbine and compressor inertia, reportedly promote excellent load acceptance. A similar concept is applied to the V8-cylinder model but with the Garrett turbochargers replaced by dual ABB RR151 units.

V-cylinder models use similar line components to the in-line cylinder models: the cylinder head, piston, liner and connecting rod are the same, as are the cam followers, pushrods and camshaft drive. The crankcase and crankshaft, of course, are different.

To ensure that the underslung main bearing cap is rigidly located it is provided with cross-bolts in addition to the hydraulically tensioned main studs. V-engines use side-by-side connecting rods, and the crankshaft—while having the same size of crankpin as the in-line engine—has a 25 per cent larger main bearing journal diameter. The resulting overlap between pin and journal imparts strength and rigidity to the crankshaft.

RK280 ENGINE

A completely new medium speed engine, the RK280 design, joined the Ruston programme in 2001, offering a considerably higher output than the long-established RK270 series for commercial and naval propulsion and genset drive applications (Figure 24.5). The 280 mm bore engine was introduced as the world's most powerful 1000 rev/min design, released with an initial rating of 450 kW/cylinder. The programme of V12, 16 and 20-cylinder models thus covers a power range from 5400 kW to 9000 kW. An output of 500 kW/cylinder was specified for the design, however, equating to a future rating of 10 000 kW for the 20RK280 engine.

Figure 24.5 V16-cylinder version of the RK280 engine

RK280 design data

Bore	280 mm
Stroke	330 mm
Cylinders	V12, 16, 20
Output*	450 kW/cyl
Rated speed	1000 rev/min
Mean effective pressure	26.5 bar
Mean piston speed	11 m/s
Power range	5400–9000 kW
Power density	< 5.1 kg/kW
Fuel consumption	< 190 g/kWh

* Continuous rating

A competitive specific fuel consumption of under 190 g/kWh is reported, with NOx emission levels of less than 10 g/kWh from primary reduction measures alone. Compactness and lightness are reflected in a dry weight of 46 tonnes and a power density of 5.1 kg/kW for the V20-cylinder version, which measures 7.33 m long × 2.1 m wide × 3.18 m high.

The RK280 design benefited from experience with RK270 engines powering fast ferries, as well as feedback from RK215 engine installations. Advantage was also taken of advanced software packages, new development methodologies and project management techniques. Areas of the RK270 design that had proved particularly successful—such as the integrity of the bottom end in a dynamic fast vessel application exploiting an elastic mounting system—were thus translated into the design parameters of the RK280 engine. At the same time the opportunity was taken to address other features in eliminating the possibility of problems in service: for example, minimizing overhangs and induced stress in the engine, and reducing pipework on and off the engine. Among the measures adopted to enhance performance, fuel economy, reliability and serviceability were:

- Electronically-controlled pump-pipe fuel injection system to foster a low specific fuel consumption and reduced NOx and particulates emissions throughout the load range.
- Simple, single-stage high efficiency turbocharging based on ABB Turbo Systems' TPL 65 turbocharger for improved specific fuel consumption and lower operating temperatures; and a simple exhaust system with rigid heat shields helps to minimize and ease maintenance.
- Integrated pipe design with four-pipe connection to reduce vibration and ease installation and serviceability.
- Two-stage fresh water-cooled high efficiency intercooler in the centre of the engine to improve specific fuel consumption and curb NOx emissions while reducing mechanical loading.
- Reduced number of components (40 per cent fewer than the RK270 engine) for higher reliability and lower servicing costs.
- Rigid dry crankcase design for reduced weight and easier servicing.
- Low inertia/low friction two-piece pistons with alloy steel crowns to reduce weight and improve specific fuel consumption; three-ring pack.
- Simplified gear train of robust construction; water and lube oil pumps driven from the free end of the engine through gears housed in the pump drive casing (Figure 24.6).
- External viscous damper for ease of maintenance.

Key to main components

A Turbocharger
B Air filters
C Fresh water pumps
D Sea water outlet
E LT water outlet
F Fuel inlet
G Lubricating oil filter & dipstick
H Fuel lift pump
I Lubricating oil pump
J Fuel filters
K HT water outlet
L Duplex lubricating oil filter
M Lubricating oil cooler
N Heat shields
O Intercoolers

The New RK 280
up to 9000 kWb

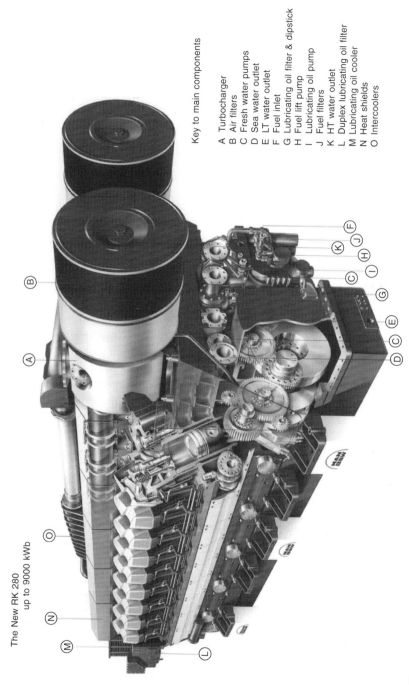

Figure 24.6 Cutaway of RK280 engine showing the gear train for the camshaft and service pump drives

RK 280 engine details

Crankshaft: manufactured from a high tensile NiCrMo continuous grain flow steel forging.

Crankcase: machined from spheroidal graphite cast iron and featuring underslung main bearings retained by two vertical studs and two cross bolts per side for overall stiffness. The main bearing caps are secured by hydraulically-tensioned studs to ensure maximum integrity of the crankcase system. A 52-degree V-angle minimizes torsional effects and allows location of the intercooler between the cylinder banks, reducing overhang loadings while minimizing engine height. Inspection covers on both sides of the engine provide access to internal components, and selected covers carry the explosion relief valves. Engine mountings on either anti-vibration mountings or solid seating is by separate bolt-on feet.

Cylinder liners: individual units incorporating deep flanges; strategically cooled by a separate water jacket enabling a dry crankcase to be used, reducing overall weight. The running surfaces are plateau honed and finished to improve oil retention throughout the liner life; a cutting ring is fitted at the top of the liner to eliminate the build-up of carbon on the piston crowns and minimize lube oil consumption.

Bearings: the generously dimensioned main bearings feature easily replaceable thin wall, steel-backed aluminium-tin shells.

Camshafts: modular design with one cam element per cylinder; they are hollow and form the main lube oil supply to the engine. Optimized cam profiles for electronically-controlled fuel injection minimize Hertzian stresses, enhancing reliability and extending component life. The camshaft drive is located at the free end of the engine; the crankshaft gear drives via a compound idler gear for each camshaft.

Piston: two-piece design with a lightweight body and alloy steel crown. A three-ring pack comprises two chrome-ceramic compression rings and an oil control ring. The case-hardened gudgeon pin is fully floating and retained by a circlip at each end. Lube oil is fed from the connecting rod through drillings in the gudgeon pin and piston to a cooling chamber in the piston crown. The oil is then discharged through drillings in the underside of the piston crown back to the sump.

Connecting rods: of forged high tensile alloy steel, the rods have obliquely split large ends that carry fully grooved bearings with the cap secured by four hydraulically-tensioned studs.

Cylinder heads: the individual heads have a thick combustion face incorporating coolant drillings. The two inlet and two exhaust valves, all with cooled seats, surround the central fuel injector. Twin inlet ports connect directly to the air manifold, and there is a single tandem exhaust port outlet in the top face for ease of maintenance.

Valve gear: Each pair of valves, operated via pushrods and rockers, is driven from the camshaft via bucket tappet-type followers mounted in a separate housing bolted to the crankcase.

Air manifold: these are modular castings mounted down the vee of the crankcase and incorporating passages for the lube oil and water systems.

Exhaust system: modular and compact, the system comprises single cylinder units bolted to the cylinder head and connected to the next unit with expansion bellows. The whole exhaust system is arranged in a lagged enclosure made up of two-cylinder units for ease of maintenance.

Charge cooler: the cylindrical two-stage charge cooler is contained in a housing that includes part of the inlet ducting. The assembly is mounted directly on top of the air manifold to provide good support. Special attention was paid to minimizing overhangs on external brackets to reduce the impact of shock loadings in fast commercial vessel or naval applications.

Turbochargers: twin high efficiency axial turbine turbochargers are mounted on a cast bracket at the free end of the engine.

Fuel system: an electronically-controlled pump-pipe injector system is used, with the fuel pump mounted in the cam follower housing which forms part of the body of the pump. Low pressure, modular fuel supply and return rails connect each pump to the next, and the short high pressure pipes to the injectors are double skinned. The electronic control for the fuel pump and injector is mounted locally on the engine.

Lubricating oil system: all contained on the engine. The lube oil pump is mounted directly on the free end of the crankcase and driven from the camshaft gear drive. The plate-type oil cooler is mounted horizontally on top of the filter housing at the free end of the engine, the duplex filter incorporating an integral oil thermostat.

Cooling system: a twin-circuit cooling system is used, with both pumps mounted on the free end of the engine and driven by the camshaft gear, and provision for a seawater pump. The charge cooler thermostat is integrally mounted in the turbocharger bracket.

Starting system: the air starting motor incorporates a control valve, pressure regulator and strainer, and engages with a gear ring on the flywheel. A barring motor can be supplied as a service tool or fitted as standard when it is fully protected against inadvertent starting of the engine.

Governor: the engine is served by a digital engine management system which controls its operation and communicates via a CAN bus link to a set of intelligent cylinder control modules that drive the pump and injector solenoids. The system dictates the fuelling, timing and pressure base upon pre-set mapped information.

25 SEMT-Pielstick (MAN B&W)

SEMT-Pielstick of France can claim a pioneering role in the development and promotion of medium speed engines for propulsion through in-house production and licensees in key shipbuilding countries, notably Japan. The PC medium speed design interests of Paris-based SEMT (Société des Études de Machines Thermiques) are now under the umbrella of Germany's MAN B&W Diesel group.

The Pielstick marque originated in the early 1950s as part of a family of monobloc multiple-crankshaft engines designated the PC1 series. Further development resulted in more conventional single-crankshaft designs: the 400 mm bore PC2 series, introduced in the mid-1960s; the 480 mm bore PC3 series, introduced from 1971; and the 570 mm bore PC4 series which was launched in the late 1970s. The PC3 series is no longer produced but successive modernization and upgrading programmes for the PC2 and PC4 designs have resulted in the PC2-6B and PC4-2B versions providing the main Pielstick thrust in recent years. The rise in maximum firing pressure and reduction in specific fuel consumption for successive generations of 400 mm bore PC2 designs is illustrated (Figure 25.1).

Experience gained with the PC2-2 design showed that the one-piece light alloy piston had to be abandoned in favour of the first two-piece design with steel crown and forged light alloy skirt. Developed by SEMT-Pielstick in conjunction with a leading piston manufacturer, the new concept allowed higher mechanical and thermal loads, and secured better production quality. The PC2-5 engine, launched in 1973, benefited from this development in delivering a 30 per cent increase in power over the original PC2-2. Other refinements included a bore-cooled cylinder liner and a bevel-cut connecting rod allowing the crankpin diameter to be enlarged.

New turbocharging and fuel-optimized injection systems were applied to the PC2-6 engine, launched in 1982. The modular pulse converter (MPC) turbocharging system features an almost constant pressure exhaust gas feeding to each turbine through a single exhaust pipe.

Longer stroke derivatives arrived in the mid-1980s to supplement the PC2 and PC4 series, these PC20 and PC40 models retaining the

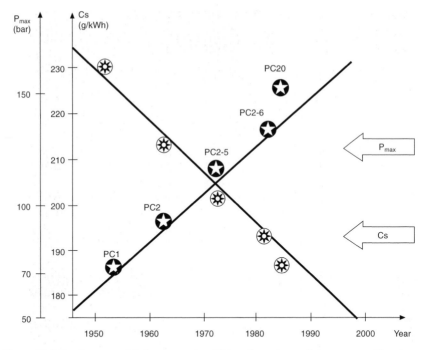

Figure 25.1 Development of fuel consumption (Cs) and maximum firing pressure (P_{max}) ratings for successive SEMT-Pielstick 400 mm bore PC engine designs

400 mm and 570 mm bore sizes but exploiting higher stroke–bore ratios, respectively increased from 1.15 to 1.375:1 and from 1.08 to 1.31:1. Significantly lower specific fuel consumptions were achieved. The PC20, with a stroke of 550 mm and rating of 607 kW/cylinder at 475 rev/min, has since been dropped from the programme.

In 1996 SEMT-Pielstick's Japanese licensee Diesel United delivered the world's most powerful medium speed engines which were specified for propelling longhaul coastal ferries. Each V18-cylinder PC4-2B model in the twin-engine installations yielded a maximum continuous output of 23 850 kW at 410 rev/min (Figure 25.2).

PC2-6B DESIGN

The most recent manifestation of SEMT-Pielstick's 400 mm bore PC2 medium speed engine is the 500 mm stroke PC2-6B design which develops 615 kW/cylinder at 500–520 rev/min. An output range up to 11 340 kW is covered by six, seven, eight and nine in-line and V45-

Figure 25.2 SEMT-Pielstick 18PC4-2B propulsion engine (23 850 kW at 410 rev/min) on resilient mountings

degree 10-, 12-, 14-, 16- and 18-cylinder models (Figures 25.3 and 25.4); a V20-cylinder version was also planned. The PC2-6B engine retains the same overall envelope for propulsion installations as the shorter stroke (460 mm) PC2-6 predecessor as well as its overhaul space requirements.

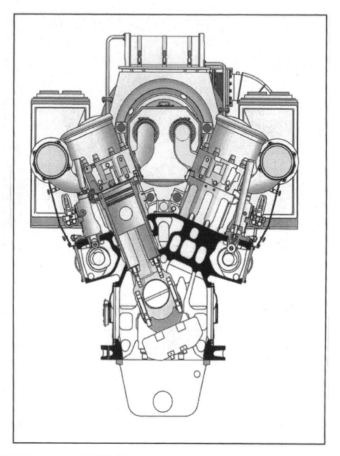

Figure 25.3 Cross-section of PC2-6B engine

Planning the PC2-6B engine, the designers aimed to increase power output by at least 10 per cent over the established PC2-6 model without penalising specific fuel consumption and making maximum use of experience from the PC20 (400 mm bore/550 mm stroke) series which was introduced in 1985 (Figure 25.5). A rise in power through a higher mean effective pressure rating was limited by:

• The pressure ratio of contemporary turbochargers.

Figure 25.4 SEMT-Pielstick 18PC2-6B engine, rated at 11 340 kW

• A peak combustion pressure deliberately limited to 150 bar and a compression ratio set at 12.5 to allow trouble-free burning of heavy fuels. The 150 bar limit also reflected a commitment to a short development period for the new model (around one year) and the avoidance of costly production plant investment.

A second path to increased power is to raise the mean piston speed, which, for the PC2-6 engine, was conservative compared with its rivals. Two routes are available: increasing the rotational speed (the smallest possible step to 600 rev/min is dictated by the synchronous speed requirement of genset applications); or increasing the stroke.

The latter alternative was pursued, seeking the longest possible stroke that still allowed the main dimensions of the PC2-6 engine crankcase to be retained and the cylinder heads to remain at the same position. These considerations determined a stroke of 500 mm. Engine component validation work was reduced since few new parts were used, mechanical and thermal stresses remained at the same levels as the PC20L, and the running speed was unchanged (hence also the dynamic behaviour of the valve train and injection pumps).

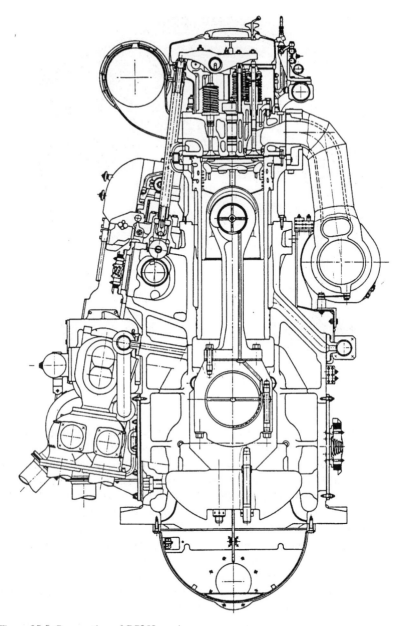

Figure 25.5 Cross-section of PC20L engine

Manufacturing costs were cut by specifying a cast iron crankcase (Figure 25.6) featuring an increased distance between seating paths to achieve high rigidity and a modified water jacket to avoid cooling

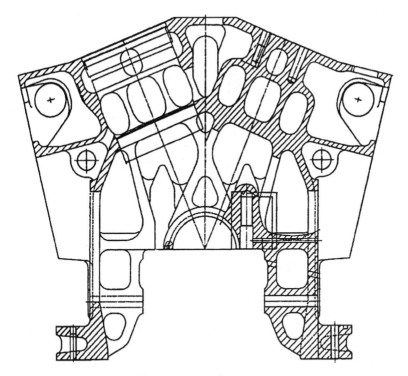

Figure 25.6 Cross-section of PC2-6B engine crankcase

water circulation in the crankcase. Strain measurements made on the first V18-cylinder PC2-6B industrial engine reportedly verified the predicted values from a comprehensive 3D FEM investigation. An interchangeable welded crankcase was proposed as an alternative.

The cylinder liner (Figure 25.7) is similar to that of the PC2-6, the smaller water jacket avoiding the direct contact of cooling water with the engine block and hence the possibility of corrosion.

The composite piston design (Figure 25.8) is the same as that of the PC20. Its steel head is without valve recesses to promote a uniform temperature distribution and thus limit thermal stresses; and the light alloy skirt has stepped bosses to reduce mechanical stresses under combustion pressure.

Only two compression rings (formerly three) are featured in the five-ring pack which, in conjunction with a honed liner, was designed to foster low lubricating oil consumption. An inspection of the pistons after endurance testing of a three-cylinder prototype engine revealed a 'very good' bearing picture for the skirt, without any contact marks

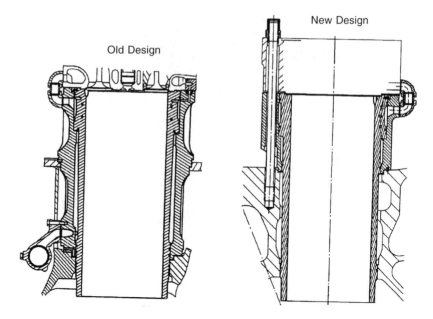

Figure 25.7 Comparison of crankcase, water jacket and cylinder liner arrangement for PC2-6B engine and its predecessor

between head and liner despite the reduction in skirt length under the piston pin.

An exhaustive study of the connecting rod design (Figure 25.9) was dictated by the increased inertia forces resulting from the longer stroke. The key concerns were reliability and the prevention of fretting on the serration. The main differences with the PC2-6 rod design are a higher tightening force at the big end cap and larger bolts which have been moved 10 mm outwards from the bore. The effectiveness of the modifications was confirmed by an endurance test on the prototype engine under extremely severe conditions (600 rev/min). A subsequent inspection revealed no trace of fretting on the serration.

A modified crankshaft—with a main journal diameter increased to 350 mm and crankpin diameter increased to 330 mm—was also deemed necessary in view of the impact of a longer stroke and higher peak pressure on the bending and torsional stresses as well as on the bearing operating conditions. The larger diameter main journal maintains the oil film thickness and specific pressure at similar levels to the PC2-6. Despite the larger crankpin diameter, the connecting rod big end bearings are slightly more loaded than those of the PC2-6: the 15 per cent higher loading is largely compensated by improved bearing shell technology (Rillenlager-type shells).

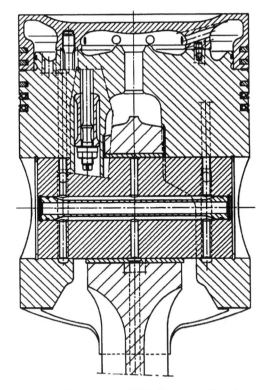

Figure 25.8 Two-part stepped boss piston of PC2-6B engine. Note the two compression rings

A modified cylinder head design was introduced to improve the pressure distribution on the gasket, the main modifications involving fireplate local reinforcements and repositioning of the side wall closer to the gasket circle. The highest temperature measured on the fireplate (between the exhaust valves) during tests of the prototype was, at 280°C, identical to that found on the PC20 engine.

PC2-6B design details

Frame: a stiff nodular cast iron crankcase construction.

Cylinders: each cylinder, including a water jacket and bore-cooled liner, is fitted into the crankcase. Only the upper part of the liner is cooled, avoiding any flow of water in the crankcase and hence possibility of corrosion.

Crankshaft: the one-piece forged unit rests on underslung main bearings fitted with thin bearing shells. A temperature sensor is fitted

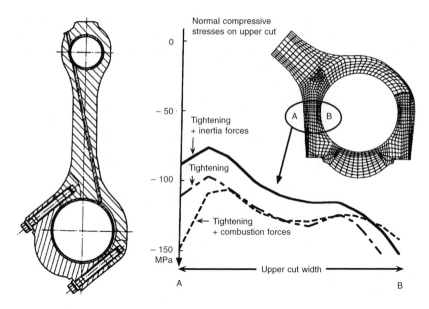

Figure 25.9 PC2-6B engine connecting rod design and big end calculations

on each main bearing, preventing the development of shell bearing abnormality.

Connecting rods: the bevel cut big end cap is secured by serrations on the rod. In V-type engines the connecting rods of a cylinder pair are mounted side-by-side on the same crankpin.

Piston: the composite-type piston comprises a steel crown ('shaker' oil cooled) and a light alloy skirt of the stepped boss type to reduce stresses under peak combustion pressure. The floating piston pin permits free rotation. The five-ring pack includes two compression rings while two spring-loaded scraper rings control lube oil consumption, their position in the upper part of the skirt facilitating easier piston and liner lubrication.

Cylinder head: a vermicular cast iron component attached to the water jacket and to the liner by eight tie-bolts anchored in the crankcase bosses.

Valves: both inlet and exhaust valves have a tight guide bush and the exhaust valves feature a turning device. The entire water flow across the cylinder head passes through the exhaust valve cages, substantially decreasing the valve seat temperature.

Fuel injection: the injectors are cooled by a separate freshwater system. The fuel pumps provide a pressure exceeding 1000 bar.

Camshaft: the bearings are secured with four bolts directly below the injection pump supports, an arrangement which avoids the transmission of injection stresses to the crankcase. Each camshaft and its bearings can be removed laterally from the side of the engine. The fuel pump driving gear design secures a low contact pressure between cam and roller.

Turbocharging: a patented modular pulse converter (MPC) system represents a compromise between impulse and constant pressure systems, and also fosters easier maintenance of piping and expansion bellows.

COMPARISON OF PC2.6-2 AND PC2-6B PARAMETERS

	PC2.6-2	PC2-6B
Bore (mm)	400	400
Stroke (mm)	460	500
Swept volume (dm^3)	57.8	62.8
Max. combustion pressure (bar)	137 (V-cyl)	150 (V-cyl)
	142 (L-cyl)	150 (L-cyl)
Nominal power (kW/cyl)	550	615
Speed (rev/min)	500/514/520	500/514/520
Mean piston speed (m/s)	7.7–8.0	8.3–8.6
Mean effective pressure (bar)	22.9	23.5
Specific fuel consumption (g/kWh)	183	182

PC2-6B LOW POLLUTANT VERSION

In creating a low pollutant version of the PC2-6B engine, SEMT-Pielstick aimed to find a compromise between low emission levels, first cost and running cost using proven solutions. The following nitrogen oxides (NOx) reduction techniques, investigated earlier on other prototype engines, were applied:

- Increased compression ratio: raised to 14.8 instead of 12.7:1 on the standard engine.
- Retarded fuel injection: 4 degrees before TDC instead of 11 degrees.
- Water/fuel emulsion injection: 30 per cent/70 per cent ratio.
- Injection rate modification: an injector nozzle with nine 0.76 mm diameter holes instead of the standard nine 0.72 mm holes.
- Turbocharger matching modification: rematched to increase slightly the combustion air excess.

PC4-2B DESIGN

The latest version of the 570 mm bore PC4 series, the PC4-2B design, has a 660 mm stroke and develops 1325 kW/cylinder at 430 rev/min on a mean effective pressure of 22 bar. The production programme embraces V10-, 12-, 16- and 18-cylinder models covering a power band up to 23 850 kW (Figures 25.2 and 25.10). A specific fuel consumption of 176 g/kWh is reported. Progressive improvements have benefited the PC4 design since the first example entered service in 1977, the refinements focusing on the exhaust and inlet manifolds, exhaust valves, fuel injection system, crankshaft and camshafts. The PC4-2 featured a 620 mm stroke and the PC40L (Figure 25.11) a stroke of 750 mm and a running speed of 375 rev/min.

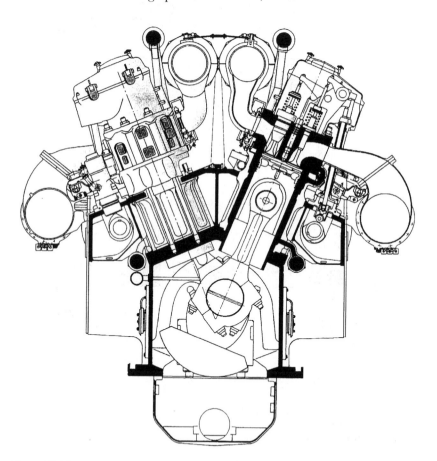

Figure 25.10 Cross-section of PC4-2B engine

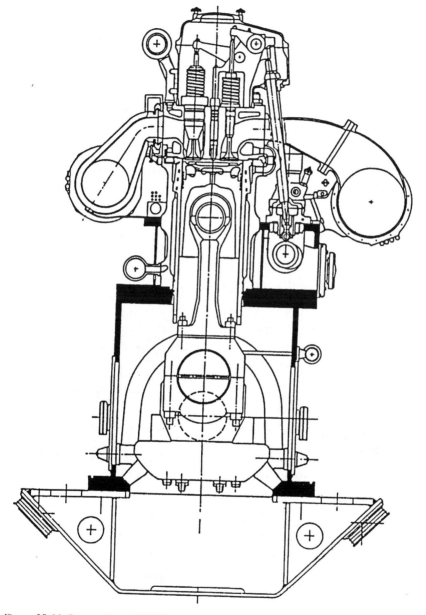

Figure 25.11 Cross-section of PC40L engine

The key features of the PC4-2B engine include:

• A one-piece welded steel engine frame with a steel plate oil sump mounted on the bottom.

- A bore-cooled cylinder liner of special cast iron located inside a cast iron water jacket, avoiding contact between the cooling water and the engine frame. Cooling of the liner's upper part was calculated to reduce thermal stresses.
- A one-piece underslung crankshaft of chromium molybdenum forged steel; each main bearing is fixed by two vertical and two horizontal tie-bolts, fostering a weight reduction for the frame. A temperature sensor is installed on each main bearing to prevent shell bearing abnormalities.
- Composite-type pistons embrace a steel crown and light alloy skirt, with a floating-type piston pin. The crown is oil cooled by the shaker effect.
- Forged alloy steel connecting rods with a wide big end and stepped boss small end to secure low bearing pressures.
- The cast iron cylinder head is fixed to the water jacket and to the liner by eight studs screwed into the frame bosses. It incorporates two inlet and two exhaust valves, a fuel valve, an indicator valve, a safety valve and a starting valve.
- Heat-resistant steel inlet valves exploit the two-guides solution while the Nimonic exhaust valves have sealed guides. All water flowing across the cylinder head runs through the exhaust valve cages, significantly reducing the valve seat temperature. All the valves are provided with Rotocap rotating devices.
- Camshaft bearings are fixed directly under the fuel injection pump brackets, avoiding the transmission of injection stresses to the frame.
- A large fuel injection pump diameter and short piping underwrite a high injection pressure achieving fine fuel pulverization even at low load; complete and clean combustion of the heaviest residual fuels with a high asphaltene content is promoted. The fuel pump plunger and barrel can be exchanged through the pump's upper part without dismantling the pump body. The weight of the dismountable equipment is 20 kg. The fuel injectors are cooled by a separate freshwater circuit.
- The patented modular pulse converter (MPC) turbocharging system is the best compromise between pulse and constant pressure systems, according to SEMT-Pielstick. Air coolers may be supplied with two banks of tubes, one of which can be deployed to heat air when the engine operates at low load.

See Chapter 22 for MAN B&W Diesel's V40/50 design, a derivative of the SEMT-Pielstick PC2.6B engine; and Chapter 30 for SEMT-Pielstick PA series high speed engines.

26 Sulzer (Wärtsilä)

A successful medium speed engine programme embracing the 200 mm bore S20, 400 mm bore ZA40S and 500 mm bore ZA50S designs was offered by New Sulzer Diesel before becoming part of the Wärtsilä Corporation. Only the ZA40S series now remains in the Wärtsilä portfolio. The three designs, covering an output range from 460 kW to 21 600 kW for propulsion and auxiliary power duties, were released for operation on fuels with viscosities up to 700 cSt (Figure 26.1).

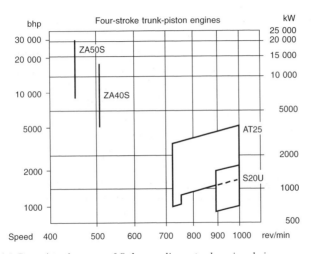

Figure 26.1 Power/speed ranges of Sulzer medium speed engine designs

The modern four-stroke Z-engines matured from a 400 mm bore family founded by the ZH40 two-stroke medium speed engine, designed in the early 1960s and installed in icebreakers, ferries and cruise ships. The uniflow-scavenged two-stroke design was succeeded in 1972 by a four-stroke version (the Z40) which was in turn completely redesigned and replaced in 1982 by the ZA40 engine. A longer stroke (560 mm instead of 480 mm) ZA40S engine was introduced in 1986 to succeed the Z40 and ZA40 designs. The performance development steps over the years are illustrated in Figure 26.2.

641

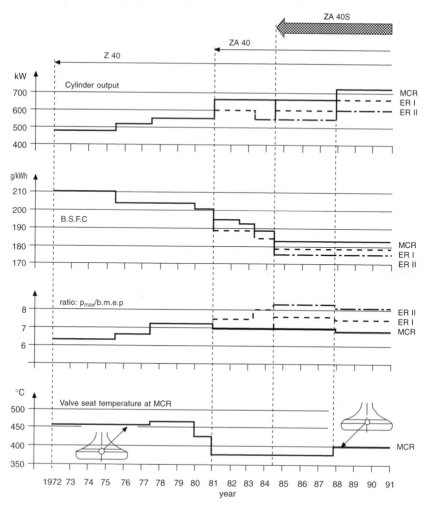

Figure 26.2 Development steps of successive Sulzer Z40, ZA40 and ZA40S designs illustrated by key performance parameters

In autumn 1988, three years after its announcement with a cylinder output of 660 kW, the ZA40S was uprated to 720 kW/cylinder following successful service experience and heavy fuel endurance tests. The new rating was associated with an increased maximum cylinder pressure of 165 bar and mean effective pressure of 24.1 bar, underwritten by thermodynamic and mechanical optimizations. Some design modifications were also introduced to improve product reliability and durability in general (Figure 26.3):

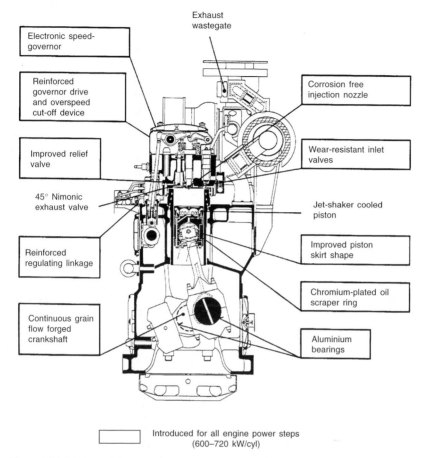

Figure 26.3 Main modifications for uprated ZA40S engines

- An exhaust valve with a 45 degree seat angle, this refinement achieving: better cooling without reduction of mechanical and thermal safety at the higher engine load and maximum cylinder pressure; and more efficient seating due to a wider seat area.
- Sulzer's new combined 'jet-shaker' piston cooling principle in which the conventional shaker effect in the piston crown is supplemented by oil jets sprayed through nozzles. This offers advantages in lowering piston crown temperatures, thereby preventing carbon formation, and gaining an improved washing effect.
- A new version of the waste gate in the exhaust gas manifold before the turbine, securing a better match between the engine and turbocharger at the increased power rating.

Rotating piston

A classic feature of Sulzer Z-type engines retained throughout successive generations is the rotating piston. Sulzer patented the concept in 1937 after tests on a number of research engines with cylinder bores from 90 mm to 420 mm which confirmed the special benefits offered for highly loaded trunk piston engines. A ratchet mechanism transforms the swinging motion of the connecting rod into a smooth rotation of the piston (Figure 26.4). The connecting rod has a spherical small end which allows some 40 per cent more bearing area than a gudgeon pin bearing.

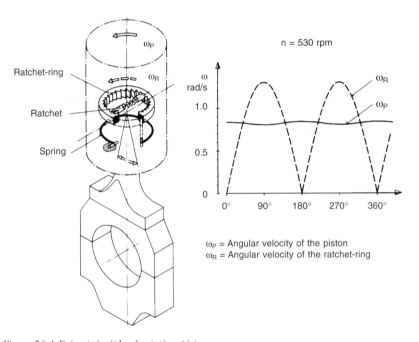

Figure 26.4 Drive principle of rotating piston

The rotating piston was eventually adopted for the Z-type design, first introduced in two-stroke form in 1964. The concept remained unique to Z-engines until its adoption in 1995 for GMT's upgraded 550 mm bore VA55 medium speed model. (The Italian designer benefited from a technology transfer arrangement with Sulzer; both later became members of the Wärtsilä Corporation.)

For rotating the piston, the connecting rod is provided with two pawls positioned slightly out of the centre of the spherical small end bearing. When the connecting rod performs its swinging movement relative to the piston the pawls impart an intermittent rotating motion to a toothed rim, from which it is transmitted to the piston by an annular spring. The flexible connection maintains the forces necessary for rotating the piston at a constant low level.

Sulzer cites the following merits of the rotating piston:

- Even temperature distribution around the piston crown as there are no particular inlet and outlet zones.
- Small and symmetrical deformation from a top end bearing with a relatively large area and of spherical design (Figure 26.5).
- Optimum sealing and working conditions for the piston rings because the small, symmetrical deformations of the piston allow the use of the smallest running clearance between piston and cylinder liner.
- Low and stable lubricating oil consumption because the small piston running clearances minimize piston slap and obviate the need for the traditional oil cushion, thereby allowing the oil scraper ring to be located at the lower end of the piston skirt.
- Good margin for unfavourable running conditions with the smallest risk of seizure because the grey-iron piston skirt is always turning to a fresh part of the cylinder liner surface.

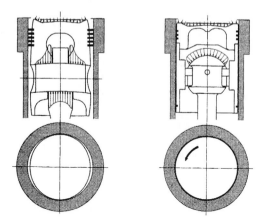

Figure 26.5 Symmetric design, with rotation, ensures that Sulzer Z-engine pistons (right) keep their circular form. Gas and mass forces are transmitted centrally through the spherical bearing with no load on the piston skirt

ZA40 AND ZA40S ENGINES

The current ZA40S programme is released with a rating of 750 kW/ cylinder at 510 rev/min on a mean effective pressure of 25.1 bar and with a mean piston speed of 9.5 m/s. Six, eight and nine in-line and V12, 14, 16 and 18-cylinder models cover a power band from 4500 kW to 13 500 kW. (Figure 26.6). The crankcase is fabricated for in-line cylinder engines while V-form versions feature a cast iron monobloc. The originally fabricated two-part design of the Z-engine frame was superseded by a monobloc casting for the ZA40 and ZA40S in-line models, as had always been standard for the V-engines. (Figure 26.7).

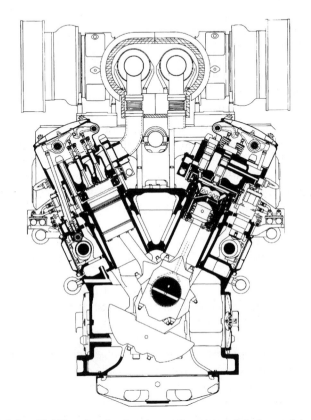

Figure 26.6 Sulzer ZA40S engine showing the rotating piston, fully bore-cooled combustion space and single-pipe exhaust system

Uneven cylinder liner wear was initially encountered in the mid-1970s by the first Z40 engines burning blended or heavy fuel oils with increased sulphur content, at a height equal to the top dead centre of the piston rings. This wear could be defined as local low temperature corrosion which was countered by a four-step solution: ensuring adequate alkalinity of the lubricating oil; modified cylinder lubrication;

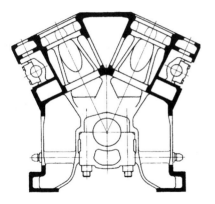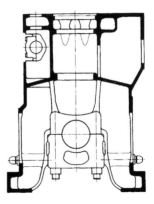

Figure 26.7 Engine blocks of ZA40S vee-form and in-line cylinder versions

higher cooling water temperature; and increased cylinder liner wall temperature from a modified bore-cooling arrangement with local insulating tubes. (Operators of earlier Z40 engines were offered insulating tube sets, together with appropriate fitting devices, for retrofit to original cylinder liners during a routine overhaul.) Subsequent service results demonstrated that these measures reduced low temperature corrosion to a negligible (if any) amount.

A modified O-ring material (changed to a silicon quality and an appropriate fluor elastomer quality, such as Viton) was adopted for the cooling water space of the cylinder liner after leakage was experienced with some units. The dry lower half of the liner almost entirely eliminates the risk of water leakage into the crankcase and contamination of the lubricating oil.

The double-bottomed water-cooled cylinder heads of Z40 engines, made of cast iron, performed well. Some cracks resulted, however, either from casting and machining faults or from the supply of spares not intended for engines with the later, higher outputs. The material for the O-rings of the valve seats and guides was also changed to an appropriate fluor elastomer quality, both with good results.

The cylinder heads of the ZA40 and ZA40S engines are bore cooled (Figure 26.8). The measured bottom deformations under gas load are only one-third those of the conventional double-bottom head, and the mechanical stresses were also found to be low in the same proportion. The drilled cooling passages permitted a uniform temperature adjustment within a few degrees at the desired low level. The resulting low thermal distortion as well as the high mechanical stiffness foster good valve seat sealing with optimum heat transfer to the exhaust valve seats.

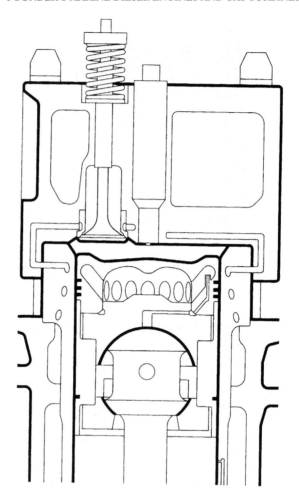

Figure 26.8 Fully bore-cooled combustion space of Sulzer ZA40S engine

The well-proven rotating piston concept was retained for the ZA40 and ZA40S engines but benefited from simplification; bore cooling of the crown was also introduced. An optimized arrangement of the cooling bores reduced stresses by around 10 per cent with practically no change in crown temperatures. The substantially stiffened periphery secured by bore cooling also resulted in reduced thermal deformation of the crown and hence improved working conditions for both piston and rings (Figure 26.9).

The top ring of the ZA40S piston features a Sulzer plasma-coated running face, while the running faces of the other two rings are chromium plated. All three rings are barrel shaped to promote initial

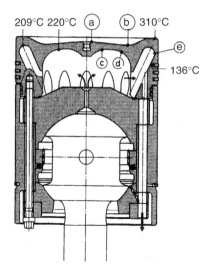

Figure 26.9 Measured temperature in ZA40S rotating piston

running in and to avoid a need to hone smoothened cylinder liners at routine overhauls. (The complete set of piston rings and scraper rings should therefore be replaced during a piston overhaul.) The flanks of the rings and their grooves are also chromium plated. A combination of bore-cooled rotating pistons and plasma-coated piston rings promotes low wear rates and long piston life, Sulzer reports. Rotation and negligible distortion due to the spherical top end, as well as no transmission of gas forces through the cast iron piston skirt, underwrite a small skirt clearance.

The connecting rod with marine-type bottom end features large crankpin diameters and hydraulically tightened studs for easy maintenance; no interference with the bottom end bearing is necessary at piston overhauls. Shims can be applied to achieve the different compression ratios for three engine tuning options.

The well-proven exhaust valve arrangement refined over many years in the Z40 and ZA40 engines was retained for the ZA40S: Nimonic exhaust valves with rotators and water-cooled valve seat inserts but no valve cages (Figure 26.10). The higher rigidity of the bore-cooled cylinder head and the generally lower temperature levels from the enhanced turbocharging system further improved the sealing conditions of the exhaust valves in the ZA40S. Longer times-between-overhauls were thus secured, the intervals coinciding with those for the piston.

A new generation of aluminium/tin alloy bimetal-type bearings was introduced for both main and bottom end bearings of ZA40 and

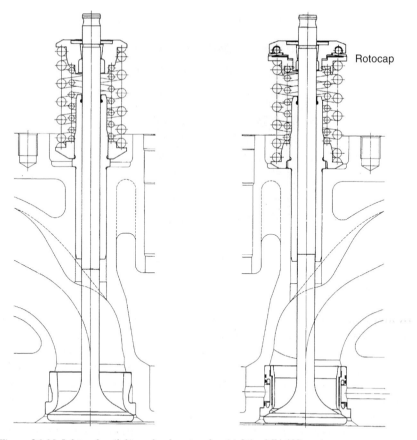

Rotocap

Figure 26.10 Inlet valve (left) and exhaust valve (right) of ZA40S engine

ZA40S engines (Figure 26.11). These 'Rillenlager' and bimetal types dispensed with the traditional overlay of the previous trimetal type to give the following reported benefits: sustained fatigue resistance and embeddability; better wear resistance for heavy fuel operation; improved local self-healing and emergency running behaviour; and better adaptability after inspections.

The loading of the spherical top end bearing remained some 15–20 per cent lower than that of a conventional gudgeon pin design. The big end of the ZA40 and ZA40S connecting rod was also enlarged from the Z40 design, resulting in a reduction of the dynamic stresses of around 20 per cent. The bolts are hydraulically tightened for easier and faster maintenance procedures.

Helix-controlled fuel injection pumps (Figure 26.12) are provided for each cylinder, with fuel leakage into the lubricating oil prevented

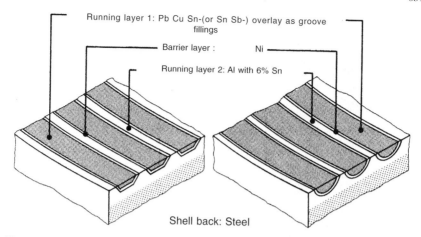

Figure 26.11 Features of 'Rillenlager'-type bearings: thickness of the Al/Sn running layer No. 2 is a few tenths of a mm; depth of grooves is a few hundredths of a mm; grooves are laid as a helical thread; shape and dimension of grooves may vary, depending on manufacturer

by an oil barrier at the lower end of the plunger sleeve. Depending on the engine application, helix control edges can be chosen for constant or variable injection timing (CIT or VIT). The simple arrangement of short high pressure fuel pipes to the injection valves, together with the circulation of preheated fuel oil up to the level of the pumps, underwrites routine pier-to-pier operation on heavy fuel (with appropriate circulation and preheating).

The injection valves (Figure 26.13) are equipped with sleeve-type nozzles cooled by fresh water and provided with rounded-off inner edges of the spray holes (Figure 26.14). Rounded-off spray holes are traditional for Sulzer nozzles and reportedly result in a more stable spray pattern over longer service periods with a narrower scatter of injection rates between individual cylinders. A more corrosion-resistant sleeve material was introduced to replace the original material whose punishment threshold was not large enough to cope with corrosion attacks by insufficiently treated cooling water (as a result, some tip breakages occurred).

Plunger seizures initially occurred on Z40 engines due to insufficient clearance (either through manufacturing with too small a clearance or reduction of clearance in heavy fuel service as a result of the tendency of the material used earlier to grow at higher temperature levels). An increased plunger clearance and a changeover to a new material specification for the plungers, supported by stricter quality control, proved to be design remedies. Operationally, shipyards and operators

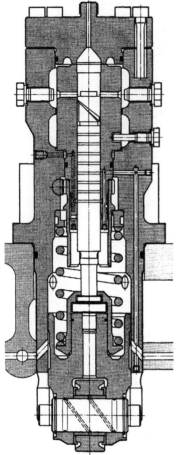

Figure 26.12 Fuel injection pump for ZA40S engine

must take care that no undue impurities are present in the fuel and lubricating oil systems which may lead to avoidable plunger seizures. Slow heating up of the fuel system and the maintenance of the correct temperature level for the heavy fuel oil used are also advisable.

Some sticking of fuel pump plungers was reported from a growing lacquer formation on the guide part side, depending on the combination of heavy fuel oil and lubricating oil used. As a solution, the lubrication and oil barrier arrangement was redesigned to increase the oil flow and thus reduce the tendency for oil ageing, if any.

With regard to the hydraulically shrunk-on fuel cams and rollers, some failures initially occurred on Z40 engines as a result of misaligned drive housings or insufficient case hardening of the cam surfaces. Stricter specifications and quality control, together with wider cams

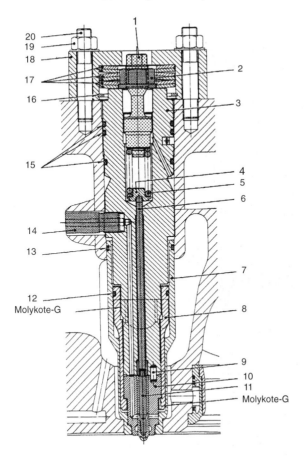

Figure 26.13 ZA40S engine fuel injection valve

and rollers, remedied the problem. Two-piece replacement cams were introduced for non-reversible engines to simplify renewal.

A variable fuel injection timing (VIT) system is standard for ZA40S engines not equipped with a charge air waste gate (see below). The helix of the pump plunger is shaped so as to raise Pmax throughout the part-load range and thereby reduce part-load brake specific fuel consumption.

The single-pipe exhaust system serving the ZA40S engine yields a reasonable combination of the advantages of the pulse system (partially retaining the kinetic energy of the exhaust gases) with the simplicity of a constant pressure system. This had been recognized in the 1950s but could only be realized when modern high efficiency turbocharger types became available. Turbocharger matching, however, can be

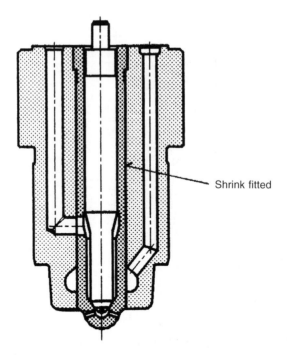

Shrink fitted

Figure 26.14 ZA40S engine fuel injection nozzle

demanding with the short valve overlap necessary for low fuel consumption at full load; it may lead to surging at part loads and, furthermore, a high efficiency turbocharger shows a pronounced decrease in charge air pressure at part load.

These difficulties were overcome in the ZA40S by using charge air bypasses and charge air waste gates to create load-adaptable supercharging systems that often enable the engines to surpass the performance of conventional supercharging systems (Figure 26.15). The ZA40S system, Sulzer asserts, can be tailored to give low fuel consumption through the normal engine load range but can also be adapted to yield the high torque required at part load in the propulsion of icebreakers and shortsea shuttle ferries. Progressively opened above about 85 per cent load, the optional waste gate valve allows higher boost pressures to be used at part load for a substantial improvement of part-load brake specific fuel consumption while avoiding excessive cylinder pressures at higher loads.

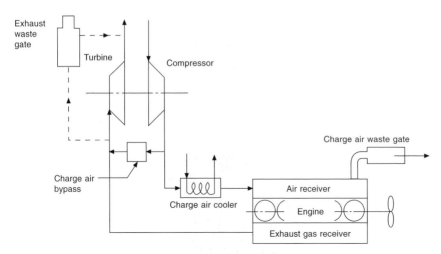

Figure 26.15 ZA40S engine turbocharging system with charge air bypass and waste gate

ZA50S ENGINE

Rising demand for compact, high output medium speed diesel propulsion installations—particularly from cruise and ferry sectors—stimulated Sulzer to introduce in 1995 a larger bore derivative to complement the successful ZA40S engine. This 500 mm bore/660 mm stroke ZA50S design extended the power range of the programme to 21 600 kW at 450 rev/min from six, eight and nine in-line and V12-, 14-, 16- and 18-cylinder models (Figures 26.16 and 26.17). The ZA50S engine is no longer in the Wärtsilä programme, its output range covered by the Wärtsilä 46 design.

A shorter and lighter engine than rivals in the same power class was sought by the ZA50S designers, and attention was paid to reducing the engineroom headroom requirement and the minimum distance necessary between engines in multi-engine plants. The connecting rod design contributed to a particularly low dismantling height.

The ZA50S engine inherited the main features of the ZA40S, notably the rotating piston, fully bore-cooled combustion space, high stroke–bore ratio, cylinder head without valve cages, and load-adaptable turbocharging with single-pipe exhaust system. The opportunity to innovate was taken, however, by adopting hydraulic actuation for the gas exchange valves, standard for many years in low speed two-stroke engine practice but introduced here for the first time to a medium speed four-stroke design (Figure 26.18).

Hydraulic actuation–in conjunction with pneumatically controlled

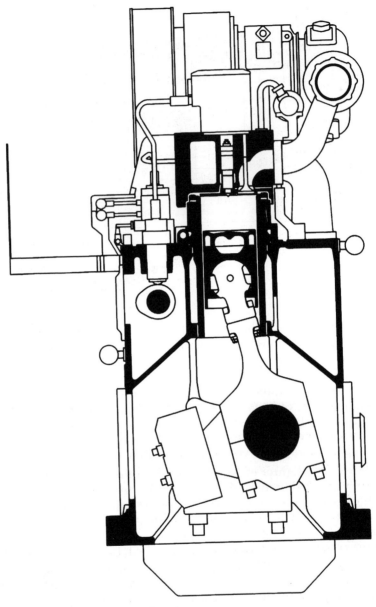

Figure 26.16 Cross-section of ZA50S engine

load-dependent timing to provide variable inlet closing–yields flexibility in valve timing, fostering lower exhaust gas emissions and improved fuel economy. Variable inlet closing, together with optimized super-

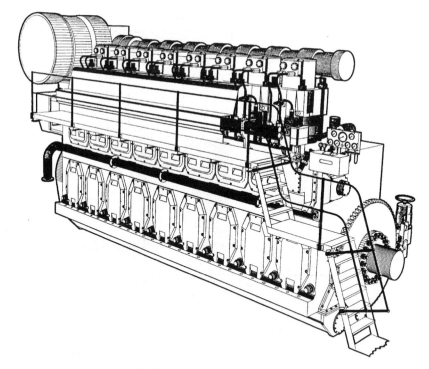

Figure 26.17 Eight-cylinder ZA50S engine

charging, delivers a very flat fuel consumption characteristic across the whole load range of the engine. A considerable reduction in smoke levels in part-load operation is also secured.

A high stroke–bore ratio, similar to that of the ZA40S engine, underwrites a compression ratio promoting clean and efficient combustion over the whole engine load range. The combustion chamber is nevertheless sufficiently deep for the clean combustion of heavy fuel, with low smoke emissions and low thermal loadings of the surrounding components. The standard ZA50S engine meets the nitrogen oxide emission limits set by the IMO; further reductions are possible by engine tuning refinements.

A robust engine block with underslung crankshaft combines maximum rigidity with low stresses, the designer reports. The spheroidal cast iron block is designed for direct installation on resilient mountings. The main and bottom end bearings follow experience with the ZA40S engine, featuring bimetal aluminium/tin-alloy running surfaces on thin-walled shell bearings. The bimetal-type bearings, which have demonstrated fatigue resistance and a long life expectancy, also exhibit

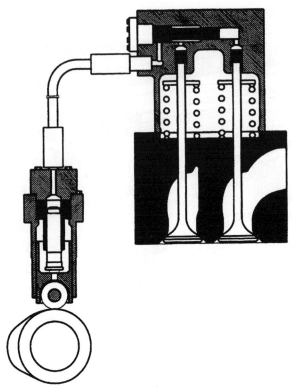

Figure 26.18 Hydraulic exhaust and inlet valve drive for Sulzer ZA50S engine. The merits cited include: constant valve timing for the exhaust valves; variable valve timing for the inlet valves (improving fuel economy, emissions and part-load operation); no valve clearance adjustment; and low noise operation

high resistance against wear and corrosion, together with good embeddability and emergency running behaviour.

The symmetrical design of the rotating piston achieves even distributions of thermal load and wear around the piston. The piston comprises a steel crown and spheroidal graphite cast iron skirt. The crown is bore cooled with oil jet-shaker cooling; and the skirt is provided with inner lubrication, whereby lube oil is fed to the cylinder liner outwards from within the piston. Effective lubrication of the liner, rings and pistons, with low lube oil consumption, is reported from ZA40S installations. Full bore cooling of the combustion chamber—the cylinder head, liner and piston crown—fosters optimum temperatures and temperature distribution, low mechanical stresses and low thermal strains.

The ZA50S engine inherited the proven exhaust valve concept of earlier Z-type engines, based on water-cooled seat inserts, Nimonic valves and mechanical rotators but no valve cages. The inlet valves also feature mechanical rotators. The times between valve overhauls are arranged to coincide with those for the pistons.

A key contribution to economy and performance derives from another feature shared with the ZA40S engine: optimized supercharging incorporating a simplified single-pipe exhaust system. Appropriate use is made of waste gates and charge air bypasses to suit the individual application. The reported benefits are realized in low fuel consumption at both full and part loads, improved load acceleration behaviour and reduced thermal loading of engine components.

MAIN PARAMETERS OF ZA40S AND ZA50S ENGINES

	ZA40S	ZA50S
Bore, mm	400	500
Stroke, mm	560	660
Speed, rev/min	510	450
Mep, bar	25.1	24.7
Mean piston speed, m/sec	9.5	9.9
Cylinders	6, 8, 9L/12, 14, 16, 18V	
Output/cylinder, kW	720	1200
Power range, mcr, kW	4500–13 500	7200–21 600
Specific fuel consumption, g/kWh	183–185	181

S20 ENGINE

Introduced in 1988 for genset drives and small-ship propulsion, the Sulzer S20 engine was distinguished by a significantly higher stroke/bore ratio (1.5:1) than established contenders in its class (typically with ratios between 1 and 1.2). The 200 mm bore/300 mm stroke design was available in four-, six-, eight- and nine-cylinder in-line versions with a maximum rating of 160 kW/cylinder at 1000 rev/min on marine diesel oil or heavy fuel oil. A power band from 460 kW to 1440 kW was covered over a speed range from 720 rev/min to 1000 rev/min. Engines for genset duty (designated S20U) were offered with a higher rating to cover power demands up to 1575 kW at 1000 rev/min (Figure 26.19). The S20 engine was displaced from the joint programme when Wärtsilä acquired New Sulzer Diesel in favour of the Wärtsilä 20 series.

The high stroke/bore ratio of the S20 engine allowed a greater compression ratio while keeping the combustion space high enough

Figure 26.19 Sulzer 6S20 engine

to foster an undisturbed fuel spray pattern for clean combustion. The height of the combustion space also gave freedom in the choice of valve timing. A high compression ratio further yields the high temperatures at the end of compression that are necessary for quick ignition and clean combustion of heavy fuel oils, especially at low loads. A pier-to-pier capability burning 700 cSt fuels is claimed, allowing auxiliary engines to operate in uni-fuel installations.

The S20 has almost the same overall dimensions as its predecessor, the A20 engine, the height being uncompromised thanks to a short connecting rod design. The engine block combines crankcase and cylinder housing in one rigid casting with an integral camshaft housing and charge air manifold. The main bearing caps of the underslung crankshaft are secured by hydraulically tightened studs both vertically and horizontally to increase the structural stiffness of the block. Large crankcase doors on both sides of the engine offer good access for maintenance. As far as possible, lubricating oil and cooling water are distributed through drillings in the block to reduce pipework. A deep sump provides a large oil capacity to reduce ageing of the lubricating oil and extend intervals between oil changes.

The underslung crankshaft is a single-piece forging, machined all over, with large diameter crankpins and journals, and fully balanced with bolted-on counterweights at each crank throw. Both main and big end bearings have thin-walled aluminium/tin shells, a type which has demonstrated excellent running and wear behaviour and has become standard for Sulzer four-stroke engines. Provision is made at the free end of the crankshaft for a gear drive for pumps and a shaft extension for power take-off drive.

High tensile forged steel connecting rods have a diagonally split big end with a serrated joint and hydraulically tightened studs. The rod is drilled longitudinally to convey lubricating oil to the gudgeon pin bearing and piston.

A two-part piston–comprising bore-cooled steel crown and lightweight nodular iron skirt–was specified for maximum reliability in heavy fuel service. The corrosion-resistant steel crown addresses the high combustion pressures and surface temperatures associated with heavy fuel operation; and the cast iron skirt allows smaller running clearances at every load compared with aluminium skirts. This enables the lateral thrust of the short connecting rod to be distributed evenly over the full depth of the skirt. Better lubrication conditions are also fostered over the entire power/speed range. Three piston rings and a scraper ring are specified.

The cylinder liners are centrifugally cast from iron whose composition and structure were specially developed to yield high mechanical and fatigue resistance as well as good corrosion and wear resistance. The wall temperatures obtained prevent low temperature corrosion and minimize abrasive wear.

Individual cylinder heads, incorporating a thick bore-cooled flame plate, are held down by four hydraulically tightened studs. Each head accommodates a central fuel injector, two inlet valves, two exhaust valves, a starting air valve and an indicator cock. The exhaust valves and their housings were designed to achieve high reliability and durability in heavy fuel operation, the specification calling for:

- Nimonic alloy valves with a high corrosion and heat resistance.
- Intensively cooled valve seat inserts in the cylinder head to allow efficient cooling with even temperature distribution around the valve seat.
- Valve rotators.
- A cylinder head flame plate which is particularly rigid and bore cooled for minimum deformation under severe load conditions.

A helix-controlled fuel injection pump for each cylinder has a lubricating oil barrier on the plunger to prevent fuel oil dilution of the system oil. The regulating linkage for the pumps is inside the camshaft housing. The nozzles of the fuel injection valves are cooled by lubricating oil.

Assembled from individual cylinder segments to facilitate overhaul, the camshaft is driven from the crankshaft via a train of spur gear wheels.

A forced lubrication system for the engine is based on a deep sump. Before the engine is started oil is automatically delivered by a motor-driven pre-lubrication pump.

Engine-driven pumps are arranged as standard for lubricating oil, jacket cooling water and, in the case of marine diesel oil installations, fuel supply. If required, a circulating pump can also be mounted for sea water or low temperature cooling water in a central cooling installation. The pumps are driven from gearing at the free end of the engine.

Standardized ratings for engines operating on marine diesel oil or heavy fuel oil, introduced in 1993, were accompanied by a unified turbocharging match and a new cooling arrangement. The latter featured a compact, two-stage charge air cooler and an oil cooler in series to create a cost-effective configuration with reduced cooling water flow rates.

Turbocharging is based on a high efficiency radial-type turbocharger in conjunction with a single-pipe exhaust system. The pipe is assembled from individual nodular cast iron segments for the cylinders, linked by bellows. As an option, engines can be tuned for higher charge air pressures at part load to give enhanced fuel economy in that operating range. Overpressure at full load would then be avoided by fitting a charge air waste gate to blow off excess air. For gensets, such tuning has the added benefit of achieving a quicker response to changes in load. In propulsion engines, an automatically controlled bypass is fitted to connect the charge air manifold to the exhaust pipe at part load.

Maintenance is facilitated by hydraulic tightening devices for the cylinder head holding-down studs, the main bearing studs and the connecting rod big end studs.

27 Wärtsilä

The Wärtsilä Corporation grew from a modest Finnish base in the 1970s to become a world force in four-stroke engine design and production through the development of its own medium speed programme and international acquisitions. Enginebuilders absorbed into the group over the years include Nohab Polar of Sweden, Stork-Werkspoor Diesel of The Netherlands, Wichmann of Norway, the French companies Duvant-Crepelle and SACM, Echevarria of Spain, and GMT of Italy.

In 1997 Wärtsilä Diesel's Finnish parent group Metra took a dominant holding in the newly formed Wärtsilä NSD Corporation, with Fincantieri of Italy taking the remaining minority share. The new group embraced Wärtsilä Diesel, New Sulzer Diesel and Diesel Ricerche (Fincantieri's R&D arm) as well as Grandi Motori Trieste (GMT), the enginebuilding division of Fincantieri, which became a fully owned subsidiary in January 1999. The New Sulzer Diesel element brought the significant Swiss-designed low and medium speed engine portfolio under the Wärtsilä umbrella. Wärtsilä Corporation's medium speed engine production is now based in Finland and Italy.

Wärtsilä Diesel laid the groundwork for its international growth in the early 1970s by deciding to develop a 320 mm bore medium speed engine designed from the outset for operation on heavy fuel. The resulting Vasa 32 engine was tested at end-1976 and the first production model delivered in 1978. The design supplemented the company's established smaller bore marine diesel oil-burning Vasa 22 series and helped to pioneer heavy fuel-burning medium speed engines for propulsion and auxiliary power.

Underwriting the Vasa 32 design's pier-to-pier heavy fuel burning capability were a number of innovations which have since become standard medium speed engine practice: nodular cast iron pistons; forced exhaust valve seat cooling to combat valve burning; an inverse cooling system to foster heavy fuel operation in the low-load range; a high injection pressure for improved combustion; and a cylinder liner design with temperature control preventing corrosion and deposit formation.

Wärtsilä Diesel's medium speed engine programme was progressively extended in the 1980s and 1990s with smaller and larger bore heavy fuel-burning models to create the current programme outlined in Table 27.1. The more recent designs are the Wärtsilä 32 (a more modern longer stroke derivative of the Vasa 32), the Wärtsilä 64 (the world's most powerful medium speed engine) and the Wärtsilä 50DF dual-fuel engine (derived from the Wärtsilä 46 design). Wärtsilä 32 and Vasa 32 engines are produced at the Vaasa factory in Finland, while another Finnish factory, at Turku, manufactures Wärtsilä 46 engines; the Wärtsilä 26, 38 and 64 engines are assigned to Wärtsilä Italia in Trieste.

Table 27.1 Wärtsilä medium speed engines

Design	Bore/Stroke (mm)	Output (kW/cyl)	Speed (rev/min)	Cylinders
W20	200/280	200	720–1000	3, 4, 6, 8, 9L
W26	260/320	325	900/1000	6, 8, 9L/12, 16, 18V
W32	320/400	500	720/750	6, 8, 9L/12, 16, 18V
W38B	380/475	725	600	6, 8, 9L/12, 16, 18V
W46C	460/580	1050	500/514	6, 8, 9L/12, 16, 18V
W64L	640/900	2010	333	6, 7, 8, 9L
W64V	640/770	1940	428	12V

The various designs introduced features now widely used by other four-stroke enginebuilders, Wärtsilä citing: side-mounted one-cylinder camshaft sections; four cylinder head bolts for large engines; 1500 bar-plus fuel injection pressure; and a 'hot box' for fuel injection pumps and pipes. Advanced machining systems allowed comprehensive integration of fluid systems within the engine block and the creation of multi-functional structural elements, significantly reducing the overall number of components.

Maximum cylinder pressure limits beyond 200 bar have been pursued by the company (Figure 27.1), exploiting nodular cast iron piston skirts with forced piston skirt lubrication: part of the oil for cooling the piston crown is forced out through four nozzles leading to an annular oil distribution groove, fostering well-controlled lubrication and hence increased reliability and reduced liner wear. Thick-pad bearing technology has benefited engines operating with high cylinder pressures.

A three-ring piston pack is standard for Wärtsilä engines as is an anti-polishing ring at the top of the cylinder liner (Figure 27.2). The function of the latter ring is to calibrate the carbon deposits on the

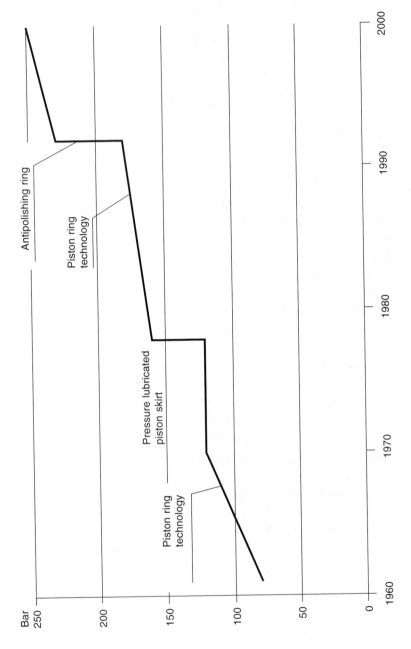

Figure 27.1 Contributions to the development of the maximum cylinder pressure limit in Wärtsilä engines

Figure 27.2 Anti-polishing rings have benefited all Wärtsilä engines (W32 engine liner shown)

piston top land to a thickness small enough to prevent contact between the liner inner wall and the deposits in any position of the piston. Bore polishing is reportedly eliminated and liner lifetimes can be doubled, accompanied by cleaner piston ring areas and significantly reduced lubricating oil consumption.

Low NOx combustion technology (whose principles are detailed in the Wärtsilä Vasa 32 section below) can be applied to all current Wärtsilä models, reducing noxious exhaust emissions without undermining fuel economy. The engines can also be arranged for gas-diesel and dual-fuel burning installations (see Chapter 2) and offered as part of group-supplied integrated Propac propulsion packages embracing reduction gearing, propeller and control systems.

Wärtsilä has pioneered computer-based systems for engine control and monitoring, condition monitoring and fault diagnosis, as well as multi-media documentation and remote expert communication systems to support servicing and maintenance (Figure 27.3).

Figure 27.3 The engine computer is the 'brain' of the Wärtsilä Engine Control System (WECS), shown here on a Wärtsilä 32

WÄRTSILÄ VASA 32

The 320 mm bore/350 mm stroke Vasa 32 engine (Figure 27.4) was designed in the 1970s but benefited from refinements yielding higher power outputs, enhanced reliability and serviceability, and the acceptance of fuels with higher CCAI numbers until the arrival in 1997 of the new generation Wärtsilä 32 (see page 672). By that time

Figure 27.4 The Wärtsilä Vasa 32 engine founded a family of heavy fuel-burning designs

around 1900 Vasa 32 engines had been sold for marine service since the seagoing debut of the design in 1978. Sustained demand nevertheless dictated continued production, overlapping with the Wärtsilä 32 engine. The Vasa 32LN is still produced in four, six, eight and nine in-line and V12-, 16- and 18-cylinder models to cover propulsive power demands from 1480 kW to 7380 kW at 720/750 rev/min. The D-rated versions offer 375 kW/cylinder while E-rated models yield 410 kW/cylinder with respective mean effective pressures of 21.9 bar and 24 bar.

The specification comprises the following main elements:

Engine block: designed for maximum overall stiffness and cast in one piece for all cylinder numbers; arranged for underslung crankshaft; and direct installation on resilient mountings possible.

Crankshaft: forged in one piece; fully balanced; and optional torsional vibration dampers.

Connecting rod: forged in alloy steel and machined; and diagonally stepped split in the big end.

Bearings: designed for maximum wear resistance.

Cylinder liner: special cast iron; and bore cooling for efficient control of liner temperature.

Piston/rings: composite piston with steel top and nodular cast iron skirt; forced skirt lubrication; and ring set comprising three compression rings and one oil scraper ring, all chromium plated.

Cylinder head/valves: grey cast iron; absorbs the mechanical load with a box section formed by an intermediate and an upper deck; mounted on the engine block with hydraulically tensioned studs; twin inlet and twin exhaust valves, all equipped with rotators; and water-cooled exhaust valve seats.

Camshaft: drop-forged one-cylinder shaft segments with integrated cams; and bearing journals fitted to the camshaft segments with flange connections.

Fuel injection: one-cylinder pumps with built-in tappets; through-flow type pumps for heavy fuel operation; uncooled nozzles; and all fuel-carrying equipment located in a drained and insulated space (the hot box) which keeps the system at operating temperature and ensures safety in the event of leakage.

Exhaust pipes: nodular cast iron, the entire system enclosed in an insulating box for safety.

Low NOx Vasa 32

The Vasa 32 engine in 1994 became the first in the Wärtsilä Diesel programme to be released with low NOx combustion technology, a key measure calling for the compression ratio to be increased from 12 to 14:1 to secure a sufficiently high compression temperature. The smaller combustion space dictated reshaping of the piston crown and the cylinder cover flame plate to allow for the fuel jets and good air/fuel mixing.

A new piston was developed for the higher maximum firing pressure (raised by 10 bar to 165 bar). Efficient cooling of the alloyed steel crown of the composite piston was secured by applying the cocktail shaker principle. The piston pin diameter was increased by 10 per cent to match the higher cylinder pressure. The three-ring concept (as opposed to the original four-ring set) was also adopted following good experience on the Wärtsilä 46 engine, the two compression rings benefiting from a special wear-resistant chrome ceramic coating (Figure 27.5).

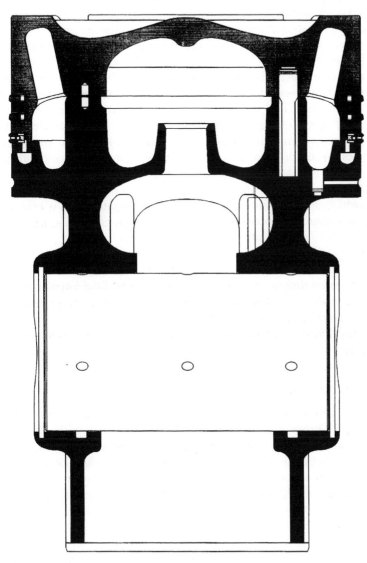

Figure 27.5 Composite piston of Vasa 32 Low NOx engine designed for increased combustion pressure and low friction

The stiff bore-cooled cylinder liner was equipped with an anti-polishing ring to keep carbon and ash build-up on the piston top land thin enough to prevent contact between the liner inner wall and the deposits, regardless of the piston's position. The liner lifetime is reportedly doubled and the lubricating oil consumption halved to 0.6 g/kWh or less.

The increased cylinder pressure also required the connecting rod to be changed from the original diagonally-stepped two-piece component to a fully machined three-piece design with a horizontally split big end bearing. The change yields greater safety at increased load and eliminates the need to interfere with the big end bearing assembly when overhauling the piston; a disadvantage is the increased number of parts. The load-carrying capability of the big end bearing was raised by switching from an SnSb or SnAl to a BiAl material type.

Only a minor modification to the cylinder head was necessary: a slight reshaping of the flame plate to provide more space for the fuel jets.

The fuel injection system was redesigned to handle demands for an increased injection rate and improved atomization. As a result, the fuel cam is faster and provides higher lift to compensate for the reduced plunger bore/stroke ratio. The injection valve opening pressure was raised from 350 to 600 bar, mainly to improve atomization at the start and stop of injection, and to maintain the higher injection pressure needed for a shorter injection period and reduced ignition delay. Improved part- and low-load performance was apparent, along with reduced NOx formation and fuel consumption. Test results showed it was possible to reach an NOx level of 11 g/kWh, a 30 per cent reduction compared with a standard Vasa 32 engine (IMO's limit is 12 g/kWh NOx at 720/750 rev/min). At the same time, the specific fuel consumption was lowered from 187 to 180 g/kWh while maintaining a reasonable combustion pressure of 165 bar, only 10 bar higher than that of a standard Vasa 32 engine.

Wärtsilä Diesel also developed a planetary gear with controls to replace the standard intermediate gear of the camshaft transmission. The arrangement allows a retarded fuel injection mode to be temporarily engaged while the engine is in operation and meet modest local NOx emission limits (up to 30 per cent reduction on normal emission levels at the expense of a rise in fuel consumption).

WÄRTSILÄ 32

A new generation 320 mm bore model was introduced to the market in 1997, the Wärtsilä 32 exploiting a longer stroke (400 mm) than its

Vasa 32 predecessor and incorporating design features established on the Wärtsilä 20, 26, 38 and 46 engines (detailed below). A higher output than its predecessor was offered (460 kW/cylinder at 750 rev/min on a mean effective pressure of 22.9 bar) to give a power range up to 8280 kW from in-line six, eight and nine and V12, 16 and 18-cylinder versions.

Scope for future specific power rises was demonstrated by the test engine operating comfortably on an mep of 31 bar. A higher rating was subsequently released (500 kW/cylinder at 750 rev/min on an mep of 24.9 bar) to extend the power range to 9000 kW. A specific fuel consumption of 177–179 g/kWh (without engine-driven pumps) is quoted, depending on nominal output or optimal point operation, with a lubricating oil consumption of 0.5 g/kWh.

Although slightly taller and heavier than the Vasa 32, the Wärtsilä 32 is more compact and has a better power/weight ratio. Integration reduced the number of components in the new engine by 40 per cent compared with its forerunner and contributed to a neater look and lower production costs (Figure 27.6). Ducts for water and oil are

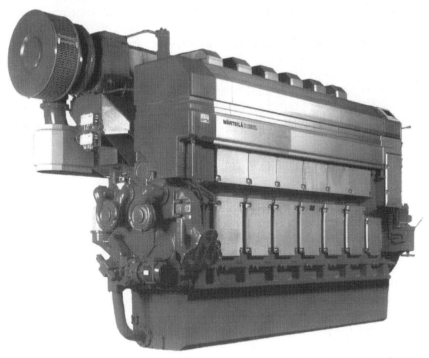

Figure 27.6 Engine-driven pumps are standard for the Wärtsilä 32. Note the clean lines achieved by integrated duct systems which eliminate external piping

incorporated in the engine block, and low pressure fuel pipes were replaced by ducts integrated in the injection pump design.

A substantial reduction in the number of joints to be broken during engine overhauls resulted. Further attention was paid to easing maintenance. The multi-duct (Figure 27.7) containing the water system and the exhaust and air piping, for example, stays in place when the cylinder head is lifted; the joints for starting air, oil and return fuel are slide-in connections; and, as there is no real jacket water, only a limited amount of water has to be drained when maintaining the cylinder unit. Ergonomic refinements were also pursued: the rocker arm cover is suspended on hinges and stays in place when the rocker arms of the injector are attended. Service pumps, including the pre-lubricating pump, are arranged on the engine along with a lubricating oil module which incorporates an automatic back-flushing filter (Figures 27.8 and 27.9).

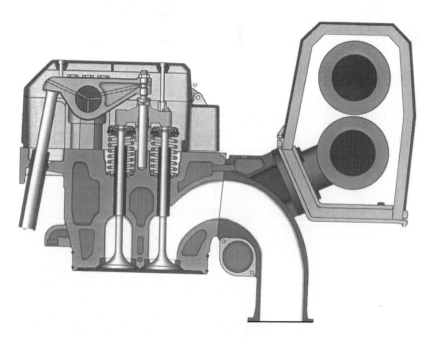

Figure 27.7 The multi-duct remains in place when the cylinder head is lifted (Wärtsilä 32 engine)

The low pressure fuel line comprises drilled channels in cast parts clamped firmly on the engine block, these parts consisting of the pump housing, the tappet housing, the fuel transfer housing and the multi-cover. For easy assembly/disassembly the parts are connected to

Figure 27.8 Lubricating oil module for mounting on the Wärtsilä 32 engine

each other with slide connections. Both the whole low pressure and high pressure systems are housed in a fully covered compartment. The high pressure system is designed and endurance tested at 2000 bar, and the injection pressure is around 1800 bar. With a wear-resistant low friction coating on the fuel plunger, no lubricating oil is required for the pump element. The profiled plunger geometry allows the clearance between plunger and barrel to be kept small, thereby allowing only a minimum of oil to pass down the plunger; this small leakage is collected and returned to the fuel system. Any likelihood of fuel mixing with the lube oil is eliminated.

The turbocharging system is specified to suit the requirements of each engine application from these standard options: pulse system, single pipe exhaust (Spex) system and Spex system with exhaust wastegate and air bypass. The systems are designed for minimum flow losses on both exhaust and air sides, the interface between engine and turbocharger streamlined to avoid all the adaption pieces and piping often used in the past.

The Wärtsilä Engine Control System (WECS) serving the Wärtsilä 32 measures a range of parameters and undertakes safety monitoring of the main bearing and cylinder liner temperatures, as well as individual exhaust valve monitoring. In the case of the V18-cylinder model, for

Figure 27.9 Lubricating oil module mounted on the side of a Wärtsilä 32 in-line cylinder engine

example, the computer-based system measures 97 temperatures, 10 pressures, four speeds, three positions and seven levels. It also controls starting and stopping, the low temperature thermostatic valve, wastegate and bypass valves, and slow turning, and interfaces with external systems.

WÄRTSILÄ 20

Introduced in 1992, the 200 mm bore/280 mm stroke Wärtsilä 20 engine (Figure 27.10) was designed primarily as a genset drive to replace the long-established Wärtsilä Vasa 22 but small-ship propulsion installations have also been logged. An output per cylinder of 130 kW to 165 kW was initially offered, depending on the nominal speed (720–1000 rev/min), but subsequent power rises took the rating to

Figure 27.10 Three Wärtsilä 6L20-powered gensets in the engineroom of a feeder containership

180 kW/cylinder (end-1998) and 200 kW/cylinder at 1000 rev/min (2003). The rating increases were underwritten by designing all key components for a maximum cylinder pressure of 200 bar; engines are currently released with a Pmax of 170 bar.

The W20 series now embraces four-, five-, six-, eight- and nine-cylinder models covering a maximum continuous output range from 720 kW to 1620 kW (to be extended to 1800 kW by the latest uprating). Up to 24 000 hours between overhauls have been achieved by engines running on light fuel oils.

A flat fuel consumption curve, with the lowest point preferably at part load, is desirable for an auxiliary engine which seldom operates at full load and may often run at a very low load. This was addressed in the Wärtsilä 20 design by optimizing the cylinder dimensions in conjunction with modern turbocharging technology; adopting a pulse charging system; and exploiting a high cylinder pressure.

Physically, the Wärtsilä 20 was intended to be shorter and lower than any existing engine in its performance class, with advantages in ease of installation and maintenance. The nodular cast iron engine block was designed for maximum overall stiffness and local stiffness around the upper part of the cylinder liner, and to incorporate a number of cast-in or machined water and oil channels in pursuit of component integration. The camshaft bearings are directly housed in the block as is the camshaft gear train at the flywheel end of the engine. Provision is made for five engine-driven pumps (three are standard). A four-point support configuration underwritten by the overall stiffness of the engine is especially attractive for resilient mounting, and the screwed-on feet arrangement offers considerable freedom in foundation design.

The four-screw cylinder head and box-cone design is another Wärtsilä tradition retained to achieve even sealing pressure and prevent liner deformation under high cylinder pressures. It also secures ample space for large inlet and exhaust channels which are necessary for good gas exchange. The head is mounted on the engine block with hydraulically tensioned studs, and is arranged to house two inlet and two exhaust valves, all equipped with rotators; the exhaust valve seats are water cooled. A starting air motor for all engine cylinder configurations was specified in contrast to traditional direct starting by air injection into the combustion chamber, contributing to optimized cylinder head stiffness and valve performance.

An innovation in the W20 design was a multi-functional duct between the cylinder head and the engine block (Figure 27.11). The multi-duct houses the air duct from the air receiver to the cylinder head (the air duct creates the initial swirl which helps to achieve optimum

Figure 27.11 The multi-duct of the Wärtsilä 20 engine facilitates maintenance of the cylinder components

part-load combustion) and incorporates the exhaust gas channel. The cooling water from the cylinder head is led to a cast-in channel in the engine block through the multi-duct. The multi-duct is also the support for the exhaust pipes, remaining in place when the cylinder head is removed to ease maintenance.

A 'distributed water flow' pattern is used to effect cooling of the exhaust valves and cylinder head flame deck. An even temperature distribution results from increasing the flow velocities on the exhaust side in general and around the exhaust valve seat rings in particular.

The wear-resistant cast iron cylinder liner was designed for minimum deformation thanks to its own stiffness and support from the engine block. It was also important to tune the inner wall temperature to a level just above the sulphuric acid dew point; this was addressed by creating an increased tangential cooling water flow velocity just below the liner collar. The risk of bore polishing is eliminated by an anti-polishing ring at the upper part of the liner.

A traditional Wärtsilä concept was retained for the composite piston (Figure 27.12) but refined for a smaller bore engine operating on heavy fuel: a nodular cast iron skirt, steel crown and pressurized skirt

Figure 27.12 Wärtsilä 20 engine composite piston with nodular cast iron pressure-lubricated skirt and steel crown

lubrication. Two compression rings and a spring-loaded oil scraper ring are located in the crown, the efficiency of a three-ring concept having been proven in the Wärtsilä 46 engine. (Most of the frictional loss in a reciprocating combustion engine originates from the piston rings; a three-ring pack is considered optimal by Wärtsilä with respect to function and efficiency.) Each ring is chromium plated and dimensioned and profiled for its individual task. The top ring benefits from a special wear-resistant coating.

A connecting rod of forged and machined alloy steel with a stepped split line lower end similar in design to that used in the Vasa 22 and 32 engines allows the maximum pin diameter while still making it possible to pull the rod through the cylinder liner (Figure 27.13). The advantage cited over the conventional straight, oblique split line is an increase in pin diameter of at least 15 per cent, a key factor in an engine designed for high combustion pressures. The fully balanced crankshaft is forged in one piece and amply dimensioned for reliable bearing performance. A small cylinder distance facilitates thin webs and large bearing surfaces. A low bearing load and optimized balancing of the rotating masses secure good oil film thicknesses at all engine loads and speeds.

Figure 27.13 The connecting rod of the Wärtsilä 20 engine incorporates a stepped split

The low pressure fuel system is an integrated part of the injection pump achieved via a multi-functional housing which acts as a housing for the single-cylinder pump, a fuel supply line along the whole engine, a fuel return line from each pump, a housing for the pneumatic overspeed trip device, and guidance for the valve tappets. The high pressure fuel line is also fully enclosed with an alarm for possible leakages, and all fuel-carrying parts are housed in a fully covered compartment (hot box). The high pressure side is designed for a 2000 bar injection pressure although the pressure used is around 1500 bar. The pump element itself features a solution to eliminate cavitation erosion. The fuel injection valve is designed for uncooled nozzles which are heat treated for low wear and high heat resistance.

A traditional two-circuit cooling water system incorporates a hot circuit for jacket water and a cold circuit for air cooling and oil cooling. The inverse cooling principle introduced earlier by Wärtsilä on the cold circuit side has reportedly proven valuable for engine operation on low quality heavy fuels at part load. A stepless electronically controlled version was developed for standard fitment to the W20 engine. The two pumps serving the cooling systems are engine driven. The high temperature (HT) cooling system is fully integrated into associated engine components, virtually eliminating the use of pipes. Supply and return channels are cast into the engine block while the pump cover and turbocharger support hold the connections between pump and engine block. The system temperature is kept at a high level (about 95°C) for safe ignition/combustion of low quality heavy fuels at low loads. An engine-driven pump of the cassette type is standard. The low temperature (LT) cooling system is also fully integrated with the engine components (engine block, pump cover, air and lubricating oil cooler housings) and served by an engine-driven pump of the same type as in the HT-system. System temperature is electronically controlled as a function of engine load. The charge air temperature is optimized at every operating point by heating the charge air with heat from the lubricating oil according to a pre-set curve programmed into the electronic board.

The lubricating oil system is grouped compactly into two adjacent modules at the flywheel end: one contains the cooler thermostatic valves and filters, and the other is a pump unit with a pre-lubricating pump and an engine-driven main pump. The filter arrangement comprises three individual filter chambers which can be switched off one by one when replacing the filter cartridge. A 'safety mesh' is also incorporated in each chamber. The paper-type filter cartridges are horizontally positioned to facilitate proper cleaning of the chamber before a new cartridge is installed. Centrifugal filters fitted for further filtration of the lubricating oil also serve as the oil condition indicator. A pulse turbocharging system was considered the best choice for an engine designed for installations normally subject to frequent load variation and sudden load application. All cylinder configurations are served by an optimized pulse system; no additional solutions are required for part-load operation.

The control and safety system is based on two cabling modules, one on each side of the engine, and sensors of the 'low mass' type to avoid possible vibration damage. The sensors are prefabricated with the correct cable length and a connector for connection to the cabling module. The modules in turn are connected with multi-pin type connectors to the engine control unit (ECU) which is resiliently mounted on the engine. Incorporated in the ECU is instrumentation for local reading of engine and turbocharger speed, exhaust temperature after each cylinder and after the turbocharger, an operating hour counter, and electronic boards for speed measuring and charge air temperature control. Plug-in connectors are provided for connecting the ECU to an external control system.

New W20 layout

Seven years after Wärtsilä 20 engine production started some design revisions were made, mainly involving the ancillary equipment and with the aim of improving environmental friendliness and adaptability to customer needs. The main components, such as pistons, bearings and crankshaft, remained untouched. The new layout was introduced at end-2000.

A drawback of the original layout was the paper cartridges used for lube oil filtration. The number of filter cartridges per engine varied between two and four, depending on the cylinder configuration. A typical genset installation of three 9L20 engines running for 5000 hours annually on heavy fuel oil would therefore generate hazardous waste in the shape of 100 used cartridges a year. The main objective in revising the engine layout was to eliminate this waste and the high cost of new filters for shipowners.

A completely new oil module was designed to enable the use of a fully automatic back-flushing lube oil filter incorporating elements of seamless sleeve fabric with high temperature resistance. An overhaul interval of one year is recommended, and the expected filter lifetime is four years. The channels directing the fluids to the thermostatic valves for the LT-water and lubricating oil are integrated in the bracket for the filter and the cooler. A horizontal position for the filter was chosen because of space limitations and maintenance benefits. The filter housing drains automatically when the engine is stopped so that any overhaul inside the filter is easier, and the risk of foreign particles entering the clean side is minimal.

The oil cooler design was also changed from a tube-type to a brazed plate heat exchanger. The cooler needs no outer casing and its cooling capacity varies with the number of plates (the height of the cooler); this makes it possible to use one bracket for all engine cylinder configurations. The same basic oil module is used on both LF (auxiliary) and LD (main) engines; it is simply rotated 180 degrees.

The number and size of non-machined parts after the oil filter were reduced significantly to minimize the risk of foreign particles entering the system. Since the Wärtsilä 20 cannot be equipped with running-in filters, to prevent damage to the bearings or crankshaft during test runs, this redesign was necessary to allow uninterrupted production. A dual drive, turbine-driven centrifugal oil filter is used for removing particles from the flush oil coming from the automatic filter. Both dry and wet oil sumps are available. The main oil pump was moved to the front of the engine to optimize the suction arrangement for every installation type.

The charge air cooling module was redesigned to fit both versions of the engine. A self-supporting flange sealing was selected to eliminate the problems with sealing the cooler insert on the original design. The same cooler and the outlet duct are used on both LD and LF engines. The compressor washing arrangement is integrated in the water box for the cooler on the LD engine. Also integrated in the design was the possibility to have a water mist catcher. The water and fuel connections were moved to the sides of the engine to improve the pipe layout and the support for the pipes. No piping disturbs the overhaul space since all the pumps are arranged at the front of the engine.

Among the options are a direct-driven seawater pump, 100 per cent power take-off at both ends, WECS digital control system, fully integrated standby connections for water, fuel and lube oil, and fuel flexibility (light and heavy) possible for all engine types.

The old engine block design with loose feet (to make installation more flexible) had proved unnecessary in practice as it was useful in only a handful of applications. The feet were thus integrated with the engine block casting to reduce cost; the option to use loose feet was nevertheless retained and wider feet were designed to create a more effective resilient mounting. The main oil channel is machined in the engine block to reduce both labour and material costs. The charge air receiver wall was provided with two openings since the charge air duct can be placed at either end of the engine.

A major disadvantage of the Wärtsilä 20 engine for propulsion applications was that the turbocharger could only be mounted at its free end. It was therefore decided to make it possible for the turbocharger to be located at either end on the six- to nine-cylinder models, the main contenders in the small ship propulsion market.

WÄRTSILÄ 26

The 260 mm bore/320 mm stroke Wärtsilä 26 engine (Figure 27.14) was launched in 1995 as an eventual replacement for the SW28 model in the Stork-Wärtsilä Diesel medium speed programme. The design is

Figure 27.14 The W26 engine in V18-cylinder form. It is possible to change the lube oil filter while the engine is running

characterized by a high power density and the capability to run on heavy fuel up to 730 cSt viscosity. An advanced crankshaft design, significantly reducing the distance between cylinders, and a short connecting rod limiting the height of the engine contributed to a short and low overall configuration.

Propulsion and genset drive applications are targeted by six, eight and nine in-line and V12-, 16- and 18-cylinder models covering an output band from around 1800 kW to 5850 kW at 900–1000 rev/min. Mean effective pressures range from 22 to 24 bar, depending on heavy fuel or marine diesel oil operation, with power outputs up to 325 kW/cylinder. A more powerful version yielding up to 395 kW/cylinder on gas oil was planned for a later launch with high performance turbocharging and a mean effective pressure of 28 bar to serve the fast ferry propulsion sector (see page 689).

Most of the design work was executed at Stork-Wärtsilä Diesel in The Netherlands. The designers defined five key objectives: high reliability; low fuel consumption and emissions; simple, space-saving installation; multiple applications and fuel choices; and low operating cost per kW. The goals were addressed by combining technology developed specifically for the application with design features of three earlier 'new generation' engines in the Wärtsilä Diesel programme: the medium speed W38 and W20 series and the high speed W200 series (Chapter 30).

High reliability was sought from the incorporation of structural and component designs proven in these and other Wärtsilä engines, such as thick-pad bearing technology and pressurized skirt lubrication. Extended times between overhauls and enhanced lifetimes were also targeted.

Compact, more simple and accessible installations are addressed by integrating ancillary support equipment (pumps, filters, coolers and thermostats) on the engine. Piping is eliminated from outside the engine; the channels for lubricating oil and cooling water are all incorporated in the block and other castings; and the inlet air receiver is also completely integrated in the engine block casting. A multi-duct (Figure 27.15) combines several functions, providing a connection between the inlet air receiver and the inlet port of the cylinder head, a flow channel for exhaust gases and a cooling water channel. It also supports the exhaust system and facilitates dismantling of the cylinder head without disturbing that system.

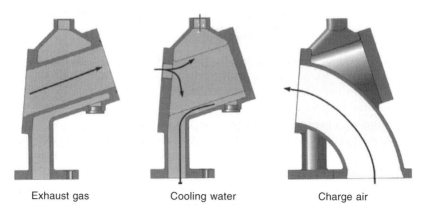

Exhaust gas   Cooling water   Charge air

Figure 27.15 Schematic diagram of multiduct connected to the cylinder head (Wärtsilä 26 engine)

The pursuit of integration reduced the overall number of Wärtsilä 26 engine components by around 30 per cent compared with the SW28 design.

Engine length was saved by giving the crankshaft undercuts. The radii in the transition from the crankweb to the crankpin and from the crankweb to the journal are included in the web itself. As a result, it was possible to keep the distance between the cylinders short and, at the same time, specify an optimum bearing length. A beneficial effect on the oil film thickness of the bearings, and hence their durability, was gained. The fully balanced crankshaft, suitable for full power output to be taken off either end, was designed for maximum overall rigidity and moderate bearing loads.

The highest possible rigidity was also sought from the cylinder head design which features a conical intermediate deck and provides optimum support to the flame deck. The flame deck has drilled coolant channels to inhibit large thermal stresses. The exhaust valve seats are water cooled and the valve temperatures measured by sensors.

Flanged cylinder liners with tangential cooling water flow are installed. The liner is supported symmetrically at the top by the cylinder block and sized so that the temperature on the inner wall remains high enough to prevent low temperature corrosion yet low enough to ensure good lubrication. This design, proven on the Wärtsilä 20 engine, makes drilled coolant holes unnecessary; as a result, a thinner liner wall can be used. An anti-polishing ring at the top of the liner is specified to prevent contact between the piston crown and the liner, thus avoiding any bore polishing by combustion residues. High resistance to wear is further fostered by elements in the alloy material of the liner. The anti-polishing ring also contributes to a reported lubricating oil consumption of under 0.5 g/kWh.

The composite piston consists of an oil-cooled steel crown with three rings (two ceramic chromium-plated compression rings and an oil distributor ring) and a nodular cast iron skirt. Crown and skirt are fastened to each other by one central bolt to foster an even, symmetrical load distribution (another feature whose effectiveness had been demonstrated in the Wärtsilä 20 engine). Pressurized skirt lubrication secures excellent tribological behaviour at high cylinder pressures.

A connecting rod with a horizontal split was developed for the Wärtsilä 26 engine, exploiting the finite element method to optimize the component with regard to maximum stiffness and minimum mass. Key merits cited for the resulting design are a short length, high rigidity, reliability and low mass (and consequently a smaller load on the bearings). The rod is drilled with a single hole without plugs for delivering lubricating oil to the piston.

An injection pressure of 2000 bar, necessary for fine atomization of the fuel and a short injection duration, is attained by closed barrel-type fuel pumps. Fuel feed lines are integrated in the pump housing, and the fuel system is designed to keep pressure pulses low. The high injection pressure dictated a very rigid camshaft made of single-cylinder sections connected to each other by bearing journals.

Design details facilitate easier inspection and maintenance routines. The stud bolts for the cylinder head, connecting rod and main bearing are hydraulically tensioned (Figure 27.16). When the head is removed the multi-duct supports the exhaust gas piping and the insulation cover.

Figure 27.16 The stud bolts for the cylinder head (shown here) as well as those for the connecting rod and main bearing of the W26 engine are hydraulically tensioned

The seals of the cooling water pump can be replaced without dismantling the pump itself. Replacement or inspection of the piston rings is possible without disassembly of the liner: the piston can be partially removed from the liner for this purpose.

The fuel pumps can be replaced quickly and simply without taking apart the low pressure piping and the tappets; the pump can be taken from the engine and overhauled under clean workshop conditions. Both air cooler and lubricating oil cooler are of the insert type, allowing convenient removal for cleaning, and it is possible to change the oil filters while the engine is running.

A Wärtsilä-designed computer-based engine control unit (ECU) protects the engine, warning against possible problems before damage occurs. Signals from sensors are transmitted to the unit via a CAN bus connection comprising a single cable from one end of the engine to the other.

W26X version

A high performance variant, the Wärtsilä 26X engine, was developed for fast commercial and naval propulsion applications with respective outputs of 4800 kW, 6400 kW and 7200 kW at 1000 rev/min from V12, 16 and 18-cylinder models on a mean effective pressure of 28.2 bar. Key components of the standard engine—notably the crankshaft, cylinder heads and liners—were redesigned and reinforced to suit the load profile imposed by fast ferry operation. Weight reductions in structural elements were also sought without compromising reliability; other design targets included compactness, installation friendliness and ease of accessibility for servicing. A specific fuel consumption of 195 g/kWh (including engine-driven pumps) was achieved by optimized injection and sequential turbocharging systems. ABB Turbo Systems' TPL turbochargers (model 65 sets in the case of the V18-cylinder engine) were specified in association with a Spex (single pipe exhaust) system, an arrangement standard for all cylinder numbers.

WÄRTSILÄ 38

The 380 mm bore/475 mm stroke W38 design was developed by Wärtsilä Diesel's Dutch subsidiary Stork-Wärtsilä Diesel as a replacement for its ageing TM410 engine (see Chapter 28), filling a power gap in the group portfolio between the Vasa 32 and W46 designs. The W38 series was introduced in 1993 with a specific rating of 660 kW/cylinder at 600 rev/min to cover a power band from 3960 kW to 11 880 kW with six, eight and nine in-line and V12-, 16- and 18-cylinder models (Figure 27.17).

The design was distinguished by a new combustion philosophy applied to achieve a high thermal efficiency and reduced NOx emissions (down by 50–70 per cent) without compromising fuel economy. The key parameters are a stroke/bore ratio of 1.25:1, a maximum cylinder pressure of up to 210 bar and a maximum fuel injection pressure of 1500 bar. Such parameters provide the flexibility to select a compression ratio, injection timing and injection rate that allow the desired low NOx values to be attained.

A completely new fuel injection system was developed with the following characteristics: suitable for continuous 1500 bar injection pressure; closed barrel; optimized constant pressure valve design for

Figure 27.17 V18-cylinder version of the Wärtsilä 38 engine

short injection period and high brake mean effective pressure; no mixing of lubricating oil and fuel between plunger and barrel to avoid lacquering and sticking; separation of leakage fuel from the oil sump to avoid lubricating oil contamination; integrated low pressure fuel supply lines; fuel injection valves with uncooled nozzles; and shielded high pressure lines and connections. The fuel injection system is fully covered by the hot box, and possible low and high pressure leakages are detected by alarms.

The fuel pump has an integrated low pressure channel which means there is no low pressure pipe to fit to the pump because the fuel pipes are fitted directly to the fuel pump foundation. Integrated fuel pulse dampers in the fuel supply lines ensure low pressure pulses on the system side. Each fuel pump is equipped with a flow distribution throttle. The high pressure fuel pipe is shielded and fitted with a leak alarm. Maintenance is facilitated by a locking device flange on the cylinder head for the high pressure pipe.

The engine starting system incorporates a starting fuel-limiter to secure a smooth and safe start. Each fuel pump is equipped with an overspeed trip cylinder to ensure that the fuel rack is pushed to zero in the event of overspeed. Additional safety is provided by an extra stop cylinder turning the regulating shaft to zero in the event of overspeed. Overall compactness and lightness were prime targets for the W38 engine designers. Component integration resulted in 40 per cent fewer parts than the older generation TM410 engine, with many of the traditional bolted-on elements now incorporated in the main block casting (for example, lines and piping for lubricating oil, cooling water and fuel).

Simplicity of maintenance and extended overhaul intervals were other design goals, with only a modest envelope required for servicing and dismantling. Engine-driven cooling and lubricating oil pumps as well as engine-mounted coolers and filters simplify installation and release machinery space. Integrated brackets are provided for resilient mounting systems. The bending stiffness and torsional rigidity of the engine block and crankshaft allow savings in the specification of flexible couplings, torsional vibration dampers and elastic elements for resilient mounting.

W38A specification

Engine block: underslung crankshaft; integrated air receiver; single-piece casting for all cylinder numbers; and arranged for resilient mounting with screwed-on feet.

Crankshaft: fully balanced and designed for space saving, maximum overall rigidity and moderate bearing loads. The crankshaft of the V-cylinder models has undercuts for both the main and pin fillet, the design combining a short cylinder distance and maximum bearing length. Normal fillets are applied in the in-line cylinder engines.

Connecting rod: marine head design, completely machined and dimensioned to achieve the lowest dismantling height (Figure 27.18).

Bearings: Thick-pad bearing technology and moderate main bearing load.

Cylinder liner: high collar design with optimized operational roundness and straightness; bore cooling for effective control of liner temperature; and anti-polishing ring.

Piston and rings: composite piston with forged steel crown and nodular cast iron skirt; pressurized skirt lubrication; two ceramic chromium-plated directional barrel-shaped compression rings in the crown; a ceramic chromium-plated taper-faced compression ring and an oil distributor ring in the skirt.

Cylinder head: stiff box with four screws; flame plate supported by intermediate deck, allowing no deformation of valve seat environment.

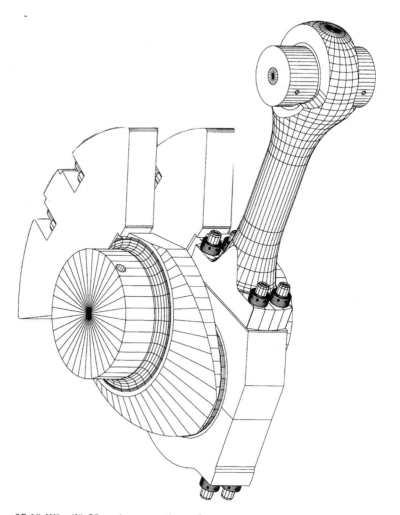

Figure 27.18 Wärtsilä 38 engine connecting rod

Camshaft: single-cylinder shaft segments with integrated cams; and built-in valve tappet modules screwed onto the engine block.

Turbocharger/charge air cooling: modular exhaust gas pipes in insulated hot box; two-stage air cooler; and integrated, rigid turbocharger support/charge air cooler.

W38B version

The popularity of the Wärtsilä 38 encouraged the development of a B-version offering higher powers with lower fuel consumption and reduced emissions. The new model also benefited from an advanced control and monitoring system, and full modularization to ease assembly. The B-series was fully released in 2001 with in-line six, eight and nine and V12, 16 and 18-cylinder models covering a power band from 4350 kW to 13 050 kW at 600 rev/min. The original engine, re-designated the Wärtsilä 38A, remained in the programme. Both versions are assigned for production at Wärtsilä Italia in Trieste.

Careful selection of thermodynamic processes and progress in component design contributed to the achievement of the development goals. The refinements included a higher maximum cylinder pressure, compression ratio and mean effective pressure, and advanced fuel injection timing (see table). Special attention was paid to optimizing the combustion chamber characteristics, the fuel injection profiles and the piston shape. The air flow system was fully revised (inlet bends to the cylinder head, flow channels at the head, port design, exhaust bends and turbocharger). State-of-the-art valve timing (including application of the Miller cycle) was also introduced. New generation turbochargers were adopted to further boost overall performance and economy, with integration of the bypass and waste gate valves. The position of the valves and their controls was selected to secure low vibration levels, low local temperatures and short pipework.

These measures resulted in a significant reduction in specific fuel consumption (6 g/kWh below the 38A engine) and opened the way to further improvements in fuel economy. The limits on pollutant emissions were also comfortably met; the NOx emission levels can be lowered further by up to 50–60 per cent by incorporating direct water injection (see Chapter 3).

Critical components were given greater strength to support the higher power levels and raise the safety margins for further rating growth; examples include additional strengthening of the connecting rod and geometrical modifications at the interface of the cylinder liner with the cylinder head (area of the gas sealing ring). All bearing loads are kept low and an ample oil film thickness ensures safe bearing operation.

A new three-piston ring pack and an anti-polishing ring incorporated at the top of the cylinder liner foster low lube oil consumption. The 'hot box' environment was completely redesigned to secure high reliability and functional integration; externally, there are swinging hot box covers and new cylinder head covers enclosing the hot box area; internally a new LP and HP fuel system is based on a new generation pump.

The lubrication module incorporates a cooler, an automatic back-flush filter, a centrifugal filter in the back-flush line, and thermostatic valves. The overall system includes main and pre-lubrication pumps and the regulating valve. Lubricating oil flow through the engine was revised and based on optimized bearing clearances to allow large safety margins. Running-in filters of the spin-on type can be mounted on the non-operating side of the engine and removed after the factory approval test. The LT and HT cooling water systems embrace pumps and thermostatic valves built on the engine.

Compactness and easier installation were promoted by 'all on' engine support systems, with complete lubrication and cooling water modules mounted on the engine (Figure 27.19). All piping connections are located at the free end to allow simplified interfacing with external systems. Among the advanced automation features is Wärtsilä's WECS 7000 engine-integrated system, which is tailored for the group's medium speed designs.

Wärtsilä 38A and 38B differences

	38A	*38B*
Bore	380 mm	380 mm
Stroke	475 mm	475 mm
Output/cylinder	660 kW/cyl	725 kW/cyl
Cylinders	6, 8, 9L/12, 16, 18V	same
Speed	600 rev/min	600 rev/min
Compression ratio	12.7	15
Maximum combustion pressure	180 bar	210 bar
Mean effective pressure	24.5 bar	27 bar
Piston speed	9.5 m/s	9.5 m/s
Injector holes	11	10
Injector position	Ref	6 mm higher
Fuel cam	Ref	low speed
Cylinder head	Ref	optimized air flow
Piston shape	Ref	optimized
Inlet valve timing	Ref	Miller cycle
Injection timing	Ref	advanced
Exhaust system	Ref	optimized flow
Turbocharger efficiency	65 per cent	70+ per cent
Specific fuel consumption	Ref	– 6 g/kWh
NOx emissions	IMO	IMO-50 per cent*

* Prepared for Direct Water Injection

WÄRTSILÄ 46

Several innovative features were introduced by Wärtsilä Diesel with the launch of the W46 engine in 1988 which has since found favour in diverse passenger and cargo ship propulsion sectors (Figures 27.20

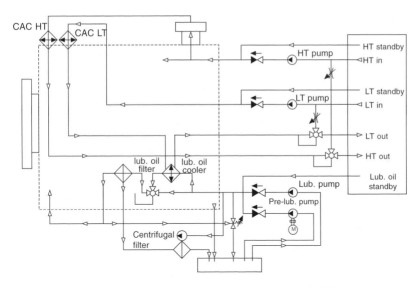

Figure 27.19 Integrated cooling water and lubrication oil systems of W38B engine

and 27.21). The designer's prime development goal in creating a large bore medium speed trunk piston engine to compete with low speed crosshead machinery was the highest reliability. A significant

Figure 27.20 Six-cylinder version of the Wärtsilä 46 engine

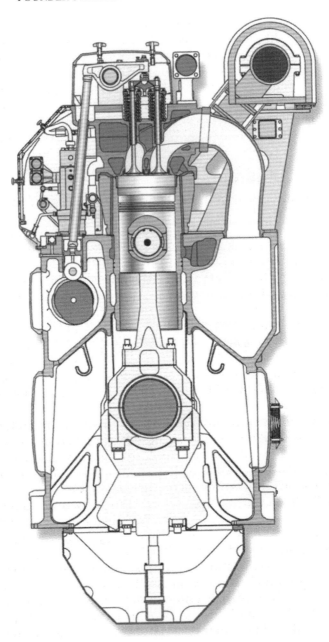

Figure 27.21 Cross-section of Wartsila 46 engine

contribution was sought from thick-pad bearing technology, exploiting large bearings with thick oil films, and pressure-lubricated piston skirts.

During its first four years of production the 460 mm bore/580 mm stroke design was offered with an output of 905 kW/cylinder at 450/500/514 rev/min as the Wärtsilä 46A. Experience with pilot installations encouraged the release in 1992 of a B-version with an output of 975 kW/cylinder at 500/514 rev/min. The C-output version followed in autumn 1995, yielding 1050 kW/cylinder at 500/514 rev/min with reduced thermal load and lower emissions thanks to advances in turbocharging and fuel injection systems and the adoption of low NOx combustion principles.

The three ratings allow the 46 series to cover a power band from 5430 kW to 18 900 kW with six, eight and nine in-line and V12-, 16- and 18-cylinder models. The C-output versions operate with a mean effective pressure up to 26.1 bar and the engine is designed for a maximum combustion pressure of 210 bar.

A nodular cast iron engine block (Figure 27.22) is configured to achieve a rigid and durable construction for flexible mounting (the W46 engine was perhaps the first of its size designed for elastic and super-elastic mounting, Figure 27.23). The main bearings are of the underslung type with hydraulically tightened bolts; side bolts add further rigidity to the main bearing housing. In-line cylinder engines are equipped with an integrated air receiver fostering rigidity, simplicity and cleanliness. A welded block of steel castings and plates is offered as an alternative.

A rigid box-like cylinder head design aims for even circumferential contact pressure between the head and cylinder liner, with four fixing bolts simplifying maintenance procedures. No valve cages are used, improving reliability and increasing the scope to optimize exhaust port flow characteristics. Water-cooled exhaust valve seat rings are specified. Both inlet and exhaust valves receive a forced rotation from Rotocaps during every opening cycle, fostering an even temperature distribution and wear of the valves, and keeping the sealing surface free from deposits. Good heat conduction results.

Cylinder liner deformations are normally caused by cylinder head clamping and thermal and mechanical loads. A special liner design with a high collar-to-stroke ratio for the W46 engine minimizes deformation, the round liner bore in combination with efficient lubrication enhancing conditions for the piston rings and reducing wear. The liner material is a special grey cast iron alloy developed for wear resistance and high strength. An anti-polishing ring in the upper part of the liner prevents the bore polishing that leads to local liner wear and increased lubricating oil consumption. The simple yet highly

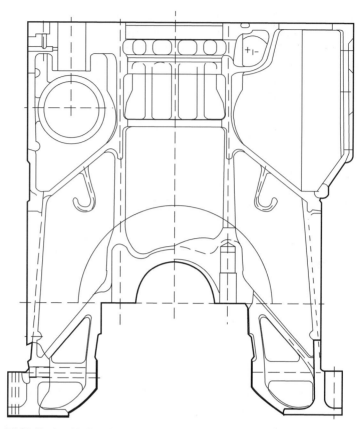

Figure 27.22 Engine block of Wärtsilä 46 engine

effective device was introduced as standard on later engines and retrofitted to existing installations.

A composite low friction piston with a nodular cast iron skirt and steel crown incorporates a special cooling gallery configuration to secure efficient cooling and high rigidity for the piston top. It is designed to handle combustion pressures beyond 200 bar. A long lifetime is sought from hardened top ring grooves. The three-ring set includes a top ring with a special wear-resistant coating; all the rings are dimensioned and profiled for maximum sealing and pressure balance. Low friction is addressed by a skirt lubrication system delivering a well-distributed clean oil film that eliminates the risk of piston ring scuffing and reduces the wear rate, and fostering cleaner rings and grooves free from corrosive combustion products. Noise and wear are reduced by hydraulically damped tilting movements provided by an oil pad between liner and piston.

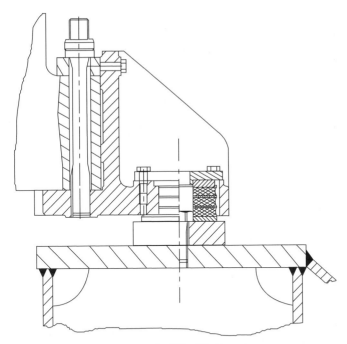

Figure 27.23 Flexible mounting system for Wärtsilä 46 engine

Hard skirt contact against the liner experienced with the first seagoing W46 engine installation was solved by introducing a piston with increased compression height, a slightly increased skirt length and an optimized skirt form.

A three-piece marine design connecting rod was designed for distribution of the combustion forces over a maximum bearing area with relative movements between mating surfaces minimized. Piston overhaul is facilitated without touching the big end bearing, and the bearing can be inspected without removing the piston. The three-piece design also reduces the piston overhauling height.

The one-piece forged crankshaft is designed to accept a high combustion pressure while maintaining a conservative bearing load; rigidity is underwritten by a moderate bore/stroke ratio and large pin and journal diameters. Counterweights fitted on every crankweb provide 95 per cent balancing. Reliability is sought from the thick-pad bearing design and bearing loads reduced by increasing crankshaft journal and pin diameters as well as length. Low bearing loads allow for softer bearing materials with greater conformability and adaptability, underwriting a virtually seizure-free bearing, according to Wärtsilä. A thick corrosion-resistant overlay of tin–antimony is used in the big end bearings.

The key elements of Wärtsilä's thick-pad bearing technology are: ample oil film thicknesses achieved by proper overall dimensioning; corrosion-resistant bearing materials with excellent imbeddability properties; radially rigid bearing assemblies during both mass and gas force loading conditions; conformable bearing materials and axially conformable housings, which minimise edge pressure; and oil grooves and oil holes located where no harm can be caused to the oil film.

The camshaft is built of single-cylinder sections with integrated cams. The shaft sections are connected through separate bearing journals which make it possible to remove the sections sideways from the camshaft compartment. The valve follower is of the roller tappet type, the roller profile being slightly convex for good load distribution. The valve mechanism includes rocker arms working on yokes guided by pins.

W46 fuel injection

An innovation on the original W46 engine was a twin injection fuel system exploiting a pilot valve and a main injection valve (Figures 27.24 and 27.25). The initial aim of the system was to secure safe combustion with the poorest fuels on the market; the very fast main injection process facilitated by the system also contributed to a low

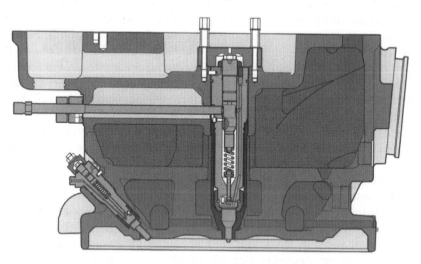

Figure 27.24 Wärtsilä 46 twin injection system, showing the pilot fuel valve (left) and the main fuel valve (centre)

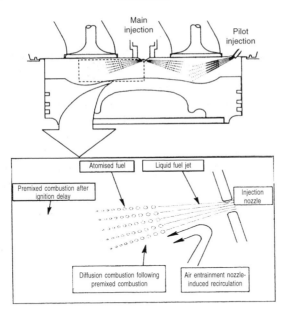

Figure 27.25 Twin injection system principle for original Wärtsilä 46 engines

fuel consumption. It became evident too that the twin injection configuration yielded a positive effect on NOx emissions, stimulating further refinement of the system.

The first seagoing W46 engine installation was later optimized with a reduction of the pilot fuel amount to below 3 per cent. Both pilot and main injection valves were designed with a small heat absorbing surface facing the combustion space and arranged for efficient heat transfer to the cooling water, eliminating the need for a separate nozzle temperature control system. Fuel was transported over the shortest distance from the pump to the valve via a high pressure pipe in the cylinder head.

Wärtsilä cited the following merits of twin injection through individually optimized nozzles: excellent pilot fuel atomization and combustion over a wide load and speed range; reasonable injection rates even under idling conditions when the combustion process is often a poor compromise; facility to burn fuels of low ignition quality with a minimum of ignition delay; low combustion noise level thanks to controlled pre-burning, resulting in a lower rate of increase in cylinder pressure and consequently lower mechanical load; reduced delay in the pre-burning phase, which lowers peak pressure and temperature and hence curbs NOx emission formation; improved starting ability; and increased thermal efficiency.

The mono-element injection pump (Figure 27.26) was designed to serve the twin injection system. A constant pressure relief valve eliminates the risk of cavitation erosion by maintaining a residual pressure at a safe level over the full operating field. A drained and sealed-off compartment between the pump and tappet prevents leakage fuel from mixing with lubricating oil. The pre-calibrated pumps are interchangeable.

Fuel system pipes and main components are located in a hot box to secure safety at high preheating temperatures; fuel pipes outside the hot box are also carefully shielded. The fuel feed pipes on the engine are equipped with pressure pulsation dampers. Any fuel leakage from pipes, injection valves and pumps is collected in a closed piping system which keeps the hot box and the engine dry and clean.

The twin injection system was later replaced by a single injection configuration as higher pressure traditional layouts became available (Figure 27.27).

Spex turbocharging

Swirl-Ex turbocharging—a constant pressure system with an in-built diffusor—was specified for early Wärtsilä 46 engines. It demonstrated high efficiency but service experience showed up a disadvantage due to the large volume in the exhaust pipe which made the load pick-up slower than the standard in other Wärtsilä engines. A faster turbocharging system based on a single pipe with a much smaller volume was therefore developed, resulting in the Spex (single pipe exhaust) system. This offers a swift load pick-up and almost the same efficiency at high load as the Swirl-Ex system.

The aim of an exhaust system is to transfer the energy released from the cylinder, after the exhaust valve has opened, to the turbine of the turbocharger in the most effective way. This could easily be done by making the pipes as short as possible and the areas wide without restrictions. There are, however, other parameters that must be taken into account, such as uniform scavenging. Since all the cylinders are connected to the common system there is an obvious risk that a pulse coming from one cylinder at the beginning of the blow-down phase disturbs the scavenging of another cylinder. To avoid this, the design has to be suited to all the possible cylinder configurations and speed ranges that the engine type will cover.

Another parameter affecting scavenging is the pressure drop between inlet and exhaust manifolds. The higher the pressure difference, the better resistance the cylinder has against the disturbing pulses in the exhaust manifold; the pressure difference is largely a function of the

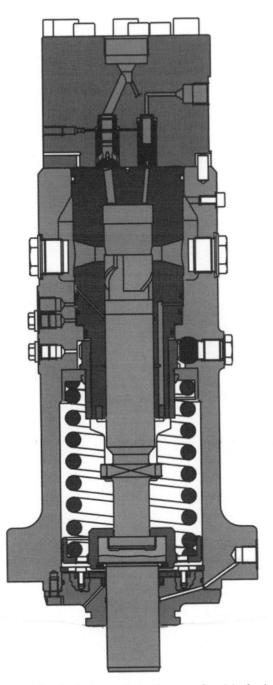

Figure 27.26 Wärtsilä 46 engine fuel pump designed to serve the original twin injection system

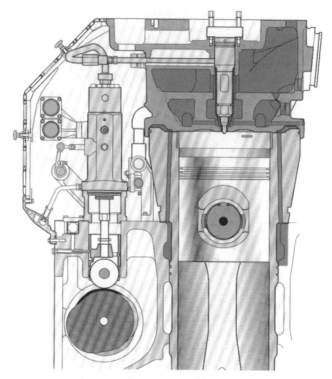

Figure 27.27 Single injection system for later Wärtsilä 46 engines

turbocharger efficiency. Thus there are many parameters affecting the performance of the exhaust system, and the final design is a compromise between various factors.

Turbocharging of the W46 engine is now optimized for the given application exploiting the Spex system as standard (original Swirl-Ex engine installations in service were reconfigured). For special applications, where low load running accounts for a large proportion of the total operating time, a three-pulse system is also offered for appropriate cylinder numbers. Pulse charging is reportedly excellent at partial load operation and for fast load response, while Spex charging is said to be excellent at high and steady loads, with the load response remaining good. Spex is based on an exhaust gas system that combines the advantages of both pulse and constant pressure charging. Compared with a constant pressure system, the ejector effect of the gas pulses provides better turbine efficiency at partial loads. Very small deviations in the scavenging between the cylinders result in an even exhaust gas temperature.

Modular-built exhaust gas systems on the W46 engine are designed to handle high pressure ratios and pulse levels while being elastic enough to cope with thermal expansion in the system. Exhaust wastegate, air bypass and air wastegate arrangements can be specified to satisfy specific operating requirements, such as load response or partial load performance.

A load-dependent cooling water system is used for the low temperature (LT) water circuit. At low load the water temperature is elevated by transferring heat from the lubricating oil to the charge air. This eliminates the risk of cold corrosion in the combustion space and also improves the ability to burn fuels of low ignition quality at low loads. The high temperature (HT) cooling water system operates constantly at a high temperature to make the temperature fluctuations in the cylinder components as small as possible and to prevent corrosion due to undercooling. The charge air cooler is split into a high and low temperature section to maximize heat recovery. Engine-driven pumps are available as an option.

A wet or dry sump lubricating oil system can be specified. On its way to the engine the oil passes through a full flow automatic filter unit and a safety filter for final protection. A small proportion of the flow is diverted to a centrifugal filter which acts as a lubricating oil condition indicator and warns the operator of excessively dirty oil and wear. Both the safety filter and the indicator filter are designed for cleaning while the engine is running. An engine-driven lubricating oil pump can be provided optionally.

WÄRTSILÄ 64

The most powerful medium speed design ever brought to the market— and reportedly the first to exceed the 50 per cent thermal efficiency barrier (Figure 27.28)—the Wärtsilä 64 engine was prototype-tested in six-cylinder form in September 1996 (Figure 27.29). Production models became available from autumn 1997 and have since logged propulsion applications in containerships, multipurpose cargo vessels and chemical/product tankers. The engines are assigned for manufacture by Wärtsilä Italia in Trieste.

A rated nominal output of 2010 kW/cylinder at 333 rev/min is quoted for the 640 mm bore/900 mm stroke in-line cylinder design on a mean effective pressure of 25 bar and a maximum cylinder pressure of 190 bar (Figure 27.30). Power demands up to 23 280 kW can be met by a programme of six, seven, eight and nine-cylinder in-line models and a V12-cylinder model. The V-engine has the same bore

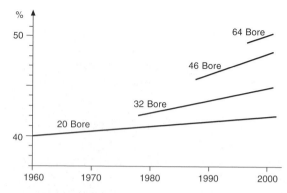

Figure 27.28 Rise in thermal efficiency of successive Wärtsilä medium speed engine designs

Figure 27.29 The Wärtsilä 64 engine is the world's largest and most powerful medium speed design

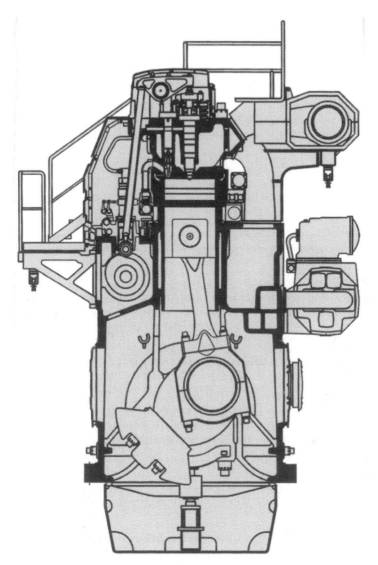

Figure 27.30 Cross-section of Wärtsilä 64 design

but a shorter stroke (770 mm) and lower specific power rating (1940 kW/cylinder at 400–428 rev/min on an mep of 22–23.5 bar). Performance is boosted by a high efficiency TPL80E turbocharger, one of ABB Turbo Systems' latest TPL series (see Chapter 7).

Reliability is sought from traditional Wärtsilä solutions, notably a pressure-lubricated piston skirt, large bearings with thick oil films, thick collar cylinder liners fitted with an anti-polishing ring, high fuel injection pressures for optimized combustion, and camshafts with high torque capacity and low Hertzian pressures. Easier installation is addressed by built-on cooling water pumps and lubricating oil system (including automatic filters). Simplicity of maintenance is reflected in a maximum overhaul time of three hours for the 'strategic' components (main and big end bearings, pistons, cylinder heads and injection pumps) thanks to design measures and special tools.

W64 engine details

Engine block: nodular cast iron is the natural choice for modern engine blocks, Wärtsilä suggests, because of its strength and stiffness properties and the freedom that casting offers. Optimum use was made of current foundry technology to integrate most oil and water channels, resulting in a virtually pipe-free engine with a clean outer exterior. Resilient mounting, now common, calls for a stiff engine frame; integrated channels designed with this in mind thus serve a dual purpose.

Crankshaft and bearings: advances in combustion development require a crank gear which can operate reliably at high cylinder pressures. The crankshaft must be robust and the specific bearing loads kept at an acceptable level; this is achieved by optimizing the crankthrow dimensions and fillets. The specific bearing loads are conservative and the cylinder spacing (important for the overall length of the engine) is minimized. Apart from low bearing loads, the other crucial factor for safe bearing operation is oil film thickness. Ample film thicknesses in the main bearings are ensured by optimized balancing of rotational masses and in the big end bearing by ungrooved bearing surfaces in the critical areas. All these features ensure a free choice of the most appropriate bearing material. The other thick-pad bearing technology concepts proven on the Wärtsilä 46 engine (see page 694) are also applied.

Piston and rings: a rigid composite piston with a steel crown and nodular cast iron skirt has been adopted for highly rated diesel engines for years to secure reliability in a high cylinder pressure and combustion temperature environment. Wärtsilä's patented skirt lubrication is applied to minimize frictional losses and ensure appropriate lubrication of piston rings and skirt. Each ring of the three-ring pack is dimensioned and profiled for a specific task. The pressure balance above and below each ring is crucial in avoiding carbon deposits in the ring grooves of a heavy fuel engine. Experience has shown that this effect is most likely achieved with a three-ring pack, Wärtsilä reports. Another factor favouring a three-ring pack is that most frictional losses in a reciprocating combustion engine originate from the rings. An extended lifetime is sought from the piston ring pack and ring grooves through wear-resistant coating and groove treatment (Figure 27.31).

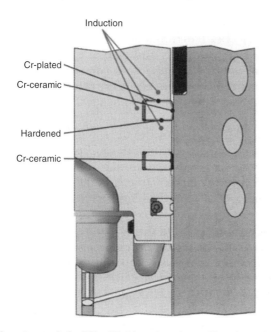

Figure 27.31 Three-ring pack for Wärtsilä 64 engine piston. Note the anti-polishing ring incorporated in the upper cylinder liner (top right)

Cylinder liner and anti-polishing ring: the thick high-collar type liner is designed with the stiffness necessary to withstand both pre-tension forces and combustion pressures with virtually no deformation. Its temperature is controlled by bore cooling of the upper part of the collar to achieve a low thermal load and avoid sulphuric acid corrosion. The cooling water is distributed around the liners with simple water distribution rings at the lower end of the collar. At its upper end the liner is fitted with an anti-polishing ring to eliminate bore polishing and reduce lube oil consumption. The ring's function is to calibrate the carbon deposits formed on the piston top land to a thickness small enough to prevent any contact between the liner wall and the deposits at any piston position. When there is no contact between liner and piston top land deposits no oil can be scraped upwards by the piston; at the same time, liner wear is also significantly reduced.

Connecting rod: a three-piece rod with all highly stressed surfaces machined is the safest design for engines of this size intended for continuous operation at high combustion pressures, according to Wärtsilä. For easy maintenance and accessibility the upper joint face is placed right on top of the big-end bearing housing. A special hydraulic tool was developed for simultaneous tensioning of all four screws. An intermediate plate with a special surface treatment is arranged between the main parts to eliminate any risk of wear in the contact surfaces.

Cylinder head: high reliability and ease of maintenance were sought from a stiff cone/box-like design able to cope with the high combustion pressure and secure both cylinder liner roundness and even contact between the exhaust valves and their seats. The head design is based on the four-screw concept developed by Wärtsilä and applied for over 20 years. Such a design also provides the freedom needed for designing inlet and exhaust ports with minimal flow losses. The port design was optimized using computational fluid dynamics (CFD) analysis in conjunction with full-scale flow measurements. Wärtsilä's extensive heavy fuel-burning experience contributed to the exhaust valve design, the basic criterion for which is correct temperature; this is achieved by carefully controlled cooling and a separate seat cooling circuit to secure long lifetimes for valves and seats.

Fuel injection system: the split-pump technology first introduced on the W64 engine offers advantages in terms of operating flexibility, mechanical strength and cost effectiveness. Fuel injection timing can be freely adjusted independently of the injected quantity, and tuning of the injection parameters according to the engine's operational conditions improves engine performance and reduces exhaust emissions. Smaller closed-type pump elements—derived from high volume production of smaller engines—reduce mechanical stresses and enhance reliability, while lower loads on rollers, tappets and cams improve pump drive reliability.

This new solution was dictated when injection pump manufacturers suggested that for such a large medium speed engine it would be very difficult to produce pump plungers of the size and accuracy required to secure the reliability associated with smaller engine designs. Since the output of the Wärtsilä 64 is approximately double that of the established Wärtsilä 46, it was decided to use two plungers (each of roughly W46 size) per engine cylinder.

The two plungers have slightly different functions (Figure 27.32). Both pump fuel at each stroke and are connected to the same line, from where fuel is led to the nozzle via a single high pressure line. Although both plungers pump the fuel in a similar way, only one of them needs to be controlled to adjust the fuel quantity. This made it possible to reserve the other plunger for another task: turning it to control the injection timing during engine operation. New possibilities

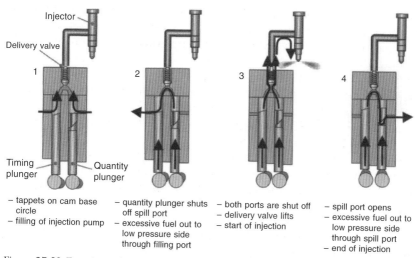

Injector

Delivery valve

1   2   3   4

Timing plunger   Quantity plunger

| – tappets on cam base circle
– filling of injection pump | – quantity plunger shuts off spill port
– excessive fuel out to low pressure side through filling port | – both ports are shut off
– delivery valve lifts
– start of injection | – spill port opens
– excessive fuel out to low pressure side through spill port
– end of injection |

Figure 27.32 Functions of twin plungers of fuel pump for the Wartsila 64 engine

thus opened up for controlling different load modes and fuel qualities, even allowing injection retardation when lower NOx emission values are demanded.

A contribution to reliability by the fuel pump design comes from splitting the plunger loading between two cams and rollers, thus lowering the loading on these components and underwriting safe operation at injection pressures up to 2000 bar. The associated tappets for these components are both integrated in the same housing as the tappets for the inlet and exhaust valves.

The high pressure fuel system was designed and endurance tested at 2000 bar; an actual injection pressure of around 1400 bar thus represents a substantial safety margin. No lubricating oil is required for the pump element since the plunger has a wear-resistant low friction coating. A profiled plunger geometry keeps the clearance between the plunger and barrel small, allowing only minimal oil to pass down the plunger; the small leakage is collected and returned to the fuel system. Any chance of fuel mixing with the lube oil is eliminated. Both the nozzle and nozzle holders are made of high grade hardened steel to withstand the high injection pressures and, combined with oil cooling of the nozzles, to foster extensive nozzle lifetimes.

Low pressure fuel system safety is underwritten by the Wärtsilä-patented multi-housing concept. The fuel line consists of channels drilled in cast parts which are clamped firmly on the engine block and connected to each other by simple slide-in connections for ease of assembly and disassembly. The pumps are connected together to form the complete low pressure fuel line with both feed and return channels; any need for welded pipes is eliminated. Safety is further enhanced by housing the entire low pressure and high pressure systems in a fully covered compartment.

Turbocharging system: based on non-cooled turbochargers with inboard plain bearings lubricated from the engine's lube oil system. The Spex (single pipe exhaust) turbocharging system is standard, with the option of exhaust waste-gate or air bypass according to the application. Spex, which exploits the pressure pulses without disturbing the cylinder scavenging, is described in the Wärtsilä 46 engine section above. The interface between the engine and turbocharger is streamlined, eliminating all the adaptation pieces and piping formerly used.

Cooling system: split into separate high temperature (HT) and low temperature (LT) circuits (Figure 27.33). The cylinder liner and cylinder head temperatures are controlled through the HT circuit; the system temperature is kept at a high level (around 95°C) for safe ignition/combustion of low quality heavy fuels, including operation at low loads. An additional advantage is maximum heat recovery. To further increase the recoverable heat from this circuit it is connected to the high temperature part of the double-stage charge air cooler. The HT water pump is integrated in the pump cover module at the free end of the engine; the complete HT circuit is thus virtually free of pipes.

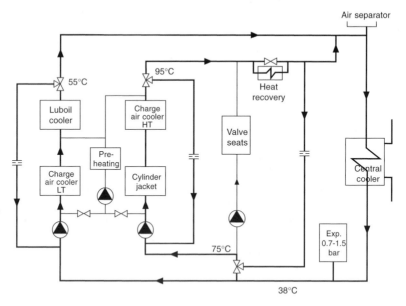

Figure 27.33 Cooling water system of Wärtsilä 64 engine

The LT circuit serves the low temperature part of the charge air cooler and the built-on lube oil cooler. It is fully integrated with engine parts such as the LT water pump with pump cover module, the LT thermostatic valve with the lube oil module, and transfer channels in the engine block. In addition, the LT circuit provides separate cooling of the exhaust valve seats and a lower seat/valve temperature, thus promoting long lifetimes for these components. Directly-driven pumps ensure safe operation even during a short power cut.

Lubricating oil system: all W64 engines are equipped with a complete built-on lube oil system comprising:

- Pump cover module: engine-driven main screw pump with built-in safety valve; pre-lubricating module; electrically-driven pre-lubricating screw pump; pressure regulating valve; and centrifugal filter for lube oil quality indication.
- Lubricating oil module: lube oil cooler; oil thermostatic valves; full flow automatic filter; and special running-in filters before each main bearing, camshaft line and turbocharger.

On in-line cylinder engines the lubricating oil module is always located at the back side of the engine, while on V-engines it can be built on the engine at the flywheel or free end, depending on the turbocharger position. The lube oil filtration is based on an automatic back-flushing filter requiring minimal maintenance and no disposable filter cartridges.

Automation system: an engine-integrated system, WECS, is standard and has these main elements:

- The Main Control Unit (MCU) cabinet, which comprises the MCU itself, a relay module with back-up functions, a Local Display Unit (LDU), control buttons and back-up instruments. The MCU handles all communication with the external system.
- The Distributed Control Unit (DCU) handling signal transfer over a CAN bus to the MCU.
- The Sensor Multiplexing Units (SMU) transferring sensor information to the MCU.

Software loaded into the system is easily configured to match the instrumentation and the safety and control functions required for each installation. The MCU cabinet is well protected and built into the engine; most of the rest of the hardware is housed in a special electrical compartment alongside the engine.

W64 in-service experience

Observations made at the 12 000 hour overhaul of the first Wärtsilä 64 engine in marine service (a seven-cylinder model) revealed no evidence of heavy carbon deposits on the cylinder head flame deck. Excellent behaviour was indicated by the exhaust valves, with no failure or burn marks on the contact surface with the seat ring and no sign of wear or corrosion on the valve stem. The wear on the ring grooves fell within the permitted limit, promising an expected lifetime of at least 86 000 hours; only minor carbon deposits were noted on the ring grooves (averaging 60 microns on the first groove, and no deposits on the other two grooves).

Lube oil consumption and its quality, especially in an engine running on heavy fuel oil, is a useful indicator of wear. The anti-polishing ring used on all Wärtsilä engines ensured that consumption and wear of the piston top and liner upper part remained low. Good lube oil quality also results in less carbon deposits on the piston crown cooling gallery, and consequently the absence of hot corrosion problems on top of the piston crown.

The connecting rods inspected showed no fretting marks on the mating surface and no cavitation marks on the oil feeding channels. The big end bearings (tri-metal type) proved to work very well without cavitation or fretting; some scratches were found on the running surface but the overlay had worked well. The extensive use of integrated channels adopted in the W64 design contributed to a lack of oil leakage anywhere around the installation; similarly, the absence of piping reduces problems during overhauls.

A few failures were reported, however, some resulting from a combination of second order causes:

Fuel injection equipment: some problems emerged in the sealing and minor components on the injection valve and were addressed in co-operation with the supplier, allowing the nozzle lifetime to be extended up to 7000 hours.

Cylinder head stud: failures on two installations were caused by poor quality in the material and the production process (rolling of the large M110 × 6 size thread) combined with high flexural vibration. It is unlikely that any of these factors alone would have caused the failure. A new damping device tested on the prototype engine yielded lower vibration levels and was adopted.

Intermediate gear stud: three cases of failure were logged, the problem solved by reducing the free length of the stud stem. A stem support was integrated on the structure and helped to achieve the targeted reduction in stud stress.

See Chapter 2 for Wärtsilä 32 and 50 dual-fuel engines; Chapter 3 for Enviroengine versions of W32 and W46 designs with common rail fuel injection and/or direct water injection systems; Chapter 26 for Sulzer ZA40S engine, which is part of the Wärtsilä portfolio; and Chapter 30 for W200 high speed engine.

28 Other medium speed engines

ABC

Two medium speed engine series are fielded by the Anglo Belgian Corporation (ABC) for propulsion and auxiliary power applications. The 242 mm bore/320 mm stroke DX type is produced in three, six and eight in-line cylinder versions to cover an output band up to 883 kW at 720 or 750 rev/min. DX models are naturally aspirated; DXS models are turbocharged; and DXC models are turbocharged and intercooled.

The DZ type is a 256 mm bore/310 mm stroke design offered in six and eight in-line and V12 and V16-cylinder form to serve power demands up to 3536 kW at 720/750/800/900 and 1000 rev/min. The turbocharged/intercooled engine is designed for burning heavy fuel with suitable adaptation. Its nodular cast iron piston features a Hesselman-type head which is oil cooled by the shaker effect, and four rings: a chrome fire ring, two compression rings and an oil scraper ring located above the floating piston pin.

The cylinder liner is produced from a wear-resistant centrifugal cast iron slightly alloyed with chromium. The slightly alloyed cast iron cylinder head is fitted with two inlet and two stellited exhaust valves. An intermediate deck assures a forced water circulation around the valve seats and the centrally mounted fuel nozzle. The one-piece crankshaft, forged from 42CrMo4 steel and with induction hardened running surfaces, is arranged in a box-type crankcase of nodular cast iron. An additional bearing on the flywheel side limits crankshaft deflection and bearing wear.

ABC effectively doubled the power threshold of its programme in 2000 with V-cylinder variants of the in-line DZ design, the new V12 and V16 models offering outputs of 2650 kW and 3536 kW at 720–1000 rev/min (Figure 28.1). A 45-degree V-bank was selected, fostering an overall width of 1.65 m; an engine height of 2.5 m is quoted, with lengths of 4.18 m and 4.94 m respectively for the V12 and V16 models. As much commonality of parts as possible with the in-line cylinder models was sought in the development, but new components were

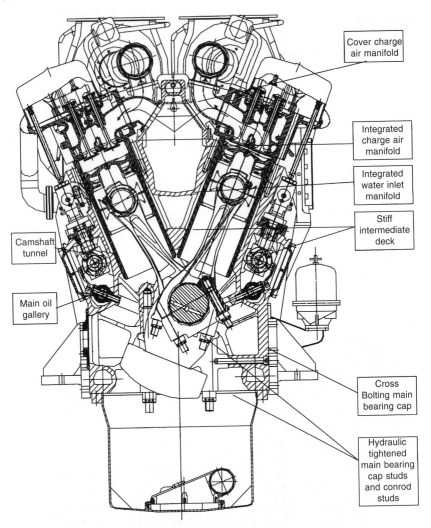

Figure 28.1 V-version of Anglo Belgian Corporation's DZ engine

designed to permit a 20 per cent rating rise in the future and to support a peak cylinder pressure of 150 bar. Easy accessibility to parts needing frequent control or regular preventative maintenance was also pursued.

DAIHATSU

A Japanese contender in the domestic and international small bore medium speed engine arenas, Daihatsu evolved its current programme from the PS series which was developed in 1950 and helped establish the company as a major supplier of auxiliary diesel engines. Some 10 000 examples of the PS-26 model, for example, were sold worldwide.

The PS was succeeded in 1967 by a second generation engine, the DS series, whose development goals focused on high output, low specific fuel consumption and compactness for genset and main engine roles. Market demands for higher fuel economy with the capability to burn lower grade bunkers led to the introduction of a third generation engine, the longer stroke DL series, which includes eight medium speed models with bore sizes ranging from 190 mm to 400 mm and covering an output band up to 4415 kW. The DL series remains in the production programme alongside the fourth generation DK series, designed to address requirements in the 1990s for high reliability, durability and extended service intervals.

The DK series (Figure 28.2) embraces 200 mm, 260 mm, 280 mm and 360 mm bore medium speed models covering power demands up to 6825 kW at 720/750/900 rev/min from in-line and V-cylinder models.

Figure 28.2 Daihatsu DK-28 engine in six-cylinder form

The engine is based on a one-piece cast iron frame designed and manufactured for high rigidity to minimize noise and vibration. Large crankcase windows are arranged on both sides to ease access for maintenance. The thick bore-cooled flame plate of the cylinder head seeks improved mechanical and heat resistance, and an extended service life. The head is secured by four hydraulically tightened studs. The crankshaft incorporates hardened large diameter pins and journals forged in one piece and supported by heavy duty bearings for durability. The connecting rod is formed from three pieces with horizontal joints to reduce piston overhauling height and yield high rigidity.

A gap between the DK-20 and DK-28 models was filled by the 260 mm bore/380 mm stroke DK-26, which is intended as a genset drive or propulsion plant for small ferries, cargo vessels and tugs. Produced as an in-line six-cylinder model, the engine develops up to 1620 kW at 720/750 rev/min.

Similar design principles to the previous DK models are exploited by the DK-26, notably a high stroke/bore ratio of 1.46:1, but the engine features an improved combustion chamber profile and higher compression ratio. Combustion is claimed to be stable over the entire load range when operating on heavy fuel, without compromising NOx emission and fuel consumption performance. The engine frame is fitted with deep mounting feet with I-contour sections so that rubber elements for resilient supports can be attached directly to the feet. The charge air manifold, lubricating oil gallery and cooling water passages are all integrated in the frame, creating a double-wall configuration to help reduce noise and vibration.

A simple yet efficient exhaust layout designated DSP (Daihatsu single pipe exhaust system) recovers gas energy effectively by allowing a good match to the turbocharger. An NOx emission level of 9.4 g/kWh and a specific fuel consumption of 185 g/kWh at full load are claimed, with temperatures of key combustion chamber components well within the optimum range.

Supplied with matching gearboxes, the DKM series is produced in 200 mm, 250 mm, 260 mm, 280 mm and 360 mm bore sizes in six- and eight-cylinder models covering an output band from 880 kW to 4413 kW at speeds from 900 rev/min to 600 rev/min.

A new auxiliary engine, the DC-17, was introduced in 2001 in five- and six-cylinder versions offering generator outputs of 440 kW and 560 kW at 900/1000 rev/min. The 170 mm bore/270 mm stroke design is released for heavy fuel burning applications.

GMT

A wide range of medium speed engines was designed and produced over many years by Grandi Motori Trieste (GMT), which was founded in the late 1960s to pool the diesel engine expertise and resources of Fiat, Ansaldo and CRDA, and rationalize the Italian enginebuilding industry. GMT later became part of the state-owned Fincantieri shipbuilding group's diesel engine division and in 1997 entered a new era as a member of the Wärtsilä Corporation (see Chapter 27), after which its own-design portfolio was phased out.

GMT's last large engine, the A55 series, was the final evolution of a 550 mm bore model first designed and produced in Torino by Fiat Grandi Motori in the early 1970s. The original A550 design with a 590 mm stroke and a rating of 880 kW/cylinder was developed into a B550 version with an output of 1080 kW/cylinder at 450 rev/min. The BL550 derivative (Figure 28.3), launched in 1986, retained the same bore size and running speed but exploited a longer stroke (630 mm) to yield a 10 per cent higher output and a reduction in specific fuel consumption on low grade fuel. The power uprating—to 1213 kW/ cylinder—was achieved with a modest increase in mean effective pressure (from 20.6 to 21.6 bar) and a rise in mean piston speed from 8.85 to 9.45 m/sec.

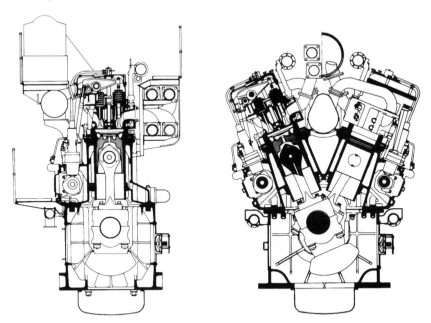

Figure 28.3 GMT's BL550 design in in-line and V-cylinder form

Higher mechanical and thermal loads dictated some changes in the design of combustion chamber elements and running gear but the engine frame and bedplate remained separate units of nodular cast iron for the in-line cylinder engines. Unlike the original design, the eight- and nine-cylinder bedplates and frames are monobloc castings secured by tie-rods; the V-type engines, however, feature a welded steel bedplate and a special cast iron frame for which tie-rods are unnecessary and replaced by stud bolts.

The cylinder head for the BL550 engine (Figure 28.4) was redesigned to assure a more even distribution of temperatures between the exhaust valves and the fuel nozzle, and additional bore cooling introduced in the gas side walls for better distribution. A new exhaust valve cage (Figure 28.5) with a thinner seat wall and more efficient cooling from improved water circulation was specified to reduce the valve seat temperature and extend valve life when the engine burns fuels with high sodium and vanadium contents. Better cooling of the cage seat and effective pulse turbocharging lowered the temperature of the continuously rotated composite exhaust valves which comprise a Nimonic 80A alloy head and anti-corrosion coated steel stem.

A   B

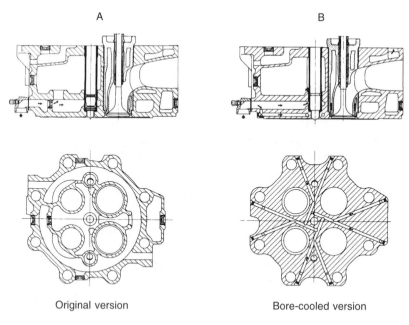

Original version   Bore-cooled version

Figure 28.4 Sections through GMT B550 cylinder head (left) and the redesigned BL550 component

1—Original valve cage   2—Valve cage with improved cooling

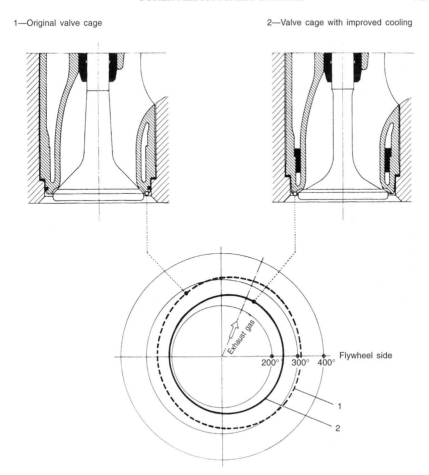

Figure 28.5 Temperature distributions of the GMT B550 valve cage and the improved design for the BL550 engine (21 bar mep and 450 rev/min)

A much stiffer cylinder liner unit (Figure 28.6) was designed, without a heat shield and bore cooled at the top near the gas-side wall. The cooling bores are of the blind type in which the inlet and outlet water flows are separated by diaphragms, allowing the radial outlet holes to be positioned in the lower part of the liner and thus eliminate a concentration of stresses in those areas most affected by dynamic load fluctuations below the cylinder head gasket. Tests showed that the bore-cooling arrangement secures a lower peak temperature and a more uniform distribution on the piston ring rubbing surface. Load-dependent cooling was specified as standard.

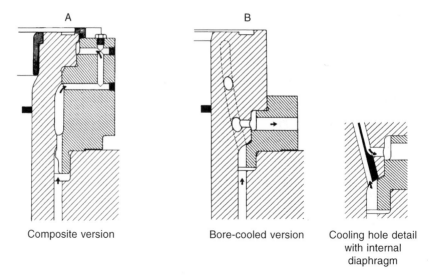

Composite version   Bore-cooled version   Cooling hole detail
                                                    with internal
                                                    diaphragm

Figure 28.6 Improved cooling arrangement for GMT BL550 cylinder liner (B) over original configuration

The composite pistons (steel crown and aluminium alloy skirt) carry three compression and one oil scraper ring in chromed grooves. The upper two compression rings were increased in thickness to withstand the higher combustion pressure and reduce torsional deformation.

A subsequent development programme in 1992/93 resulted in the A55 design (Figure 28.7) which retained the 550 mm bore of its precursors but featured an extended stroke (680 mm). An eight per cent power increase over the BL550 series was derived—taking the specific output to 1250 kW/cylinder at 425 rev/min—along with enhanced fuel economy. The redesign also sought simplified maintenance procedures and higher reliability from a range comprising six, eight and nine in-line and V12-, 14-, 16- and 18-cylinder models covering a power band up to 22 500 kW. Special propulsion market targets included cruise ships and high powered ferries, GMT citing the low height of the A55 engine for its power output: 5.765 m for the in-line version and 5.67 m for the V-cylinder model.

The A55 benefited from GMT's technical collaboration with New Sulzer Diesel, permitting elements of the Swiss designer's successful ZA40S medium speed engine to be exploited. The longer stroke took the stroke/bore ratio from 1.15 to 1.24:1, giving a better thermal efficiency. This was achieved without raising the cylinder head; a more compact piston allowed the same head casting to be used. A reduction

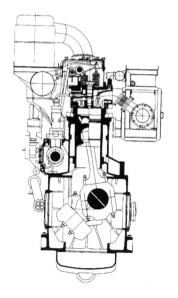

Figure 28.7 GMT's A55 design

in engine speed compared with the BL550 limited the mean piston speed and yielded the increased output with virtually the same brake mean effective pressure as the earlier model.

Fuel injection system improvements were also made. Fuel enters the valve laterally rather than vertically, underwriting a reduced fuel pipe length and a shorter fuel valve. The injection itself takes place over a shorter period, covering around 30 degrees of the cycle, and at a higher pressure (around 1200 bar).

The crank mechanism and other components were also redesigned, mainly to simplify maintenance. The connecting rod, for example, is in three parts (a shank and two caps) connected by hydraulically tightened studs. Despite the longer stroke, it was possible to house the crank mechanism in the same crankcase as the BL550 by redesigning the crankshafts and connecting rods. The bedplate and cylinder block remained virtually unchanged.

Most notable of the A55 refinements was the adoption of Sulzer's unique rotating piston concept, previously unique to the ZA40S engine and its predecessors (Chapter 26). A bore-cooled steel crown is mated with a cast iron skirt. Cast iron rather than aluminium alloy was specified to allow the operating clearance of the skirt to be reduced; this, together with the inertia from the rotating piston, reduces impact velocities with the liner resulting from piston slap. The rotating piston (Figure 28.8) underwrites reliability and a lower lubricating oil consumption. Simpler and more efficient turbocharging was introduced. In-line and

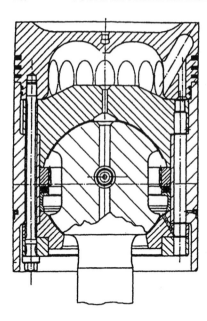

Figure 28.8 GMT's A55 design benefited from a Sulzer-derived rotating piston

V-cylinder propulsion engines are served by a single pipe exhaust system associated with high efficiency and compact configuration; an air/gas bypass is also provided.

GMT was also active in the popular 320 mm bore class, its A32 design having emerged in the 1990s as a longer stroke derivative of the A320 engine introduced in 1986. The 320 mm bore/390 mm stroke A32 series (Figure 28.9) had a maximum continuous rating of 440 kW/cylinder at 750 rev/min and was produced in six, eight and nine in-line and V12-, 14- and 16-cylinder versions to cover a power range up to 7040 kW.

A monobloc casting of high strength iron arranged for an underslung crankshaft forms the structure of both in-line and V-cylinder engines. The high strength lamellar iron cylinder head (Figure 28.10) is secured by four stud bolts for ease of dismantling and has a thick bore-cooled bottom plate for low thermal loading and high stiffness. The inserted thin-walled exhaust valve seat rings are directly cooled by water circulation to achieve a low seat temperature and extended time-between overhauls on heavy fuel.

High strength cast iron cylinder liners have a thick bore-cooled collar for low mechanical and thermal stresses. The composite pistons feature thin forged steel crowns and aluminium alloy skirts. The connecting rods are drop forged in two pieces and withdrawable from the top along with the piston.

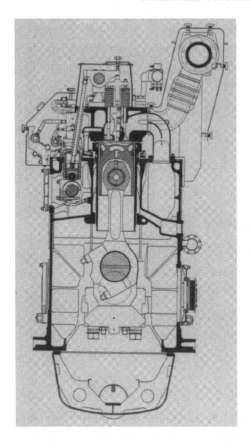

Figure 28.9 GMT A32 design

Reliable operation on residual fuels was addressed by basic design parameters (mean piston speed, compression ratio, combustion pressure and air/fuel ratio) and attention to key component and system design, GMT citing:

- A facility to provide automatic load-dependent control of jacket cooling water and charge air temperatures, gradually raising them with diminishing load to improve combustion, reduce fouling and maintain the combustion chamber walls above the acidic condensation temperature to prevent cold corrosion.
- Exhaust valves of Nimonic 80A material with corrosion-resistant coated stems and rotating devices to prevent hot spots.
- Forged steel piston crown carrying corrosion- and wear-resistant plasma-coated rings.
- High pressure fuel injection system ensuring fine atomization for better ignition; and availability of a remote-controlled facility

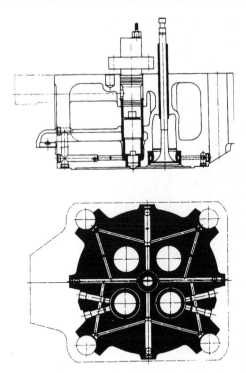

Figure 28.10 Section and plan of GMT A32 cylinder head

for recirculation of fuel through the injectors at engine standstill for starting on fuels of the highest viscosities.
- Bore cooling of the hot components of the combustion chamber (cylinder liner, head and exhaust valve seats) through machined passages, allowing precise temperature control in inhibiting both hot and cold corrosion.

GMT also developed 420 mm bore engines for a number of years. The A420 had a stroke of 500 mm and ran at 500 rev/min to deliver 515 kW/cylinder, while the shorter-stroke (480 mm) A420H ran at 600 rev/min for an output of 588 kW/cylinder.

Higher speed 230 mm bore GMT engines are covered in Chapter 30.

HYUNDAI (HIMSEN H21/32 AND H25/33)

A prolific South Korean builder of two-stroke and four-stroke diesel engines under licence, Hyundai Heavy Industries developed its first own-design engine, the HIMSEN H21/32 series, for launching in 2001. Initially targeting marine genset drives, the medium speed engine was also intended for propulsion markets (Figure 28.11).

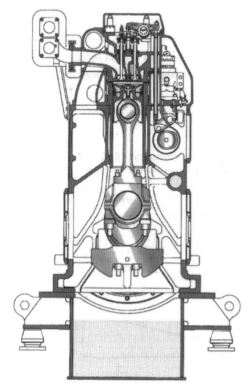

Figure 28.11 Cross-section of Hyundai H21/32 design

A heavy fuel-burning 210 mm bore/320 mm stroke design was selected for development, the high stroke/bore ratio (1.52) and high compression ratio (17) considered appropriate for fuel economy and low NOx emissions. A simplified configuration with a minimal number of components was pursued to enhance reliability and ease production and maintenance. Two sets of prototypes were built and tested to confirm engine performance and reliability. The resulting production engine programme covers a power band from 800 kW to 1800 kW with in-line five to nine-cylinder models; specific ratings from 150 kW to 200 kW per cylinder at 720/750/900 and 1000 rev/min are delivered.

Key structural and running elements of the design were optimized for Hyundai's production facilities, with generous dimensions underwriting moderate fatigue safety factors. Ease of maintenance was addressed by adopting a cylinder unit concept, allowing the cylinder head, valve train, fuel injection equipment, cylinder liner, piston and connecting rod to be withdrawn as an assembly for servicing onboard or ashore (Figure 28.12).

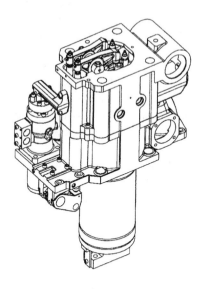

Figure 28.12 H21/32 engine cylinder unit

The engine block is a simple and robust structure of nodular cast iron. A large volume of combustion air chamber and lubricating oil channel is incorporated but water spaces are excluded to avoid any risk of corrosion and flooding into the oil chamber. The nodular cast iron cylinder head features a V-type fire deck support to combat the high maximum firing pressure (200 bar); the inlet and exhaust ports were evolved in-house by Hyundai using its own flow test rig and computational fluid dynamics (CFD) analysis. The structural design of the head and the cylinder liner was optimized based on the CFD analysis of coolant flow as well as finite element analysis.

The two-piece piston comprises a steel crown and box-type steel skirt, delivering lower deformation under high cylinder pressure. Two compression rings and one oil scraper ring are fitted, the top ring coated with chromium-ceramic material on its running surface to curb wear. The connecting rod is a three-piece marine head-type component, shot peened to prevent any fretting from relative movement of the mating surfaces of rod and big end. The crankpin bearing is an aluminium tri-metal Rillenlager type, and the main bearing an aluminium bi-metal component.

A special chromium molybdenum steel camshaft incorporates a fuel cam and intake and exhaust cams arranged as a one-piece element serving each cylinder. The cam profiles were optimized using the Hermite Spline Curve method developed by Hyundai. The robust fuel injection system is designed for pressures up to 2000 bar. The injection pump has a roller tappet with generous dimensions, the injection pipe is of the short block type and the injection valve has a high opening pressure capability with an oil-cooled nozzle body for heavy fuel operation.

An innovative design of air cooler cover controls the cooling water flow and hence the combustion air temperature at low load; the air cooler is arranged to function as a heater at low loads. ABB Turbo Systems' compact TPS turbocharger is specified to boost engine efficiency, power density and durability, offering the potential to support future rating rises.

Hyundai's portfolio also includes the 250 mm bore H25/33 medium speed engine developed jointly with Rolls-Royce group member Bergen of Norway, which markets the design as the C25:33 series (see Chapter 23 for details).

MIRRLEES BLACKSTONE

An impressive medium speed enginebuilding tradition was established by Mirrlees Blackstone of the UK which traced its roots back to the first British licence arranged with Dr Diesel at the end of the 19th century. The Stockport-based company was acquired by the MAN B&W Diesel group from Alstom Engines in June 2000, along with Paxman and Ruston.

Mirrlees Blackstone's final programme—now phased out—was headed by the 800 kW/cylinder MB430L engine, a 430 mm bore/ 560 mm stroke in-line cylinder derivative of the MB430 design which was launched in V45-degree cylinder form in 1985 with a shorter stroke of 480 mm. The two types were respectively produced in six, eight and nine in-line and V12-, V16- and V18-cylinder versions to cover power demands up to 13 144 kW, the in-line engines running at 514 rev/min and the V-engines at 600 rev/min.

The MB430 (Figure 28.13) was introduced with a rating of 665 kW/ cylinder (later increased to 730 kW) and the following key component specification:

Crankshaft: underslung, thin web and high overlap.
Cylinder liners: centri-cast grey iron of heavy section with intensively cooled flanges.

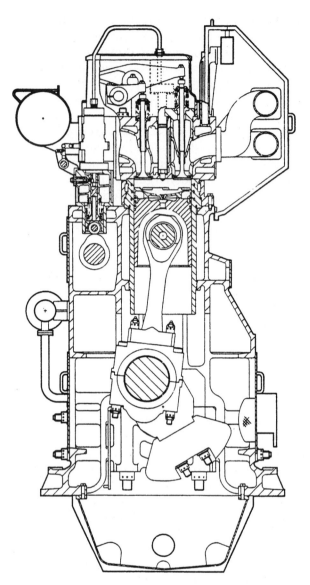

Figure 28.13 Mirrlees Blackstone MB430 design

Cylinder heads: individual, with six holding-down bolts, housing caged exhaust valves and air inlet valves seated on pressed-in seat inserts; valve operation is by Y lever-type gear.

Connecting rods: three-piece construction with a palm-ended shank secured to the large end block with four alloy steel studs, and the halves of the large end block secured together by alloy steel bolts. Large end bearing shells are of steel-backed lead bronze design. Oilways are provided in the connecting rod to facilitate lubrication of the piston pin bearings and deliver a copious flow of cooling oil to the piston.

Pistons: two-piece construction with heat-resistant alloy steel crown and aluminium alloy skirt.

Camshafts: driven by a crankshaft gearwheel via compound intermediate gears; all meshing points are lubricated and cooled by oil sprays; separate cylinder casings house the camshafts and high level fuel pumps.

Engine-driven lubricating oil and water pumps are mounted at the free end of the engine; air manifolds are mounted on the outside of the vee; and turbochargers and intercoolers can be located at either end of the engine.

The MB430 retained many basic features of Mirrlees Blackstone's long-established 400 mm bore/457 mm stroke K Major series which remained in the programme in Mark 3 form, delivering an output of 545 kW/cylinder at 600 rev/min. The design matured from the K-engine of the 1950s, one of the first large medium speed engines in the world, which in its previous Mark 2 form featured a 381 mm bore (Figure 28.14). The connecting rod assembly of a Mark 3 engine is shown in Figure 28.15.

The MB430 design also exploited some concepts introduced for the MB275 series in 1979. This 275 mm bore/305 mm stroke engine was available in six and eight in-line and V12- and 16-cylinder versions covering a power range up to 5070 kW at 1000 rev/min.

Lower power requirements were served by Mirrlees Blackstone's 222 mm bore/292 mm stroke ESL series (Figure 28.16) whose upper output limit was doubled in 1986 by V12- and V16-cylinder models, the latter developing 2985 kW at 1000 rev/min. The programme also embraced five, six, eight and nine in-line cylinder models.

The V-versions tapped the pedigree of the former E range but incorporated several interesting features, including an underslung crankshaft and a belt drive for the lubricating oil and water pumps. The V-models have a 45-degree cylinder configuration, deep main bearing caps, side bolts for added structural stiffness and a central air intake. A simplified rocker arrangement operates four valves (two inlet and two exhaust) through two levers and bridge pieces enclosed within an aluminium cover. The valve gear is actuated by side-loaded camshafts driven from the flywheel end through nitride-hardened gears. The three-piece construction of the connecting rod allows piston withdrawal without disturbing the large end bearing.

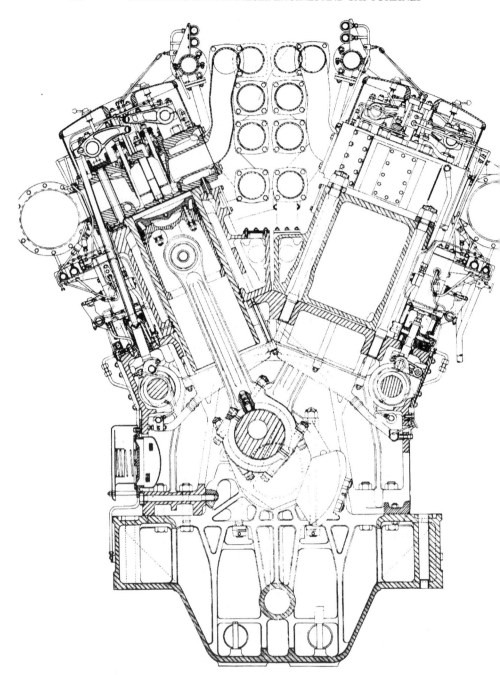

Figure 28.14 Cross-section of Mirrlees Blackstone KV Major Mk 2 engine

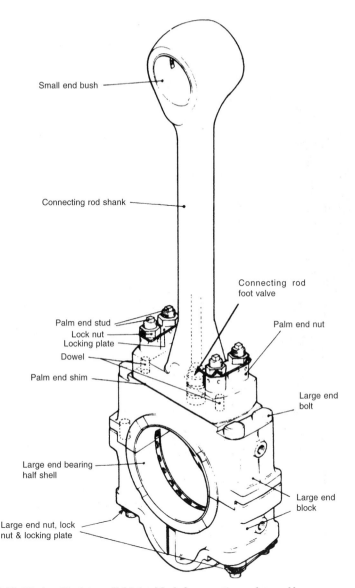

Small end bush

Connecting rod shank

Connecting rod
foot valve

Palm end stud
Lock nut
Locking plate
Dowel

Palm end nut

Palm end shim

Large end
bolt

Large end bearing
half shell

Large end
block

Large end nut, lock
nut & locking plate

Figure 28.15 Mirrlees Blackstone K Major Mark 3 connecting rod assembly

A pulley and flat belt arrangement for the service pumps, mounted on the engine front, represented a departure from the traditional gear drive used by most enginebuilders. Its adoption was considered beneficial in lower production costs, pump speed flexibility, easier installation and maintenance, and weight and power savings. The belt

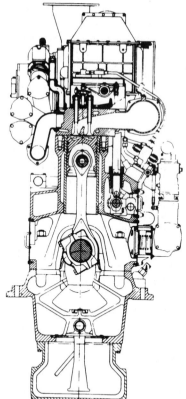

Figure 28.16 Mirrlees Blackstone ESL Mk 2 design

comprises a core of endless polyester cords helically wound with a chrome leather friction face covering. A minimum belt life of 10 000 hours was anticipated.

MITSUI

A major licensee of MAN B&W Diesel in the production of MC low speed engines, Mitsui also offered its own 420 mm bore/450 mm stroke medium speed design whose last version was designated the L/V42MA series. The launch of the Japanese group's earlier venture in the four-stroke sector, the 600 mm bore 60M design with a rating of 1120 kW/cylinder, unfortunately coincided with the 1973 oil crisis which undermined its market.

The smaller bore L/V42M design, however, benefited from that ill-fated project and became available in in-line and V-cylinder form with an output per cylinder of 558 kW at 530 rev/min. The frame is in two parts, comprising a welded steel crankcase structure made from cast sections and plate, and a cast iron cylinder housing held together with tie-bolts. The underslung crankshaft bearings are secured by vertical and lateral bolts, all hydraulically tightened. The designers sought a stress-free structure, controlling the lateral fit between bearing keep and frame, for example, by tapered adjusting liners before tightening to avoid stressing the frame unduly.

For the crankshaft bearings, Mitsui was able to keep within the stress capacity of thin whitemetal (on a lead–bronze backing in the case of the crankpins) with an overlay. While this is a function of the control of maximum bearing pressures, it also means that damage to a bearing does not necessarily lead to shaft damage.

The combustion area and cylinder head configuration can be appreciated from Figure 28.17. Note the composite piston, bore-cooled cylinder liner, three-deck head, caged exhaust valves and rigid high pressure fuel line, as well as the camshaft bearing arrangement. In addition to the four compression rings in the piston crown (the second is Ferrox-filled to improve running-in) there are two conformable SOC rings at the top of the cast iron piston skirt and two broad lead–bronze bands above the gudgeon and one below; these also assist during the running-in phase.

A three-part marine-type connecting rod was selected with serrations at the big end joint. The exhaust valve cage can be removed easily by taking out the pushrod and lifting clear the rocker. Rotators are applied on the exhaust valves, and guide vanes are formed in the inlet passages of the head to foster swirl in the charge air. A mist catcher is installed after the intercooler to reduce the water loading of the charge air in humid ambient conditions.

The fuel injector sleeve is provided with small water passages at the seat. These passages communicate with the cylinder head water passages and foster a convection flow which gives adequate cooling to the nozzle for operation on all fuels without the complication of water cooling the nozzles.

Subsequent refinements led to the Mitsui L/V42MA design with an output raised to 625 kW/cylinder at 600 rev/min, allowing 6L-to- V18-cylinder models to cover a power band up to 11 250 kW.

NIIGATA

A wide range of own-design high and medium speed engines from Niigata Engineering of Japan is supplemented by licences to build

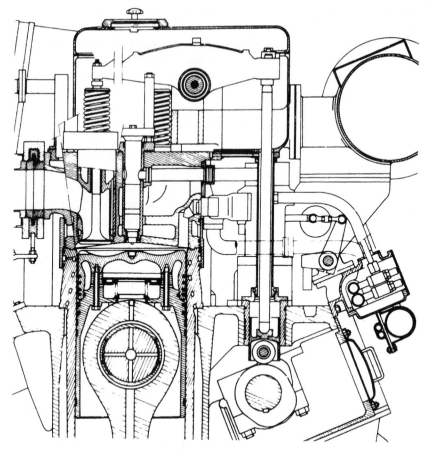

Figure 28.17 Cross-section through cylinder head and piston of Mitsui 42M engine

SEMT-Pielstick PA and PC series and MAN B&W Holeby 23/30 and 28/32 series engines. Niigata's own MG (Marine Geared) medium speed programme embraces designs with 190 mm, 220 mm, 250 mm, 260 mm, 280 mm, 320 mm, 340 mm, 400 mm, 410 mm and 460 mm bore sizes; the high performance FX series includes 320 mm and 410 mm bore versions. (*See Chapter 30 for Niigata high speed engines.*)

Its largest medium speed engine, the MG46HX, is a 460 mm bore/ 600 mm stroke design with an output per cylinder of 823 kW at 450 rev/min and the capability to operate on 700 cSt fuel. Power demands up to 13 180 kW are met by 6/8L and V12/16-cylinder models, the targeted propulsion plant applications including medium-to-large ferries, containerships and RoRo freight vessels. The crankcase is built as a box structure to secure rigidity in handling a maximum combustion pressure of 180 bar; engine noise and vibration levels are also claimed to be low. Easier maintenance was addressed by simplified component designs and hydraulic tools. Turbocharging is based on MAN B&W turbochargers produced under licence by Niigata.

Niigata also developed the HLX series, which embraces 220 mm, 280 mm and 340 mm bore sizes covering a power band from 1392 kW to just under 10 000 kW from in-line and V-cylinder models.

NOHAB

The Nohab engine business was acquired by Wärtsilä, which focused on the Swedish designer's Nohab 25 series before phasing out the programme. This 250 mm bore/300 mm stroke design was offered in six in-line and V12- and 16-cylinder versions covering output demands up to 3680 kW at 720/750/825/900 and 1000 rev/min for auxiliary and propulsion plant markets. The engine proved particularly popular in the offshore power arena. A cross-section of the 6R25 model is shown in Figure 28.18.

SKL

Four-stroke engines in a number of bore sizes—160 mm, 200 mm, 240 mm and 320 mm—are produced by SKL whose pedigree dates back to 1905 when the Magdeburg, Germany-based company's forerunner (Lokomobilenfabrik R Wolf) took out a licence to build diesel engines. Now a subsidiary of the Laempe group, SKL Motor's medium speed engine programme includes the VDS 29/24 AL series for propulsion and genset duties which is offered in six, eight and nine in-line cylinder versions to cover an output range up to 2350 kW at 750/900 and 1000 rev/min. Its brake mean effective pressure rating is 23.9 bar.

The 240 mm bore/290 mm stroke design (Figure 28.19) is based on a rigid nodular cast iron cylinder block housing a suspended crankshaft. The block is free of cooling water thanks to a special cylinder liner design which incorporates a raised, rigid and separately cooled collar; this feature protects against corrosion and deposits, and reduces maintenance costs. The alloyed flake-graphite cast iron liner benefits from all-round nitration to enhance its wear resistance. Four cylinder head screws with long anti-fatigue shafts transmit forces via intermediate walls and ribs in the block and via the main bearing screws to the main bearing covers.

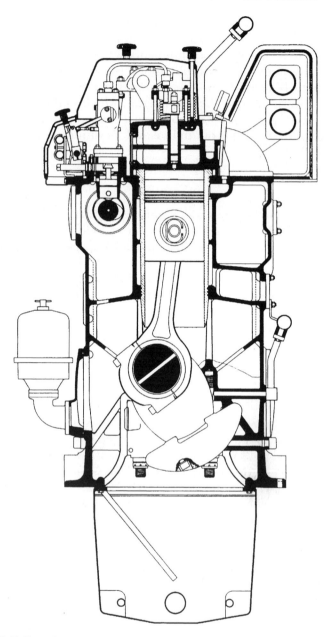

Figure 28.18 Wärtsilä Nohab R25 design

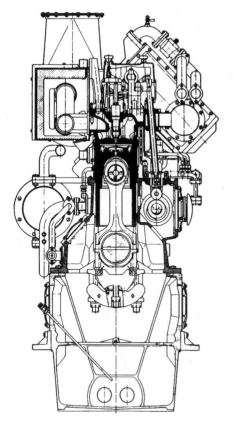

Figure 28.19 SKL's VDS 29/24 AL design

A deformation-resistant bedplate forms the bottom of the housing and can, in appropriately extended form, also serve as the mounting for an alternator. The camshaft is carried by bearings in the cylinder block and driven by gearing at the opposite end to the coupling. A hardened and tempered crankshaft, upset forged from C-steel, is supported by interchangeable thin-walled multi-layer slide bearings. A viscosity-type torsional vibration damper is arranged at the free end of the crankshaft. The control gear and vibration damper flange are attached to the crankshaft by a cone interference fitting and hydraulically mounted. Up to 60 per cent of the rated output can be taken off at the opposite end to the coupling. Also hardened and tempered and forged from alloyed C-steel, the connecting rod is divided horizontally at its large end and supported by interchangeable thin-walled multi-layer slide bearings.

The composite piston consists of a steel crown and aluminium alloy skirt, bolted from the top and supplied with cooling oil through a longitudinal bore in the connecting rod. The ring pack comprises three compression and one oil scraper rings with chromium-plated surfaces.

Four necked-down bolts secure the cylinder head to the block, the head incorporating two inlet and two exhaust valves. Engines destined for heavy fuel burning are equipped with turning devices for the exhaust valves. The hard facing of the valve seat surfaces in combination with the cooling of the exhaust valve seat rings, mounted in the cylinder head, foster extended operational lifetimes for the components. The fuel injection valve is located in the centre of the cylinder head. The valve drive (lubricated by pressure oil and arranged in an oil-tight enclosure) is effected from the camshaft via roller tappets, pushrods and forked rockers.

Turbocharging is based on a three-pulse system with a claimed maximum efficiency of at least 64 per cent. A compressor washing device and, for heavy fuel service, a turbine washing device are specified to reduce maintenance demands.

Fuel is injected by individual top-mounted pumps at maximum injection pressures of 1400 bar-plus which, in conjunction with an optimized cam geometry, ensures a short combustion time. Ideal combustion with minimal noxious emissions is further pursued by short injection pipes (incorporating double walls for safety) and optimized nozzle spray angles and injection hole diameters. In heavy fuel engines the injection nozzle is continuously cooled with lubricating oil from the engine circuit and fuel pipings sheathed with heating piping are isolated. The quantity of fuel injected is controlled by a mechanical–hydraulic speed governor via a compound control rod arrangement.

Engine cooling is effected by two circuits. The cylinder liner, head and turbocharger are cooled via a high temperature circuit while a low temperature circuit serves the oil and charge air cooler as well as the heat exchanger (water). Both circuits are designed for the direct fitting of centrifugal pumps. Engine lubrication is carried out by a pressure circulation system served by a lubricating oil pump, filter and cooler mounted on the engine. The system also supplies the piston with cooling oil.

STORK-WERKSPOOR DIESEL

Formed in 1954 by the merger of Stork and Werkspoor, along with other interests in the Dutch enginebuilding industry, Stork-Werkspoor Diesel (SWD) was acquired in 1989 by the Finland-based multi-national

Wärtsilä Diesel group, now the Wärtsilä Corporation. SWD's own-design medium speed engines were gradually phased out of the production programme and replaced by the Wärtsilä 26 and 38 series (Chapter 27), largely designed by SWD staff and subsequently manufactured in Zwolle, The Netherlands. These models were later assigned for production by Wärtsilä Italia in Trieste.

Not long after its formation SWD decided to develop a high powered medium speed engine to contest diverse coastal and deepsea propulsion arenas, the resulting TM410 series appearing on the market in the later 1960s. The 410 mm bore/470 mm stroke design (Figure 28.20) was uprated a number of times in its career to a level of 564 kW/cylinder at 600 rev/min, the series embracing six, eight and nine inline and V12-, 16-, 18- and 20-cylinder versions.

A scaled-up version—the TM620 series—was introduced in the mid-1970s to complement the TM410, the 620 mm bore/660 mm stroke design enjoying a status for many years as the world's largest and most powerful medium speed engine (Figure 28.21). Examples of both types—in in-line and V-cylinder form—remain at sea (Figure 28.22).

The bedplate of the TM410 engine is a U-shaped iron casting in which the main bearing caps are fitted with a serrated joint to form, with the bearing saddles, a rigid housing for the thin-wall steel main bearing shells. For the TM620 engine what is in effect a single tooth serration is used for the main bearing cap joint: the joint faces are inclined towards the abutment.

Bearing shells have copper–lead lining and lead–tin plating. Both upper and lower shells can be removed easily by lifting the bearing cap. Large inspection openings in the bedplate facilitate access to the crankcase, the covers being provided with relief valves. Integrally cast columns in the in-line engine bedplate accommodate alloy steel tierods to connect bedplate and cylinder block tightly together in a rigid construction preventing high tensile stresses from cylinder pressures being transmitted to these major cast iron elements. V-type engines are fitted with two rows of tie-rods. As the rods are in an oblique position, a number of short bolts are additionally fitted to prevent relative movement of cylinder block and bedplate. A thick upper bridge between pairs of cylinders on opposite V-banks also contributes to rigidity.

The crankshaft is a fully machined high tensile continuous grain flow forging in one piece. The large diameter journals and crankpins are provided with obliquely drilled holes for the transmission of lubricating and piston cooling oil. Counterweights fitted on all crankwebs are secured by two hydraulically stressed studs and two keys.

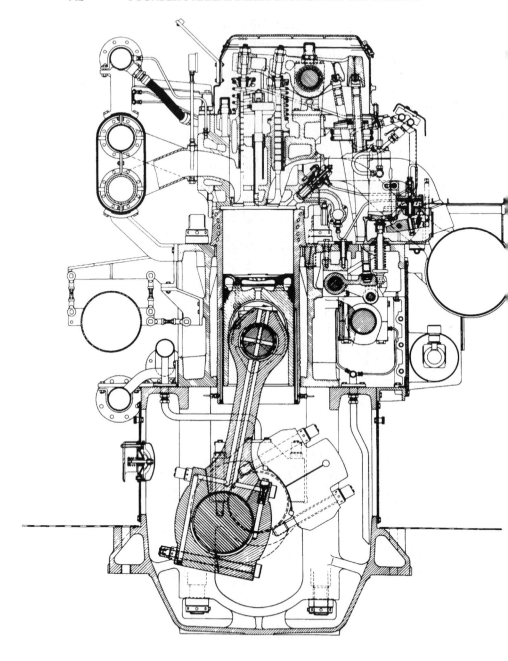

Figure 28.20 Cross-section of in-line SWD TM410 engine

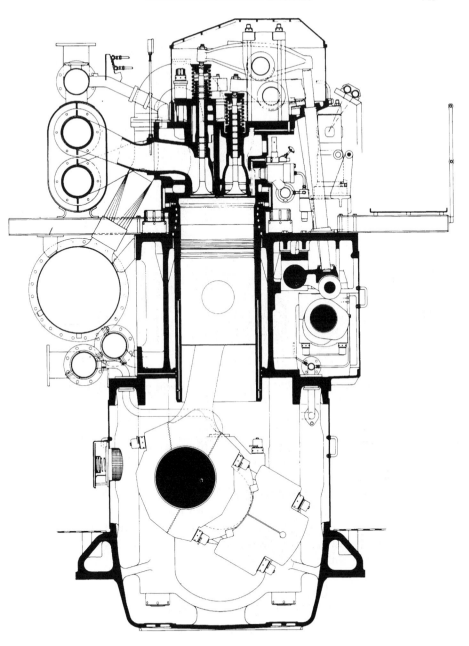

Figure 28.21 Cross-section of in-line SWD TM620 engine

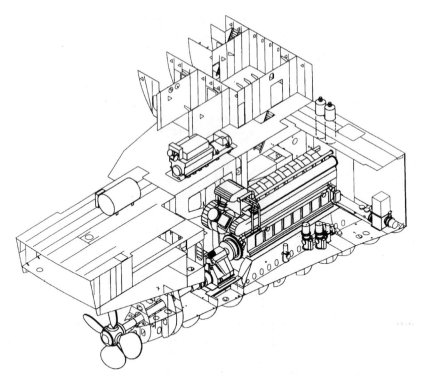

Figure 28.22 An eight-cylinder Stork-Werkspoor Diesel TM620 engine installed to power a containership

A rigid iron casting cylinder block incorporates the cooling water jackets and the camshaft space. Covers are arranged on the camshaft side of the block to allow inspection of the cams and rollers. Three roller levers are fitted per TM410 engine cylinder: one each for the inlet and exhaust cams, and one for the fuel pump drive. Only two levers per cylinder—for the valves—are arranged on the TM620 engine. The pushrod passages are equipped with seals to prevent any oil leakage from entering the camshaft space.

The camshaft is ground to a single diameter over its entire length, with the hardened steel cams hydraulically shrunk on with the aid of tapered bushes. When necessary, the complete camshaft can be removed sideways; and cams can be exchanged without removing the camshaft from the engine since they can be removed from the shaft at the forward end of the section. The camshaft is driven from the crankshaft by nodular cast iron gearwheels and runs in thin-wall bearings. The direct-reversible engine is provided with double cams with oblique transition faces and a pneumatically controlled hydraulic reversing gear to move the camshaft in an axial direction.

An extremely heavy big end is a feature of the connecting rod, forming a rigid housing for the copper–lead lined lead–tin plated thin-wall steel bearing shells. To allow the rod to be removed through the cylinder liner, the big end is split by serrated joints in two planes. The design fosters a low headroom and easy dismantling. The big end bolts are hydraulically stressed. The TM620 engine connecting rod is similar in principle to that of the TM410 component but uses three pairs of studs, each set normal to the face of the joint it closes rather than bolts at right angles to the axis of the rod.

Lower rated TM410 engines were specified with a light alloy piston with a cast-in top ring carrier and a cast-in cooling oil tube; the higher rated models and the TM620 engine exploited a two-piece piston. Both feature one chrome-plated top ring, three compression rings with bronze insert and an oil control ring above the gudgeon. The TM410 engine additionally uses a second oil control ring at the base of the skirt when the single-piece piston is fitted.

The gudgeon pin is fully floating and the small end bearings have different widths in order to provide a greater area to withstand better the combustion loads. The small end bearing material is of the same specification as the large end. The small end and piston cooling oil supply passes through drillings in the connecting rod; the spent oil from the pistons and gudgeon pins is expelled from the piston. In the case of the one-piece piston, however, cooling oil passes through a separate drilling to the bottom of the big end. Both engines deploy restrictors in the big end oil passages incorporating non-return valves. Despite the three-part large end configuration, both engines use horizontally split two-part big end shells. The V-engine big ends run side-by-side on the crankpin.

The special cast iron cylinder liner is provided with cooling water passages drilled to a hyperboloid pattern in its thick upper rim. This feature was designed to secure intense cooling of the upper liner part and also to ensure, by equalizing the temperatures of the connected liner rim and the cylinder block, a perfectly circular liner when the engine is operating.

Two holes for cylinder lubrication are drilled in each liner from the bottom, thus avoiding oil pipes through the cooling water spaces. Each hole feeds one lubricating oil quill arranged halfway along the liner length. A normal Assa cylinder lubricator is provided. There are four holes per cylinder in the case of the TM620 engine.

Both engine types are equipped with four-valve cylinder heads with an unusual two-bearing exhaust rocker design. In the case of the

TM620 the inlet valve is shorter than the exhaust valve and two fulcrum shafts are used with different lengths of Y-shaped rockers to operate each pair of valves.

Cooling water is directed over the flameplate of the head by the intermediate deck which also serves the function of a strength member so that the flameplate itself can be thinner and thus better able to withstand thermal load. The exhaust valves work in water-cooled cages and are Stellited on the seating face. Careful cooling of the seat area, and hence of the sealing face, reportedly gave the valves a lifetime between overhauls when burning heavy fuels not greatly inferior to that experienced by valves in marine diesel-burning engines. The inlet valves seat on hardened cast iron inserts mounted directly in the cylinder head.

Cam profiles were designed to avoid rapid acceleration changes and thus to minimize noise and dynamic stresses. The valve rockers are fitted with needle bearings that require no oil supply; all other contact faces within the cylinder head covers are lubricated by separate impulse oiling equipment. The associated lubricator is driven, in parallel with the cylinder lubricator, from the camshaft gear. The oil used for these purposes is drained separately to prevent contamination of the crankcase oil should a fuel leak occur.

A dedicated fuel injection pump serves each cylinder. The injector is equipped with an easily replaceable nozzle which is water cooled to prevent carbon deposits forming when handling residual fuels. The TM410 fuel pump is actuated by a spring-loaded pushrod which itself follows a lever-type cam follower similar to those that actuate the valve pushrods. On the TM620 engine the pump has an integral roller.

The mechanism for the starting air pilot valves is also used to stop the engine at overspeed. In such an event the mechanical overspeed trip actuates a pneumatic valve to allow air pressure into an auxiliary cylinder which then keeps every fuel injection pump plunger lifted with the roller free of the cam (Figure 28.23). Safety devices for low lubricating oil pressure and for low cooling water flow are integrated in the engine systems and are thus independent of the external electrical alarm system.

A more modern smaller bore design, the SW 280 engine, was introduced by SWD in 1981 to fill a gap between the TM410 and the SW 240 models and remained in the Stork-Wärtsilä Diesel production programme until 1997 when the Wärtsilä 26 series was launched (see Chapter 27). The 280 mm bore/300 mm stroke design (Figure 28.24) developed 300 kW/cylinder at 720–1000 rev/min and found favour as a genset drive in deepsea tonnage as well as a propulsion unit for smaller ships. It was available in six, eight and nine in-line and

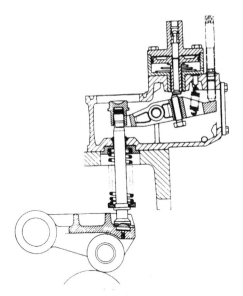

Figure 28.23 Fuel pump drive and cutout on SWD TM410 engine

V12-, 16- and 18-cylinder versions to cover power demands up to 5400 kW.

A short and compact engine was sought by the Dutch designers, with power available from both ends of the crankshaft. Engine-driven pumps could be installed on a sideways-mounted gearcase. The design allowed for combustion pressures up to 150 bar and associated high fuel injection pressures. Well-controlled wall temperatures in the combustion space addressed heavy fuel operation.

A reduced number of overall components was achieved by replacing most pipes for water and oil passage with drillings through the structure. This, and other measures, lowered the risk of fluid leakages and eased maintenance procedures. A rigid design was secured by incorporating the air duct, water and lubricating oil galleries in the cylinder block casting.

The cylinder liner features a high collar and drilled cooling channels. The piston comprises a pressed aluminium alloy body and a forged steel crown which is cooled by lubricating oil and provided with hardened ring grooves. The cylinder head is relatively high and very rigid, its stiffness largely achieved by a special shaped intermediate deck. The head incorporates four valves whose seats are detachable; the exhaust valve seats are water cooled.

A separate lubricating system serves the fuel pumps to prevent contamination of the main system oil. Pulse turbocharging was specified to secure the maximum air supply over all loads. The scavenge air is heated for starting and low-load operations on heavy fuel.

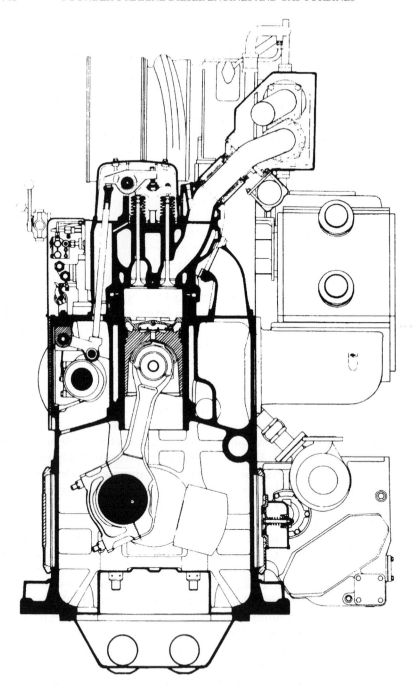

Figure 28.24 Stork-Werkspoor Diesel SW 280 design

YANMAR

Like Daihatsu above, Yanmar Diesel is a prolific Japanese producer of high speed and smaller medium speed engines for genset and propulsion drives. Its medium speed propulsion engine programme includes the S185, M200, M220, T240, T260, N260, Z280, N280 and N330 series, the designation signifying the bore size in mm. Most are built in six-cylinder in-line form, with the larger bore models also offering an eight-cylinder version. The 8N330-EN engine delivers 3310 kW at 620 rev/min.

Development goals pursued by Yanmar to benefit the designs included enhanced compatibility with heavy fuels and strengthened but lighter cylinder blocks, bedplates and cylinder heads to lower overall engine weight. A compact and lightweight reverse-reduction gear and compact arrangement of turbocharger and air cooler contribute to reduced overall engine length and height.

Measures to ease engine installation and servicing are also cited. Large inspection windows on both sides of the cylinder block facilitate dismantling and re-assembly of the main bearing and connecting rod. All pumps and filters are arranged above engine installation level to simplify access. Full power take-off is available from the front of the engine.

TWO-STROKE MEDIUM SPEED ENGINES

The medium speed engine market has long been dominated by four-stroke, uniflow-scavenged designs but at one time a number of two-stroke designs enjoyed popularity, notably the Polar loop-scavenged type which ceased production when Nohab introduced its four-stroke F20 range, and Sulzer's uniflow-scavenged ZH40.

WICHMANN's commitment to the two-stroke loop-scavenged trunk piston concept was renewed in 1984 with the launch of its 295 kW/cylinder WX28 engine (Figure 28.25). The simple 'valveless' approach had been proven in service by the Norwegian company's earlier AX, AXG and AXAG designs. The 280 mm bore/360 mm stroke WX28 covered an output band from 1180 kW to 4735 kW at 600 rev/min with four, five and six in-line and V8-, 10-, 12- and 16-cylinder models.

Development focused on low fuel and maintenance costs with high reliability. The engine was also claimed to be one of the lightest and most compact in its power class. The ability to operate on heavy fuel (180 cSt) under all conditions was another goal. A specific fuel consumption of 188 g/kWh resulted from enhanced scavenging and

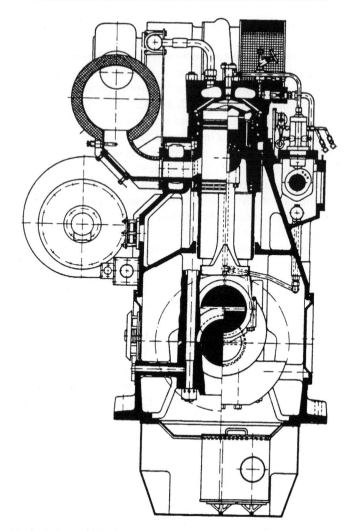

Figure 28.25 Wichmann WX28L design

fuel injection systems, and a maximum combustion pressure of 140 bar is underwritten by rugged construction. The mean effective pressure is 13.5 bar.

The valveless cylinder cover is of simple construction, the lack of ducts for hot exhaust gas promoting uniform temperature distribution and low stress. Fastened by eight hydraulically tightened nuts, the cover can be removed in a few minutes and the piston withdrawn in 10 minutes. The connecting rod can be disconnected while leaving the big end bearing on the crankshaft; this feature reduces the necessary removal height.

Wichmann stressed the overall simplicity of the engine and the impact on reliability and serviceability, citing fewer moving parts and hence fewer wearing parts. Separate cylinder lubrication—a standard feature—permits matching of the lubricating oil total base number to the fuel sulphur content. The oil is distributed over the cylinder surface by a hydraulic lubricator via four bores and quills.

Wichmann engines—including the earlier 300 mm bore/450 mm stroke AXAG design—found particular favour in the Norwegian fishing and offshore vessel propulsion sectors. A/S Wichmann became part of the Finland-based Wärtsilä Diesel group in 1986 and changed its name from Wärtsilä Wichmann Diesel in January 1994 to Wärtsilä Propulsion A/S (now Wärtsilä Propulsion Norway A/S).

The Wichmann 28 engine, which remained in production until 1997, was released with the following specification:

Cylinder block: cast iron monobloc design with integrated crankcase, scavenging air receiver, water manifold and camshaft box; underslung type of crankshaft support.

Crankshaft: fully forged and machined in Cr–Mo steel; dimensionally laid out for 50 per cent power growth potential.

Cylinder liner (Figure 28.26): wear-resistant cast iron alloy; bore cooled with strong backed top section; balanced cooling water flow for efficient temperature control; separate cylinder lubrication through four quills.

Cylinder head: cast iron, valveless, simple design; bore cooled with strong backing to secure efficient cooling and low stress level.

Piston (Figure 28.27): oil-cooled composite design with cast iron skirt and steel crown; ring grooves hardened for low wear rate in heavy fuel operation; integrated small end bearing in full gudgeon pin length.

Connecting rod: drop forged and fully machined; separate large end bearing unit for easy piston withdrawal and low removal height.

Bearings: three-metal steel-backed type, interchangeable with main and crank journal.

Turbocharging: constant pressure system with auxiliary blower in series; the moderate speed auxiliary blower boosts the turbocharger effort to ensure an adequate air supply under all load conditions; the blower is engine driven via low pressure hydraulics using the engine lubricating oil and pump.

Fuel injection system: individual high pressure monobloc pumps with built-in roller tappet; short high pressure pipes and temperature-controlled nozzles for heavy fuel operation.

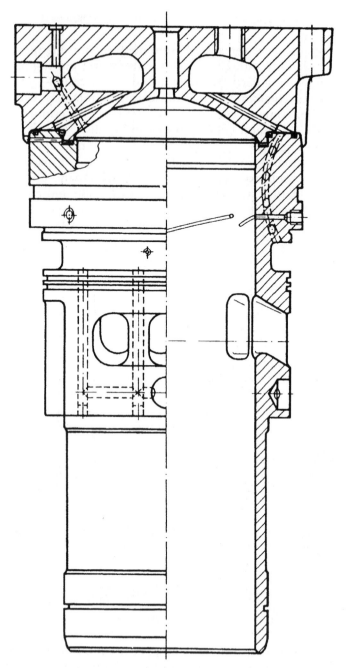

Figure 28.26 Bore-cooled cylinder liner and cover of Wichmann WX28 engine; separate cylinder lubrication is standard

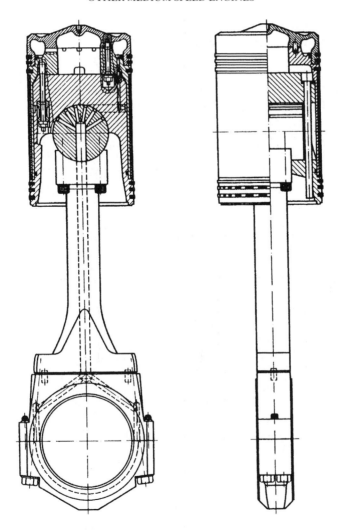

Figure 28.27 Composite piston (high alloy steel crown, cast iron skirt and light alloy gudgeon pin support) and connecting rod of Wichmann WX28 engine

Auxiliary pumps: engine gear-driven units for lubricating oil, fresh water and sea water.

Loyal to the two-stroke uniflow-scavenged principle for medium speed trunk piston engines—exploiting air inlet ports in the cylinder liner and exhaust valves in the head—is General Motors' Electro-Motive Division (**EMD**). The US designer argues better component wear life, reliability and serviceability over four-stroke designs serving the same power range.

EMD's current 645 series and 710 series cover an output band from 785 kW to 3730 kW at 900/1000 rev/min from V8-, 12- and 16-cylinder models (Roots blown) and V8-, 12-, 16- and 20-cylinder turbocharged models. The 645 design has a 230 mm bore/254 mm stroke while the 710 design has the same bore size but a stroke of 280 mm; each is produced in a V45-degree cylinder bank configuration.

The 710G series (Figure 28.28) was launched in 1986 as a longer stroke derivative of the established 645FB range. A more advanced turbocharger (yielding a 10 per cent increase in the overall air–fuel ratio) and larger fuel pump plunger diameter contributed to the higher power rating and lower fuel consumption.

A number of improvements benefited the current 710GB series which offers outputs up to 187 kW/cylinder at 900 rev/min:

- An L-11 liner design for enhanced durability and performance, reduced scuffing and higher wear resistance, and better fuel economy.
- A Duracam camshaft, extending valve train component life and lowering valve vibration.
- A Diamond six-cylinder head with tangent flow fireface, yielding improved cooling and better valve sealing, eliminating core plugs and hence water leaks, and providing hardened valve guides for improved valve and valve guide life.
- Improved unit fuel injector, with new diamond seal design, enhanced check valve and stiffer follower spring.
- A four-pass aftercooler achieving improved thermal efficiency and hence fuel economy, and exhaust emissions reduction.
- A new turbocharger incorporating an external clutch for easier maintenance.
- Lower vibration levels from a new crankshaft and coupling disc balancing technique.

Another medium speed two-stroke design philosophy was pursued for many years by **BOLNES** of The Netherlands until its acquisition by the Wärtsilä group and the subsequent cessation of engine production. The company produced the world's smallest two-stroke crosshead engine, the last manifestation being the 190/600 series. The earlier 150/600 design is illustrated (Figure 28.29).

The 190 mm bore/350 mm stroke design delivered a maximum continuous output of 140 kW/cylinder at 600 rev/min on a mean effective pressure of 14.1 bar. The range embraced 3–10 in-line models (excluding a four-cylinder version) and V10–20 cylinder models covering

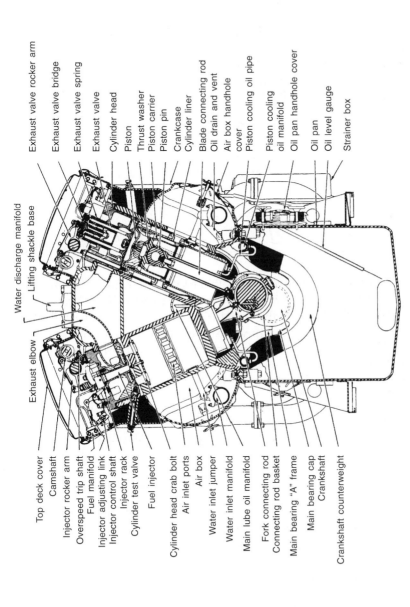

Top deck cover
Camshaft
Injector rocker arm
Overspeed trip shaft
Fuel manifold
Injector adjusting link
Injector control shaft
Injector rack
Cylinder test valve
Fuel injector

Cylinder head crab bolt
Air inlet ports
Air box
Water inlet jumper
Water inlet manifold
Main lube oil manifold

Fork connecting rod
Connecting rod basket
Main bearing "A" frame
Main bearing cap
Crankshaft
Crankshaft counterweight

Water discharge manifold
Lifting shackle base

Exhaust elbow

Exhaust valve rocker arm
Exhaust valve bridge
Exhaust valve spring
Exhaust valve
Cylinder head
Piston
Thrust washer
Piston carrier
Piston pin
Crankcase
Cylinder liner
Blade connecting rod
Oil drain and vent
Air box handhole cover
Piston cooling oil pipe
Piston cooling oil manifold
Oil pan handhole cover
Oil pan
Oil level gauge
Strainer box

Figure 28.28 General Motors EMD 710G two-stroke medium speed engine with overhead camshafts and unit injectors

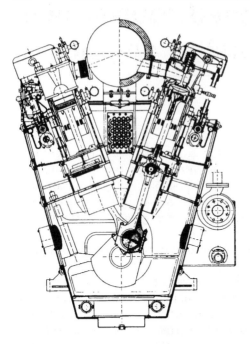

Figure 28.29 Cross-section through Bolnes VDNL 150/600 two-stroke engine

an output band from 400 kW to 2800 kW. The crosshead took the form of a lower piston/scavenge pump, with uniflow scavenging arranged via a single valve in the head. Air is drawn in by the turbocharger, passes through a first-stage air cooler to the scavenge pump and then fed to the cylinder for the compression stroke via a second-stage air cooler.

Bolnes cited the following merits of the design:

- Very low lubricating oil consumption due to the complete separation of combustion space and crankcase.
- Unique air control due to the design of the 260 mm bore crosshead scavenge pump which was claimed to secure completely smokeless combustion under all load conditions.
- High performance at low speed (for example, 110 per cent torque at 70 per cent rev/min).
- Separate lubrication systems promoting good heavy fuel-burning operation.
- Simple maintenance.

Bolnes engines had a faithful following in fishing, dredger, coaster and survey vessel propulsion sectors. The design was also valued as a test engine in fuel and lubricating oil research laboratories.

29 Low speed four-stroke trunk piston engines

The marine diesel propulsion market is dominated by direct-coupled low speed two-stroke crosshead engines and geared high/medium speed four-stroke trunk piston engines but some Japanese and east Asian regional operators of coastal/shortsea, fishing and small oceangoing vessels appreciate the merits of a 'hybrid' alternative: the low speed four-stroke trunk piston engine.

Low speed four-stroke engine designs, a traditional Japanese speciality, are characterized by simple and rugged construction and comparatively long strokes. Nominal operating speeds, in some cases less than 200 rev/min, allow direct coupling to the propeller. In contrast to the equivalent European/US engine sector, served by high/medium speed machinery with a fixed bore size and variable number of cylinders, the Japanese designs are invariably based on a six-cylinder in-line configuration with a range of bore diameters to meet the desired power output.

As an example, Akasaka Diesels' programme has over the years embraced a dozen bore sizes from 220 mm to 510 mm, all offered in six-cylinder form only. Other company ranges have extended up to 580 mm bore with outputs as high as 735 kW per cylinder. Research and development efforts in the 1970s addressed higher specific ratings, with some designers introducing two-stage turbocharging systems to boost power. Enhanced operating economy, heavy fuel-burning capability and reliability were pursued in the 1980s.

The drawbacks of the direct-coupled low speed four-stroke engine concept compared with medium speed rivals are the fixed propeller speed (though some models are arranged for reduction gearing), higher weight and increased space demands. But these are outweighed, according to proponents, by advantages: improved reliability and reduced maintenance derived from a fewer number of components; less noise and vibration; better fuel and lubricating oil consumption figures; and a greater ability to handle low grade fuels.

With few exceptions, however, the Japanese designs have made little impact in Western markets where the geared medium speed engine maintains its supremacy in the small-ship propulsion arena.

This failure to break through decisively reflects the difference in traditional vessel designs and operating philosophies between the Asian and European/US sectors. And not all the Japanese enginebuilders have made special export efforts to the latter territories, acknowledging that a substantial spares and service support network is required for credibility. The doorstep markets of Indonesia, South Korea, the Philippines and Taiwan offer the best export potential. The domestic market represents the baseload for production, the enginebuilders enjoying longstanding relationships with small-to-medium sized yards.

Hanshin Diesel is perhaps the most well-known designer/builder of the hybrids outside Japan, the Dutch shipping company Spliethoff having standardized on its engines for deepsea cargo ships in the 1970s/1980s. The EL design (illustrated in Figure 29.1), evolved by Hanshin since the late 1970s, indicates the construction and

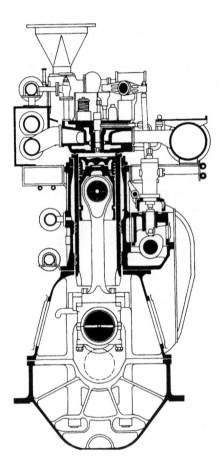

Figure 29.1 Hanshin Diesel's EL design

performance characteristics of a contemporary low speed four-stroke design.

The current EL programme is reduced to 380 mm and 400 mm bore models but past portfolios have also included 300 mm, 320 mm, 350 mm and 440 mm versions, all produced in six-cylinder form only and featuring a stroke/bore ratio of 2:1. An output band from 2058 kW to 2430 kW at 240 rev/min is covered by the 6EL38 and 6EL40 engines on a mean effective pressure of around 20 bar.

The bedplate, crankcase and cylinder block are of cast iron and form a highly rigid construction. The exhaust valve is provided with a cage and rotator, and the valve seat faces are stellited. The cylinder liner is supported at its shoulder by an annular piece to prevent it from deforming due to thermal expansion. No separate cylinder lubrication device is necessary.

The piston crown is of forged steel and cooled by the shaker method using the lubricating system oil; the skirt is of cast iron; and the piston ring set consists of two barrel face, chromium-plated inner cut rings, two conventional cast iron compression rings and two oil scraper rings with expanders. The connecting rod is comparatively short to reduce the overall height of the engine, and the integral crankshaft is made of continuous grain flow forged steel. Both main and crankpin bearings are of the precision shell type.

Hanshin also offers its LU and LF series, the latter extending to a 580 mm bore/1050 mm stroke LF58A model with an output per cylinder of up to 772 kW at 190 rev/min. A new LHL series design benefits from experience with the LU, EL and LF engines. The range embraces 280 mm, 300 mm, 320 mm, 340 mm, 360 mm, 410 mm and 460 mm bore models which feature a stroke/bore ratio of around 1.9:1. Again, all are produced in six-cylinder form to cover a power band from 1176 kW to 3309 kW at speeds from 380 rev/min down to 220 rev/min.

Akasaka Diesels' low speed trunk piston engine equivalent to the Hanshin EL programme, the A series, is produced in six-cylinder 280 mm, 310 mm, 340 mm, 370 mm, 380 mm, 410 mm and 450 mm bore versions with a stroke/bore ratio of around 1.95:1. The A45S model develops 3309 kW at 220 rev/min. The overall Akasaka portfolio embraces designs with bore sizes ranging from 220 mm to 500 mm and offering outputs from 375 kW to 6066 kW. The 500 mm bore/ 620 mm stroke U50 model is available in six, eight and nine-cylinder forms developing 674 kW/cylinder at 380 rev/min.

Among other Japanese designers contesting the market are Makita, Matsui Iron Works and Niigata Engineering.

30 High speed engines

High speed four-stroke trunk piston engines are widely specified for propelling small, generally specialized, commercial vessels and as main and emergency genset drives on all types of tonnage. The crossover point between high and medium speed diesel designs is not sharply defined but for the purposes of this chapter engines running at 1000 rev/min and over are reviewed.

Marine high speed engines traditionally tended to fall into one of two design categories: high performance or heavy duty types. High performance models were initially aimed at the military sector, and their often complex designs negatively affected manufacturing and maintenance costs. Applications in the commercial arena sometimes disappointed operators, the engines dictating frequent overhauls and key component replacement.

Heavy duty high speed engines in many cases were originally designed for off-road vehicles and machines but have also found niches in stationary power generation and locomotive traction fields. A more simple and robust design with modest mean effective pressure ratings compared with the high performance contenders yields a comparatively high weight/power ratio. But the necessary time-between-overhauls and component lifetimes are more acceptable to civilian operators.

In developing new models, high speed engine designers have pursued essentially the same goals as their counterparts in the low and medium speed sectors: reliability and durability, underwriting extended overhaul intervals and component longevity and hence low maintenance costs; easier installation and servicing; compactness and lower weight; and enhanced performance across the power range with higher fuel economy and reduced noxious emissions.

Performance development progress over the decades is highlighted by considering the cylinder dimension and speed of an engine required to deliver 200 kW/cylinder (Figure 30.1). In 1945 a bore of 400 mm-plus and a speed of around 400 rev/min were necessary; in 1970 typical medium speed engine parameters resulted in a bore of 300 mm and a speed of 600 rev/min, while typical high speed engine parameters were 250 mm and 1000 rev/min to yield 200 kW/cylinder.

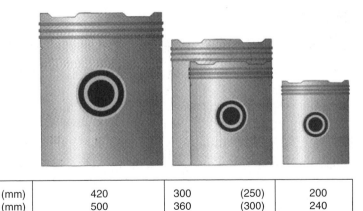

Bore	(mm)	420	300	(250)	200
Stroke	(mm)	500	360	(300)	240
Speed	(rpm)	428	600	(1000)	1500
Year		1945	1970		1995

Figure 30.1 Cylinder dimensions and speeds for medium and high speed engines delivering 200 kW/cylinder (1945, 1970 and 1995). (Reference Wärtsilä)

Today, that specific output can be achieved by a 200 mm bore high speed design running at 1500 rev/min.

Flexible manufacturing systems (FMS) have allowed a different approach to engine design. The reduced cost of machining has made possible integrated structural configurations, with more functions assigned to the same piece of metal. The overall number of parts can thus be reduced significantly over earlier engines (by up to 40 per cent in some designs), fostering improved reliability, lower weight and increased compactness without compromising on ease of maintenance. FMS also facilitates the offering of market-adapted solutions without raising cost: individual engines can be optimized at the factory for the proposed application.

A widening market potential for small high speed engines in propulsion and auxiliary roles encouraged the development in the 1990s of advanced new designs for volume production. The *circa*-170 mm bore sector proved a particularly attractive target for leading European and US groups which formed alliances to share R&D, manufacture and marketing—notably Cummins with Wärtsilä Diesel, and MTU with Detroit Diesel Corporation.

High speed engine designs have benefited from such innovations as modular assembly, electronically controlled fuel injection systems, common rail fuel systems and sophisticated electronic control/monitoring systems. Some of the latest small bore designs are even released for genset duty burning the same low grade fuel (up to 700 cSt viscosity) as low speed crosshead main engines.

Evolving a new design

An insight into the evolution of a high speed engine design for powering fast commercial vessels is provided by MTU of Germany with reference to its creation of the successful 130 mm bore Series 2000 and 165 mm bore Series 4000 engines, which together cover an output band from 400 kW to 2720 kW.

MTU notes first that operators of fast tonnage place high value on service life and reliability, with fuel economy and maximized freight capacity also important. In the fast vessel market, conflicting objectives arise between key parameters such as low specific fuel consumption, low weight/power ratio and extended engine service life. If one parameter is improved, at least one of the others is undermined. The engine designer's aim is therefore to optimize co-ordination of the parameters to suit the application.

Knowledge of the anticipated service load profile is vital for determining the specific loads that must be addressed during the engine design stage so that the required maintenance and major overhaul intervals can be established. Load acceptance characteristics and performance map requirements have a strong influence on turbocharging and the maximum possible mean pressures.

Specifying performance map requirements is simultaneously connected with the selection of the lead application, in this case high speed tonnage. The maximum possible mean pressures are determined on the basis of the power-speed map requirements of various vessel types (for example, air cushion, hydrofoil and planing hull types) and the form of turbocharging (sequential or non-sequential, single or two-stage). With increasing mean pressures (higher power concentration), the weight/power ratio of the engine can be improved.

The maximum mean piston speed is derived from a service life requirement (time-between-overhaul) and the target for the weight/power ratio. With increasing mean piston speed, a greater power concentration in a given volume is achieved, thus improving the weight/power ratio. For fast vessel engines, mean piston speeds of 11–12 m/s and mean effective pressures up to 22 bar (single-stage turbocharging) or 30 bar (two-stage turbocharging) are typical. Figure 30.2 shows the correlations of four-stroke engines for determining bore diameter, stroke and speed. The output per cylinder (Pe) is known from the power positioning of the proposed new engine, and the maximum mean pressure and maximum mean piston speed have already been established. The required minimum bore diameter (D) can thus be determined.

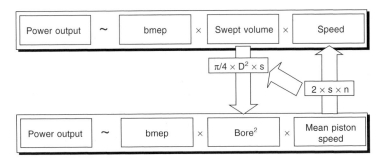

Figure 30.2 Calculation of power output

The appropriate stroke (s) is determined using the specified stroke/bore ratio (s/D). If large s/D ratios are selected, a large engine height and width results (V-engine); small s/D ratios are associated with somewhat reduced fuel efficiency. For relevant MTU engines with good weight/power ratios, the stroke/bore ratio lies within a range of 1.1 to 1.25. The engine speed appropriate to the established stroke is determined via the resulting mean piston speed. Engine speed is an important factor for the customer as the size of the gearbox required is based on the speed and torque.

The peak firing pressure or peak firing pressure/mean pressure ratio is the most important factor influencing the specific effective fuel consumption. For high efficiency, MTU suggests, peak firing pressure/mean pressure ratios of around 8 should be targeted. With the mean pressure already specified, the peak firing pressure can be established. If, for example, a mean pressure of 22 bar (single-stage turbocharging) is selected, the peak firing pressure should lie in the 160–180 bar range. For two-stage turbocharged engines with correspondingly high mean pressures, peak firing pressures of above 200 bar must be targeted. With two-stage turbocharging, the potential for fuel consumption reduction can be increased using charge air cooling.

CATERPILLAR

A wide programme of high speed engines from the US designer Caterpillar embraces models with bore sizes ranging from 105 mm to 170 mm. The largest and most relevant to this review is the 170 mm bore/190 mm stroke 3500 series which is produced in V8-, V12- and V16-cylinder versions with standard and higher B-ratings to offer outputs up to around 2200 kW. The engines, with minimum/maximum running speeds of 1200/1925 rev/min, are suitable for propelling workboats, fishing vessels, fast commercial craft and patrol boats. Genset applications can be covered with ratings from 1000 kVA to 2281 kVA.

The series B engines (Figure 30.3) benefited from a number of mechanical refinements introduced to take full advantage of the combustion efficiency improvement delivered by an electronic control system. Electronically controlled unit fuel injectors combine high injection pressures with an advanced injector design to improve atomization and timing. Outputs were raised by 17 to 30 per cent above earlier 3500 series models.

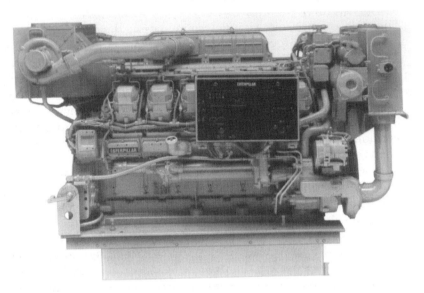

Figure 30.3 Caterpillar 3512B engine with electronic control system

A special high performance variant of the V16-cylinder 3500 series model was introduced to target niche markets, the refinements seeking increased power, enhanced reliability and lower fuel and lubricating oil consumptions without undermining durability. This Phase II high performance version of the 3516 has an upper rating of 2237 kW at 1925 rev/min. It was released for fast passenger vessels with low-load factors with a standard maximum continuous rating of 1939 kW at 1835 rev/min and a 'two hours out of 12' rating of 2088 kW at 1880 rev/min. Optional higher ratings up to 2205 kW at 1915 rev/min can be specified for cooler climate deployment, with revised turbocharger, fuel injector and timing specifications.

Key contributions to higher performance came from high efficiency ABB turbochargers, a seawater aftercooler to supply colder air to the combustion chambers, larger and more aggressive camshafts, and a new deep crater piston design. The fuel is delivered through strengthened unit injectors designed and manufactured by Caterpillar to secure injection pressures of 1380 bar.

An optimum air-fuel mixture which can be burned extremely efficiently is fostered by the combination of a denser air intake and the high injection pressure. The reported result is a specific fuel consumption range at full load of 198–206 g/kWh with all fuel, oil and water pumps driven by the engine. Modifications to the steel crown/aluminium skirt pistons and rings lowered lubricating oil consumption to 0.55 g/kWh.

A particularly desirable feature for fast ferry propulsion is underwritten by the high efficiency combustion and low crevice volume pistons which help to eliminate visible exhaust smoke at all steady points along the propeller demand curve. The rear gears were widened and hardened to serve the higher pressure unit injectors. New gas-tight exhaust manifolds with bellow expansion joints and stainless steel O-rings improved engineroom air quality by eliminating exhaust gas leaks.

A longer-stroke variant of the Cat 3500 series B engine was introduced after marine field tests undertaken from early 1998, these 3512B and 3516B models offering as much as 13 per cent higher powers than their standard counterparts, with respective maximum commercial ratings of 1380 kW and 1864 kW at 1600 rev/min. Seven per cent improvements in power-to-weight ratio and fuel economy were reported, along with lower emission levels than the standard engines. The higher output was achieved by enlarging the cylinder displacement (increasing the stroke by 25 mm) and without raising cylinder pressure or undermining bearing life or the durability of other key components.

A new single-piece forged crankshaft has more mass and is made from a stronger steel alloy than before to handle the higher loads. The connecting rods are longer and feature stronger shaft geometry; and a more robust rod pin end enhances the durability required for the increased piston speeds and higher inertia loads. The pistons are of the same two-piece design proven in standard Cat 3500 series B engines, a steel crown and aluminium skirt securing high strength and reduced weight. The engine footprint of these more powerful variants remained unchanged; only the dimensions of the higher capacity aftercooler and turbochargers were increased.

All Caterpillar 3500 series-B engines are controlled by a microprocessor-based electronic control module (ECM). Information is collected from engine sensors by the ECM which then analyses the data and adjusts injection timing and duration to optimize fuel efficiency and reduce noxious exhaust emissions. Electronic control also supports onboard and remote monitoring capabilities, the ECM reporting all information through a two-wire Cat Data Link to the instrument panel. The panel records and displays faults as well as operating conditions. An optional Customer Communications Module translates engine data to standard ASCII code for transmission to a PC or via satellite to remote locations.

Caterpillar's Engine Vision System (EVS) is compatible with the high performance 3500 series-B engines and the company's other electronically controlled engines. The EVS displays engine and transmission data, vessel speed, trip data, historical data, maintenance intervals, diagnostics and trouble-shooting information. Up to three engines can be monitored simultaneously, the system transferring between the vision display and individual ECMs via the two-wire data link.

An upgrade announced in 2002 introduced the 3500B series II engines with enhancements to their electronic control, monitoring, display and cooling systems as well as new derating and operating speed options designed for specific applications. New electronics included the latest Caterpillar ADEM III control system, allowing more engine parameters to be controlled and monitored, with more accuracy and fault-reporting capability. A new 'programmable droop' capability allows precise governor control for load-sharing applications. A combined cooling system, rather than two separate circuits, became an option. A higher maximum continuous rating of 2000 kW from the 3516B series II engine was offered to yield more power and bollard pull capacity for larger harbour tugs; the higher rating also addressed some types of ferries and offshore service vessels.

CUMMINS

The most powerful own-design engine in Cummins' high speed programme, the KTA50-M2 model, became available from early 1996 (Figure 30.4). The 159 mm bore/159 mm stroke design is produced by the US group's Daventry factory in the UK in V16-cylinder form with ratings of 1250 kW and 1340 kW for medium continuous duty and 1030 kW and 1180 kW for continuous duty applications. The running speeds range from 1600 rev/min to 1900 rev/min, depending on the duty; typical applications include fishing vessels, tugs, crewboats and small ferries.

Figure 30.4 Cummins' largest own-design engine, the KTA50-M2 model

The KTA50-M2 engine benefited from a new Holset turbocharger, low temperature after-cooling and gallery-cooled pistons. Cummins's Centry electronics system contributes to enhanced overall performance and fuel economy, providing adjustable all-speed governing, intermediate speed controls, dual power curves, a built-in hour meter and improved transient response. Diagnostic capabilities are also incorporated.

DEUTZ

A long tradition in high speed engine design is maintained by Deutz of Germany (formerly Deutz MWM) whose current programme is focused on the 616 and 620 series with an upper output limit of 2336 kW for propulsion and genset drives. Other high speed engines produced under the Deutz banner cover a power range down to 14 kW.

The 616 series is a 132 mm bore/160 mm stroke design covering propulsion plant applications from 320 kW to 1360 kW at speeds up to 2300 rev/min by V8-, 12- and 16-cylinder models (Figure 30.5).

Figure 30.5 V12 cylinder block of Deutz 616 series engine

The 170 mm bore/195 mm stroke 620 series design also embraces V8, 12 and 16-cylinder versions covering a power band from 829 kW to 2336 kW at speeds up to 1860 rev/min for propulsion duty (Figure 30.6). Good exhaust gas exchange and optimum combustion with minimal emissions are fostered by cross-flow cylinder heads and four valves per cylinder, while a relatively high compression ratio promotes excellent cold-start characteristics and low fuel consumption. High pressure fuel injection further contributes to these operating qualities. Correct swirl under both low load and full load conditions is secured by the boost pressure-controlled throttle of the HALLO-Swirl system, a variable swirl-supporting air intake system. Outputs from 1035 kW to 3600 kW at 1000 rev/min are delivered by six, eight and nine in-line and V12- and V16-cylinder versions of the 240 mm bore/280 mm stroke 628 series, which can be operated on heavy fuels (RMK 35) as well as marine diesel oil. (*See Chapter 20 for details of Deutz 628 series.*)

Figure 30.6 V16-cylinder version of Deutz 620 series engine

GMT

Grandi Motori Trieste (GMT) of Italy—now part of the Wärtsilä Corporation—focused its high speed engine developments on a 230 mm bore design offered in several versions for commercial and naval propulsion applications as well as for genset drive duties (Figure 30.7). A heavy fuel-burning model was available for unifuel machinery installations while a non-magnetic version could be specified for minehunters.

The series was produced in standard (B230) form with a stroke of 270 mm and in BL230 long stroke (310 mm) form, later versions benefiting from improved cooling and turbocharging arrangements for higher outputs at speeds up to 1200 rev/min. The B230 had a rating of 210 kW/cylinder and the BL230 a rating of 225.5 kW/cylinder. Both types were produced in in-line four to V20-cylinder configurations. A two-stage turbocharged version with variable compression ratio, the BL 230 DVM, developed 283 kW/cylinder at 1050 rev/min on a mean effective pressure of 25.1 bar to provide a compact plant for corvettes and frigates. The V20-cylinder model delivered 5660 kW.

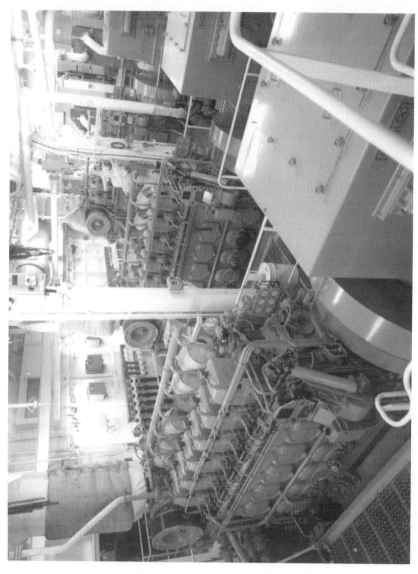

Figure 30.7 GMT BL230 engines in a genset installation

A BL230P version was developed for operation on heavy fuel up to IFO 500 with a maximum speed of 1000 rev/min and a power rating— 190 kW/cylinder—lower than the gas oil-fuelled models. Its special features included: composite pistons with forged steel crowns and plasma-coated rings in the first groove for wear resistance; Nimonic A material exhaust valves; valve rotators; cooled fuel injection valves; and oversized fuel injection pumps for enhanced atomization and combustion. Other variants in the engine programme targeted natural gas and LPG burning installations.

A highly rigid cast iron main structure for the 230 mm bore engine incorporates the water, oil and air manifolds, virtually eliminating external pipes and fostering compactness. Camshafts, oil cooler, oil, water and fuel pumps, and filters are arranged for accessibility; and pistons and connecting rods are withdrawable from the top.

GMT's high speed portfolio also included the A210 series, the 210 mm bore/230 mm stroke design developing 170 kW/cylinder at 1500 rev/min on a mean effective pressure of 16.2 bar. A programme embracing V6- to V20-cylinder models covered output demands up to 3400 kW. A special version designed for submarine propulsion—the compact and lightweight A210SM—benefited from anti-noise, shock and vibration characteristics. An ability to operate in severely inclined positions was also addressed, along with ease of access for maintenance.

ISOTTA FRASCHINI

A family of high speed engines from Isotta Fraschini, part of Italy's Fincantieri group, is headed by the 170 mm bore/170 mm stroke 1700 series which is available in different versions for light, medium and heavy propulsion duties in commercial and military vessels, as well as for genset drives. A non-magnetic variant was developed for mine warfare vessels.

The 1700 series embraces V8, 12 and 16-cylinder models, all with a 90-degree configuration arranged on a high tensile alloy iron crankcase and featuring a direct fuel injection system and four valves per cylinder. Recent upgrades raised the power output, the largest model—the V1716 T2—now offering up to 2595 kW at 2100 rev/min, compared with its previous maximum rating of 2350 kW at the same speed.

MAN B&W HOLEBY

In 1995 MAN B&W Diesel's genset engine specialist Holeby Diesel of Denmark supplemented its popular L23/30H and L28/32H medium speed auxiliary prime movers (see Chapter 18) with the innovative high speed L16/24 series. The 160 mm bore/240 mm stroke design (Figure 30.8) was conceived as a new generation 1000/1200 rev/min engine dedicated to genset drives and capable of operating on an unrestricted load profile on heavy fuel up to 700 cSt/50°C viscosity. The programme embraces five-, six-, seven-, eight- and nine-cylinder models covering a power range from 450 kW to 900 kW.

The main problem in burning heavy fuel oil in small high speed engines is ignition delay. Smaller quantities of the volatile, easily combustible hydrocarbon fractions are present in such fuels than in lighter diesel fuels. The L16/24 designers addressed the problem by adopting a number of measures to achieve excellent heavy fuel combustion even at part- and low-load operation: a higher injection pressure (1500 bar), a higher opening pressure for the fuel valve, smaller nozzle hole diameters in the valve, and a higher compression ratio (15.5:1). The maximum combustion pressure is 180 bar. Another contribution came from a cylinder head design refined to improve the swirl of fuel in the combustion chamber.

Support-function components were traditionally distributed around the engine block and connected with externally mounted supply pipes. The practice was reversed for the L16/24 engine: all support elements— oil and water pumps, coolers, filters, and safety and regulator valves— are arranged in a single front-end box for ease of accessibility and maintenance (Figure 30.9); and the supply channels are cast into the block for maintenance-free operation. Virtually all the engine's internal supply lines are channelled through the cooling water jacket and cylinder head. The arrangement considerably simplifies the overall design, reducing the overall number of components by some 40 per cent compared with earlier engines. The front-end box system components can be exchanged using clip-on/clip-off couplings without removing any pipes.

High rigidity was sought from a monobloc cast iron engine frame whose elements are all held under compressive stress. The frame is designed to accept an ideal flow of forces from the cylinder head down to the crankshaft and to yield low surface vibrations from the outer shell. Two camshafts are located in the frame: the camshaft for the inlet/exhaust valves is arranged on the exhaust side in a very high position; and the fuel injection camshaft is on the service side of the engine.

Covers in the frame sides offer access to the camshafts and crankcase; some of the covers are arranged to act as relief valves. There is no cooling water in the frame. The framebox is designed to minimize noise emission, the inner frame absorbing all the engine forces and

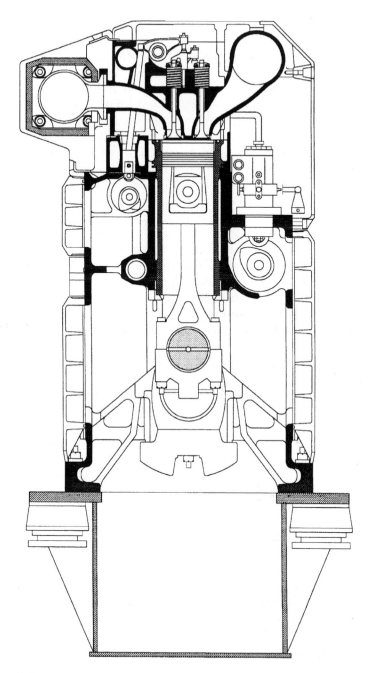

Figure 30.8 Cross-section of MAN B&W Holeby L16/24 genset engine. Note the separate camshafts for the gas exchange valves (left) and fuel injection (right)

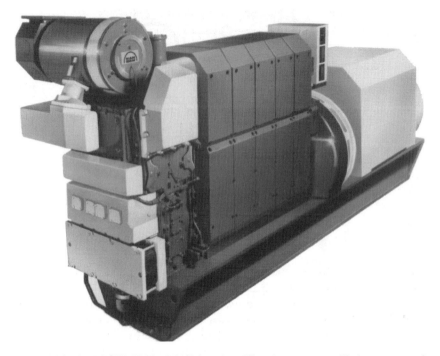

Figure 30.9 MAN B&W Holeby L16/24 engine. All main support ancillaries are grouped in a single front-end box for ease of access

the outer frame forming a stiff shell with minimal vibration (Figure 30.10). The main bearings for the underslung crankshaft are carried in heavy supports by tie-rods from the intermediate frame floor and secured by bearing caps. The caps are provided with side guides and held in place by studs with hydraulically tightened nuts. The main bearing features replaceable shells which are fitted without scraping.

Both engine and alternator are mounted on a rigid baseframe that acts as a lubricating oil reservoir. Specially designed engine mounts reduce noise and vibration.

The centrifugal cast iron cylinder liner, housed in the bore of the engine frame, is clamped by the cylinder head and rests on its flange on the water jacket; it can thus expand freely downwards when heated during engine operation. The liner is of the high flange type, the height of the flange matching the water-cooled area to give a uniform temperature pattern over the entire liner surface. The liner's lower part is uncooled to secure a sufficient margin for cold corrosion at the bottom end. There is no water in the crankcase area. Gas sealing between liner and cylinder head is effected by an iron ring. The liner

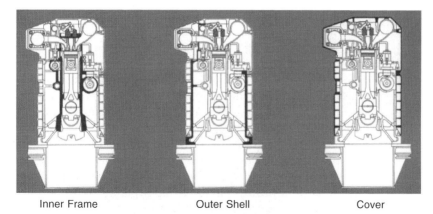

Inner Frame   Outer Shell   Cover

Figure 30.10 Noise emission and vibration from the MAN B&W Holeby L16/24 engine are minimized by the inner frame, outer shell and cover arrangement

is fitted with a slip-fit-type fire ring at its top to reduce lubricating oil consumption and bore polishing.

The cast iron cylinder head, with integrated charge air receiver, is made in one piece and incorporates a bore-cooled thick-walled bottom. It has a central bore for the fuel injection valve and a four-valve cross-flow configuration with a high flow coefficient. The valve pattern is turned about 20 degrees to the axis to achieve an intake swirl promoting optimized combustion. The head is hydraulically tightened by four nuts acting on studs screwed into the engine frame. A screwed-on top cover for the head has two main functions: oil sealing the rocker chamber and covering the complete top face of the head.

Air inlet and exhaust valve spindles are of heat-resistant material and their seats armoured with welded-on hard metal. All the spindles are fitted with valve rotators to ensure an even temperature on the valve discs and to prevent deposits forming on the seatings. The cylinder head is provided with replaceable valve seat rings of heat-resistant steel, and the exhaust valve seat rings are water cooled. The seating surfaces are hardened to minimize wear and prevent dent marks.

The valve rocker arms are actuated through rollers, roller guides and pushrods. The roller guides for the inlet and exhaust valves are mounted in the water jacket part. Access for dismantling is provided by a side cover on the pushrod chamber. Each rocker arm activates two spindles via a spring-loaded valve bridge with thrust screws and adjusting screws for valve clearance. The valve actuating gear is pressure feed-lubricated from the centralised lubricating system, through the water chamber part and from there into the rocker arm shaft to the rocker bearing.

The oil-cooled pistons comprise a nodular cast iron body and forged steel crown with two compression rings and one scraper ring fitted in hardened grooves. The different barrel-shaped profiles of the compression rings and their chromium-plated running surfaces aim to maximize sealing and minimize wear. The piston has a cooling space close to the crown and the ring zone which is supplied with oil from the engine lubricating system. Heat transfer, and thus cooling, is promoted by the shaker effect stimulated by the piston movement.

Oil is supplied to the cooling space through channels from the oil grooves in the piston pin bosses. Oil is drained from the space through ducts located diametrically to the inlet channels. The piston pin is fully floating and kept in position in the axial direction by two circlips.

The die-forged connecting rod has a big end with a horizontal split and bored channels to transfer oil from the big end to the small end. The big end bearing is of the tri-metal type coated with a running layer. The bearing shells are of the precision type and can therefore be fitted without scraping or any other adaptation. The tri-metal small end bearing is pressed into the connecting rod. The bush is provided with an inner circumferential groove and a pocket for distributing oil in the bush itself and for supplying oil to the pin bosses.

A one-piece forged crankshaft with hardened bearing surfaces is suspended in underslung, tri-metal main bearings coated with a running layer. To attain a suitable bearing pressure and vibration level the crankshaft is equipped with counterweights which are attached to the shaft by two hydraulic screws. At its flywheel end the crankshaft is fitted with a gearwheel which, through two intermediate wheels, drives the twin camshafts. Also mounted here is a coupling flange for the alternator. At the opposite (front) end is a gearwheel connection for the lubricating oil and water pumps.

Lubricating oil for the main bearings is supplied through holes drilled in the engine frame. From the main bearings the oil passes through bores in the crankshaft to the big end bearings and thence through channels in the connecting rods to lubricate the piston pins and cool the pistons.

Separate camshafts for the inlet/exhaust valves and the fuel pump facilitate adjustment of the gas exchange settings without disturbing the fuel injection timing (Figure 30.8). Likewise, it is possible to adjust fuel injection without disturbing gas exchange parameters. The resulting flexibility allows engine operation to be adjusted and optimized for fuel economy or low NOx emissions. The camshafts are mounted in bearing bushes fitted in bores in the engine frame. The valve camshaft is arranged in a very high position on the engine exhaust side to secure a short and stiff valve train and reduce moving masses.

The fuel injection camshaft is arranged on the service side of the engine.

Both camshafts are structured in single-cylinder sections and bearing sections in such a way that disassembly of individual cylinder sections is possible through the side openings in the crankcase. The camshafts and governor are driven by the main gear train at the flywheel end of the engine, rotating at a speed half that of the crankshaft. The lubricating oil pipes for the gearwheels are equipped with nozzles adjusted to apply lubricant at the points where the gearwheels mesh.

All fuel injection equipment is enclosed securely behind removable covers. Each cylinder is individually served by a fuel injection pump, high pressure pipe and injection valve with uncooled nozzle. The injection pump unit, mounted on the engine frame, comprises a pump housing embracing a roller guide, a central barrel and a plunger. The pump is activated by the fuel cam and the volume injected controlled by turning the plunger. The fuel injection valve is located in a valve sleeve in the centre of the cylinder head, its opening controlled by the fuel oil pressure and closure effected by a spring.

The high pressure fuel pipe is led through a bore in the cylinder head surrounded by a shielding tube. The tube also acts as a drain channel to ensure that any leakage from the fuel valve and the high pressure pipe is drained off.

A lamda controller ensures that all injected fuel is burnt, countering the internal engine pollution and increased wear that might otherwise result from genset step loading.

A constant pressure turbocharging system embraces an MAN B&W NR/S turbocharger purpose-designed for the L16/24 engine, charge air cooler, charge air receiver and exhaust gas receiver. The charge air cooler is a compact two-stage tube unit deploying a large cooling surface.

A patented 'intelligent' cooling water system was designed to secure an optimized temperature across the engine operating band from idling to full load. The system, which accepts fresh water within the 10–40°C temperature range, has one inlet and one outlet connection. Its two pumps, in combination with thermostatic valves, continuously regulate cooling water temperature to achieve the optimized operating condition. Since charge air from the turbocharger never falls below the dew point there is no danger of water condensation in the cylinders.

The cooling water system comprises a low temperature (LT) system and a high temperature (HT) system, each cooled by fresh water. The LT circuit is used to cool charge air and lubricating oil. The HT circuit cools the cylinder liners and heads, fostering optimized combustion conditions, limiting thermal load under high load conditions, and preventing hot corrosion in the combustion area. Under low load, the system is designed to ensure that the temperature is high enough for efficient combustion and that cold corrosion is avoided.

Water in the LT system passes through the low temperature circulating pump which drives the water through the second stage of the charge air cooler and then through the lubricating oil cooler before the water leaves the engine together with the high temperature water. The amount of water passing through the second stage of the charge air cooler is controlled by a three-way valve dependent on the charge air pressure. If the engine is operating at low-load condition the temperature regulation valve cuts off the LT water flow, thus securing preheating of the combustion air by the HT water circuit in the first stage.

The HT cooling water passes through the high temperature circulating pump and then through the first stage of the charge air cooler before entering the cooling water jacket and the cylinder head. It then leaves the engine with the low temperature water. Both LT and HT water leaves the engine via separate three-way thermostatic valves that control the water temperature.

All moving parts of the engine are lubricated with oil circulating under pressure, the system served by a lubricating oil pump of the helical gear type. A pressure control valve built into the system reduces the pressure before the filter with a signal taken after the filter to ensure constant oil pressure with dirty filters. The pump draws oil from the sump in the baseframe, the pressurized oil then passing through the lubricating oil cooler and the filter. The oil pump, cooler and filter are all located in the front box. The system can also be provided with a centrifugal filter. Lubricating oil cooling is carried out by the low temperature cooling water system, with temperature regulation effected by a thermostatic three-way valve on the oil side (see above). The engine is equipped as standard with an electrically driven prelubricating pump.

The L16/24 engine is prepared for MAN B&W Diesel's CoCoS computerized surveillance system, a Microsoft Windows-based program undertaking fully integrated monitoring of engine operation, maintenance planning, and the control and ordering of spares. The four CoCoS software modules cover engine diagnosis, maintenance planning, spare parts catalogue, and stock and ordering (see Chapter 22).

Each cylinder assembly (head, piston, liner and connecting rod) can be removed as a complete unit for repair, overhaul or replacement by a renovated unit onboard or ashore. Replacing a cylinder unit

(Figures 30.11, 30.12 and 30.13) is accomplished by removing the covers and high pressure fuel injection pipe, and disconnecting a snap-on coupling to the exhaust gas pipe. The only cooling water connections are to the cylinder unit as there is no cooling water in the baseframe. Inlet and outlet cooling water passes between cylinder units via bushes which are pushed aside in disassembling a unit. The charge air connections are dismounted in the same way. The four hydraulically fastened cylinder head nuts and the two connecting rod nuts (all six are of the same size) are then removed, allowing the 200 kg unit to be withdrawn from the engine.

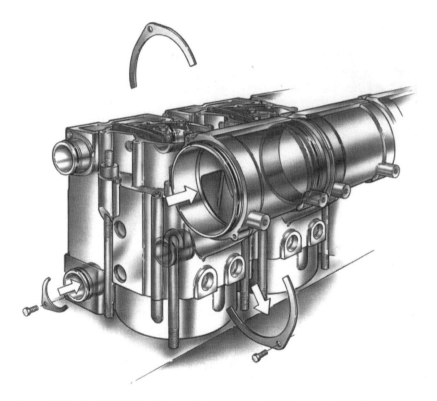

Figure 30.11 MAN B&W Holeby L16/24 engine: preparing for cylinder unit removal

The design principles of the L16/24 engine were later applied to the larger L27/38 and L21/31 medium speed engines, respectively introduced in 1997 and 2000 for both genset and propulsion applications (*see Chapter 22*).

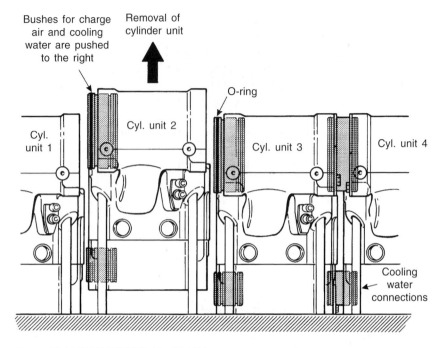

Figure 30.12 MAN B&W Holeby L16/24 engine: preparing for cylinder unit removal

MITSUBISHI

Fast ferry propulsion business potential stimulated the development in the early 1990s of the Mitsubishi S16R-S engine by the Japanese group's Sagamihara Machinery Works. The higher performance V16-cylinder model was evolved from the established S16R design which had been in production since 1989 as a general purpose marine engine. A constant pressure turbocharging system based on a newly developed turbocharger and a revised fuel injection system contributed to a 20 per cent rise in the power output. The 170 mm bore/180 mm stroke design has a maximum continuous rating of 2100 kW at 2000 rev/min with overload ratings up to 2300 kW.

The weight was reduced to 89 per cent of the original engine, primarily through the adoption of aluminium alloy components optimized in size for the duty. An overall weight of 5500 kg underwrites a power-to-weight ratio of 2.62 kg/kW.

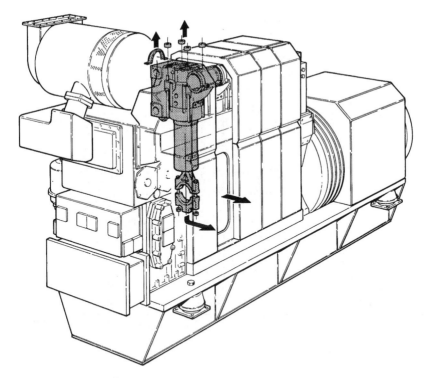

Figure 30.13 MAN B&W Holeby L16/24 engine: removing a complete cylinder unit

MTU

A portfolio of high performance high speed designs with an upper output limit of 9000 kW is offered by MTU (Motoren- und Turbinen-Union) of Germany, which was created in 1969 when Daimler-Benz and MAN consolidated the development and production of relevant engines from MAN and Maybach Mercedes-Benz. MTU Friedrichshafen became part of the Deutsche Aerospace group in 1989, and is now a DaimlerChrysler company.

Series 396

Addressing lower power demands, MTU's Series 396 engine is long established as a propulsion and genset drive, the 165 mm bore/185 mm stroke design delivering up to 2560 kW at 2100 rev/min from V90-degree configuration 8-, 12- and 16-cylinder models (Figure 30.14). Engines are available in three different versions:

- TB04, with external charge air cooling (intercooler in raw water circuit).

Figure 30.14 MTU 16V 396 TE94 engine with split-circuit cooling system

- TC04, with internal charge air cooling (intercooler in engine coolant circuit).
- TE04, with internal charge air cooling (intercooler in engine coolant circuit; split-circuit coolant system).

TE split-circuit cooling system: the Series 396 engine—and other MTU designs—can be supplied with the TE split-circuit coolant system with a power-dependent sub-circuit to cool the combustion air. Optimum performance is fostered throughout the engine's power range: at idling speed the air supply is heated to achieve complete fuel combustion; in the medium power and full load range conditions are optimized for high output while keeping thermal stress on engine components at a low level.

Coolant flow from the engine is split in two. Approximately twothirds of the flow passes through a high temperature (HT) circuit and returns directly to the engine inlet, while the remainder is fed into a thermostatically controlled low temperature (LT) circuit. During engine idling or low-load operation the thermostat allows heated coolant to bypass the recooler on its way to the intercooler in order to warm up the combustion air and prevent white smoke in the exhaust.

An annular slide valve in the thermostat remains in its initial position until increasing power raises the coolant temperature, causing the wax pellet in the thermostat to expand. Gradual closing of the bypass line now directs the coolant stream through the recooler. As a result, coolant entering the intercooler is at a low temperature which, in turn, underwrites a high combustion air volume and, consequently, maximum engine power. After flowing through the intercooler and the oil heat exchanger, 'cold' coolant rejoins the uncooled HT circuit, thereby cooling the total volume flow before it re-enters the engine.

Sequential turbocharging is exploited for propulsion engines required to deliver high power in the lower and medium speed ranges. The system incorporates two or three turbochargers with automatic on/off control as a function of engine speed, power demand and turbocharger maximum efficiency. In addition to increased torque the system yields reduced fuel consumption and lower exhaust temperatures.

Cylinder cut-out (no fuel injection into selected cylinders) is adopted for engines operating under varying speed conditions (low idling speed). The system enhances combustion at idle, thus shortening the warm-up phase and avoiding white smoke from unburnt fuel, and generally contributing significantly to environmental compatibility.

Series 595

A power range from 2000 kW to 4320 kW at 1500–1800 rev/min is covered by the 190 mm bore/210 mm stroke Series 595 engine, introduced in 1990 in V12- and V16-cylinder versions for propulsion and genset applications (Figure 30.15).

Contributing to a compact design yielding power-to-weight ratios of under 3 kg/kW and a power-to-space ratio of 250 kW/m^3 (including all ancillaries) are the following features: a V72-degree cylinder configuration with all ancillaries arranged for space saving within the engine contour; nodular cast iron crankcase extending below the crankshaft centreline to maximize rigidity; crankshaft with hardened main and connecting rod bearing radii, designed to withstand firing pressures of up to 180 bar; individual fuel injection pumps designed for injection pressures of up to 1500 bar; plate-type coolant heat exchanger integrated with the engine; split-circuit coolant system; and two-stage sequential turbocharging system with charge air cooling for boost pressures of up to 4.8 bar. The turbocharging sequencing is electronically controlled.

Components requiring regular servicing are located to provide good access at the engine's auxiliary power take-off end. Maintenance is also smoothed by the modular arrangement of functionally interlinked components, plug-in connections and the omission of complex

Figure 30.15 V16-cylinder MTU Series 595 engine

pipework, hose couplings and cabling. A new electronic control system (ECS) was purpose-developed by MTU for the Series 595 engine to provide information for the operator, promote easier operation and enhance safety, reliability and economy. The safety functions embrace engine and plant monitoring, overload protection, diagnostics, automatic start-up and load control.

Series 1163

The highest power outputs from the MTU programme were offered for many years by the Series 1163 engines (Figure 30.16). The 230 mm bore/280 mm stroke design is produced in V60-degree configuration 12-, 16- and 20-cylinder versions delivering up to 7400 kW at 1300 rev/min for commercial and naval propulsion installations. The key elements of the design are summarized as:

Figure 30.16 MTU Series 1163 engine in V20-cylinder form

Crankcase: nodular cast iron structure with access ports on both sides and a flange-mounting facility for alternators or other driven machinery; a welded sheet steel oil pan is provided.

Cylinder liner: replaceable, wet-type centrifugally cast.

Cylinder head: cast iron component arranged with two inlet and two exhaust valves, all equipped with Rotocap rotators, a centrally located fuel injector separated from the rocker area, and a decompression valve.

Valve actuation: by two camshafts, roller tappets, pushrods and rocker arms. Valve clearance adjustment is performed via two screws located in the rocker arms.

Crankshaft: a single-structure forged component finished all over and featuring bolted-on counterweights; axial crankshaft alignment is effected by a deep-groove ball bearing.

Bearings: thin-walled, two-piece, replaceable steel-backed tri-metal sleeve bearings for the crankshaft and connecting rod, with cross-bolted bearing caps.

Connecting rods: forged and finished all over and grouped in pairs to serve two opposite cylinders.

Piston: composite-type with light alloy skirt and bolted-on steel crown, cooled by oil spray nozzles; three compression rings are fitted in the crown, with one oil control ring between crown and skirt; all rings replaceable after piston crown removal.

Fuel injection: direct injection by individual pumps and via short high pressure lines; pump replacement requires no readjustment; gear-type fuel delivery pump; two duplex fuel filters; cylinder cut-out facility available (see under Series 396 engine section).

Governor: hydraulic MTU unit mounted on the gearcase, with the linkage between governor and fuel injection pumps accommodated in the gearcase and camshaft space. Engine shutdown solenoids act on the fuel rack; in addition, independent emergency air shut-off flaps are arranged to block the engine's air supply.

Lubricating oil system: self-contained forced feed system, with oil flow progressing through gear-type pumps, heat exchanger and filters to the engine lubrication points. Oil for piston cooling is tapped off after the heat exchanger and filters; the engine-mounted oil heat exchanger is integrated in the engine coolant circuit (Series 1163-02 model) or raw water circuit (Series 1163-03 model).

Cooling system: two-circuit system; closed engine coolant circuit using two centrifugal pumps; thermostatic control; recooling provided by raw water heat exchanger or fan cooler.

Turbocharging: the Series 1163-02 model exploits pulse charging based on two turbochargers (one turbocharger for the 12-cylinder engine) with intercoolers incorporated in the raw water circuit. The Series 1163-03 model exploits constant pressure turbocharging with two-stage air compression and interstage cooling; four or five charger groups (depending on the application) are under sequential turbocharging control; the constant pressure exhaust manifold is located in the V-bank, and charge air pipework mounted externally; high pressure (HP) and low pressure (LP) intercoolers are incorporated in the raw water circuit. For start-up and low-load operation the Series 1163-03 model features a charge air preheater, fed with engine coolant, in each HP intercooler outlet. Coolant jackets for the exhaust manifolds and turbocharger turbines of this model secure reduced heat rejection.

Like the Series 595 engine, the Series 1163 design features inboard high pressure fuel injection lines and the enclosure of all hot exhaust components in water-cooled and gas-tight casings: valued for unmanned enginerooms by helping to reduce the possibility of fire in the event of fuel or lubricating oil leakages. A triple-walled insulation design also maintains surface temperatures well within the limit of 220°C dictated by classification societies, as well as considerably reducing radiant heat in the engineroom.

Sequential turbocharging (output-dependent control of the number of turbochargers deployed) fosters high torque at low rev/min (wide performance band) and hence good acceleration. Optimization during sea trials, with assistance from an electronic control system, can eliminate black smoke emissions.

An uprated Series 1163 engine for fast ferry propulsion benefits from:

- An improved fuel injection system to optimize consumption and exhaust emissions reduction.
- A higher cylinder head fatigue strength, thanks partly to bore cooling.
- Improved power and torque characteristics achieved by modifying the two-stage sequential turbocharging system.
- Adaptation and optimization of the turbochargers to match the increased cylinder output (325 kW).
- Intelligent electronic engine management.

Series 8000 engine

MTU's thrust in commercial and naval propulsion sectors was strengthened in 2000 with the introduction of its most powerful-ever engine, the advanced 265 mm bore Series 8000 design (Figure 30.17). Innovative fuel injection, turbocharging and electronic management systems proven on earlier MTU engines are exploited, while a new modular power unit-based structure helps to ease maintenance and minimise life-cycle costs.

The Series 8000 engine was conceived with the aim of consolidating the company's position in the market for fast ferries and extending opportunities to mainstream commercial arenas, such as cruise liners. Outputs up to 9000 kW for naval vessels and megayachts were initially sanctioned from the V20-cylinder model which launched the series, with an 8200 kW rating at 1150 rev/min approved for high speed commercial tonnage. A potential for firing pressures up to 230 bar underwrites future power rises, projected after successful long term testing of engines in service. Extension of the programme with V12 and V16 models was planned to allow the series to target power demands down to 5000 kW; and in-line cylinder and heavy fuel-burning versions were anticipated for large workboat propulsion applications. The first four production 20V 8000 engines were installed as a megayacht propulsion plant, subsequent orders calling for units to power a large catamaran fast ferry, frigates and naval supply vessels.

Figure 30.17 MTU Series 8000 engine in V20-cylinder form. Note the fuel pumps (lower right) for the common rail injection system

A V48-degree cylinder configuration contributes to a narrow engine (1.9 m wide), well suited for fast catamaran and monohull ferries where machinery space is at a premium. The specific power output related to volume and weight is reflected in figures of 200 kW/m^3 and just over or under 5 kg/kW, depending on the application.

Common rail fuel injection, proven on MTU's 1996-launched 165 mm bore Series 4000 engine (see page 795), allows all injection parameters affecting combustion to be independently controlled; these include such variables as the timing, period and pattern of injection as well as the injection pressure. Independent control fosters reduced fuel consumption and exhaust emission levels across the entire engine power curve. The common rail system also benefits engine noise and vibration levels, substantially lowering them at idling and mid-range running speeds compared with conventional systems.

Two high pressure fuel pumps mounted at the free end of the engine develop pressure levels of up to 1800 bar. The fuel is pumped through double-walled delivery lines arranged longitudinally down the engine and directed to the cylinders via distributors into individual accumulators. The accumulators are large enough to prevent any drop in pressure before the fuel injector when maximum fuel quantities are being injected, and to prevent pressure oscillations in the system. The fuel is injected via electronically-controlled injectors arranged centrally in the cylinder head.

Experience gained from numerous Series 4000 engines in service moulded the common rail system created by L'Orange for the new larger engine, which is detailed in Chapter 8/Fuel Injection. Considerable scope for further development of the system promised further advances in fuel economy, emission levels and noise reduction in future production engines.

The Series 8000 design's sequential turbocharging system had been successfully applied by MTU to its Series 396, 538, 595 and 1163 engines for many years to secure a large performance map width. First fitted on production models in 1982, the system marries a series of individual turbochargers that can be switched in and out while the engine is running to match the load. High brake mean effective pressures can thus be sustained even at low engine speeds. Among the merits cited for the sequential turbocharging system are excellent acceleration and substantially lower fuel consumption and smoke emission levels at low and medium power rates. Additionally, in twin-engine plants with common transmission and fixed pitch propellers or waterjets, the system can be run on one engine virtually up to its rated power. Quadruple MTU ZR 265 turbochargers are specified for the single-stage turbocharging system serving the 20V 8000 engine, Figure 30.18.

Both common rail fuel injection and sequential turbocharging systems contribute to a claimed specific fuel consumption of below 195 g/kWh across a broad power range, and less than 190 g/kWh at the most economical points. These figures are achieved while maintaining NOx emissions within IMO limits. Smoke indices are reportedly <0.4 Bosch throughout the entire engine performance map and <0.2 Bosch at full load.

Electronics are increasingly exploited by engine designers, MTU addressing the full potential from the conceptual stage of the Series 8000. The resulting MDEC (MTU Diesel Engine Control) engine management system allows all functions and parameters to be precisely monitored and controlled for optimum effect. The system also incorporates the basic functions for trend analysis and diagnostics, and the facility for integrating the engine in a comprehensive control and monitoring package for the entire propulsion system as well as other ship systems. MTU offers these advanced options in the shape of its RCS (remote control system) and MCS (monitoring and control system) products.

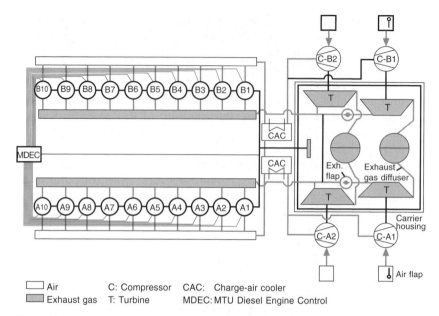

☐ Air   C: Compressor   CAC: Charge-air cooler
▨ Exhaust gas T: Turbine   MDEC: MTU Diesel Engine Control

Figure 30.18 Thermodynamic model of the MTU 20V 8000 M70 engine and its sequential turbocharging system

Structurally, the Series 8000 is the first MTU engine to feature modular 'power units', which can be quickly removed and replaced, Figure 30.19. Each integrated module comprises:

- A cylinder head, including valve gear, fuel injector, exhaust T-piece and coolant line segment.
- Cylinder liner.
- Piston and connecting rod.
- Spacer (for channelling coolant to the upper section of the liner and to the coolant pipes).

Each power unit is attached to the crankcase by four hydraulically-tightened stud bolts which have to withstand the high ignition pressures but do not perform a sealing function. A good seal is effected by securing the cylinder liner directly to the cylinder head from below with 24 bolts. By separating the functions of sealing and retention, MTU explains, wider design scope was created for optimizing the channels, valve gear and cylinder head cooling.

High design rigidity and endurance strength were sought from the transverse-flow nodular grey cast iron cylinder head. Its exhaust ports and particularly the inlet ports are said to yield excellent flow coefficients which establish the fundamental conditions for fuel economy and low

Figure 30.19 All the cylinder components of the MTU Series 8000 engine are combined in a functional module termed a 'power unit' for ease of withdrawal and replacement

emission levels. Circularity of the cylinder liner when the engine is running is sustained by the 24 evenly-spaced bolts around its perimeter and by which it is attached to the cylinder head. That circularity underwrites highly efficient sealing by the piston rings and, combined with a special liner honing process developed by MTU, promotes smooth running and durability. The high sealing efficiency of the rings and liners is also the basis for the 'exceptionally low blow-by levels' demonstrated by the Series 8000 engine, the designer reports. This significantly extends the life of the lubricating oil by reducing the amount of dirt escaping past the rings.

A robust nodular cast iron crankcase extending well below the centreline of the crankshaft is described as particularly rigid, the structure offering large inspection ports to ease checking and maintenance. Integrated in the crankcase are large lateral charge air ducts to the cylinders, the central bore for the camshaft operating the valve gear, and a centrally located main oil gallery. The cylinder liner is cooled only in its upper region and the crankcase is therefore completely coolant free.

Wide, cross-bolted bearing caps retain the crankshaft with its bolt-on counterweights. Both big end and crankshaft main bearings are of the micro-groove type, and the latter can be replaced in-situ. Other key elements of the running gear are fully machined connecting rods with diagonally-split big ends, and composite pistons comprising a forged steel skirt (for the first time in an MTU engine) bolted to a steel crown carrying two compression rings and an oil control ring. The pistons are cooled by nozzles spraying cooling oil.

Exceptional engine rigidity imparted by the power unit concept, and consequently low-wear characteristics, were expected to help keep lube oil consumption at a constantly low level throughout the period between major overhauls. Confidence in a promised overhaul interval of up to 24 000 running hours—depending on the engine's load profile and power rating—was supported by trials during which the test engines were subjected to an extremely severe alternating load sequence as well as continuous operation at full and overload powers.

Minimizing the number and complexity of necessary interfaces, the pumps for engine coolant, raw water, fuel (high and low pressure) and lube oil, as well as the associated filters and oil coolers, are mounted on the engine. The electronic MDEC engine management unit is also arranged here. Locating all these components together as a 'service block' on the free end of the engine fosters accessibility and simplifies integration of the engine with the ship's systems. Installation work is further reduced by an engine mounting concept proven with the Series 1163 engine in many high speed ferries, the torsionally resilient, offset-accommodating shaft couplings and a special arrangement of the air intake and exhaust connections.

Engine starting is effected by an air starter, the sequence incorporating an initial slow-turn phase which allows engine functioning to be quickly checked without having to open the decompression valves. This is followed immediately by the main starting sequence.

Contributing to trouble-free, low maintenance operation are: an automatic oil filter and two oil centrifuges in the secondary oil circuit (an indicator filter is fitted in the return line of the automatic filter for monitoring purposes); fuel filter cartridges with paper elements that can be replaced while the engine is running; and an engine coolant filter. The circulation systems are also fitted with running-in filters which trap any dirt that has found its way into the engine fluids during the installation process.

Adopted from previous MTU engines is the TE cooling system. Raw water never comes into contact with the engine components and the inlet temperature of the coolant at the engine can be controlled to suit the operating conditions. At low ambient temperatures and/or low engine loads, for example, the recooler is bypassed so that the coolant remains at a high temperature and heats up the combustion air. 'White smoke' emissions produced by cold engines can thus be eliminated. At full power, however, the system provides highly efficient cooling and creates the optimum conditions for combustion, both in terms of fuel economy and NOx emission levels, according to MTU.

MTU 20V 8000 engine data

Bore	265 mm
Stroke	315 mm
Cylinders	V20
Output/cylinder	410 kW
Power, max	9000 kW
Power, continuous	8200 kW
Rated speed	1150 rev/min
Idle speed	380 rev/min
Mean effective pressure	24.6 bar
Mean piston speed	12.1 m/s
Weight-power ratio	<5.3 kg/kW
Weight	43 tonnes
Specific fuel consumption	190–195 g/kWh
Dimensions (L × W × H)	7.4 m × 1.9 m × 3.3 m

MTU/DDC designs

An alliance formed by MTU in 1994 with Detroit Diesel Corporation (DDC) resulted in the creation of two advanced high speed designs for joint marketing by the German/US partners, the Series 2000 and Series 4000 engines, which were launched in 1996. MTU was responsible for the basic design development of both engines and the variants for propulsion applications.

Series 2000

The Series 2000 design, with a 130 mm bore/150 mm stroke, is based on the Mercedes-Benz 500 commercial vehicle engine (Figure 30.20). Models with V90-degree 8, 12 and 16 cylinders cover a propulsion band from 400 kW to 1343 kW, while genset drive applications start from 270 kW. The most powerful version, intended for yachts, has a maximum speed of 2300 rev/min. A single-stage turbocharging system is specified as standard but, when special requirements regarding power band width and acceleration characteristics are dictated, Series 2000 engines can benefit from MTU's sequential turbocharging system. Fuel injection, adapted from the commercial vehicle engine, is a solenoid valve-controlled pump/fuel line/injector system, each cylinder having its own separate injection pump.

Figure 30.20 MTU V16-cylinder Series 2000 engine

Marine versions are fitted with MTU's TE twin-circuit cooling system (detailed above in the Series 396 engine section) which fosters an optimum temperature for all operating conditions: intake air can be heated at idling speeds or at partial throttle and cooled at full throttle.

Series 4000

A number of notable design and performance features were highlighted by MTU for the 165 mm bore/190 mm stroke Series 4000 engine whose designers focused on high reliability and ease of maintenance without compromising compactness (Figures 30.21 and 30.22). A propulsion power band from around 840 kW to 2720 kW at 2100 rev/min is covered by the V90-degree 8-, 12- and 16-cylinder versions.

Figure 30.21 MTU Series 4000 engine in V16-cylinder form

Testing confirmed fuel economy figures that reportedly set a new standard for compact engines in the design's performance class and speed range. A specific fuel consumption of 194 g/kWh was considered exceptional for an engine with a power-to-weight ratio of between 2.7 and 3.5 kg/kW.

A key contribution, along with a high peak firing pressure potential, is made by the common rail fuel injection system—an innovation in this engine category—whose development was carried out in conjunction with MTU's specialist subsidiary company L'Orange: see also Chapter 8/Fuel Injection.

The system, which allows infinite adjustment of fuel injection timing, volume and pressure, embraces a high pressure pump, a pressure accumulator, injectors and an electronic control unit. The tasks of pressure generation and fuel proportioning are assigned to different components: pressure generation is the job of the high pressure pump while metering of the fuel relative to time is the job of the injectors.

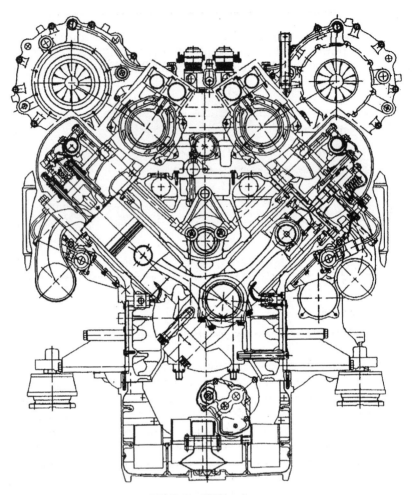

Figure 30.22 Cross-section of MTU Series 4000 engine

The flexibility of the common rail system across the engine power band enables it to deliver the same injection pressure (around 1200 bar) at all engine speeds from full rated speed down to idling (Figure 30.23). The high pressure fuel pump and electronically controlled injectors are fully integrated in the electronic control system. Peak pressures with the common rail system are around 20 per cent lower than with conventional systems, reducing stress on the high pressure components. The pump is also simpler and its plunger has no helices. Since the pressure-relief requirement of conventional systems is eliminated the mechanical effort is reduced, further enhancing fuel economy.

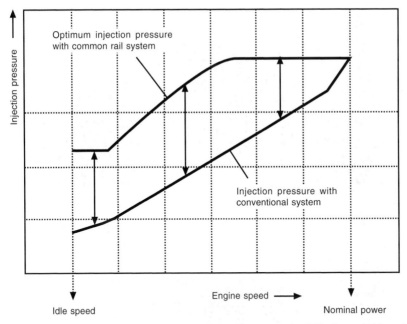

Figure 30.23 Performance of common rail fuel injection system of MTU's Series 4000 engine

One fuel rail is provided for the injectors and their respective solenoid valves on each bank of cylinders. The fuel pressure is generated by the high pressure pump driven via a gear train at the end of the crankcase. The electronic system controls the amount of fuel delivered to the injectors by means of the solenoid valves, while the injection pressure is optimized according to the engine power demand. Separate fuel injection pumps for each cylinder are thus eliminated, and hence the need for a complicated drive system for traditional pumps running off the camshafts. A reduced load on the camshaft and gearing system is therefore realized, and the removal of the mounting holes for separate injection pumps enhances the rigidity of the crankcase.

Maximum strength and optimized rigidity were sought from the crankcase, with integration of the main lubrication and coolant circulation channels reducing the number of separate components (Figure 30.24). The crankshaft, machined all over and with bolted-on counterweights, is designed to withstand maximum combustion pressures of up to 200 bar. Wear-resistant sleeve bearings contribute to longer service intervals. The cylinder head design, with integrated coolant channels and stiffened base, also addresses a 200 bar pressure. Provision is made for two inlet and two exhaust valves, and a fuel injector located in the centre of the head (Figure 30.25).

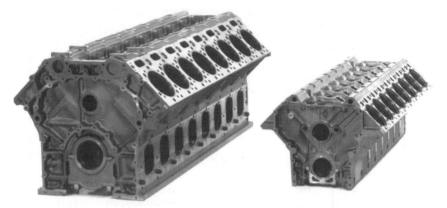

Figure 30.24 The crankcases of MTU's Series 2000 (right) and Series 4000 engines

Figure 30.25 The four-valve cylinder heads of MTU's Series 2000 engine (right) and Series 4000 engine (left) with central fuel injector

A piston comprising an aluminium skirt and a bolted-on steel crown is provided with chrome-ceramic coated rings which, in combination with the plateau-honed cylinder liner, foster low lubricating oil consumption and an extended service life. An optimized combustion chamber shape and the bowl of the composite piston promote low fuel consumption and low emission levels.

Exhaust and turbocharging systems on the Series 4000 engine benefit from established MTU practice. A triple-walled exhaust system, with an outer water-cooled aluminium casing, ensures that the surface temperatures do not exceed permissible levels at any point while also securing gas tightness. Heat dissipation via the cooling system is reduced. The exhaust pipework is positioned centrally between the V-cylinder banks. The turbochargers are mounted in a water-cooled housing. A sequential turbocharging system is exploited to deliver high torque at low engine speeds. A wide power band at high fuel economy is also claimed, along with excellent acceleration capabilities without black smoke emission.

A split-circuit cooling system for the engine and intake air complements the sequential turbocharging system. It acts as an 'intelligent' cooling system to maintain engine coolant, lubricating oil and intake air at an optimum temperature for all operating conditions. When the engine is idling or running at low load the temperature of the intake air is raised in order to ensure smooth and complete combustion without generating white smoke emissions.

On the front end of the engine is an integral service block providing ease of inspection and maintenance of the seawater cooler, oil cooler, oil filter, oil centrifuge and fuel filter. The turbochargers and charge air coolers are located at the flywheel end.

Access openings in the crankcase are said to be large enough to allow all running gear servicing to be performed without removing the engine, even under restricted machinery space conditions. The application-tailored service block integrated in the auxiliary power take-off end also simplifies routine maintenance tasks. A variety of ancillary mounting options enable the engine to be matched to specific customer requirements.

Early indication of necessary maintenance, based on the actual duty profile, is provided by an electronic engine management system designed to foster reduced downtime periods and lower servicing costs. A time-between-major-overhauls of 18 000 hours was anticipated by MTU.

NIIGATA

The FX series from Niigata Engineering was completed in 1996 with a 205 mm bore model to complement the established 165 mm and 260 mm high speed engines in a programme developed for fast commercial and military vessel propulsion. The resulting 16FX, 20FX and 26FX models offer maximum continuous outputs from 1000 kW to 7200 kW in commercial service, with slightly higher ratings available for naval propulsion. The 16FX engine is produced in in-line eight, V12- and V16-cylinder versions running at 1950 rev/min, the V20FX in V12 and V16 versions running at 1650 rev/min, and the V26FX (Figure 30.26) in V12-, 16- and 18-cylinder versions running at 1300 rev/min.

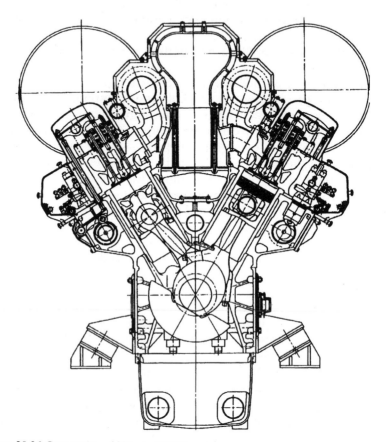

Figure 30.26 Cross-section of Niigata V26FX engine

Compact and highly rigid engines were sought by the Japanese designer from a monobloc structure fabricated from nodular cast iron, with the extensive use of light alloy elements contributing to a modest overall weight. Engine width was minimized by adopting a V60-degree cylinder bank and overall length reduced by minimizing the distance between cylinders while maintaining the required bearing width.

A one-piece nodular cast iron piston was specified for the V16FX engine, but the higher mechanical and thermal stresses imposed by the cylinder pressure parameters of the V20FX 'Blue Arrow' engine influenced the selection of a component with a steel crown and aluminium skirt for the larger design. A bore cooling-type oil gallery structure on the crown reduces the temperature of the piston at higher mean effective pressures. Higher cylinder pressures also foster increased piston blow-by, calling for a high sealing capability and wear resistance from the piston rings. A chromium-ceramic coated top ring contributes to these properties, while an anti-polishing ring in the top of the cylinder liner yields a lower lube oil consumption rate.

An output ceiling of 7200 kW at 1300 rev/min from the Niigata high speed engine portfolio is provided by the 18-cylinder version of the 260 mm bore V26FX design. A monobloc frame of high tensile strength ductile cast iron and the use of light alloy parts where appropriate contribute to rigidity and light weight from a compact envelope.

Niigata 16V20FX engine data

Bore	205 mm
Stroke	220 mm
Cylinders	V16
Output, mcr	4000 kW
Speed	1650 rev/min
Mean piston speed	12.1 m/s
Specific fuel consumption	210 g/kWh
Weight	12 100 kg
Weight/power ratio	3.15 kg/kW

Miller system for 32FX

The potential of the 32FX design was demonstrated by a development project to create a lightweight, high performance 320 mm bore medium speed engine series with a weight/power ratio of 4 kg/kW yielding outputs from around 8000 kW to 13 000 kW. Using the established 32CX engine as a basis, Niigata reportedly achieved a rating of 769 kW/cylinder at 1030 rev/min from the 32FX variant.

High output achieved by traditional development routes, Niigata explains, is accompanied by an increase in mechanical load attributable to the higher maximum pressure. Coping with the increased loading normally dictates the specification of thicker and hence heavier components. Thermal loads also rise, resulting in elevated NOx emissions. Reducing the compression ratio to maintain Pmax at a normal level, however, lowers efficiency and undermines overall performance and starting capability.

Given these constraints, Niigata opted for the Miller system to keep the Pmax and temperatures in the combustion chamber at a normal level, and reduce NOx emissions, while increasing the brake mean effective pressure. The system is designed to reduce charge air temperature and pressure before compression (thus lowering the combustion temperature) by subjecting the charge air in the cylinder to adiabatic expansion.

Miller systems come in two variants: an advanced closure timing, which closes the intake valve before bottom dead centre; and a delayed closure timing, which closes the intake valve after BDC. Based on an engine performance simulation analysis, Niigata selected the advanced closure timing route for the 6L32FX engine, adopting a closing timing of 10 degrees before BDC in contrast to the 35 degrees after BDC applied in the 6L32CX engine (Figure 30.27).

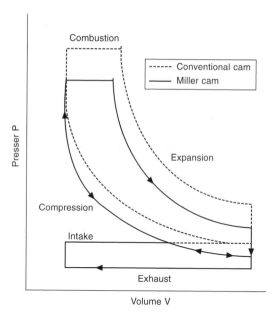

Figure 30.27 P-V diagram for Miller system (Niigata)

In the Miller system, with the intake valve closed before BDC at the normal boost pressure, the exhaust temperature rises as the intake air volume is reduced. This effect makes it necessary to increase the boost pressure to achieve the targeted output, which lowers the exhaust temperature. Securing a higher boost pressure than that obtainable with single-stage turbocharging called for the adoption of a two-stage system in which the low pressure and high pressure turbochargers are connected in series with two intervening air coolers (Figure 30.28).

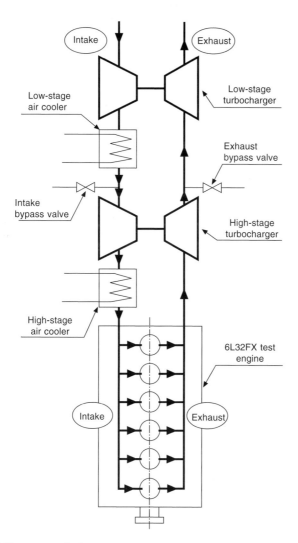

Figure 30.28 Two-stage turbocharging system (Niigata)

Applying the Miller system in conjunction with two-stage turbocharging—low pressure and high pressure turbochargers connected in series with two intervening coolers—secured the desired goal of a power rate of 38.2 (3.09 Mpa brake mean effective pressure × 12.4 m/s mean piston speed) at 110 per cent maximum load—claimed to be the highest of any engine in the circa-320 mm bore class. In addition, the NOx emission level with the Miller cam was 20 per cent lower than with a conventional cam, equating to 5.8 g/kWh compared with the IMO limit of 11.2 g/kWh at 1030 rev/min in E3 mode. Fuel consumption with a conventional cam was lower than with the Miller cam at low loads (which could anyway be improved by using a variable geometry cam) but it rose rapidly at higher loads and exceeded the Miller cam at a load of around 75 per cent.

Supporting the performance boost, Niigata also investigated new materials for key components to enhance engine reliability and reduce weight, notably: a ceramic-sprayed bimetal cylinder liner; tri-alloy intake valves and pushrods; and sodium-sealed and Waspaloy Inconel 718 exhaust valves. The ceramic-sprayed coating was applied to the sliding part of the liner surface to improve resistance to abrasion and seizure.

PAXMAN (MAN B&W)

Over 60 years experience in high speed diesel design was exploited by Paxman Diesels in creating the VP185 engine, launched in 1993 to join the UK-based company's established 160 mm bore Vega and 197 mm bore Valenta series (Figure 30.29), which respectively offered specific outputs of 107 kW/cylinder and 206 kW/cylinder. The new 185 mm bore design was introduced initially in V12-cylinder form with outputs up to 2610 kW and complemented in 1998 by a V18-cylinder version extending the power limit of the series to 4000 kW at a maximum speed of 1950 rev/min. Paxman became a member of Germany's MAN B&W Diesel group in mid-2000.

In designing the 12VP185 engine (Figure 30.30), Paxman sought improvements in compactness and lightness over its earlier models, goals dictating a smaller swept volume, slightly higher ratings and a review of piston speeds. The first parameter to be fixed was the stroke which was set at 196 mm; this, linked to a crankshaft speed of 1800 rev/min, gave a mean piston speed of 11.8 m/sec for continuous duty. An improved speed platform for marine applications was derived, the edge of which is bounded by a maximum speed of 1950 rev/min with a mean piston speed of 12.8 m/sec.

The cylinder bore was fixed at 185 mm which, coupled with a design maximum mean effective pressure of 25.3 bar, delivered a maximum power of 2610 kW at 1950 rev/min from the V12-cylinder configuration: a rating primarily targeting high speed military and commercial marine

Figure 30.29 Paxman V16-cylinder Valenta engine

markets, as well as megayacht propulsion. An unrestricted rating of 2180 kW at 1800 rev/min is quoted (at 45°C air and 32°C sea water conditions) for continuous marine duties (for example, fast ferry propulsion) where the engine's high torque capability over a broad speed range is considered especially attractive for hydrofoil service.

The 18-cylinder VP185 model offers a maximum output of 4000 kW at 1950 rev/min, a 3300 kW continuous power rating for fast ferries and patrol craft, and 3000 kW at 1770 rev/min for unrestricted marine duty.

A smaller bore and stroke than the Valenta design, combined with a switch from a 60-degree cylinder bank angle to a 90-degree configuration, achieved a layout which was short and with a width under 1.5 m while providing an adequate platform for the charge air system mounted above the engine. The 90-degree bank angle itself also lowered the profile of the charge air system to foster space efficiency (Figure 30.31).

A higher mean effective pressure normally leads to a lower compression ratio which in turn promotes poor starting characteristics,

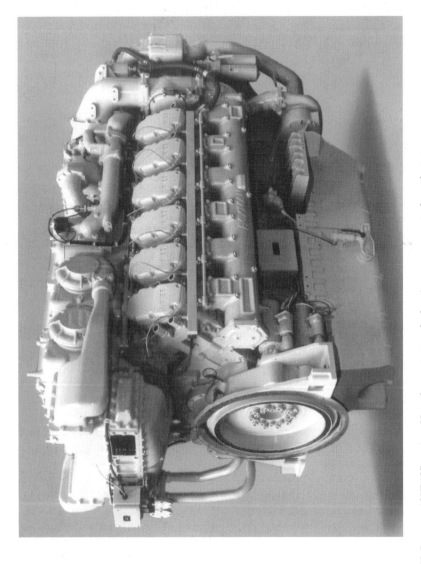

Figure 30.30 Paxman 12VP185 engine. Note the compact turbocharging arrangement along the top

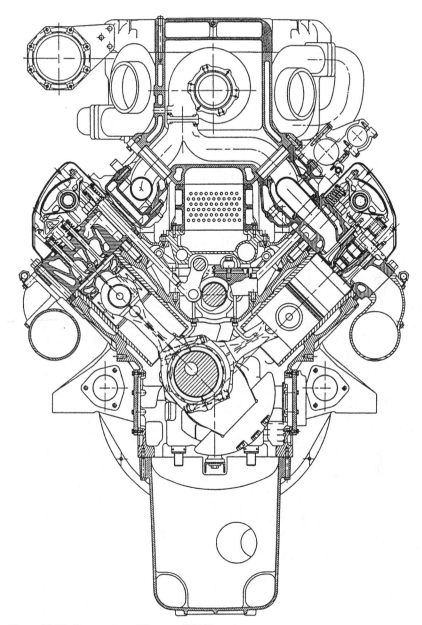

Figure 30.31 Cross-section of Paxman VP185 engine

reduced thermal efficiency and possible exposure to ignition delay damage. Some of these problems can be addressed by the use of inter-cylinder charge air transfer arrangements but Paxman decided to adopt a compression ratio in excess of 13:1. The resulting VP185 engine was claimed to be simple and easy to start under cold conditions, with good thermal efficiency and the potential to satisfy low NOx emission limits. The combination of high boost levels and compression ratios demands a robust construction but Paxman sought to avoid a heavy and cumbersome engine by securing robustness within a small and hence a stiff envelope.

The reciprocating assembly of the engine was designed for high strength and rigidity in handling a high maximum cylinder pressure and securing high reliability with a long lifetime. The backbone is a compact crankcase cast in high strength spheroidal graphite iron to provide a stiff and solid support for the underslung crankshaft. Crankcase doors along each side of the engine give access to the connecting rod large ends for *in situ* servicing and piston removal.

The crankshaft is a fully machined steel forging, fully nitrided to yield strength and durability. It is secured by main bearing caps which are drawn internally against the deep-fitting side faces of the crankcase by high tensile set screws and hydraulically tensioned main bearing studs. Such a configuration, the designer asserts, delivers good strength and stiffness characteristics to the bottom end of the crankcase and provides solid support for the crankshaft. A generous overlap between the crankpin and main bearing journals adds to stiffness and strength. Both main and big end bearings are of steel-backed aluminium–tin type with thrust washers controlling axial location. The crankshaft is provided with a viscous torsional damper totally enclosed within the gearcase.

Connecting rods of side-by-side design are forged in high tensile steel, fully machined and ferritic nitro-carburised. The large end is obliquely split to allow the rod to pass through the cylinder liner and the joint faces are serrated. The one-piece nodular cast iron pistons incorporate a large oil cooling galley above the ring belt, oil being supplied by accurately aligned standing jets mounted in the crankcase. The pistons run in wet cylinder liners made of centrifugally cast high grade iron.

The cast iron cylinder heads are designed to withstand the high firing pressures with an internal configuration ensuring maximum flame face stiffness and high cooling efficiency. They incorporate two exhaust valves, two inlet valves and a centrally mounted unit fuel injector (combined pump and injector) which is fixed with a single screw. The valves and injectors are actuated through a pushrod arrangement from a single large diameter camshaft mounted centrally in the vee of the engine. The stiffness of the actuating system reportedly achieves valve control similar to that of an overhead camshaft, while the specially designed unit injectors yield very high rates of fuel injection with clean cut-off, contributing to low NOx generation and good fuel efficiency. The fuel cam design was based on an optimized width for long life and a large base circle, coupled with high rates of lift to secure the key injection characteristics.

A maximum injection pressure of 1400 bar promotes good combustion characteristics with low emissions and fuel consumption. The unit injector is rack controlled, the rack itself being controlled by a linkage operated from twin shafts in the centre vee of the engine. The shafts in turn are coupled to the governor via an overspeed protection assembly which closes the fuel racks independently of the governor in the event of the engine overspeeding.

A safety merit of the unit injector (which combines fuel pump and injector in one unit) is that it eliminates the need for high pressure fuel pipes. Low pressure fuel oil is delivered to the injectors by a gear-driven lift pump mounted on the main gearcase. The fuel is fed to a sub-assembly—comprising cooler, filter (duplex optional), reservoir and solenoid-operated shut-off valve—mounted high at the free end of the engine for convenient access.

A single camshaft reflects the policy adopted by the designers of minimizing the number of components. This, allied to the 90-degree cylinder bank angle, allowed the camshaft gear to mesh directly with the crankshaft gear and so eliminate the need for idlers in a critical area and reduce the component count even further.

Charge air is delivered by a 'valve-less' two-stage turbocharging system designed for simplicity and ease of maintenance. Preliminary explorations of single-stage high pressure ratio turbocharging showed poor low speed torque characteristics due to surge limitation, along with poor acceleration and load acceptance. Sequential turbocharging was considered but the necessary valve gear and associated control system conflicted with the primary aim of keeping the engine as simple as possible, Paxman explains. Another factor was that few turbocharger designs in the smaller size range were capable of meeting the pressure ratio required.

A two-stage turbocharging arrangement with intercooling and aftercooling was finally adopted for the VP185 engine, based on six Schwitzer automotive-type turbochargers with broad and stable operating characteristics: four turbochargers provide low pressure charge air and two provide the high pressure air.

Inlet air compressed in the low pressure turbochargers is fed through a raw-water intercooler to the compressors of the high pressure turbochargers; high pressure air is then passed through the jacket water aftercooler into the air manifolds on either side of the engine.

The high pressure stage exploits a pulsed exhaust system to give good low-end performance without recourse to complex valve systems associated with sequentially turbocharged configurations. The turbocharging system's high air/fuel ratios, coupled with high pressure/high rate fuel injection, address low NOx emission requirements. A highly responsive performance with good engine torque characteristics is reported throughout the speed range, Figure 30.32.

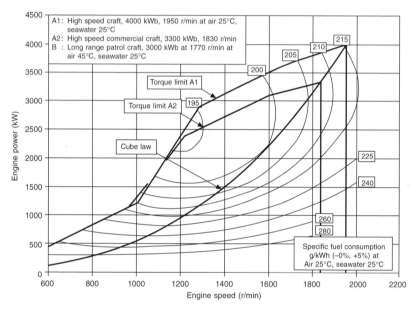

Figure 30.32 Performance map for Paxman 18VP185 engine

All six turbochargers are mounted in the walls of a water-cooled gas-tight casing, the turbine sides arranged on the inside and the compressors on the outside. The rotating assembly and compressor casing of each turbocharger form a cartridge which can be replaced quickly without the need for lifting gear and without disturbing the rest of the engine. The exhaust manifolds feature sliding joints and, like the turbochargers, are housed in water-cooled gas-tight aluminium casings; together with the turbocharger enclosure, they form a single, cool gas-tight shell around the hot parts of the engine. The casings foster low radiant and convective heat losses from the engine.

The gear train is mounted at the engine's free end and the camshaft gearwheel meshes directly with the crankshaft gearwheel. Auxiliary drives are provided for the externally mounted lubricating oil and jacket water pumps, the governor, overspeed governor and fuel lift pump; further auxiliary power take-off capability is incorporated for different applications. The PTO at the free end will accept either gear-driven or belt-driven pumps and alternators. The VP185 is supplied as a complete assembly with engine-mounted jacket water and lubricating oil heat exchangers to simplify installation procedures. A choice of air, electric or hydraulic starting systems is available.

Marine engine versions are controlled by a Regulateurs Europa digital Viking 2200 governor and 2231 actuator with the option of ball head back-up.

Multiple Schwitzer turbochargers in a passive two-stage configuration with intercooling and aftercooling were retained for the V18-cylinder model, albeit with an increased number of sets. Three high pressure turbochargers each expand into a pair of similar low pressure turbochargers, giving a total of nine turbochargers. Unlike the 12VP185 engine, the turbochargers are packaged in groups of three (one HP-two LP), each group having its own water-cooled gas-tight housing. The hot turbine volutes are fixed to the inside of each housing, with the rest of the turbocharger, the 'cartridge', plugged in from the outside: an arrangement which enables the turbochargers to be replaced quickly. The hot exhaust manifold pipes are similarly housed in their own water-cooled gas-tight casings, so eliminating unwanted heat and gas around the engine (an important safety aspect).

SEMT-PIELSTICK

The Paris-based designer SEMT-Pielstick fields three long-established high speed programmes, the PA4, PA5 and PA6 series. The PA4 is produced in 185 mm and 200 mm bore versions (both having a stroke of 210 mm) and features a variable geometry (VG) pre-combustion chamber; and the PA5 is a 255 mm bore/270 mm stroke design.

Development has benefited the 280 mm bore PA6 design, introduced in 1971 with a stroke of 290 mm and a maximum continuous rating of 258 kW/cylinder at 1000 rev/min. The output was raised to 295 kW/cylinder in 1974 and to 315 kW/cylinder in 1980. Extensive service experience in naval vessels moulded progressive refinements over the years, including the creation of a BTC version in 1980 which yielded 405 kW/cylinder at 1050 rev/min through the adoption of a reduced compression ratio together with two-stage turbocharging. This PA6 BTC model was released in 1985 with a higher output of 445 kW/cylinder. The series was extended in 1983 by a longer stroke

(350 mm), slower speed PA6 CL variant with a rating of 295 kW/cylinder at 750 rev/min.

A sequential turbocharging system (STC) was introduced to the 290 mm stroke engine in 1989, this PA6 STC model developing 324 kW/cylinder at 1050 rev/min. Performance was enhanced in a 330 mm stroke B-version from 1994, an STC system and a cylinder head with improved air and gas flow rates contributing to a nominal maximum continuous rating of 405 kW/cylinder at 1050 rev/min. A maximum sprint rating of 445 kW/cylinder at 1084 rev/min equates to an output of 8910 kW from the V20-cylinder PA6B STC engine which targets high performance vessels.

At the nominal mcr level the V12-, 16- and 20-cylinder models, respectively offer 4860 kW, 6480 kW and 8100 kW for fast ferry propulsion, and are released for sustained operation at these ratings. The weight/power ratio of the 20PA6B STC model at 8100 kW is 5.2 kg/kW (Figure 30.33).

Figure 30.33 SEMT-Pielstick V20-cylinder PA6B STC engine with sequential turbocharging system. A rating of 8100 kW is offered for fast ferry propulsion

PA6B STC design

High rigidity without compromising overall engine weight is achieved by a stiff one-piece crankcase (Figure 30.34) of nodular cast iron specially treated for shock resistance, with transverse bolt connections between both crankcase sides through the underslung-type bearing caps. Integrated longitudinal steel piping supplies lubricating oil to the main bearings.

Figure 30.34 SEMT-Pielstick PA6B STC crankcase

Large dimension main journals yield a particularly large bearing surface and conservative pressures which underwrite prolonged bearing life, as do the large surface area connecting rod bearing shells. The alloy steel one-piece forged crankshaft (Figure 30.35) has high frequency hardened crankpins and journals, and is bored to feed lubricating oil to the connecting rods.

Figure 30.35 SEMT-Pielstick PA6B STC crankshaft

A nodular cast iron cylinder head was configured to achieve improved air and gas flow rates for the PA6B STC version. The composite piston (steel crown and aluminium skirt) is fitted with five rings. The forged steel connecting rod is fitted with large surface bearing shells to yield high durability under high firing pressures. Special low wear characteristics were sought from the centrifuged cast iron material specified for the cylinder liner (Figure 30.36). The camshaft is formed from several sections for ease of dismantling.

Figure 30.36 SEMT-Pielstick PA6B STC cylinder liner

Operational flexibility and power output are fostered by the single-stage sequential turbocharging system which is based on two turbochargers and delivers a large combustion air excess at partial loads. Supercharging is effected by one turbocharger for engine loads up to 50 per cent of the nominal power, its effort boosted by the second identical turbocharger at higher loads. Switching from single to twin turbocharger mode is performed automatically by opening two flap valves. Engine performance at prolonged low load is improved with respect to fuel consumption, smoke emission, fouling resistance and transient performance. Additionally, engine utilization is expanded towards the high torque/low rev/min area.

Developing the high performance derivative

Refining an established engine for high speed vessel propulsion duty must focus on achieving a lighter and more powerful package within a more compact envelope, while also enhancing operating flexibility. The resulting PA6B STC design was released with a nominal maximum continuous rating of 405 kW/cylinder at 1050 rev/min, and a sprint rating of 445 kW/cylinder at 1084 rev/min (Figure 30.37).

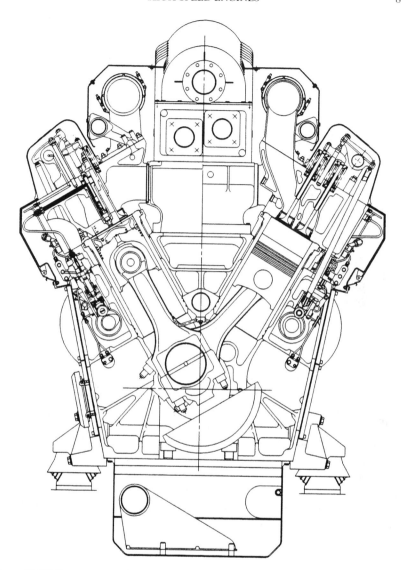

Figure 30.37 Cross-section of SEMT-Pielstick PA6B STC engine

SEMT-Pielstick's development goals aimed to:

- Increase the power rating by 25 per cent.
- Use an operating field allowing operation following a double propeller law.
- Maintain, and if possible reduce, the specific fuel consumption.

- Reduce the weight/power ratio and enhance engine compactness.
- Reduce installation costs.
- Retain the maximum number of existing components.

Power can be raised in two ways: by increasing the brake mean effective pressure and by increasing the piston speed. Increasing piston speed can be effected by increasing the rev/min and by lengthening the stroke. Increasing the rev/min was not attractive because of a need to maintain an acceptable synchronized speed for land power plant applications but also because of these drawbacks: increased specific fuel consumption; higher wear rate; and higher noise levels. Raising piston speed by lengthening the stroke is considered preferable because it reduces specific fuel consumption and fosters optimum performance at starting and low load (the compression ratio is secured without resorting to a combustion chamber that is too flat). This route was therefore selected but with the piston speed limited to 11.5 m/s (representing an increase of 14 per cent with a stroke of 330 mm). To reach the required power rating, the brake mean effective pressure had to be increased by 10 per cent; this was achieved by adopting high performance turbochargers. Attaining the desired specific fuel consumption called for the maximum cylinder pressure to be raised to 160 bar.

Coping with these new parameters dictated redesigning some key components. The existing cylinder head, for example, was incompatible with the targeted specific fuel consumption level. This was partly because its mechanical strength was insufficient in relation to the necessary peak combustion pressure, and partly because the pressure loss through the inlet and exhaust ports would be unacceptable in relation to the intervening gas flow dictated by the increased power. A new cylinder head was therefore designed with a reinforced bore-cooled fire plate and incorporating inlet and exhaust ports with large dimensions. In addition, cooled seats were specified for the inlet valves as well as the exhaust valves to maximize seat reliability and avoid risks of burning.

A redesigned connecting rod was also necessary to address the increased peak compression pressures and the inertia efforts linked to the lengthening of the stroke. The bevel cut design of the original component was abandoned in favour of a straight-cut rod to avoid weak points, such as the bevel cut serrations and the threading anchorage in the shank. Piston cooling by jet replaced the traditional oil supply through the connecting rod from the crankpin, simplifying machining and allowing the bearing shell grooves to be eliminated and yield these benefits: increasing capacity of the bearing shells; stopping cam wear of the crankpin (differential wearing of the pins between the side areas of the plain bearing shell and the central area; that is, the area including the groove); and cutting out the risk of cavitation erosion on the bearing shells at the end of the grooves. The new connecting rod is 10 per cent lighter than its bevel cut forerunner. The weight reduction, along with an improved bearing shell capacity, partly compensates for the increased inertia efforts.

Temperature measurements on the jet-cooled piston head indicated similar levels to those of the original piston, and even lower in some areas. Critical points of the piston were modified to cope with the higher peak combustion pressure: a spherical shape was given to the support spot face used for the piston head/skirt tightening spacers; and the radius under the skirt vault was increased.

Crankshaft dimensions were modified to target the same reliability from the component as before, despite the increased stroke and peak combustion pressure. Finite element analyses of the crankwebs and hydrodynamic calculations led to an increase in the journal diameter from 230 mm to 250 mm, and in the crankpin diameter from 210 mm to 230 mm.

High performance turbochargers were necessary to secure the brake mean effective pressure increase with the required efficiency, a model from MAN B&W's then new NA series—the NA 34S—being selected to meet the performance and compactness parameters. Sequential turbocharging was applied, based on the principle of reducing the number of turbochargers in operation as the engine speed and load fall. The speed of the turbochargers still operating consequently rises and significantly larger quantities of air are thus delivered to the engine.

A simple system was adopted using only two turbochargers, one being switched off at below approximately 50 per cent of the nominal engine power rating. This is effected by closing two flap valves located at the compressor outlet and at the turbine inlet of one of the turbochargers. The designer cites the following benefits from the PA6 engine's sequential turbocharging (STC) system:

- High torque and power ability at reduced engine speed.
- A gain in fuel economy at low and part loads.
- Capability to run the engine at very low loads for extended periods with minimal fouling (the light deposits can be cleaned out by running for half an hour at 50 per cent load).
- Invisible smoke emissions over a wide working range.
- Reduced exhaust temperature.
- Lower thermal stresses in the combustion chamber components at part loads.

A higher output rating naturally reduced the engine's weight/power ratio but other measures were pursued to trim overall weight. The scope for using aluminium alloy was explored for all components where operating stresses (particularly thermal) and class rules allowed, leading to the engine supports, turbocharger support, air manifolds (after the air cooler), lube oil and water cooler support, and lube oil filter support being designed in cast aluminium. Studies assessed other components for which aluminium could not be considered, either to use alternative materials with higher mechanical properties and so reduce thickness, or simply to optimize existing shapes and thicknesses. As an example, specifying high yield point steel sheet for the manifolds connecting the turbocharger to the air cooler allowed a reduction in thickness from 10 mm to 4 mm. The lube oil sump plate was also modified by reducing the material thickness.

A 10 per cent reduction in the original engine weight, along with the increased power output, contributed to a weight/power ratio of 4.8 kg/kW, including all ancillaries. In parallel with the weight trimming studies, SEMT-Pielstick focused on reducing the overall dimensions of both engine and ancillaries.

A key element here is the combi-cooler, integrating one lube oil and one freshwater plate cooler circulated by a common seawater system (Figure 30.38). The combi-cooler's support is used as the rear plate of the cooler and includes as-cast part of the connections to the high temperature freshwater and lube oil systems. Its front plate incorporates as-cast the connections to the low temperature freshwater

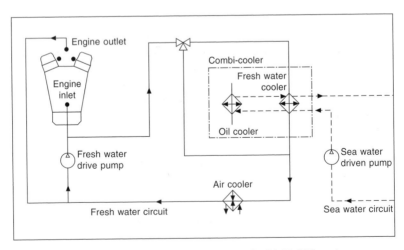

Figure 30.38 Fresh and seawater circuits integrated on the PA6B STC engine

loop and the water thermostatic valve. The main self-cleaning lube oil filter is incorporated axially in a cast support located under the combi-cooler, and includes the lube oil thermostatic valve as well as the centrifugal oil filters.

Such solutions fostered compactness, a simple pipeless configuration and good access to the main subjects of maintenance. Integration of the ancillaries on the engine further eases shipboard installation procedures. All pumps (water, oil, fuel make-up), as before, are driven by the engine upon which is also mounted the fuel filter.

A potential for burning an intermediate fuel oil grade such as IF30 was addressed in the development programme, and reflected in the specification of cooled valve seats and exhaust valve rotators. The 75°C temperature necessary to reach an adequate viscosity for its injection can be derived by taking heat from the engine's high temperature freshwater system. In such installations, however, the engine is derated to 360 kW/cylinder and the time-between-overhauls is reduced.

WÄRTSILÄ

The medium speed specialist Wärtsilä added high speed designs to its four-stroke engine portfolio with the acquisition in 1989 of SACM Diesel of France. The Finnish parent group saw the market potential for a new generation engine blending the best features of high and medium speed designs for continuous duty applications, resulting in the 1994 launch of the Wärtsilä 200 series (Figures 30.39 and 30.40).

The 200 mm bore/240 mm stroke engine is produced by Wärtsilä France for propulsion and genset drive applications, with an output band from 2100 to 3600 kW at 1200 or 1500 rev/min covered by V12, 16- and 18-cylinder models. High reliability in continuous duty (defined as 24 hours a day operation with an annual running period of over 6000 hours) was sought from an engine structure and main components designed for a maximum cylinder pressure of 200 bar. Wärtsilä's medium speed engine technology was exploited to achieve a high power density, low emissions and fuel consumption, and ease of maintenance.

A multi-functional connection piece or 'multi-duct' located between the cylinder head, engine block and exhaust manifold has the following duties:

- Combustion air transfer from charge air receiver to cylinder head.
- Introduction of an initial swirl to the inlet air for optimal part load combustion.

Figure 30.39 Wärtsilä W200 engine

- Exhaust gas transfer to the exhaust system.
- Cooling water transfer from the cylinder head to the return channel in the engine block.
- Insulation/cooling of the exhaust transfer port.
- Support for the exhaust manifold and insulation.
- Inclined face towards the cylinder head facilitates easy removal/ remounting of the cylinder head. The exhaust gas piping and insulation box, supported by the multi-duct, stay in place when the head is removed.

The nodular cast iron engine block (Figure 30.41) is designed for maximum stiffness with a V60-degree configuration optimizing balancing and hence limiting vibration. Bolted supports dictate only four to six fixing points, and provision is made for elastic mounting. The locations of the camshaft drive (integrated in the flywheel end of the block) and the air receiver channel (in the middle of the vee-bank and also integral with the block) reflect solutions adopted in Wärtsilä medium speed engines. Arranging the oil lubrication distribution

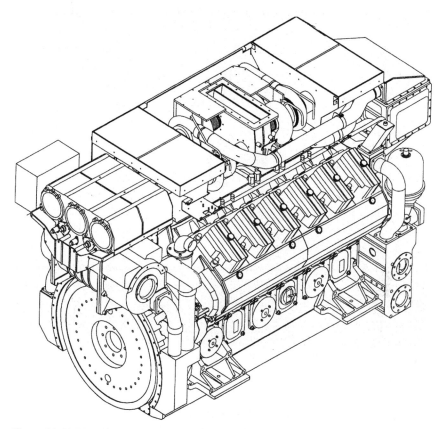

Figure 30.40 Wärtsilä W200 engine. The original automotive-type turbochargers shown here were later replaced by ABB or Holset models

through a channel in the middle of the vee, however, was a new idea. The areas of the block that support the camshafts and fuel injection pumps are designed for high strength to accept the forces created by the high injection pressures. Large openings on both sides of the block facilitate access for inspection and maintenance.

The cylinder unit also benefited from solutions validated in other Wärtsilä designs, notably: a plateau-honed cast iron cylinder liner incorporating an anti-polishing ring at the top to eliminate bore polishing; a composite piston (steel crown and aluminium alloy skirt) with a three-ring pack (two hard chromium-plated compression rings and one spring-loaded oil scraper ring) and optimized cooling; and Wärtsilä's patented piston skirt lubrication system. The piston is lubricated by an oil jet, which also supplies oil to the cooling gallery in the piston top.

Figure 30.41 The nodular cast iron block of the Wärtsilä W200 engine incorporates fluid channels

The forged high tensile steel connecting rods are fully machined and balanced for low vibration running, and designed to facilitate oil lubrication to the small-end bearings and pistons. The big end has a diagonal split with serration teeth. Two stud screws are hydraulically tensioned and can be accessed through the inspection doors at the engine sides. An off-set split makes its possible to overhaul the piston and connecting rod assembly through the cylinder liners.

The strong 220 mm diameter forged high tensile steel monobloc crankshaft (Figure 30.42) has gas-nitrided surfaces for added safety and a high degree of balancing from two bolted counterweights per crank throw. The generous diameters of both crankpin and journal achieve a large bearing surface while allowing a reduced cylinder spacing ratio (1.5 times the bore) to minimize engine length and weight. The crankshaft is fitted with counterweights and fully balanced, and the oil seal can be changed without removing the crankshaft. The running surfaces of the crankpin and main journal are hardened, and the connecting rod big end bearings and main bearings are of the CuPb type.

Figure 30.42 The Wärtsilä W200 engine crankshaft has gas-nitrided surfaces

Supporting the loads generated by the high injection pressure, the well-dimensioned camshafts are formed from modular sections bolted together, each section serving two or three cylinders; the sections can be removed axially from the engine block. The cams are surface hardened for wear resistance. Located on both sides of the engine block, the camshafts are driven by timing gearwheels arranged inside the engine block at the flywheel end.

Among the measures designed to ease inspection and maintenance are: cylinder heads hydraulically tightened with four stud screws on the cylinder block; and connecting rod big ends and main bearing caps fastened by two hydraulically tightened studs.

Fuel injection starts slightly before TDC so that combustion takes place during the beginning of the expansion phase in lower temperatures. The combination of a high compression ratio (16:1) and late fuel injection fosters low NOx generation without raising fuel consumption since the engine's injection period at 1500 rev/min is sufficiently short. Applying this low-NOx principle (see Chapter 27) requires the engine to be able to inject fuel late in the cycle and over a short duration without undermining performance: high pressure capacity injection equipment is therefore dictated. The engine is released for operation on marine diesel oil (ISO 8217, F-DMX to F-DMB).

Individual fuel injection pumps are integrated in the same cast multi-housing (Figure 30.43) as the inlet and exhaust valve tappets

Figure 30.43 The fuel injection pump of Wärtsilä's W200 engine is integrated in the multi-housing

and the inlet and outlet fuel connections, and can be removed directly without touching the housing or any piping. The coated pump plungers are designed for pressures up to 2000 bar and for a high flow capacity. Fuel is injected into the cylinder through a nitrided eight-hole nozzle designed to yield optimal performance in combination with the combustion bowl and swirl level, and to ensure good resistance to wear and thermal loads. The feeding parts are optimized to eliminate the risk of cavitation. Injection timing is adjusted from the camshaft drive. The injection pumps are equipped with pneumatic emergency stop pistons which are activated to switch the fuel racks to the 0-position by a signal from the engine overspeed detector or safety stop contacts.

An important contribution to engine performance is made by the cooling water system which can be divided into low temperature (LT) and high temperature (HT) circuits with separate outlets and inlets for the HT and LT sides, or configured as a single circuit with an HT outlet from the engine and return to the LT side. The cooling system comprises: a preheating module; double-impeller cooling water pump for LT/HT circuits; thermostatic valves and LT/HT water mixing; charge air cooler; lube oil cooler; seawater pump (optional); and cooling circuitry mostly incorporated in the engine block casting. The external system normally comprises a plate-type heat exchanger for cooling the engine water and another smaller plate-type unit for fuel cooling.

The LT system cools both the charge air and the lubricating oil; the charge air and oil coolers are placed in parallel to achieve maximum cooling efficiency. Increased engine component reliability is reported due to the low lubricating oil temperature (70°C after cooler), and reduced NOx emissions due to the cold charge air temperature (60°C after cooler). The HT system cools the engine block and also the first stage of the charge air when optional maximized waste heat recovery on the HT line is specified.

The main elements of the lubricating oil system (Figure 30.44) are: a wet oil sump; lube oil pump module (comprising engine-driven oil pump and electrically-driven pre-lubrication pump); lube oil module (comprising thermostatically controlled lube oil cooler and main filters); centrifugal filter; distribution channels integrated in the engine block; oil jet nozzles for piston cooling and lubrication; and instrumentation. Pre-lubrication can be controlled automatically or manually.

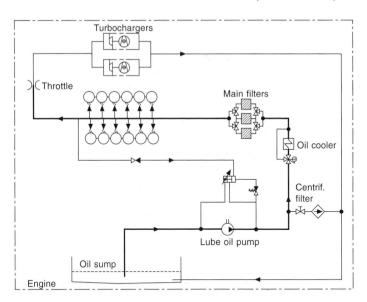

Figure 30.44 Lubricating system of W200 engine

Integration of lubricating oil, fuel oil and cooling equipment in modules mounted on the engine achieves a significant reduction in external connections: only two for water, two for fuel oil and one for starting air. The lubricating oil filters can be changed one by one, even while the engine is running, using a three-way valve to interrupt the oil flow to the filter chamber. The chamber is drained externally by a tap and cleaned without risk of contaminating the lubricating oil system. The system is equipped with a bypass centrifuge filter, located to facilitate ease of maintenance while the engine is running. The washable cell-type air filters are mounted on top of the engine and fixed with a quick clamping arrangement.

The lubricating oil pump, electrically driven pre-lubricating pump, cooling water pump, fuel pump and optional priming fuel pump, and seawater pump are all mounted externally at the free end of the engine. The fuel pump is placed on the same shaft as the lubricating oil pump. The twin water pump for both LT and HT cooling water circuits has only one shaft for both circuits; this means fewer gearwheels need checking and less changing of bearings and sealings. The pre-lubricating pump can also be used for draining the oil sump.

An engine flywheel housing of multi-functional design supports starters and turning gear, and conveys fluids between the different modules. The starting motor is of the pneumatic type, operating effectively even at low pressure.

Marine versions of the W200 engine were originally turbocharged with four automotive-derived turbochargers but these were later replaced by ABB or Holset marine models. The 12V200 models now feature twin turbochargers arranged at the free end of the engine.

The 1500 rev/min engines in service suffered some failures in 1998–99 owing to lube oil circuit problems and material issues, but design refinements and an improved manufacturing process eliminated the main causes. Lubricating circuit performance was upgraded by a better oil filter module design and main lube oil pump, allowing 'fail-safe' cartridge replacement and a constantly clean and air-free oil flow. The connecting rod big-end assembly was thoroughly re-engineered to secure the highest reliability. The engine top overhaul period for 1500 rev/min engines (50 Hz gensets) was extended from 12 000 hours to 18 000 hours in 2002.

ZVEZDA

St Petersburg-based Zvezda's portfolio includes the interesting ChN 16/17 high speed radial engine formed from 160 mm bore/170 mm stroke cylinders. Seven-cylinder aluminium monobloc banks are grouped in a star-configuration to create 42-cylinder or 56-cylinder engines (using either six or eight banks) with a steel tunnel-type crankcase.

An integral reversible gearbox or hydraulic transmission is available for parallel operation with gas turbines. Another variant offered is a 112-cylinder power unit comprising two 56-cylinder engines driving via a common reversible gearbox. Among the merits of the concept claimed for fast vessel propulsion is a weight/power ratio of less than 2 kg/kW.

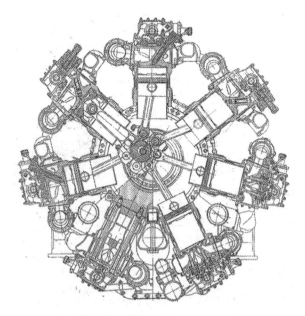

Figure 30.45 Cross-section of Zvezda radial engine

Technically more modest, the V12-cylinder ChN 18/20 engine is a 180 mm bore/200 mm stroke design based on twin six-cylinder aluminium monobloc banks. The design has benefited from a stronger crankcase and high alloy steel crankshaft and connecting rods underwriting outputs up to 1100 kW at 1600 rev/min. Both engine types feature direct fuel injection and turbocharging with or without intercooling.

AUTOMOTIVE-DERIVED ENGINES

Marinized automotive-derived high speed engines are popular for small craft propulsion and genset drives on larger ships, where their ruggedness and reliability (based on numerous truck applications) are appreciated.

Swedish truck engine specialist **SCANIA** launched its most powerful-ever marine engine in 2001, the DI16M design replacing the long-established DI14M series. The new 16-litre engine exploits an extra two litres of displacement in a package that is both shorter and lower than its forerunner (Figure 30.46). Increased torque and lower emissions are also delivered by the DI16M, a twin-turbocharged V8-cylinder design offered with commercial power ratings from 338 kW to 590 kW at speeds from 1800 rev/min to 2200 rev/min. The highest rating addresses light duty applications such as patrol boat propulsion; ratings from 440 kW to 590 kW (medium duty/high power commercial) serve pilot, police and rescue vessels; and the lower output range is applied for heavy duty commercial propulsion (tugs, workboats, fishing vessels).

Figure 30.46 Scania DI16M engine

The DI16M was the first Scania marine engine to use electronic unit injectors (EUI) and the company's new S6 fully electronic engine management control module. EUI allows fuel injection timing to be adjusted to suit all engine running conditions, from cold start to idling and throughout the power range. The system fosters improved power output, fuel economy and emission control, with reduced exhaust smoke.

Based on Scania's 16-litre truck engine, the 127 mm bore/154 mm stroke DI16M shares the same separate four-valve cylinder heads, cylinders and valve mechanism as its smaller sibling, the DI12M. New composite pistons feature aluminium bodies and steel crowns to handle the high combustion pressure. The engine measures 1.25 m long × 1.178 m wide × 1.15 m high overall, and has a dry weight of 1500–1600 kg, depending on its equipment.

Low emissions were also high priority for another Swedish specialist, **VOLVO PENTA**, when designing its 16-litre TAMD 165 automotive-based marine engine. The in-line six-cylinder model benefits from an upgraded fuel injection pump and a new injector as well as a new charge air cooling system, which reduces the inlet air temperature by around 10°C compared with earlier engines. New pistons and rings reportedly reduce lube oil consumption by 50 per cent. High reliability and low noise and vibration were also targeted, relevant measures including the cast iron engine block, trapezoidal connecting rods and a crankshaft with seven main bearings.

Volvo Penta's in-line six-cylinder D12 series is available with four power ratings—from 450 kW to 515 kW—for light duty and special light duty commercial applications. Contributing to increased power with lower emissions and fuel consumption, as well as higher reliability, are unit fuel injectors, a new electronic diesel control (EDC) system, an extremely rigid cylinder block and a seven-bearing crankshaft. Fuel is injected at a high pressure (1800 bar) through eight-hole nozzles to foster an efficient fuel/air mix. Precise control of injection is secured by the unit injectors and the EDC system, which receives continuous data from five sensors, determines the torque created by each cylinder and compensates for any difference.

31 Gas turbines

Gas turbines have dominated warship propulsion for many years but their potential remains to be fully realized in the commercial shipping sector. Breakthroughs in containerships, a small gas carrier and the Baltic ferry *Finnjet* during the 1970s promised a deeper penetration that was thwarted by the rise in bunker prices and the success of diesel engine designers in raising specific power outputs and enhancing heavy fuel burning capability.

In recent years, however, gas turbine suppliers with suitable designs have secured propulsion plant contracts from operators of large cruise ships and high speed ferries, reflecting the demand for compact, high output machinery in those tonnage sectors, rises in cycle efficiency and tightening controls on exhaust emissions. A new generation of marine gas turbine—superseding designs with roots in the 1960s—will benefit from the massive investment in aero engine R&D over the past decade, strengthening competitiveness in commercial vessel propulsion.

The main candidates for gas turbine propulsion in commercial shipping are:

Cruise ships: the compactness of gas turbine machinery can be exploited to create extra accommodation or public spaces, and the waste heat can be tapped for onboard services. In a large cruise ship project some 20–100 more cabins can be incorporated within the same hull dimensions, compared with a diesel-electric solution, depending on the arrangement philosophy. Compactness is fostered by the smaller number of prime movers and minimal ancillary systems (roughly around 50 per cent fewer than a diesel-based plant).

Large fast passenger ferries and freight carriers: the extremely high power levels demanded for such vessels is difficult to satisfy with diesel machinery alone; coastal water deployment favours the use of marine diesel oil in meeting emission controls.

LNG carriers: an ability to burn both cargo boil-off gas and liquid fuel at higher efficiency than traditional steam turbine propulsion plant should be appreciated.

Fast containerships: the compact machinery allows space for additional cargo capacity.

Among the merits cited for gas turbine propulsion plant in commercial tonnage are:

- High power-to-weight and power-to-volume ratios; aero-derived gas turbines typically exhibit power-to-weight ratios at least four times those of medium speed diesel engines; compactness and weight saving releases machinery space for extra revenue-earning activities; a General Electric LM2500 aero-derived gas turbine unit delivering 25 000 kW, for example, measures 4.75 m long × 1.6 m diameter and weighs 3.5 tonnes.
- Low noise and vibration.
- Ease of installation and servicing fostered by modular packages integrated with support systems and controls.
- Modest maintenance costs, low spare part requirements and ease of replacement.
- Environmental friendliness (lower NOx and SOx emissions than diesel engines).
- Reduced manning levels facilitated by full automation and unmanned machinery space capability.
- Operational flexibility: swift start-up: no warm-up or idling period required; idle can typically be reached within 30 seconds, followed by acceleration to full power; deceleration can be equally as rapid, after which the turbine can be immediately shut down; subsequent restarts, even after high power shutdown, can be instant with no 'cool down' restriction.
- High availability, underwritten by high reliability and rapid repair and/or turbine change options.

PLANT CONFIGURATIONS

Optimizing the combination of low power manoeuvring and high power operation is accomplished in many naval applications by using a combined diesel or gas turbine propulsion system (CODOG). Other arrangements can be configured to suit the power demands and/or operational flexibility required for a project: combined diesel and gas turbine (CODAG—Figure 31.1); combined gas turbine and gas turbine (COGAG); combined gas turbine or gas turbine (COGOG); and combined diesel and gas turbine electric propulsion (CODLAG or CODEG).

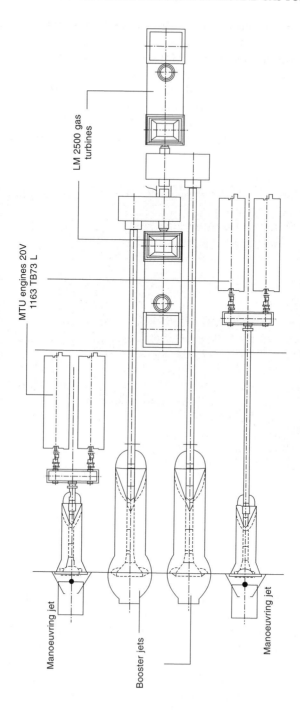

LM 2500 gas turbines

MTU engines 20V 1163 TB73 L

Manoeuvring jet

Booster jets

Manoeuvring jet

Figure 31.1 CODAG propulsion plant configuration for a large monohull fast ferry

A notable COGAG plant powers the Stena HSS 1500-class high speed passenger/vehicle ferries, whose service speed of 40 knots is secured by twin General Electric LM2500 and twin LM1600 gas turbines arranged in father-and-son configurations with a total output of 68 000 kW. All four turbines are deployed for the maximum speed mode, with the larger or smaller pairs engaged alone for intermediate speeds; this enables the turbines to operate close to their optimum efficiency at different vessel speeds, with consequent benefits in fuel economy (Figure 31.2).

Figure 31.2 Final preparation of a General Electric LM2500 propulsion module for a large high speed ferry

An example of a CODEG plant is provided by *Queen Mary 2*, the world's largest passenger ship, whose 117 200 kW power station combines twin General Electric LM2500+ sets with four Wärtsilä 16V46 medium speed diesel engines, all driving generators.

Combined-cycle gas turbine and steam turbine electric (COGES) plants embrace gensets driven by gas and steam turbines. Waste heat recovery boilers exploit the gas turbine exhaust and produce superheated steam (at around 30 bar) for the steam turbine genset. Such an arrangement completely changes the properties of the simple-cycle gas turbine: while gas turbine efficiency decreases at low load the steam turbine recovers the lost power and feeds it back into the system. The result is a fairly constant fuel consumption over a wide operating range. Heat for ship services is taken directly from steam turbine extraction (condensing-type turbine) or from the steam turbine exhaust (back pressure turbine), and there is thus normally no need to fire auxiliary boilers.

Installations supplied by General Electric Marine Engines for large Royal Caribbean Cruises' liners pioneered the COGES plant at sea, the first entering service in mid-2000 (Figures 31.3 and 31.4). Each 59 000 kWe outfit comprises a pair of GE LM2500+ gas turbine-generator sets, rated at 25 000 kWe apiece, and a 9000 kWe non-condensing steam turbine-generator. Heat recovery steam generators located in the exhaust ducts of the gas turbines produce the steam to drive the steam turbine and feed auxiliaries such as evaporators and heating systems. Since no additional fuel is consumed to drive the steam turbine, the additional power it generates represents a 15-18 per cent increase in efficiency with the gas turbines operating at rated power. The auxiliary steam represents a further efficiency improvement.

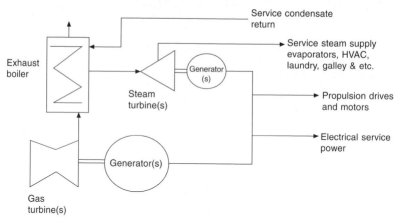

Figure 31.3 Schematic layout of COGES plant for a large cruise ship (GE Marine Engines)

Lightweight gas turbine-generator sets (weighing approximately 100 tons) have allowed naval architects to locate them in the base of a cruise ship's funnel. Such an arrangement replaces the gas turbine inlet and exhaust ducts normally running to the enginerooms with a smaller service trunk housing power lines, fuel and water supplies to the gas turbine package. A significant area on every deck between

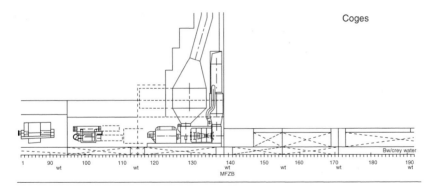

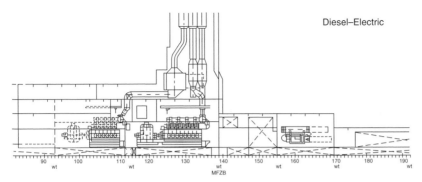

Figure 31.4 Space saving of a COGES plant compared with a diesel-electric installation for a cruise ship (Deltamarin)

the funnel and the machinery spaces is thus released for other purposes.

CYCLES AND EFFICIENCY

Significant progress has been made in enhancing the thermal efficiency of simple-cycle gas turbines for ship propulsion over the years, R&D seeking to improve part-load economy and reduce the fuel cost penalty compared with diesel engines. In 1960 marine gas turbines had an efficiency of around 25 per cent at their rated power, while second generation aero-derivatives were introduced in the 1970s with efficiencies of around 35 per cent. Subsequent advances—design refinements, new materials and cooling techniques, and the appropriate matching of higher compressor pressure ratios—have resulted in some large simple-cycle turbines achieving efficiencies of over 40 per cent, Figure 31.5.

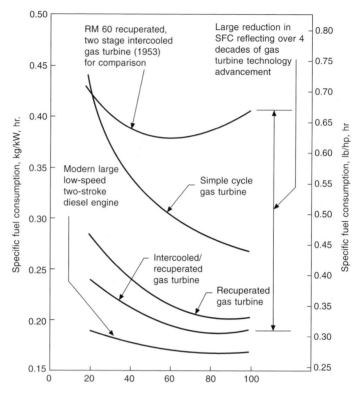

Figure 31.5 Comparison of specific fuel consumption curves against load for various gas turbine cycles and a low speed two-stroke diesel engine

More complex gas turbine cycles can deliver specific fuel consumptions closely approaching the very flat curve characteristics of larger diesel engines. Part-load efficiency can be improved in a number of ways, notably through the intercooled recuperated (ICR) cycle which uses the exhaust gas to heat the combustor inlet (Figure 31.6). This method was chosen for the WR-21 gas turbine (detailed in the Rolls-Royce section below) which has achieved a 42 per cent thermal efficiency across 80 per cent of the operating range. (The first advanced-cycle gas turbine, the Rolls-Royce RM60, was operated in *HMS Grey Goose* for four years from 1953. Unfortunately, the technical complexity of the intercooling and recuperation was far ahead of the contemporary production techniques, thwarting commercial success, but the turbine proved reliable and efficient during its comparatively brief service.)

The ICR cycle and other performance/efficiency-enhancing solutions were examined by Mitsubishi Heavy Industries, taking as an

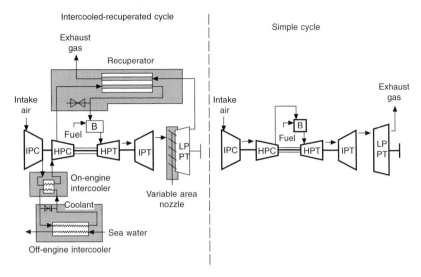

Figure 31.6 Schematic layout of intercooled-recuperated (ICR) cycle and simple-cycle gas turbines (Rolls-Royce)

example its own MFT-8 aero-derived gas turbine developed for marine use (see below). The Japanese designer investigated various methods of improving the thermal efficiency and power output to similar levels as those of the diesel engine, focusing on six gas turbine cycle configurations (Figure 31.7):

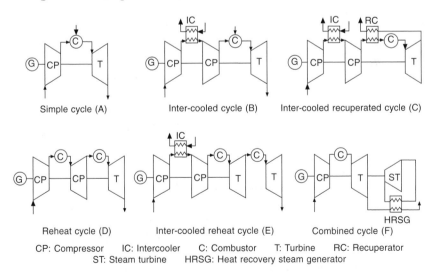

Figure 31.7 Various gas turbine cycle configurations (Mitsubishi)

— *Simple cycle*: the standard layout comprising compressor, turbine and combustor.
— *Intercooled cycle*: as for the simple cycle but with an intercooler added to increase performance through a reduction of high pressure compressor power.
— *Intercooled recuperated cycle*: as for the intercooled cycle but with a recuperator installed to recover heat from the gas turbine exhaust to the combustor inlet.
— *Reheat cycle*: as for the simple cycle but with a reheat combustor added downstream of the high pressure turbine.
— *Intercooled reheat cycle*: a combination of intercooled and reheat cycle configurations.
— *Combined cycle*: a bottoming steam turbine cycle added to a simple-cycle gas turbine.

The main conditions that are changed by varying the cycle, MHI explains, are the turbine inlet temperature and the pressure ratio (1,150°C and 20:1 respectively in the case of the MFT-8 simple-cycle gas turbine). Parametric studies were carried out to establish the optimum performance from each cycle, within the bounds of what is practically achievable in terms of temperatures and pressures. With this in mind, a maximum turbine exhaust temperature of 600°C was specified, and pressure ratios for maximum efficiency and maximum power were determined for each cycle.

The highest efficiency and power was obtained with the combined cycle option, which had the further advantage of a pressure ratio lower than even the simple cycle. The major disadvantage of the combined cycle is the fact that the heat recovery steam generator is physically large and contains very heavy components. Although this is not necessarily a problem in industrial applications, where the combined cycle is commonly used, it is a considerable drawback in marine installations.

MHI's investigations favoured the intercooled recuperated cycle, which showed high efficiency and in power output was second only to the combined cycle. Moreover, maximum efficiency and maximum power were identical, at the achievable pressure ratio figures of 20 in each case. In the other non-simple cycles the two pressure ratio values were far apart, with one or other of the values being higher than 20, a ratio that incurs many practical disadvantages. The only problem with the ICR cycle, according to MHI, is that both the cost and size of the recuperator need to be reduced to make it more applicable to a marine gas turbine.

Although atmospheric pressure variation, an important consideration for aero gas turbines, is relatively small for marine units operating at sea level both temperature and humidity can vary significantly. Both of these parameters influence the specific heat of the air as a working fluid within the gas turbine. Atmospheric variations can have a significant effect on thermal efficiency and specific fuel consumption. Increased air temperature, in particular, reduces efficiency and lowers fuel economy; relative humidity has a less significant influence, though at high inlet temperatures an increase in humidity will have an adverse effect on specific fuel consumption.

EMISSIONS

A lower operating temperature and more controlled combustion process enable gas turbines to deliver exhaust emissions with significantly lower concentrations of nitrogen oxides (NOx) and sulphur oxides (SOx) than diesel engines. Gas turbines typically take in three times the amount of air required to combust the fuel and the exhaust is highly diluted with fresh air. Along with a continuous combustion process, this yields very low levels of particulate emissions and a cleaner exhaust.

Combustion in gas turbines is a continuous process, with average temperatures and pressures lower than the peak levels in diesel engines that foster NOx emissions (see Chapter 3). The fundamental characteristic of continuous combustion in a gas turbine is that residence time at high flame temperatures (a key cause of NOx formation) is capable of being controlled. In a gas turbine a balance between smoke production and NOx generation can be easily secured.

Staged pre-mixed combustion allows NOx emission levels to be reduced without the need for expensive selective catalytic reduction systems typically required by diesel engines to meet the most stringent controls (Figure 31.8). Proven technologies are available to reduce NOx emissions from gas turbines even further:

- Wet technology based on steam or water injection into a standard combustor can lower NOx emissions to less than 1 g/kWh, exploiting a concept proven in land-based industrial applications. The availability of clean water can restrict operation at sea but the technology may be suitable for limited use in coastal areas.
- Dry Low Emissions (DLE) technology yields much lower emissions than current marine engine requirements, well below 1 g/kWh. Lean pre-mix DLE combustion technology maintains a near-

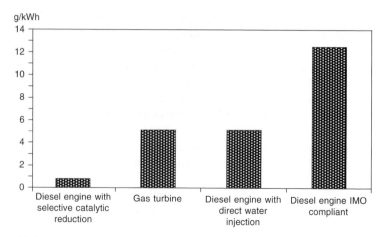

Figure 31.8 Specific NOx emissions for prime movers operating on marine gas oil

optimum fuel-air distribution throughout the combustion zone and the flame temperature in a narrow band favourable both to low NOx and low carbon monoxide production.

SOx emissions are a function of fuel sulphur content. The heavy fuel typically burned by diesel engines may have a sulphur content up to 5 per cent (3-3.5 per cent on average). In contrast, gas turbines fired by marine distillate fuels have a maximum 1–2 per cent sulphur content and the average level in gas turbine fuels is less than 0.5 per cent.

See also section on Rolls-Royce WR-21 gas turbine below.

LUBRICATION

The quantity of lubricating oil in circulation in a gas turbine is much smaller than in a diesel engine of equivalent output, and similarly the oil consumption is significantly less. (The Rolls-Royce Marine Spey, for example, has a recorded in-service average lube oil consumption of 0.1 litres/h.) The smaller charge, however, means the oil is subject to far greater stress than in a diesel engine, and the reduced consumption means it is not refreshed as often.

Apart from lubricating the bearings and other key components of a gas turbine, the bulk of the oil performs an intensive cooling function. Unlike its diesel engine counterpart, the lubricant does not come into contact with the combustion process and does not have to remove products of combustion or neutralize acids formed by burning sulphur-containing fuels.

In aero-derived gas turbines the concentration of power and heat, coupled with the comparatively small quantity of lubrication oil in circulation, results in peak oil temperatures typically over 200°C. Mineral oils are not suitable for this type of gas turbine because they would quickly oxidize at such temperatures: gums, acids and coke deposits would form and the viscosity of the oil would rapidly increase. Synthetic lubricating oils are therefore favoured because of their intrinsic ability to withstand much higher temperatures than mineral oils (see Chapter 4).

There are two types of lubricating oil generally available for gas turbine applications: 'standard' or 'corrosion inhibiting'. Most turbine operators, according to lube oil supplier BP Marine, have moved towards premium corrosion inhibiting (C/I in US military specifications) oils to take advantage of the higher protection afforded to the bearings and other key components.

A low coking propensity, good swell characteristics and compatibility with seals are cited by BP Marine as desirable properties of a synthetic lube oil for aero-derived gas turbines. Coking is the formation of hard, solid particles of carbon due to high temperatures and can result in the blocking of oil ways. The tendency of an oil to break down and produce coke may be exacerbated by turbine operating procedures commonly encountered in fast ferry service, such as rapid acceleration and sudden shut-downs, resulting in the oil being subjected to excessive temperature rises.

Swell is caused by synthetic seals coming into contact with synthetic oils and absorbing the oil. Some swelling is desirable to ensure good sealing but too much can damage the seal and result in leakage. In addition to swell characteristics, the oil and the elastomers which come into contact must be compatible in all other respects so that degradation is avoided.

BP Marine's Enersyn MGT synthetic lubricant for marine gas turbines, a corrosion inhibiting product, is claimed to offer the desirable properties outlined above, along with good thermal, oxidative and hydrolytic stability, and corrosion resistance. It also conforms to the commonly required viscosity of most synthetic oils for gas turbines, nominally 5 cSt at 100°C.

A 5 cSt synthetic lube oil from Castrol Marine, Castrol 5000, is also approved for use in a range of aero-derived marine turbines, promising excellent high temperature and oxidation stability as well as superior load-carrying capabilities. Another synthetic lubricant, Castrol 778, is claimed to exhibit excellent anti-wear and rust protection properties supported by superior oxidation stability. The turbine is protected during extreme cold weather starting and during extended high temperature operation, and deposit or sludge formation is prevented over prolonged drain interval periods.

AIR FILTRATION

The use of air filtration systems tailored to the individual application can significantly enhance the reliability and efficiency of a marine gas turbine installation. A typical filtration system comprises three stages:

- *Vane separator*: a static mechanical device that exploits inertia to remove liquid droplets from an airflow passing through it. The intended purpose of this first stage is to remove the bulk of any large quantities of water and coarse spray that may otherwise overload the remaining components of the filtration system. This type of device can remove droplets down to around 12 microns in size, beyond which inertia has little effect. Unfortunately, the damaging salt aerosol experienced within this environment is predominantly below this size and so will pass to the next filtration stage.
- *Coalescer*: this filter-type device—usually of the one-inch depth pleated variety—is specifically designed to coalesce small water (and particularly salt aerosol) droplets; in other words, to capture small liquid droplets and make them form larger droplets. This comparatively easy task for a filter can be achieved with a relatively open filter material which, although a good coalescer, only has a very limited efficiency against dust particulate.

A low efficiency open media filter has a low pressure loss and thus for this application can be run at much higher airflow face velocities than traditional filters in other applications before it has an impact on gas turbine performance. Velocities are typically between 6 m/s and 8 m/s, compared with 2.5 m/s–3.5 m/s for other applications. This enables a much smaller filter system to be designed for naval vessels, where space is at a premium.

High velocity does not come without a price, however. A coalescer operating at a low velocity is likely to be able to drain all of the liquid it collects—through its media—with carry-over into the duct beyond. But a high velocity coalescer will re-entrain the liquid captured as large droplets, and so a third stage is required, particularly as this liquid is heavily laden with salt.

- *Vane separator*: this third stage device is very similar to the first stage and in many systems is exactly the same. Its purpose is to remove droplets being re-entrained into the airflow by the second-stage coalescer; as such, it must be totally reliable since this liquid has the potential to cause significant damage to the gas turbine because of its salt content.

A successful filtration system serving the gas turbine plant of a fast ferry should have the following features and characteristics, suggests UK-based specialist Altair Filters International:

— It should be a three-stage (vane, coalescer/dust filter, vane) system.
— The filter element should have an appropriate degree of dust filtration performance.
— The filter should have a suitable lifetime.
— The system should be of the high velocity, compact type.
— The system should have a low pressure loss that is not significantly affected in wet conditions.
— The aerodynamics of the intake should be carefully considered.

MARINE GAS TURBINE DESIGNS

Most marine gas turbines are derived from aircraft jet engines, whose development from scratch can cost over US$1 billion. Such an investment cannot be justified for the maritime industry but the extensive testing and service experience accumulated by aero engines provides an ideal basis for marine derivatives. Marine gas turbines are available only in specific sizes and ratings, unlike a given diesel engine design which can cover a wide power range with different in-line and V-cylinder configurations. The economical application of gas turbines to marine propulsion therefore dictates matching the unit to the project, and hence calls for a technical and economic analysis of the vessel's proposed deployment.

Operational costs include availability for service and maintenance requirements, cost of consumables (fuel and lube oil) and manning costs. All these factors have to be considered for military ship designs but not to the extent that is called for in commercial shipping. Trade-offs among the various parameters can be conducted to determine the right fit of propulsion system to the application.

All marine gas turbines incorporate the same fundamental components (Figure 31.9):

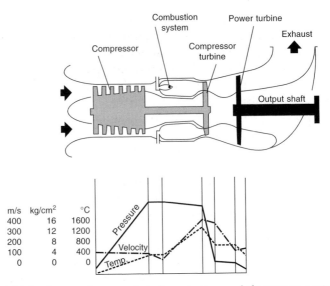

Figure 31.9 Main elements of an aero-derived gas turbine, and the pressure, temperature and air/gas speeds during the cycle

- A compressor to draw in and compress atmospheric air.
- A combustion system into which fuel is injected, mixed with the compressed air, and burned.
- A compressor turbine which absorbs sufficient power from the hot gases to drive the compressor.
- A power turbine which absorbs the remaining energy in the hot gas stream and converts it into shaft power.

Lightweight aero-derived, rather than heavyweight industrial-derived, gas turbine designs are favoured for marine propulsion applications. The aero jet engine extracts only sufficient energy from the gas stream to drive the compressor and accessories, and releases the surplus gas at high velocity through a convergent nozzle to propel the aircraft by reaction.

When converting the jet engine into a shaft drive machine it is necessary to provide an additional free power turbine to absorb the energy left in the gas stream and to transmit that energy in the form of shaft power. The original jet engine is then termed the *Gas Generator* and the whole assembly becomes a *Gas Turbine*. Using a free power turbine reduces the starting power requirement as only the gas generator rotating assemblies have to be turned during the starting cycle. The resulting prime mover benefits from the intensive development and refinement of the original aero engine, both before and during service.

Certain aero engines are already fitted with power turbines and termed turbo-prop or turbo-shaft machines; these are readily adaptable to industrial and marine use. Special technical and production techniques are applied to the re-designed gas turbine to ensure its suitability for sea level operation at high powers, and for the marine environment.

A basic gas generator has one rotating assembly: the compressor and its turbine coupled together. The characteristics of axial-flow compressors vary considerably, however, over the operating range from starting to full power. On some high compression ratio compressors it is necessary to fit automatic blow-off valves and to alter the angle of the inlet guide vanes and first stages of stator blades to ensure efficient operation.

To achieve the necessary stability in larger gas generators, the compressor is divided into two separate units: the low pressure and high pressure compressors, each driven by its own turbine through co-axial shafts. Each compressor is able to operate at its own optimum speed, giving flexibility of operation and efficient compression throughout the running range. Only the high pressure rotor needs to be turned during starting, and therefore even the largest gas generator can be started by battery.

Although such a two-spool compressor is very flexible, it is still necessary in some cases to adjust the inlet guide vanes to deal with the changing flow of air entering the compressor. This is effected automatically by pressure sensors acting upon rams which alter the angle of the guide vanes.

An axial-flow compressor consisting of alternate rows of fixed and rotating blades draws in atmospheric air through an air intake and forces it through a convergent duct formed by the compressor casing into an intermediate casing, where the compressed air is divided into separate flows for combustion and cooling purposes. A typical annular combustion chamber receives about 20 per cent of the air flow for combustion into which fuel is injected and burned.

Initial ignition is executed by electrical igniters, which are switched off when combustion becomes self-sustaining. The resulting expanded gas is cooled by the remainder of the air flow which enters the combustion chamber via slots and holes to reduce the temperature to an acceptable level for entry into the one or more axial-flow stages of the turbine. The turbine drives the compressor, to which it is directly coupled. The remaining high velocity gases are exhausted and are available for use in the power turbine.

A power turbine of the correct 'swallowing' capacity is required to convert the gas flow into shaft power, Figure 31.10. Typically of one

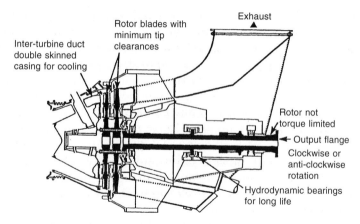

Figure 31.10 Power turbine of a Rolls-Royce Spey SM1C engine

or more axial-flow stages, the power turbine may be arranged separately on a baseframe designed to mount the gas generator, to which the turbine is linked by a bellows joint to avoid the need for very accurate alignment and to allow for differential expansion. In some cases, however, expansion is allowed for in the design of the power turbine and the gas generator is mounted directly on to the inter-turbine duct. The power turbine, usually designed to last the life of the plant, is surrounded by an exhaust volute which passes the final exhaust to atmosphere through a stack.

Most marine gas turbines incorporate a free power turbine, which is free to rotate at whatever speed is dictated by the combination of the power output of the Gas Turbine Change Unit (GTCU) and the drive train behind it. The gas turbine is designed so that there are no critical speeds in the running range, and it can operate continuously with the power turbine at any speed from stopped to the full power rev/min of the power turbine. This facility is especially important under manoeuvring conditions where the combination of a free power turbine and reversing waterjets provides a flexible and effective propulsion system.

Aero- and industrial-derived marine gas turbines are available with unit outputs ranging from around 2000 kW to over 50 000 kW. Only the larger designs (with ratings of 17 000 kW upwards) are reviewed here:

ALSTOM POWER (ABB STAL)

Lower operating costs through burning cheaper fuels than the light gas and diesel oils normally required by aero-derived gas turbines are promised by Alstom Power, which inherited the lightweight industrial-based turbine interests of Sweden's ABB Stal. Successful marine service experience on IF30 intermediate fuel is cited for its GT35 turbine, supported by a capability to burn IF180 fuel demonstrated in land-based plants.

The GT35 emerged from a post-war aero engine development programme in Sweden involving Stal Laval and Volvo Flygmotor. Small turbojets were designed and built for testing in the late 1940s/early 1950s but never entered commercial production. Stal Laval saw potential in transforming the engine into an industrial gas turbine for power generation and mechanical drive duties. The first GT35, rated at 9000 kW, was tested in 1955 and installed during the following year in a Swedish power station.

Performance and reliability were subsequently demonstrated in land-based and offshore platform power generating installations. The design has since benefited from two upgradings to attain its present output rating of 17 300 kW at 3450 rev/min for mechanical drive applications, with an efficiency of 33 per cent (Figure 31.11).

Figure 31.11 One of two 17 000 kW Alstom Power GT35 gas turbines supplied to drive the waterjets of a Stena HSS 900-class fast ferry

The gas generator of the GT35 is a two-shaft engine equipped with a 10-stage low pressure compressor, an eight-stage high pressure compressor, seven can-type combustion chambers, a single-stage high pressure turbine and a two-stage low pressure turbine (Figure 31.12). The power turbine, developed in parallel with the gas generator, is a three-shaft design. A refinement made in the 1990s enabled the power turbine to operate at variable speed and enhanced its suitability for mechanical drive and marine applications.

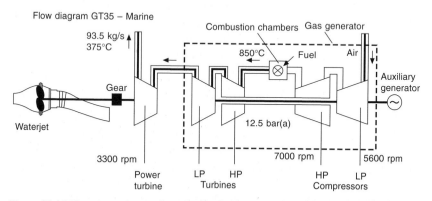

Figure 31.12 Flow diagram of Alstom Power GT35 gas turbine in a marine propulsion installation

Refining the GT35 for marine applications called for the following measures: modifying the lube oil system to allow for vessel movements; modifying the support arrangement to protect the turbine from hull deflections; and modifying the power turbine by introducing a second combined journal and thrust bearing, as well as a modified active tip clearance control mechanism (achieving a self-supported power turbine and thus eliminating the need for a thrust bearing in the gearbox).

High reliability from the marinized GT35 was sought by exploiting as many standard components as possible from the power generation and mechanical drive versions. An extensive re-calculation of low cycle fatigue and creep properties of certain critical elements of the turbine was also undertaken, leading to subsequent redesign of some parts. High availability and maintainability in arduous marine duty were addressed by adapting the installation to the specific vessel, with the aim of reducing the time for unit replacement and facilitating servicing. The maintenance programme was also adapted to operating conditions imposing only short shutdown times for routine inspections and unit change-outs.

Four key factors contribute to high fuel flexibility, underwriting operation of a GT35 turbine set on IF30 fuel (normally produced from 75 per cent IF180 fuel/25 per cent marine gas oil) after suitable treatment:

- A high viscosity tolerance due to the spill flow fuel system in which the amount of fuel injected into the turbine is governed by a regulating valve mounted in the return pipe from the injectors. The system configuration makes it possible to circulate the fuel through the injectors prior to ignition and hence warm them up sufficiently to facilitate starting on fuels with a viscosity of up to 30 cSt/50°C (equivalent approximately to 10 cSt/95°C, the temperature to which the fuel is heated on the vessel).
- A maximum inlet temperature of 880°C (fairly low compared with other gas turbine designs) makes it unnecessary to use cooled blades and vanes, yielding such benefits as insensitivity to high temperature corrosion from sodium sulphate; no need for blade coating; and no risk of plugging cooling holes and channels with ash deposits and reducing the lifetime of blading.
- Large combustion chambers (Figure 31.13) secure low sensitivity to variation in radiation heat depending on the type of fuel that is burned. The size and design of the chambers ensure their lifetime matches that of the complete turbine unit.
- A simple on-line 'soft blast' turbine cleaning procedure—with ground nutshells injected directly into the turbine section via inspection windows in the combustion chambers—results in sustained high turbine efficiency and power output combined with low outage time for cleaning. Cleaning can be effected with the turbine operating at close to full load; depending on the fuel properties, the procedure is necessary at intervals varying between 50 hours and several months' operation.

Retaining a high output to weight ratio dictated special attention to the marine GT35's auxiliary systems, the designers reportedly achieving a reduction in weight of approximately 25 per cent over a land-based unit. Among the measures was a redesign of the starting system from a pneumatic to an electric system and a reconfigured baseframe. Quiet running and high safety are fostered by solid turbine casings which allow sets to be installed without an enclosure. Packages can thus be better tailored to individual hulls, while the free-standing auxiliary systems can be located in the most convenient position.

(The gas turbine interests of Alstom Power were acquired by the Siemens group of Germany in 2003.)

Figure 31.13 Seven large can-type combustion chambers (foreground) contribute to the lower grade fuel burning capability of Alstom Power's GT35 turbine

GE MARINE ENGINES

General Electric (GE) introduced its first aero-derivative gas turbine, the LM100, in 1959 and in the same year the LM1500, derived from the successful J79 aircraft engine. The first LM1500 installation, serving the hydrofoil vessel *H.S. Denison*, was followed by over 160 applications in land-based catapults, pipeline pumping systems, marine propulsion and power generation. The mainstay of the US designer's programme, the LM2500, arrived in 1969 and has since benefited from developments which have doubled its initial rating of around 15 000 kW. Other models subsequently joined the LM-series portfolio, which now embraces five simple-cycle designs with maximum power ratings from around 4500 kW to 42 750 kW and thermal efficiencies up to 42 per cent (see table).

Valuable experience from LM-series gas turbines in naval propulsion applications ranging from patrol craft to aircraft carriers was tapped by GE Marine Engines in targeting commercial vessel projects during the 1990s, earning a number of large cruise ship and fast ferry references.

Table: GE Marine Engines' aero-derived LM gas turbine family

Model	Maximum rating	Thermal efficiency
LM500	4470 kW	32 per cent
LM1600	14 920 kW	36.7 per cent
LM2500	25 060 kW	37.1 per cent
LM2500+	30 200 kW	39 per cent
LM6000	42 750 kW	42 per cent

The LM500 is a two-shaft gas turbine derived from GE's TF34 aircraft engine; the LM1600 is a three-shaft gas turbine derived from the F404 engine; and the LM6000 is a two-shaft gas turbine derived from the CF6-80C2 engine. The most successful models in the commercial and military marine markets are the LM2500 and LM2500+ models, described below.

LM2500 and LM2500+

Derived from the GE military TF39 and commercial turbofan aircraft engines, the LM2500 marine gas turbine is a simple-cycle, two-shaft engine comprising a gas generator, a power turbine, attached fuel and lube oil pumps, a fuel control and speed-governing system, associated inlet and exhaust sections, lube and scavenge systems, and controls and devices for starting and monitoring engine operation. The four main elements (Figure 31.14) are:

- A 16-stage compressor, with an 18:1 pressure ratio; the first seven stages feature variable stators and inlet guide vanes promoting easy starting, good part-load performance, high efficiency and a high stall margin over the entire operating range.
- Fully annular combustor, with externally mounted fuel nozzles supporting liquid fuel combustion; virtually smokeless operation is assured through the complete power range even when burning heavy distillate fuels.
- Two-stage high pressure turbine, air cooled for a long life, which drives the compressor and accessory-drive gearbox.

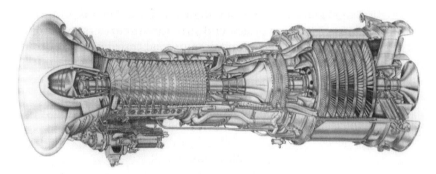

Figure 31.14 Cutaway of GE Marine Engines' LM2500 aero-derived gas turbine showing (from left to right) the compressor, annular combustor, high pressure turbine and low pressure power turbine

- Six-stage low pressure power turbine; a low speed, low stress machine aerodynamically coupled to the gas generator and driven by its high energy release exhaust flow.

Lower installed and life-cycle costs per unit kW were sought from the LM2500+ derivative released in the late 1990s with a 25 per cent higher power rating than its precursor. An introductory rating of 27 600 kW was achieved with a thermal efficiency of more than 37 per cent; release at the design rating of 29 000 kW and a thermal efficiency of 38 per cent followed early sets in service demonstrating the traditional reliability and availability of the LM2500. The current rating is 30 200 kW with a thermal efficiency of 39 per cent. The design refinements applied to the LM2500+ focused on:

— The compressor rotor: air flow was increased by approximately 20 per cent; Stage 1 blades were redesigned to a wide chord configuration eliminating the mid-span dampers; an LM6000 rotor airfoil design was added to Stages 2 and 3; LM6000 stages were incorporated to improve compressor efficiency; and a new inlet guide vane assembly was specified.
— High pressure turbine rotor and stator: redesigned to reduce maintenance costs and new materials exploited for improved oxidation life; Stage 1 and 2 contours optimized for higher flows.
— Power turbine: redesigned in line with the higher output; Stage 1 and 6 blades optimized for aerodynamic efficiency; and the rotor strengthened for the higher torque and potential energy.

No changes were made to either the standard or dry low emissions (DLE) combustion system.

Only slightly longer and heavier than the LM2500, the LM2500+
offers dual-fuel capability (distillate and gas), rapid start-up and loading,
variable speed operation and excellent part-load efficiency, GE claims.

MITSUBISHI

A Japanese challenge in the aero-derived marine gas turbine market
was spurred by the country's Techno-Superliner (TSL) fast freight
carrier project of the early 1990s which called for an indigenous
high output prime mover with a very low weight. In response,
Mitsubishi created the MFT-8 from the marriage of its own power
turbine and the GG-8 gas generator from Turbo Power and Marine
Systems, a subsidiary of Pratt & Whitney (Figure 31.15).

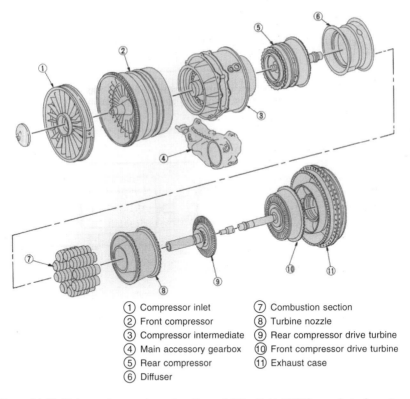

① Compressor inlet   ⑦ Combustion section
② Front compressor   ⑧ Turbine nozzle
③ Compressor intermediate   ⑨ Rear compressor drive turbine
④ Main accessory gearbox   ⑩ Front compressor drive turbine
⑤ Rear compressor   ⑪ Exhaust case
⑥ Diffuser

*Figure 31.15 Major engine groups and sections of Mitsubishi MFT-8 aero-derived marine
gas turbine*

The GG-8 is a land and marine derivative of the US group's JT8D turbofan aero engine, the design modified to replace the front fan with the forward part of the front compressor; otherwise it is fully compatible with the core engine. The key elements of the gas generator are: an eight-stage low pressure compressor; seven-stage high pressure compressor; nine burner cans; a single-stage air cooled high pressure turbine; and a two-stage low pressure turbine. Gas is fed to a high performance power turbine developed by Mitsubishi's project team as a three-stage design with lightweight, overhung ball and roller bearings.

The main modules and components are designed for ease of removal and replacement during inspection and maintenance; and provision is made for borescope inspection of the hot parts and rotating parts of the gas generator and power turbine without disassembly.

After the prototype achieved a full power rating of 24 300 kW during shop trials, two production sets were delivered for installation in the TSL craft built by Mitsubishi. The Japanese designer subsequently offered the MFT-8 turbine to the wider market with a propulsion plant rating of 25 790 kW and a thermal efficiency of just under 39 per cent.

ROLLS-ROYCE

Marine Spey

The Marine Spey first went to sea with the UK Royal Navy in 1985, and the latest C-rated version entered service in 1990. Numerous warship propulsion references have been logged and commercial opportunities targeted, notably for high speed ferry and freight vessel projects (Figure 31.16). The twin-spool design is released with

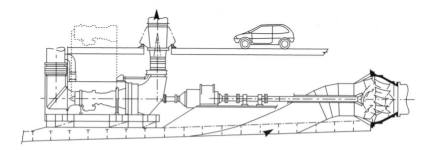

Figure 31.16 Rolls-Royce Marine Spey machinery arrangement for a waterjet-powered fast ferry

respective commercial and military marine ratings of 18 000 kW and 19 500 kW, and specific fuel consumptions of 230 and 226 g/kWh.

Gas generator, power turbine and all ancillary systems are mounted on a common bedplate to form a self-contained unit for ease of installation (Figure 31.17). Interfaces for the shipbuilder are limited to the intake and uptake flanges, the mountings to the hull and the output flange. A complete propulsion module, including power turbine, has a length of just over 7.5 m and weighs 17.5 tonnes in its commercial vessel configuration; the gas generator weighs only 1.8 tonnes. An 80 per cent weight saving and 60 per cent space saving over a medium speed diesel installation of equivalent output are reported.

Figure 31.17 A fully packaged Rolls-Royce Marine Spey propulsion module

The gas generator is a high efficiency, second generation unit of twin-spool design designated the SM1C, which is a development of the SM1A and incorporates technology from the Rolls-Royce Tay and RB211 aero engines. The core achieves a 20:1 pressure ratio and good overall efficiency from a compact unit with relatively few stages of blading. The designers sought excellent operational flexibility, rapid acceleration and good aerodynamic stability, with no visible exhaust smoke at any power level. The modular construction allows gas generator sections to be exchanged on site, the engine comprising five maintenance assembly change units (MACUs).

Specifically designed for the marine environment, the power turbine is a high efficiency, two-stage axial flow design with short and rugged shrouded blades capable of withstanding significant foreign body damage. Its rotor system is supported on hydrodynamic bearings housed in a rigid centre body, and is available with clockwise or anti-clockwise output shaft rotation. The power turbine is designed for the life of the ship but is removable in the event of serious damage (Figure 31.10).

Marine Spey data

Power, commercial service rating		18 000 kW
Intake mass flow		64.3 kg/s
Exhaust mass flow		65.5 kg/s
Exhaust temperature		480°C
Compressor stages	LP	5
	HP	11
Turbine stages	LP	2
	HP	2
	PT	2
Shaft speeds	LP	7900 rev/min
	HP	12 000 rev/min
	PT	5000 rev/min
Combustion system	Cannular	10 combustors
Number of shafts		2 + free power turbine

Marine Trent 30

A simple-cycle marine gas turbine with a thermal efficiency exceeding 40 per cent, the Rolls-Royce Marine Trent 30 was scheduled for commercial availability from early 2004 offering outputs up to 36 000 kW from a package weight of 25 tonnes. The MT30 is the eleventh engine type derived from the British designer's core aero technology.

The aero Trent 800 was selected from a diverse Rolls-Royce engine family as the parent for this new generation of marine gas turbine because it best matched current and emerging customer requirements with margins in hand. The three-spool marine derivative allows variable speed operation, and its 'turn-down' facility secures low power harbour manoeuvring capability. Furthermore, the aero engine's track record (99.9 per cent dispatch reliability) gave confidence for commercial marine applications with high annual usage.

Availability, reliability and maintainability were prime goals in setting the design parameters for the MT30, with an overhaul target of 12 000 hours for the hot end and 24 000 hours between major overhauls. The design reportedly features some 50–60 per cent fewer parts than other aero-derived turbines in its class. A sound basis is provided by an 80 per cent commonality of parts with the aero Trent. Specialized coatings are applied where necessary for protection against the marine environment. Internal temperatures and pressures are substantially lower than those at the aero engine take-off rating to foster component longevity.

A rating of 36 000 kW is available at the power turbine output shaft at ambient air temperatures up to 26°C, with a corresponding specific fuel consumption (sfc) of 207 g/kWh; under tropical conditions (32°C air temperature) the output is 34 100 kW. Competitive efficiency is sustained down to 25 000 kW and thermal efficiency is similar to that of high speed diesel engines, Rolls-Royce claims. The MT30 is designed to burn the widely available distillate marine fuel grade DMA.

Essentially the core unit of an aero Trent engine minus the fan section, the gas generator of the MT30 benefited from the latest design techniques, materials and production technology during its refinement for marine applications. An eight-stage variable geometry axial intermediate pressure (IP) compressor driven by a single-stage IP turbine feeds air to the high pressure (HP) spool, which comprises a six-stage compressor and a single-stage HP turbine. An annular combustion chamber is arranged between the HP compressor and the HP turbine. The IP and HP spools rotate in opposite directions. A two-spool compressor design yields reduced start time, good balance retention and no lock-up problems, says Rolls-Royce, resulting in greater operational flexibility and no need for in-situ balancing.

Hot gas from the gas generator section is supplied to a four-stage free power turbine whose full speed of 3300 rev/min was selected for mechanical drives to yield the lowest sfc. The design addresses the demands of waterjet operation, notably sudden large load changes if air is drawn into the jet unit in severe sea conditions.

The MT30 is supplied as a package with an acoustic enclosure housing auxiliaries and inlet and exhaust diffusers, and incorporating automatic fire detection and extinguishing systems (Figure 31.18). A complete unit—measuring 9.1 m long × 3.8 m wide × 4 m high—weighs around 22 tonnes dry, depending on the options selected. The gas turbine change unit, including power turbine, weighs 6200 kg dry and is air-freightable. Installation of the main package can be

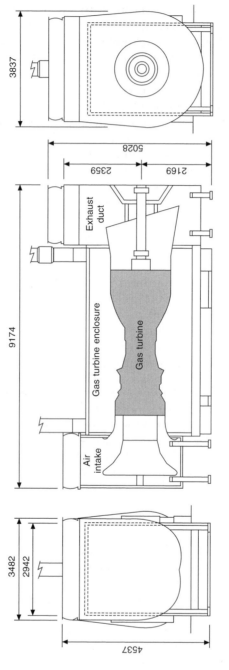

Figure 31.18 Rolls-Royce MT30 in mechanical drive package (dimensions in mm)

effected in a single lift at the shipyard, while the control panel, starter and lube oil modules can be mounted where convenient in the engineroom.

A key element of the package—the integrated control and monitoring system—is housed in a free-standing cabinet with an uninterrupted power supply for protection in the event of mains failure. An engine health monitoring system is also offered. Condition-based maintenance is facilitated by internal condition sensors; routine servicing is limited to checking fluid levels and visual examinations. Maintenance contracts covering servicing, overhauls and spares supply, long practised in the airline industry, are available to support operators and foster competitive through-life costs.

MT30 gas turbine data

Power	36 000 kW @ 25°C ambient
	30 000 kW @ 35°C ambient
Specific fuel consumption	207 g/kWh
Exhaust mass flow	113 kg/s
Exhaust temperature	466°C
Compressor stages	IPC–8
	HP–6
Turbine stages	IP–1
	HP–1
Power turbine stages	PT–4
Power turbine shaft speed	3600 rev/min

The MT30 development will serve as the basis for a more powerful aero Trent marine derivative, the 50 000 kW Marine Trent 50 (Figure 31.19). This MT50 is also a modular engine embracing eight pre-balanced units, allowing major assembly changes to be carried out locally if required. It retains the same three-spool aero core IP and HP compressors and IP and HP turbines, but with a two-stage LP compressor mounted on the third spool replacing the aero fan. The combustion system comprises an annular (donut ring) combustor with 24 individual burners through which the fuel enters the combustion chamber.

Three co-axially independent running shafts are contained within a carcass of axially-joined circular cases. The low pressure shaft features a two-stage LP compressor driven by a five-stage LP turbine; the intermediate pressure shaft features an eight-stage IP compressor driven by a single-stage IP turbine; and the high pressure shaft incorporates a six-stage HP compressor driven by a single-stage HP turbine. A full power specific fuel consumption of around 200 g/kWh is anticipated for the MT50, reflecting the decrease in the

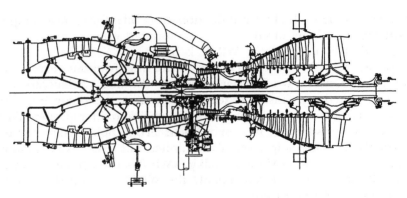

Figure 31.19 Cross-section of Rolls-Royce MT50 gas turbine

proportional blade losses in an axial turbine as the size of the turbine increases.

WR-21

The only advanced-cycle marine gas turbine currently on the market, the WR-21 was developed by Northrop Grumman in the USA and Rolls-Royce during the 1990s with funding from the US Navy, UK Royal Navy and French Navy. It is the first aero-derived gas turbine to incorporate compressor intercooling and exhaust heat recuperation (Figure 31.20) to foster a low specific fuel consumption across the

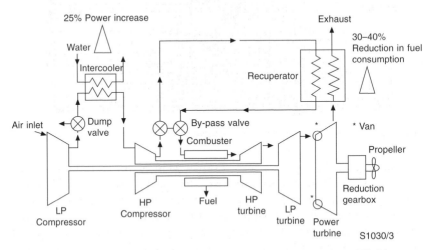

Figure 31.20 Intercooled recuperator (ICR) system applied to the Rolls-Royce WR-21 gas turbine

operating range: a fuel burn reduction up to 30 per cent over simple-cycle turbines is reported.

Rated at 25 000 kW, the WR-21 is based on the successful Rolls-Royce aero RB211 and Trent engines, with modifications to marinize the components and effectively integrate the heat exchangers and variable geometry. The first seagoing installations (featuring twin sets) will enter service from 2007 in the UK Royal Navy's new Type 45 D class destroyers but commercial propulsion opportunities, particularly cruise ships, are also targeted. The fuel consumption characteristics—approximately 205 g/kWh from full power down to around 30 per cent power—enable the WR-21 to fulfil the role of both cruise and boost engines.

Compared with simple-cycle turbines, the benefits cited for the WR-21 include an improvement in fuel efficiency over the entire operating range (with a radical improvement at low power), easier maintenance through enhanced modularization, and the facility to retrofit ultra low emission reduction systems. The intercooled recuperated (ICR) propulsion system package was designed to occupy the same footprint as existing power plant, Figure 31.21.

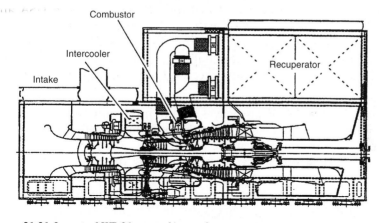

Figure 31.21 Layout of WR-21 gas turbine package

A significant fuel saving over simple-cycle turbines is achieved by using heat exchangers to improve the part-load cycle efficiency. Recuperation alone improves the thermal efficiency of low pressure ratio cycles where the exhaust temperature of the turbine is significantly higher than that of the air leaving the compressor. Heat is transferred to the compressed air before it enters the combustion system, reducing the amount of fuel required to attain the cycle turbine entry temperature.

The recuperator thus pre-heats the combustion air by recovering waste energy from the exhaust, improving cycle efficiency and reducing fuel consumption. Low power efficiency is improved still further by power turbine variable area nozzles; these maintain a constant power turbine entry temperature, which in turn maintains recuperator gas side entry conditions and improves recuperator effectiveness as power reduces.

The intercooler cools air entering the high pressure compressor, reducing the work required to compress the air; the intercooler also reduces the HP compressor discharge temperature, which increases the effectiveness of the recuperator.

Air enters the compressors via a composite radial intake designed to maintain uniform circumferential velocity at entry to the gas generator. The IP compressor (so called because it reflects commonality with the parent three-spool turbofan aero engine) includes six stages of compression; stages two to six are in common with the RB211 engine, while the first stage is modified to suit the increased flow requirements of the ICR cycle.

The intercase between the compressors transmits structural load from the engine through two support legs to the sub-base; it also houses the five intercooler segments in an outer casing and the internal gearbox within an inner casing. Both casings combine to form the air flow path between compression stages. Incorporating an intercooler between the compression stages on a twin-spool cycle increases the specific power of the engine. The amount of work needed to drive the compressor is reduced, thereby increasing the net power available. Bypass valves on the intercooler are modulated depending on the pressure and relative humidity of the air so that condensation formation can be avoided.

Cycle thermal efficiency is approximately the same as that of a simple-cycle engine, as additional fuel is required to offset the drop in compressor exit temperature. Combining an intercooler with a recuperator, however, is attractive for higher pressure ratio engines, leading to high specific power outputs and good thermal efficiency.

The HP compressor also has six compression stages and is aerodynamically identical to its aero origin. Compression is split 30:70 between IP and HP compressors, and both stages incorporate additional boroscope holes to allow greater flexibility for inspection.

The Marine Spey SM1C combustor was adopted as the basis for the WR-21 combustor design as a proven system in use worldwide. Although otherwise conventional in its construction, the combustor features a Reflex Airspray Burner method of fuel injection developed specifically for the marine versions of the Spey. This achieves a controlled mixing of fuel and air, allowing a high burner exit air-fuel ratio (AFR) to be maintained with adequate flame stability. Based on previous experience, high (lean) AFR was considered an important factor in reducing visible smoke when burning diesel fuel.

Preservation of the proven aero RB211 HP and IP spool lengths, which are characterized as short, rigid high integrity structures, was a principal design objective. The requirement to remove compressor delivery air and return recuperated air within the length constraint dictated that the annular RB211 combustor be replaced by a radially-orientated turbo-annular system.

The manifold designs underwent extensive analysis and testing to confirm structural integrity, aerodynamic flow distributions, ease of manufacture and maintainability. The inner casing carries the structural load from the HP compressor outer casing to the turbine casing; the component is designed so that it provides the minimum of blockage for HP compressor exit air entering the combustion manifold and also transfers HP compressor stage 4 air to the power turbine for sealing and cooling. In-situ removal of the combustor and burner is a key element of the maintainability strategy. The design addresses ease of life monitoring and timely repair or replacement of hardware (without engine removal) in the event of premature failure.

The HP nozzle guide vanes and rotor blade airfoils maintain commonality with the RB211-524 parent engine, with slight modifications to provide a smooth gas path from the radically-swept discharge nozzles. The disc seals and bearing arrangements are essentially unchanged from their aero origins. The IP nozzle guide vane is skewed relative to the aero RB211-535 vane and incorporates the addition of a cast boss to facilitate a boroscope inspection hole. The blade is uncooled and manufactured from a single nimonic to extend creep life.

Consistent with the pedigree of the gas generator, the power turbine also originates from an aero parent, the Trent. Stages two-through-five incorporate three-dimensional orthogonal blade geometry to maximize turbine efficiency, but the main difference is the incorporation of the Variable Area Nozzles (VAN) which control flow area. The VANs are hydraulically actuated via a single geared ring designed to maintain VAN-to-VAN throat areas within a specified tolerance. The VAN is fully open at full power and closes at part power; this has the effect of retaining the efficiency benefits across the whole power range by maintaining the high exhaust temperature at part powers. The recuperator can thus be exploited fully to give the characteristic flat fuel consumption curve for the WR-21 (Figure 31.22).

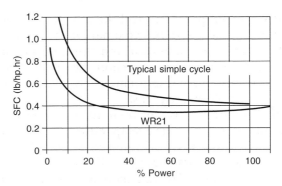

Figure 31.22 Fuel savings from ICR cycle WR-21 over the power range compared with simple-cycle gas turbine

The gas generator and power turbine comprise twelve interchangeable, pre-balanced modules, whose small size and low weight facilitate removal via simple routes; new or leased modules can be fitted in-situ, reducing maintenance costs and downtime.

The enclosure panels are structurally designed to meet blast overpressure, shock and air tightness, with insulation and acoustic damping treatment applied to satisfy thermal and noise requirements. All the panels are removable, except those welded to the recuperator housing. The engine can be withdrawn sideways by removing vertical supports at its front. The enclosure is cooled by ship's ventilation, air entering it via a connection above the recuperator ducts.

An Electronic Engine Controller (EEC), incorporating a Full Authority Digital Engine Control (FADEC) system, is responsible for controlling the system in response to commands from either the ship's monitoring and control system or from the local Man Machinery Interface (MMI). The controller covers sequencing, steady state and transient control, surveillance, fault detection and protection.

Easy control of NOx emissions when entering controlled waters is fostered by the flexibility of the ICR cycle. Punitive charges can be avoided by switching to recuperator bypass mode, lowering the combustion temperature beneath the threshold for NOx generation. Additionally, the high combustor inlet temperatures over the power range in recuperated operation ensure complete combustion with extremely low levels of carbon monoxide and unburned hydrocarbons.

GAS TURBINE MAINTENANCE

Gas turbines have nominal power ratings established for marketing purposes. A power rating at particular ambient conditions determines the internal engine temperatures and the resulting expected service life of components exposed to these temperatures. These 'hot section' components include the combustor and the high pressure turbine blades and vanes. The life of the hot section components will determine the interval between major maintenance actions, resulting in an estimated average maintenance cost.

The normal relationship between internal parts life and a representative internal gas turbine temperature, in this case the power turbine inlet temperature, is indicated in Figure 31.23. At higher internal temperatures the component life is determined by creep-rupture and decreases rapidly with increased temperature; at lower internal temperatures the component lives are determined by corrosion and do not vary as rapidly. (Turbine blades operate at high temperature and rotate at high speed. The blade material has an overall creep life which is proportional to time at a given power level.)

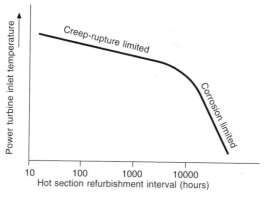

Figure 31.23 Relationship between internal parts life and power turbine inlet temperature

The nature of the curve in the figure illustrates the selection of power rating for a particular application. Operation for large percentages of time at power levels in the creep-rupture regime will severely limit component lives in the gas turbine and increase the unit's average maintenance cost. Conversely, best advantage of a gas turbine's capability can be taken with respect to maintenance near the top of the corrosion controlled regime. Establishing a rating around the 'knee' of the curve therefore tends to be a good starting point with respect to trade-off between optimum power and reasonable maintenance costs.

Vessel operations will generally involve some representative profile with a mixture of high and low power elements, and a range of ambient operating temperatures. The optimized gas turbine rating makes best use of the engine at the predominant mode of operation, while ensuring that the maintenance costs combined with other operating costs produce a minimum total life-cycle cost.

A second factor affecting maintenance costs is the number of cycles of operation per unit time. Since some internal components may be limited in life to a maximum number of thermal cycles, operation with very short mission lengths will result in higher average maintenance costs.

Figure 31.24 illustrates the nominal relationship of average maintenance cost to length of mission. For example, a fast ferry operation with mission lengths of one hour will experience a higher average annual maintenance cost than for five or six hour runs at the same power rating and ambient conditions. As the length of the mission increases, the maintenance requirements approach the level determined by the power level and ambient conditions resulting from the hot section life. For short missions, however, the replacement of cyclic-limited components can have a significant effect on average maintenance costs. Total cycle times for very short applications are usually determined by the turnaround time at each end of a route. The additional speed is less beneficial for the short cycle because a smaller percentage of time is spent in transit. Gas turbine propulsion generally yields greater benefits in longer cycle applications.

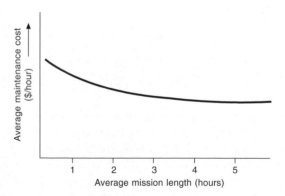

Figure 31.24 Average gas turbine maintenance cost versus length of mission

High reliability is a requirement imposed by military ship applications because of the potentially critical nature of missions. In commercial operations, however, time in service translates into revenue production; factors crucial to the commercial success of the vessel design are therefore minimal downtime and short maintenance intervals.

High availability results from two design considerations. First, the maintenance requirements must be minimized through appropriate rating of the gas turbine, and mission planning must allow for maintenance that minimizes disruption of the operating schedule. Second, appropriate installation design must foster quick removal and replacement of modular components, minimizing time out of service.

Commercial vessels are usually designed with a particular service profile in mind, but the design must be flexible enough for diverse deployment if resale to other operators is planned later. The propulsion plant must therefore allow flexibility in operation over a range of power levels while maintaining operating efficiency with respect to fuel and lube oil consumptions.

Gas turbine engine maintenance may be considered broadly in two groups: routine tasks to be carried out onboard (Figure 31.25)

Figure 31.25 Routine maintenance of the Rolls-Royce MT30 gas turbine (shown here on the testbed) is limited to checking fluid levels and visual examinations

and routines that call for service engineer support or scheduled removal of the gas turbine change unit (GTCU).

Routine onboard maintenance is limited, consisting mainly of daily visual checks of the running engine. An internal water wash is executed after a fixed number of operating hours to remove any salt deposits in the compressor which would otherwise eventually result in degraded performance.

The engine itself can be divided into two main elements: the GTCU and the power turbine with ancillaries package. The GTCU can be readily removed from the vessel, within six to 12 hours, when dictated by condition monitoring or scheduled removal. The power turbine and ancillaries are designed to remain for longer periods, up to the life of the vessel.

In general, the operator will regard the repair of the GTCU as repair by replacement; it will be removed and replaced, and the vessel returned to service in hours rather than days. An owner will therefore normally choose to own a calculated number of spare engines or major engine components (MACUs: maintenance assembly change units) to support a particular operation with a required availability.

A maintenance regime can be provided through a power-by-the-hour arrangement directly with the gas turbine manufacturer. An indicative figure for such an arrangement is often requested by operators, the figure typically including scheduled removal and overhaul of the engine and its replacement by a spare during the overhaul cycle. It is invariably misleading to quote such a figure, which is not a feature purely of the engine but highly dependent on the usage. The calculation will vary strongly with the following factors, notes Rolls-Royce:

- The operating profile of the vessel.
- The number of starts, stops and idling periods during the route.
- The ambient operating temperatures.
- The maximum and minimum powers required from the engine.

These factors are fundamental in deciding engine life and therefore the maintenance costs. The resulting calculations are based on creep and cyclic performance. With a knowledge of the operating profile an assessment can be made of the GTCU life at which the blading must be replaced. At one extreme, on a longer ferry route with operation at high power and high ambient temperatures, engine life may be creep limited.

The multiple compressor and turbine stages of the GTCU comprise individual discs with blades attached. Any metal disc that is continuously and repeatedly accelerated from rest to a high speed, then decelerated, will eventually crack. All the discs have a declared cyclic life, a full cycle being defined as from rest to full speed and back to rest. The declared life falls well within that at which a disc might be expected to crack, but when this life is reached the disc must be replaced.

Cyclic life may become a limiting factor on short ferry crossings, particularly if this is carried out at high engine speeds; it may dictate whether the engine can be shut down or should be left running at idle in harbour, largely for economic reasons based on engine lifing.

It is the combination of creep and cyclic life calculations in conjunction with condition monitoring of the engine that decides the periodicity of the required GTCU changes, and therefore impacts directly on any fired-hour maintenance support arrangement. By working with a prospective operator, the engine manufacturer is then able to make an accurate assessment of the maintenance costs. The combination of engine characteristics and selected operation determines these costs rather than the engine alone.

Gas turbine compressor washing is an important maintenance routine in sustaining power output. Since the compressor consumes up to two-thirds of the total engine power even slight blade fouling significantly reduces the net available output and hence the drive power available or increases fuel costs. Off-line washing is an effective countermeasure in restoring performance and overcoming the efficiency drop caused by compressor fouling. On-line washing is available when engines cannot easily be shut down for an off-line wash (as with cruise ships during long voyage legs).

Acknowledgements to Alstom Power, GE Marine Engines, Mitsubishi, Rolls-Royce, BP Marine and Altair Filters International.

Index

(All subentries without page numbers fall within the page range of the main entry)